THE
MANAGEMENT
OF
QUALITY
ASSURANCE

Wiley Series in Production/Operations Management

Jack R. Meredith, Advisory Editor

The Management of Quality Assurance

Madhav N. Sinha
Walter W. O. Willborn

Project Management: A Managerial Approach

Jack R. Meredith
Samuel J. Mantel, Jr.

Decision Systems For Inventory Management and Production Planning 2nd edition

Edward A. Silver

Facility Design

Stephan Konz

Modern Production/Operations Management 7th edition

Elwood S. Buffa

The Management of Operations 2nd edition

Jack R. Meredith
Thomas E. Gibbs

Quality Control

Harrison Wadsworth
Kenneth Stephens
Blanton Godfrey

THE MANAGEMENT OF QUALITY ASSURANCE

MADHAV N. SINHA, Ph.D., P.Eng.

Engineering & Management Consultant
Winnipeg, Manitoba, Canada

WALTER W. O. WILLBORN, Ph.D.

Professor of Production Management
University of Manitoba, Canada

John Wiley & Sons
New York   Chichester   Brisbane   Toronto   Singapore

Library of Congress Cataloging in Publication Data

Sinha, Madhav N.
  The management of quality assurance.

  (Wiley series in production/operations management)
  Includes index.
  1. Quality assurance—Management.   I. Willborn,
Walter W. O.   II. Title.   III. Series.
TS156.6.S48   1985   658.5′62   85-5398
ISBN 0-471-80310-3

Printed in the United States of America

10 9 8 7 6 5 4 3 2 1

To
Sharda and Vera

PREFACE

Quality in products and services is a universal concern in the modern industrial world. The great importance of quality assurance can easily be explained; only the question of how to achieve and to manage it proves rather difficult.

Quality assurance is a comprehensive management responsibility. Assurance involves both planning and control. Managers are expected to give competent and visible leadership and to commit themselves, and their organization, to effective quality assurance. This need must also be understood by students of the various management and engineering disciplines, as these students become tomorrow's managers.

This book summarizes the essential concepts, practices, and methods of modern quality assurance management. The purpose of this category of management is to lead and help employees to achieve good workmanship, and assist their companies in being productive and competitive. Benefits from such performance are shared by all as producers and consumers, and by the community at large.

The management of quality assurance must be based on interdisciplinary studies, with emphasis on management and engineering. A broad range and integrative approach best leads to modern preventive quality assurance, and ultimately to better quality of products and services.

Undergraduate and graduate programs in management and related fields normally include subject matter on quality assurance in production and operations management courses. Courses for some undergraduate and selected engineering programs cover only the statistical aspects of quality control, if it is included at all. Doubtless, students would greatly benefit from special courses in quality assurance. Such a course, however, should cover more than just traditional inspection and statistical quality control techniques. Teaching and learning about quality assurance requires a suitable textbook with a holistic approach.

In our teaching, we have found that students who are either in college or already working in a management job, learn technical and nontechnical matters of quality assurance best when concepts and methods are made easy to grasp and presented with relevance to managerial practice. This book is designed with such a goal in mind. It can also be used for self-study by practitioners, instructors, and others with interest in modern quality assurance.

This textbook has the following features.

1. As a comprehensive text, it offers more subject matter than is normally covered in one volume and course, allowing selection and special emphases.

2. Concepts and methods are presented in conjunction with practical and realistic examples. Exercises, in the text together with example cases, readings, discussion and review questions, references, and problems, aid learning and teaching.

3. In accordance with the concepts of system and life-cycle approaches, specially covered topics include product design assurance, customer and supplier relations, quality assurance standards, audits and improvement projects, motivation for quality, and community-wide quality assurance programs.

4. Analytical and quantitative methods used by managers are presented appropriately, in the sequence of chapter arrangement. However, the use of mathematics and complex statistical formulations are kept to the minimum essential to convey their relations with the managerial concepts helpful in solving real quality problems.

5.   Readers will find much direct guidance for quality planning, for preparation for new assignments in professional quality assurance examinations, for solving practical problems in the field, and for conducting research.

6.   Production and operations managers will find this book relevant and helpful when they orient their planning and control of operations towards quality improvement programs. This is true not only in manufacturing and nonmanufacturing environments, but in all industries and businesses, private or public, be they large organizations or small.

The book is divided into four parts. Each part consists of several chapters covering logically interlinked topics. A synopsis at the beginning of each part familiarizes the reader with current issues and trends, thus helping in the understanding of the importance of quality assurance as a discipline, profession, and practice. In the overall topical layout, product and service oriented planning and control is followed by discussion of company-wide management of a quality assurance program. The final part addresses the question: "where do we go from here?"

This book is the outgrowth of a long association for both of us with many outstanding managers, quality assurance professionals, and, particularly, with our students. We thank them for sharing their thoughts and experiences and for serving as a sounding board. We are also particularly grateful to Jack Meredith of the University of Cincinnati, who, as advisory editor for the Wiley series in production/operations management, guided us. Rick Leyh, editor at John Wiley & Sons, Inc., along with his assistants and team of expert reviewers, provided valuable support and encouragement.

Madhav N. Sinha
Walter W. O. Willborn

Winnipeg,
Canada

CONTENTS

BACKGROUND OF MODERN QUALITY ASSURANCE

Part I of this book presents the general historical, conceptual, human, and managerial background of quality assurance, summarized with reference to everyday experiences of the reader. Through this inductive approach, the modern concepts of quality assurance become clearly understood in the light of more recent efforts and successes in private and public businesses. Chapter 1 surveys the general meaning of quality assurance and underlines its important role and potential benefits. More precise and significant concepts that the reader must be familiarized with, are expounded in Chapter 2, in terms of quality assurance and with reference to systems concepts.

Products and services are made available through human efforts in terms of quantity (productivity), quality, time, and place. Chapters 3 and 4 in this part, follow up the conceptual framework with regard to the people involved and the managerial dimensions. Part I introduces the reader to the planning and control of product quality. This then becomes the foundation for the quality assurance management programs described in the remaining parts of this volume. These programs are discussed with particular reference to a company and, beyond that, to our society.

CHAPTER 1

CONCEPTS AND DEFINITIONS

Everybody is *for* good quality and *against* poor quality. No matter what is made, sold, serviced, or provided, every citizen in each society depends on the *quality* of such business transactions. Numerous kinds and sizes of organizations, companies, and institutions exist for the specific provision for, or solution to, human needs. All have quality objectives in common. However, many do not achieve their objectives of quality at all times and often fall below the desired expectations of their customers.

In the homestead economy, where people produced for their own consumption, quality was understood to mean well-being, survival, good simple food, and shelter—earned through one's skill, energy, and persistence, with God's help. Today, individual consumers and producers can no longer specify the quality of most complex and rapidly changing goods and services as easily, if at all. Quality awareness today differs fundamentally from in the past as does demand for its assurance. Consumers have become dependent on others in the community to an ever-increasing extent, and therefore, assurance of quality when spending money and income has become a complex task. Attaining quality assurance that people can have confidence in requires not more, but *better* goods and services that will make for better living in the future.

When people live less on the land and more in urban centers, reliance on the good will and integrity of others is necessary. However, when quality assurance is demanded and required in the marketplace, how, in fact, is quality assured?

Does quality relate just to food, shelter, clothing, and so on? Does it include the quality of an ever-increasing variety of services from private businesses and governments? At what price does the quality demanded and the one supplied meet in the marketplace? It has been

3

shown that people in different situations will have diverse understandings of quality. Is this always in fair relation to the prices they want, and can afford to pay, to meet their needs? The assurance of quality can only be articulated, demanded, and extended, once quality itself is adequately defined. Such definitions are to be found in general terms, as well as in technically oriented details.

1.1 THE QUALITY CONCEPTS BASE

To understand quality in more general terms seems to be easier, yet when explaining it to someone else, generalities do not help much, if at all. Keeping in mind the quality of this book or any other product or service,[1] the reader as consumer or producer might think of general or abstract explanations such as the following.

Quality is what people think it is, perceive it to be, or experience it to be.

Quality perceived as high, low, or negative, describes its value in utility and usefulness.

If something is of desirable value it does have a price. In a product, quality is the composite of individual attributes and measurements. A price is also paid for services, and these too have quality characteristics (or parameters) of physical (e.g., length, weight, taste, etc.), and nonphysical (e.g., courtesy, royalness, timeliness, etc.) nature. Therefore, attaining quality must involve broad responsibility for all parameters.

Quality is not always what the consumer buys and expects as fair and adequate.

It is, moreover, what the producers create in production as added value, with the expectation that consumers will be satisfied with the quality and pay a fair price. Given a product and its particular characteristics, the individual customer might find the quality good, fair, or bad. A child might use a cardboard box as a sled in winter, while a manufacturer might want to ship goods in it. Both have different uses and quality requirements for the cardboard box.

Quality is also commonly understood as "excellence," which is better than a minimum standard.

What is the dividing line between these different grades, if they are not expressed in assessable and measurable terms?

Quality's conformance with specification and price is a responsibility of the producer and a right of the consumer.

When purchasing an electric razor or washing machine, and even when purchasing groceries, we are often aware that higher quality items exist but, nevertheless, we purchase items at a lower price, confident that these will serve our requirements.

Quality means not just fitness for use but, in more specific terms, reliability, safety, maintainability, repairability, status, etcetera.

In a free-market economy, the consumer is free to determine quality, but can do this only under conditions of what is called the "buyers market." The reverse is true in a "sellers market," when, at least in the short run, producers set the quality.

Quality in business is expressed as specified standards against which actual perform-ance and conformance can be measured.

Consequently, inspection and quality control are needed, with subsequent corrective action and preventive measures that can trace the causes of defects. Therefore, via use of standards, quality attainment maintains and improves itself over the long run.

Contracted quality establishes liabilities. A specified degree of quality, spelled out un-ambiguously in contracts, leads to effective and harmonious business transactions. Product integrity based on a provable history of satisfying adequate quality requirements offers the best and most convincing defense in liability cases.

Quality, understood as product and service characteristics that are specified, standard-ized, and contracted, has a relationship to quantity, time, and space.

Often quantity and quality have certain positive and negative interrelationships. Moreover, the additional dimensions of time (e.g., reliability, maintainability, etc.) and place (delivery) influence quality with regard to actual value for the consumer.

Quality also is an expression of the people who contribute to produce. Attaining quality is thus everybody's responsibility.

Certainly a paid employee has the responsibility of meeting job performance standards, and thus, is participating in quality creation. The relation of product quality to workmanship, and high quality as the expression of good work, must not be forgotten irrespective of how impersonal and formal the work environment may be.

Quality is the totality of all attributes and characteristics of a product or service as specified, required, and expected.

In conjunction with quality assurance, quality must be measurable and verifiable in tech-nical respects by way of groupings of quality characteristics. The customer or consumer expects, and the producer assures, quality. The critical interface of the two parties that decide on quality is demonstrated in Figure 1.1. A satisfied customer is one who finds that the expected quality has been received. Otherwise, a product liability claim and rejection of trust in the producer's quality assurance might occur.

Quality can also be realized in terms of several areas of quality characteristics of a product or service. These can be technically described as *quality of design, quality of conformance,* and *quality of performance.* In response to large demands (guided by the purchasing power of consumers), the producers of goods and services intentionally evolve variations in quality (i.e., product grades). In the example of automobiles, the different makes and models clearly reflect these differences. However, the three quality areas of a product (or service) are closely interrelated.

For example, a car might have been properly designed, but the workmanship may be poor, so that the lack of quality, or *defect,* is caused by nonconformance to applicable production standards. Alternatively, if the owner does not handle the car as normally expected and does

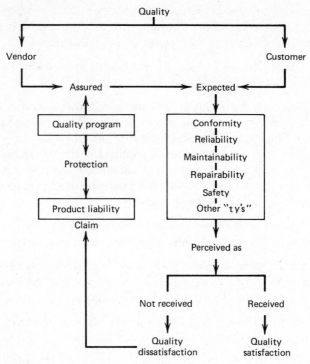

FIGURE 1.1 Critical interface of the two parties (vendor and customer) that decide on quality.

not maintain it properly, the best quality of design and conformance will not outweigh this mishandling in the long run. Thus, understanding of quality requires grasping causes of defects.

"Quality is never an accident, it is *always* the result of intelligent effort."

That was what John Ruskin, the great Victorian art critic and advocate of fine art in everyday living believed, and I think the message is as relevant and important today as it was when he wrote it.

We used to think that mass production was an alternative to quality, but today's technology teaches us how to make things quickly, cheaply and well. That's intelligent effort.

Quality isn't just for companies, either. Individuals can strive for excellence in their day-to-day life. It's a matter of personal choice: People can easily slip into just doing the job, passing the course, playing the game at a sort of 'optimum' level, or they can make a commitment to meet self-imposed quality standards and refuse to accept anything less.

I get impatient with people who complain of boredom—who are bored with bringing up children or with their job, which is 'routine.' Think of the challenge that raising 'quality children' poses! And I have seldom seen a job that couldn't be done better, no matter how routine.

Individuals can commit themselves more easily than corporations can. This commitment is as important for them as it is for companies.

It shouldn't matter that the world seems willing to accept the second rate, because it's up to each of us to challenge ourselves. The rewards for personal excellence are inside our own minds. Recognition and respect from those who notice may follow, but the most important rewards are our own feelings of self-worth and satisfaction.

Next time you read about product quality, think about your own efforts at work, at home, at school, at play and ask yourself what intelligent effort might do to upgrade the quality of your personal product.

It really is worthwhile.[2]

1.2 QUALITY AND DEFECTS

Any item we buy has numerous quality characteristics. These are specified either as non-measurable attributes or as measurements. With regard to the quality areas of a product, quality of design, of production, and actual quality of performance interrelate. In fact, the most simple product, for instance a pencil, presents a myriad of quality characteristics, when viewing not only the pencil itself, but the production stages, the design, and the pencil's use. Quality is indeed a most complex concept, but only when related to a product or service can the elementary characteristics be defined. Without a definition or *specification,* quality cannot be assured nor controlled.

Some typical quality characteristics can be perceived as *measurable.* These might be *product oriented* (e.g., length, weight, tensile strength, acidity, etc.); *time oriented* (e.g., reliability, maintainability); or *commercial* (e.g., the warranty, or the time taken in providing a service). Examples of *nonmeasurable* quality characteristics would be those that are *sensory* (e.g., beauty, taste, look, and comfort); or *ethical* (e.g., honesty, trustworthiness, or courtesy).

From a technical point of view, a given number of quality characteristics can be either functional or nonfunctional. On many occasions this forms the basis of whether or not they are critical to the product's safety, marketability, or user-affordability.

Controlling quality means to ascertain conformance with the design specification of major characteristics. For any required quality characteristic there is at least one defect, namely that the characteristic does not meet the specification. For attributes this means a go/no-go proposition, that is, the acceptance or rejection by an inspector. Measurements can have one or more common units (such as inches) expressed on a scale, and usually vary around a nominal value. For instance, the weight of a can of soup when sampled might be specified as 10 ± 2 ounces. In reality, however, this may not hold true all the time. *A quality product is the one that is defect-free.*

1.3 CONCEPTS OF DEFECTS

The concept of defect has many meanings similar to the meaning and understanding of quality. *Defects must be defined as much as the desired quality characteristics themselves.*

Defects must be controlled in order to control quality. This becomes one important prerequisite for quality assurance, namely that management must be able to prove that all due care was taken to avoid any or all defects.

Defects are like a disease against which one wants to be protected. Often they can be like an infectious disease; one defect breeds another, simply due to the close interdependencies of all the quality characteristics of an item.

Defects, once anticipated during design, defined, and then found during production, or even later, are the effect of causes. As symptoms that are undesirable and disturb quality, defects serve to provide important information. Often they can be just the tip of an iceberg which, when uncovered, reveals many hidden causes of failure and disaster.

One classic example known to us from the early 1970s is when, during one period, U.S. automobile manufacturers actually recalled more vehicles for correction of defects than they manufactured. There could not be much surprise when, for example, the Insurance Information Institute found that the average car repair claim over the 10 year period between 1962 and 1971, doubled from $176 to $345 and, in the following 5 years, doubled again.

Defects provide an opportunity to improve. Defects create unnecessary costs. A company that only sorts out the scrap without probing into the reasons for defects, especially when the scrap is more than "normal," foregoes a possible opportunity to detect and eliminate the cause. Causes such as poor material, insufficiently skilled or unmotivated workers, poor design, and so forth, not only create defects but also add on to unavoidable costs.

For example, one is reminded of the high incidence of exploding industrial boilers back in the 1880s. It was true then, as it is now—safety and good design principles are never totally overlooked. However, it took the overwhelming concern of design engineers (especially those associated with the emerging American Society of Mechanical Engineers) to create safer design standards. Manufacturers and various inspection authorities also quite clearly helped to create the very low rate of accidents found today, with an attendant increase in maximum allowable operating pressure of these boilers.

Since effective quality control involves prevention of defects, it becomes important to recognize when "normal" is normal and when tolerance limits are transgressed. It will be shown later how to determine those deviations from the norm that give rise to defects and reveal causes that must be remedied.

In all defect prevention schemes, quality characteristics help to define defects. Defects reflect undesirable causes and allow successful corrections. Only when defects do not reoccur, can quality be said to be assured. Some defects, however, are more serious than others.

It is one thing for the consumers to accept a certain inferior quality of product (early gaslighters, for example, or a set of cookware that corrodes easily) or inferior service characteristics (long waits in restaurants); it is quite another thing to accept a television set that catches fire, even when it is supposed to be turned off, as used to be the case in the beginning stage of television manufacturing.

Differences in levels of quality must be distinguished from degrees of quality that are expressed in terms of defects. Quality characteristics, individually and jointly, present certain quality levels.

For example, a luxury car is not only more costly than a compact model; its components, design, workmanship, and servicing are normally of relatively higher quality, and the price expresses this fact. However, as Figure 1.2 demonstrates, with an increasing level of quality, from a certain point and respective price the usefulness does not increase proportionally. This, together with rapidly increasing costs, describes a maximum quality level beyond which the consumer would pay a price lower than cost, and the producer would suffer a loss.

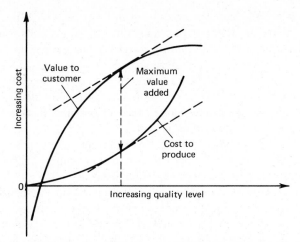

FIGURE 1.2 Cost and value of quality. The maximum difference between increasing quality cost and customer value occurs at the point where the slopes of the two curves are equal.

The degree of quality measured in number of defects, failures, repairs, and so on, relates quality and quality assurance more closely to cost than the price-related quality level.

Tracing the quality level and quality degree of a product to those of components and individual quality characteristics is rather difficult but is necessary, at least for the major characteristics. This then, is a main task of quality assurance and the various control and inspection functions.

The concept of quality cost relates to quality assurance and defects. Quality costs are, foremost, the value of resources such as material and labor that are wasted through defects. Failure costs tend to increase rapidly when the product moves through various stages. For example, a defect detected during the design stage rather than later, during production, or even later, during performance, is less costly. In this very simple fact lies the main rationale for modern quality assurance.

The strategy "get it right the first time" helps to minimize quality costs which helps to avoid costly defects and subsequent production problems. This will then raise the quality level and the product's quality image upwards. Both lower quality costs and higher quality level contribute to profit, market share, job security, and customer loyalty.

As will be shown later, quality costs do not just include defect and failure costs. Costs for quality assurance, control, and inspection have to be added as well (quality costs are described in Chapter 17). However, in a well-designed and managed quality assurance program, the overall costs for assurance remain less than the costs of defects in the long run, so that total quality costs are minimized. This positive trade-off of defect and inspection costs is often overlooked by management when it considers quality control as a cost, and not as a profit center. Reducing costs constitutes a profit increase. The contribution of quality control and the ramifications of foregoing such procedures is well understood by a competent manager. *Quality is free, yet not a gift!* It is defects that cost money.

The changing nature of defects must be thoroughly understood. When quality is seen as subject to change in the life of an item's reliability, then, logically, such change affects major quality characteristics. Quality *does* necessarily deteriorate with age, but, as the example of wine proves, it will improve before it turns sour. A used car's quality diminishes, yet, as a "classic," the car might gain in value if still performing. In more recent years quality costs have been compiled as life-cycle quality costs. This is in accordance with the definition of quality of design, conformance, and performance that encompasses the total quality lifespan and quality assurance of an item.

With this modern understanding of the life-cycle concept, quality costs, quality assurance, consumers, producers and suppliers all play their role in preventing causes of defects. Therefore, total quality assurance means more than just inspection.

Some defects are created by consumers. The consumers' own responsibility for creating defects, or making an otherwise safe product unsafe, further adds to quality problems for producers. As many design engineers would say, it is easier to make a product foolproof than to make it damn-fool-proof. What is the dividing line between the two? A good deal of damage to quality of product performance is the result of consumer neglect. Examples abound: secretaries drive staples into fingers instead of papers; hairdryers are run in refrigerators to do the defrosting; people using lawnmowers cut their toes instead of the grass, and so on.

1.4 DEFINITIONS

The technical language of quality assurance with its terms and concepts is not standardized. This means that communication requires definitions. Students who are to take on the role of quality control specialists should fully grasp the importance of various generic definitions and try to understand the meaning behind each of them. Table 1.1 gives some commonly used generic definitions along with one or more source of each.

1.5 KEY TERMS AND CONCEPTS

Quality is understood as the characteristics of a product or service that are designed to meet certain needs under specified conditions. Customers perceive quality as satisfactory or unsatisfactory. Quality is associated with all product or service stages, that is, with design, production, and performance. Quality has value not only in terms of "fitness for intended use," but also in the market, where quality is expressed as price.

Specifications describe required and expected quality. *Tolerance* is a permissible variation from a nominal value and specification. *Standards* describe normal specifications.

The degree of quality measures relative attainment of specification.

The level of quality is comparative in relation to similar items, brands, and processes. Price differences normally also reflect the quality levels. *Excellence* is a relatively high quality level, and/or degree.

Quality characteristics are individual attributes and variables (measurements) subject to planning (design, specification) and control (conformance, performance). Quality can also be understood as the totality of quality characteristics of a product or service in reference to intended use and application.

A defect is an unacceptable deviation from an expected, required, or specified quality characteristic. Different defects have different impacts on customer dissatisfaction, safety, and quality of life, with respective differences in costs. Defects are effects of causes that can be prevented by quality assurance.

TABLE 1.1
Commonly Used Generic Definitions

Definitions	Source
Quality	
"The totality of features and characteristics of a product or service that bear on its ability to satisfy given needs."	1, 4, 5
"Fitness for use."	2
"Conformance to requirements."	3
"The degree to which product characteristics conform to the requirements placed upon that product, including reliability, maintainability, and safety."	6
"The degree to which a product or service is fit for the specified use."	7
Quality Assurance	
"A planned and systematic pattern of all means and actions designed to provide adequate confidence that items or services meet contractual and jurisdictional requirements and will perform satisfactorily in service. Quality Assurance includes Quality Control."	8
"All those planned or systematic actions necessary to provide adequate confidence that a product or service will satisfy given needs."	1

1. *ANSI/ASQC Standard,* "Quality System Terminology," A3–1978, prepared jointly by the American National Standards Institute (ANSI) and the American Society for Quality Control (ASQC).
   2. Juran, J. M., editor-in-chief, *Quality Control Handbook,* 3rd ed., McGraw-Hill Book Company, New York, 1974.
   3. Crosby, P. B., *Quality Is Free,* McGraw-Hill Book Company, New York, 1979.
   4. DIN–53350, of Deutsches Institute fuer Normung, Teil 11, Beuth-Verlag, Berlin.
   5. EOQC, "Glossary of Terms Used in Quality Control," 5th ed., 1981, published by the European Organization for Quality Control.
   6. QS–Norm Draft of Swiss Standard Association, 1981.
   7. Seghezzi, H. D., "What Is Quality: Conformance With Requirements or Fitness for the Intended Use," *EOQC Journal,* 4, 1981, p. 3.
   8. Canadian Standard Association Standard, Z299.1–1978, "Quality Assurance Program Requirements."

Failure is a kind of defect related to *reliability*, which is a measure of failure-free performance. A defect that is an unacceptable deviation from a quality standard is called *nonconforming*.

Defective relates to an item or service that has one or more defects. For a given lot number, the fraction or percent defective measure the lot quality.

Quality cost, simply put, is the total cost caused by defects and the cost for preventing and correcting defects. Quality assurance must be cost-effective in the long term.

1.6 SUPPLEMENTARY READING

While Inflation Rages:—A Deflation in Quality of Goods and Services[3]

According to a survey made in the early 1970s by public opinion specialist Louis Harris: "Something went terribly wrong with the reputation for quality of American products.'' It found that almost three out of every four Americans felt something was seriously amiss in our quality controls. A 1977 report published in the *Harvard Business Review* reached similar conclusions, although the degree of dissatisfaction found was not nearly so great as in the Harris survey.

In fact, quality deterioration *is* widespread. The deterioration includes both goods and services—it represents disguised inflation. Sometimes, for example, price may not be increased (hence nothing is reflected in the price index), but quality is reduced. Often, however, price is increased while quality is simultaneously lowered, which allows some capturing of the increase in our price measures, but not the full amount. Illustrations of deterioration are not hard to find.

Not for me, at any rate. In 1968, for example, I went into a popular ice cream parlor in Tennessee, ordered a chocolate milkshake, and explained to the waitress that I wanted it all chocolate, that is, chocolate (rather than vanilla) ice cream as well as chocolate syrup. She looked up and replied, ''You want ice cream? I'm sorry, sir, but ice cream is 10 cents extra.'' I had always assumed that milkshakes were made with ice cream, but I then learned that now the base was malt. More recently, the best attraction at a fast-food establishment I frequent for lunch—the $1.29 all-you-can-eat salad bar—dropped the olives from the assortment tray. (The olives had made an otherwise bland salad rather tasty.) When I asked about the now crummy salad, I was told, ''The price of olives has really gone up, so we dropped them in order to maintain the bargain price!''

Similar examples abound. As a lifelong tennis player, I have observed two interesting phenomena. First, there has been an obvious decline in the quality of tennis balls. They do not last as long, they do not hold their bounce, and they lose the fuzz napping much more rapidly than in the past. The second phenomenon is the move away from wooden rackets. Several years ago I switched to a metal racket, although I preferred the feel of wood. I did this because, for no apparent reason, my wooden rackets had begun to break with unsettling frequency. When I mentioned this breakage to people in the business, I was told that owing to soaring lumber prices, cheaper, less resilient woods

were increasingly being used to produce the traditional racket. Another example of quality deterioration is found in book production. Coated cloth on many covers has been replaced by more fragile synthetic materials. Type size has been reduced. Indexes have often disappeared. It is estimated that in roughly 85 percent of the books manufactured today adhesive binding is used in place of the more durable Smyth sewing. Many newspapers and magazines have reduced the size of their pages to save costly newsprint. Picture sizes have shrunk.

More instances: Cosmetic firms have been experimenting with reduced fragrance levels in several products. Vegetable fat is substituted for cocoa butter in chocolate chip cookies and chocolate powder. Vending machine companies are applying the shrinking-candy-bar technique to sandwiches: you get less beef, cheese and other ingredients, yet the sandwich sells for the same price. The banana split I savored for so many years as a child is today a mere shadow of its former grandeur. Not only is its price four or five times higher but it contains far fewer frills. Often the ''standard equipment''—whipped cream, nuts, sprinkles and all that—is only available at an extra price, if at all. During the Nixon price control period a friend who frequently ate at fast-food shops complained to me that the amount of gristle in the hamburgers served had increased markedly.

The tremendous increase in housing prices in recent years has been intensely publicized. Far less attention has been drawn to the fact that, accompanying these rising prices, has been a noticeable decline in home size (lot size also). Construction quality has gone down. True, precut, prefabricated, and prefinished components have moderated price increases. But lightweight concrete floors substitute for hardwood. Pressboard replaces solid wood in most kitchens. The new home contains very few of the trim items of yesteryear, such as ceiling moldings, wallpaper, built-in bookcases, and cedar-lined closets. All this is not to be interpreted as a condemnation of new housing, nor should it be inferred that there have been no improvements. Nonetheless, it is unarguable that many features of today's home are cheesy.

In today's economy there may well be a Gresham's Law of economic goods—lower quality items tend to drive higher quality ones off the market. And that applies to services as well as goods. One investigation of substandard audits concluded that too much price competition may well hurt the quality of audits because it leads to cutting corners. While the prices of postal services rise, the number of post offices goes down and deliveries of mail are made less frequently. The gasoline service station may one day become extinct, although there are still plenty of stripped-down gas outlets. Further, many of the fringe services of past years, such as free maps, are no longer available. Since the oil embargo of 1973 more stations are closed on Sunday or open for a reduced number of hours during the week. This state of gasoline affairs has also cut down on related services such as the possibility of obtaining directions and using restrooms.

One may reflect that these examples are something less than calamitous. The matter, however, can be far more serious. Is it, for example, too bizarre to consider that the recent rash of airline crashes, mishaps, and DC-10 problems is at least partially a result of disguised inflation? While this claim might seem ridiculous to some, a leading international airline executive recently stressed the seriousness of airline safety problems, emphasizing that the "temptation to cut corners, seen or unseen" is very real. Other, less alarming examples abound. Increasing numbers of automobile owners are doing their own maintenance and repair work. A major impetus is cost saving, but a recent survey found that nearly 33 percent believe they do a better job than the typical hired mechanic. Hidden price increases can also be found in tighter warranties. In late 1976, for example, Chrysler Corporation changed its warranty on new cars from the first 12 months, regardless of mileage, to the industry standard of 12 months or 12,000 miles, whichever comes first. Similar reductions in warranty coverage have occurred on new radial tires.

At many discount stores, prices are reduced—but so are services. With fewer salespeople around, customers spend more time searching for the goods they want and wait in line longer to pay. Many food stores today have fewer baggers and bottlenecks at the checkout counter are common. At many banks there seem to be fewer tellers than in the past and the waiting lines are longer. These costs to the customer are surely real, but are not reflected in our traditional price indexes. Accordingly, our cost-of-living statistics may well understate the true level of inflation if quality deterioration is so widespread that it outweighs the qualitative advances that also occur.

For years, American economists have written about suppressed inflation in the Soviet Union. The unavailability of spare parts in the USSR, Cuba, and other communist countries has been given much notoriety. And, without question, inconveniences, queues, and waiting periods do cause a serious loss for the consumer. Such are the ways inflation expresses itself in these countries, with their controlled prices.

Unfortunately, Americans are now learning firsthand about parts problems and waiting periods. For more than two months I have been unable to get a part from the dealer who sold me my 1979 automobile. Spare parts are now a common problem in the areas of stereo equipment and household mechanisms such as central air conditioning, furnace and heating systems, and plumbing. You have to wait for long periods for several kinds of new model automobiles.

Some writers have made much of the long-standing assumption that the productivity of civil servants never changes. They then note that the introduction of computers and sophisticated office machines greatly increases the effectiveness of individual workers. This has been substantiated in a number of GAO studies of labor productivity in government. Yet there are vast areas of government services where a clear-cut measurement of productivity is not feasible (at least, not yet). For at least one important group of government employees—teachers—it is clear that their "productivity" leaves something to be desired. The Scholastic Aptitude Test scores, in a 15-year decline, indicate something about the erosion of quality in our primary and secondary schools.

I certainly have not proved statistically that quality deterioration has outweighed qualitative advances but merely have stressed that such deterioration is widespread and must be recognized. Indeed, in an inflationary age it is quite logical to expect increasing recourse to disguised measures. While the statisticians at the Bureau of Labor Statistics do make some adjustments for quality improvements, there are no similar corrections for product deterioration. For example, about two thirds of the increase in 1980 automobile prices is excluded from the Consumer Price Index on the ground that it reflects an improved product rather than simply a higher price tag for the same item. Yet no corresponding adjustment is made when, for example, renters who have to put up with sharply reduced maintenance must make the same rent payments. Or when riders travel on buses that are rattling wrecks, spewing out black clouds from their exhaust pipes.

Curiously enough, the economics profession, with precious few exceptions, has blinded itself to quality erosion possibilities. Indeed, the conventional wisdom maintains that price indexes overstate inflation rates because they fail to adequately reflect quality advances. The implicit assumption is that product evolution has been onward and upward. Each year, it is suggested, consumers do pay "nickels more," so to speak, but receive a "half-dollar's" worth of improvements. Since only the "nickels" and not the improvements are counted, the price index will rise. Thus a part of inflation is "illusory"—the product of faulty calculations. And, indeed, some very sophisticated, excellent research indicated in the past that in some instances this was exactly what has occurred. These findings, coupled with the stature of the researchers themselves, tended to support the emerging perception that there was less inflation than appeared. Such a perspective apparently took hold in the economics profession in the late 1950s. Yet a blue-ribbon panel commissioned to study price statistics cautioned in 1961 that the prevailing view was not solidly grounded. "We are impressed," they wrote "with how little empirical work has been done on so widely held a view. . . ."

Unquestionably, our traditional price measures inadequately reflect quality changes, either positive or negative. But we must recognize that the conventional wisdom concerning the quality bias in price indexes was not well-grounded to begin with and gained acceptance in a period of relative price stability. The many difficulties in the design and implementation of empirical studies to measure quality changes is an important reason why work in this area has not been definitive. Nonetheless, such work must be continued and expanded. If it is found that deterioration has outweighed improvement, the conventional wisdom would

not only be displaced but reversed. The "less inflation than appears" doctrine would give way to "more inflation than we thought."

In my opinion, consideration *must* be given to quality erosion. Indeed, in view of President Carter's anti-inflation and "price-watching" action, it is likely that inflation will more frequently go underground and quality deterioration will become even more prevalent. If so, it is distinctly possible that our traditional price level measures will increasingly *understate,* not overstate, real rates of inflation.

1.7 DISCUSSION AND REVIEW QUESTIONS

(1–1) *Quality is a complex phenomenon but still must be specified.* Discuss this from the producer's and then from the consumer's point of view.

(1–2) *Price and quality are interrelated.* Explain.

(1–3) *Contracted quality establishes liabilities.* Is this possible in view of the fact that every aspect of quality cannot be specified and standardized? Discuss.

(1–4) *A luxury car is more than a means of transportation.* Explain.

(1–5) *Quality expectations will change during the various life phases of a product, but not during the design and production phase.* Discuss.

(1–6) *Liability for quality is shared by the consumer.* Explain, using an example.

(1–7) When are quality characteristics sufficiently specified? What is the dividing line between sufficient and insufficient specification? Discuss, using an example.

(1–8) Does "quality assurance" mean "quality insurance"? Explain.

(1–9) *Both consumer and producer must agree on the meaning and degree of quality assurance.* Explain.

(1–10) *In inflationary periods, companies in most industries must compete with better quality, rather than with price.* Explain.

(1–11) *Bankruptcies and plant closures do not always mean lack of work orders.* Explain.

(1–12) *The quality you want to buy must be the one you want to produce.* Discuss.

(1–13) *To assure quality is easier in a small business than in a large one.* What are the reasons?

(1–14) *Consumers are expected to spend, rather than to save, during a period of economic recession and unemployment. But, on the other hand, high interest rates for borrowing, fear of becoming unemployed, and inflation nibbling away disposable income are powerful forces that curtail buying of new and more products and services. Furthermore, a generally high standard of living leaves people little to desire. The market seems to be saturated. Should consumers be expected to buy the third car, or to sell cars while they are still in perfect operating condition? Should manufacturers reequip their production facilities when high cost increases and dropping demand make it wiser to use the available production capacity and machines, rather than to expand and buy new ones? Consumption obviously has its limits.* What does quality assurance mean under these conditions prevailing in Western societies? Discuss.

(1–15) *Perception of quality by the consumer can be as important as reality; it has the same effect in the market.* Discuss with reference to the automobile market.

(1–16) *The principles for improving quality were developed in the United States before and during World War II—and then neglected here.* What were the main reasons?

(1–17) *Major American automakers have become tough with their suppliers and dealers in matters of quality.* What must this new strategy involve in order to bring about better cars and more satisfied customers?

(1–18) *A defect for one consumer is not a defect for another.* How can then defects be defined and prevented? Explain.

(1–19) *To prevent defects is usually more rational than to correct them, but not always.* Discuss.

1.8   NOTES

1. Note: We will frequently use the word "products" as a short generic term to designate both goods (e.g., bread, automobiles, drugs, etc.) and services (e.g., education, hospital care, restaurant facilities, etc.) respectively, as the outputs of the manufacturing and non-manufacturing (i.e., service) industries.

2. Taylor, Claude I., President and Chief Executive Officer, Air Canada, "Welcome Aboard—Message from the President," *en Route,* Air Canada's magazine, division of Southam Printing Ltd.; Weston, Ontario, March 1984, p. 1. Reprinted by permission.

3. Losman, Donald L., "While Inflation Rages: A Deflation of Quality of Goods and Services," *across the board,* 1980. Copyright © The Conference Board Inc., New York. Reprinted by permission.

1.9   SELECTED BIBLIOGRAPHY

Crosby, Phillip B., *Quality Is Free,* McGraw-Hill Book Co., New York, 1979.

Garvin, David A., "Quality on the Line," *Harvard Business Review,* September–October 1983, p. 65.

Juran, Joseph M., "Japanese and Western Quality—A Contrast," *Quality Progress,* December 1978, p. 10.

Leonard, Frank S. and Sasser, Earl W., "The Incline of Quality," *Harvard Business Review,* September–October 1982, p. 163.

Losman, Donald L., "While inflation rages—A deflation of quality of goods and services," *across the board,* July 1980, p. 36.

Peters, Thomas J. and Waterman Jr., Robert H., *In Search of Excellence: Lessons from America's Best-run Companies,* Harper & Row, New York, 1982.

"Quality: Special Report," *Business Week,* November 1, 1982.

"The Quality Concept Catches on Worldwide," *Industry Week,* April 1979, p. 125.

2 CHAPTER

ASSURANCE OF QUALITY

The assurance of quality has become increasingly important as societies and technologies have become more complex. While in many areas old approaches and knowledge have been outgrown, demand has appeared in entirely new areas. Some of the assumptions on which products were manufactured and services sold, are put into doubt by new developments demanding assurances of quality. Consumers want a claim or pledge from the producer that, by virtue of procedures and activities which they engage in and employ, does, in fact, provide confidence in the quality of the product.

Once quality has been sufficiently defined and agreed upon, responsibilities are established. The producer is now in a position to distribute areas of responsibility for the existing production capacity, which should be able to assure conformance with agreed upon and formally contracted obligations. One must keep in mind that contracts do not completely outline all such obligations. Much is left to the personal attitude and relationship when it comes to satisfying a customer's right to quality.

Quality assurance is still considered by too many as imposing additional costs, rather than being a profit center. Others have always held that quality assurance is a naturally inherent element in all business dealings and a major source of their success. These people are not much aware of modern concepts of quality assurance, yet they practice it.

Only in more recent years has a deeper appreciation of quality assurance widely evolved. The fact that better quality and greater confidence in products and services is the key to both a high standard of living and business growth is now recognized. Starting from vital areas such as nuclear, defense, and aerospace engineering, efforts are being made to introduce modern quality assurance systems for all products and services, whether the producer is a grocery store, hospital, government agency, airline, or university, quality assurance must take place. The need for such assurance keeps growing as a result of increasing technological complexities, coupled with growing quality demands by the public. Quality awareness further comes about because of difficult business conditions. Economic recession exerts pressure for improvement. Researchers begin to trace modern quality assurance back in history in order to discover trends, and deepen the understanding of present day situations and opportunities in the quality field.

We shall now highlight the evolution of fundamental concepts and practices of quality assurance. A description of the conceptual ideas and some real life examples shall follow.

2.1 THE NATURE OF ASSURANCE

The following viewpoints provide concepts of assurance that consider together all the efforts that go into production or provision of quality items.

Quality assurance, as an act by the producer means promise, conformance with quality specification, and to meet other understood requirements. Thus, quality assurance means to create confidence not only in the integrity of the product and service, but between producer and consumer; to create confidence in relationships and communications, organizations and management and, above all, to create a prosperous business future. This confidence and mutual trust more and more substitutes for direct testing and evaluation of items, particularly those of an advanced technical nature, by the customer.

Quality assurance can also mean that the producer takes every measure to ensure positive quality for the physical as well as nonphysical characteristics attached to its products and services. The traditional handshake after a business deal has become a much more formal matter in modern quality assurance. Even when a customer does not demand and expect such assurances, a prudent self-assurance of quality on the part of the producer is a very wise precautionary measure.

Quality assurance means to inform the consumer about the quality of the product and to provide sufficient help and guidance, as well as service for proper and safe usage and application. Implied in the term "quality assurance" is care in the design, production, and servicing of a product. However, "quality control" or "product inspection" mean that the producer has observed more restricted controls, which nevertheless might be absolutely sufficient. What, in fact, the producer assures is best defined not only on the basis of product specification, but also with reference to control programs and respective standards.

Quality assurance normally requires documented evidence for respective procedures and measures in production, for overall attainment of quality. While in former times quality assurance rested on the integrity and reputation of the producer, consumers today expect evidence in the form of documented and implemented quality management programs. Customers must be able to rely on such programs when the actual product quality cannot be readily determined by them. Only then is the "vendor beware" principle underlying modern quality assurance acceptable. Quality program standards assist producer and consumer alike to outline in contractual terms what quality assurance is to mean in each case.

Quality assurance is the most desired goal in quality management. Obviously, both quality and quality assurance establish consumer satisfaction and confidence. If quality assurance remains a mere lip service, it soon will undermine, rather than promote, future business.

What has always been implied is now made explicit in response to the needs of modern times. The producer ought to know what to expect and rely on, and then act on what and how to organize and implement production in accordance with extending quality assurance to the customer.

Quality assurance does not end with management that contracts with the consumer. Rather, it is an ongoing commitment that requires concern from everyone involved: staff, suppliers, governments, and, in a way, the consumer.

Total quality assurance comprehensively and continuously orients work towards quality goals. Maintaining assurance becomes a matter of decency and fairness. The benefits come not only in higher revenues, market shares, and growth, but also in the form of job security and job satisfaction. Quality assurance thus creates a quality image, not only of a product but also of its producer, staff, organization, and society as a whole.

Totality in quality assurance cannot mean total avoidance of defects and failure. Basic uncertainty, lingering unknowns, and unexpected events can still cause unavoidable defects. This is why quality assurance means to comply convincingly with established standards related to production of products and/or services as understood in a contract.

Quality Control

Quality Control, as the core activity in any quality management program, refers to the process through which actual quality performance can be measured, compared against a qualified standard, and acted upon to prevent deficiencies. Defects will occur either randomly or not randomly. Here, quality degree measured in terms of defects, must be distinguished from quality level, which is the market value of an item. For example, a Cadillac is of a higher quality level than a Volkswagen, expressed in their respective prices. Yet, the degree to which the quality levels have been achieved and measured by defects might differ in both cases. When compared with quality control, quality assurance is a more comprehensive management function that includes product design and performance.

Producers assure those quality characteristics that are specified in advertisements, orders, and contracts, and about which both consumer and producer have arrived at an understanding. As indicated in Figure 2.1, consumers have indefinite desires for products and services and for individual quality characteristics. Further, depending upon the consumer's standard of living and disposable income, only certain products and quality levels are affordable, attainable, and thus demanded. Consumer quality expectations are not entirely definable by the producers, nor even by the consumers, and the extent of quality assurance expected beyond the specified quality characteristics remains largely unsubstantiated. Here, the general quality image and experienced quality assurance stays in the mind of the public, (partly influenced by consumer reports) and substitutes for guarantees of *mere compliance* to what has been assured. In their efforts to attract customers and satisfy them more than other competitors, producers will assure more than just compliance with specifications. They will also account for the nonspecified but attainable, recommended and appreciated quality attributes.

Figure 2.1 also demonstrates that the *feasible* quality assurance a producer can extend to a consumer is limited by the *existing capacity* to create the required quality. Even when the producer has ascertained that the capacity is in line with the state of the art and that the quality assurance program is effective, the potential of staff and equipment to produce the

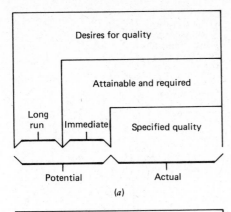

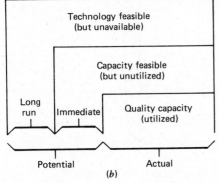

FIGURE 2.1   The quality assurance management strategy in perspective of a never-ending goal of improvement until the two areas match. (a) Consumer's quality expectations. (b) Producer's quality capability.

required and competitive quality might be further improved. This is only possible by *utilizing*, that is, motivating staff and using available equipment and organization better for the creation of quality. This aspect of effort is termed the *quality improvement program* of the company. Quality assurance then means that producers dynamically use every means at their disposal to create as high a level of satisfaction in the mind of the consumer as possible.

2.2   EVOLUTION OF QUALITY ASSURANCE

In theory and practice modern quality assurance has come about through several stages since the beginning of the Industrial Revolution. Figure 2.2 demonstrates the evolutionary steps in quality assurance as major milestones. In past centuries quality assurance remained intrinsic to the understanding of good craftsmanship and workmanship. Guilds and courts regimented work principles and quality of goods and services fairly strictly.

Good workmanship still has to be considered the foundation of sound quality and quality assurance. However, workmanship, as a concept by itself, is not restricted to the operator or worker in the conventional sense. Formerly the master, and today the foreman, set examples for good workmanship, as without doing so, they could not provide leadership and supervision. Even senior planning and controlling management must gain respect and acceptance through its own performance and leadership capabilities, and by providing an efficient system

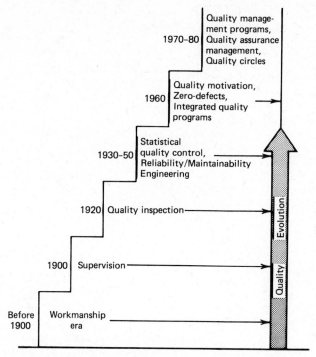

FIGURE 2.2   Quality management historical evolution. In this evolution, each new stage includes the preceding element, that is, supervision over previous independent workmanship.

to work within. Quality assurance at the worker's level depends on the sound workmanship of management. This has not changed and will not change, in spite of technological and social advancement, since the time when master craftsmen personally influenced and directed the quality workmanship of the journeymen and apprentices. The concept of modern quality management and its real application rests on the actual quality of managerial practice itself and the leadership provided in quality assurance matters. Perhaps the fact that most defects are *management-controllable* rather than caused by the workman,[1] should lead attention back to the era when the master identified personally with the work of those under his care. The purpose of modern participatory management is to restore the direct link of those who plan and control and those who are to follow the work directives.

The Taylor System and Quality Assurance

The prevalence of poorly trained workers in a growing manufacturing industry around the turn of the century led the engineer Frederick Taylor[2] to define and delineate management as the separate planning and control functions. He recognized that such specialization raised the productivity of workers. Management, as the supervisory force, was given respective training for designing the workplace, hiring and training the worker, and for supervising work performance. The former rule of thumb was replaced by a scientific approach. More importantly, the work itself became completely planned and directed, with the workers left only to execute such directives.

Management and workforce separation, however, had a different impact on quality assurance than on pure productivity. The responsibility for quality was included in the required meeting of the performance standard. However these standards, in the form of expected quantity of output per time unit, overshadowed the quality specification. Quantity, rather than quality, became stressed under a work system where one party would set standards and then solely verify conformance with these.

Advent of Specialization

The Taylor system remains unchanged even in many companies today, and the resulting dichotomy of management and labor does not always motivate better quality of performance. The risk of poor workmanship was recognized particularly during the First World War, when the scope of manufacturing increased. The division of management and labor had set in motion a specialization in the supervisory force, with quality assurance becoming one of the special functions. Quality was to be *controlled and inspected into the product,* rather than to be *built-in.* Inspection became the sole approach. Special inspectors were trained for quality assurance of work.

Inspections and Inspectors

The independent inspection, with its testing of product quality and verification of conformance to specifications, had to have some demotivating effect on the worker. Obviously, many rightly felt that quality assurance was the responsibility of the supervisor and the inspector alone. However, policing compliance with standards often resulted in punitive action in cases of nonconformance. Defects usually were not traced to their real causes. Producers mainly *weeded out* defective items, rather than to take corrective and preventive actions. This largely ineffective quality assurance, which relied heavily on the inspection force, gradually changed for the better during the late 1920s, with the advent of scientists in the field.

Effect of World War II

The difficulties of making consistent and reliable defense products, together with the inability of manufacturers to deliver in time, became the top concern of many upper managers. Following World War II, the same manufacturing difficulties became evident in nonmilitary (civilian) organizations. While managers became aware of what was needed, it remained to find ways of organizing to create an integrated and comprehensive quality assurance planning and control system.

Statistical Quality Control

Scientists, and especially statisticians, became involved and developed more reliable and valid statistical inspection methods. Walter A. Shewhart of the Bell Telephone Laboratories is generally credited with the introduction of statistical quality control methods in the United States.[3] In 1929, H. F. Dodge and H. G. Romig applied statistical methods to sampling and prepared a variety of sampling inspection plans.[4] The fact is that any process produces variation of quality attributes and variables. These variations follow certain known statistical laws of distribution. By use of control charts significant deviations of expected process and work performance can be monitored within calculated probable confidence limits.

This truly scientific method, as it is used in science to prove or disprove hypotheses, allows for inspection of quality using relatively small samples. Moreover, an objective analysis of the *causes and prevention* of poor quality turned what was formerly inspection into a more sophisticated quality control function. While control charts help to control work performance, inspection of completed lots has also been converted through statistical approaches. Acceptance sampling plans inform inspectors what sample size to draw and the number of defectives to tolerate before acceptance or rejection of the lot. Such sampling plans are based on sound sampling theory and permit reliable inferences from the findings in the sample. Statistical quality control largely improved quality control effectiveness as much more valid acceptance inspection could now be performed.

Unfortunately, many managers considered these methods too complex during the 1950s and there was not much pressure for their application. However, these statistical methods were readily accepted outside the North American Continent, in particular in Japan and Europe. Beyond any doubt, the current superior quality of these countries' products is, to a large extent, due to the mastery of statistical controls at the management and operator levels.

Reliability Engineering and Maintainability

An important development that took place in the 1950s concerned the methods of predicting unacceptable failure rates of advanced military equipment and electronic systems. Many rightly felt that quality control alone could not guarantee the ability of the design to perform under the stresses of field operation for long periods of time. This is where the discipline of reliability engineering (or product reliability) came to be known. The Advisory Group on Reliability of Electronic Equipment (AGREE) issued a report[5] in 1957 under the sponsorship of the U.S. Assistant Secretary of Defense with the primary subject being assurance of products performing adequately with a predicted probability of success for a given time. The discipline of maintainability began to appear concurrently with the movement towards reliability engineering, and with the realization that no equipment would ever be made up of ideal parts and that there would always be downtime on account of repairs and maintenance. Again, the U.S. defense forces, particularly the U.S. Air Force, were among the first to issue formal maintainability specifications.[6] One of the best sources for maintainability design principles is a widely used U.S. Navy handbook of maintainability design criteria.[7]

Motivation Programs

Meanwhile, in North America, psychologically-oriented motivation programs were devised by large corporations to bring about better workmanship and quality. This was done in the realization that the best control methods could not substitute for carelessness and apathy on the job. "Zero defect," "right the first time," and other motivational programs[8] initiated a growing quality awareness in companies, and particularly in management. However, the effect lay more in gradually overcoming Taylorism than in attaining higher quality performance. The application of modern problem solving methods (such as quality circles) for better quality in small groups, has meant a further adoption of psychologically based programs for quality improvement and control. However, a real breakthrough for quality management has not been achieved as a consequence of these programs.[9]

Quality Management Programs

Major government procurement branches, particularly in the defense area, have traditionally played a key role in modern quality assurance development. Their inspection force has helped many suppliers to establish statistical and other quality control procedures.

This strategy has stipulated that along with traditional product and contract specifications, suppliers were to have their own quality assurance programs. Governments merely described such program requirements in *quality program standards,*[10] for example, the MIL-9858A. Under this policy major customers of many companies induced the integration of individual quality controls into comprehensive quality assurance programs. Emerging quality management pioneered quality programs that not only complied with the standards but also uniquely suited the product and company conditions. This latest development stage has finally brought quality assurance as an appropriately recognized management responsibility that is on par with other managerial tasks.

Reliance of major customers on management programs for quality and reliance on management's integrity is, by necessity, a result of the ever-increasing technical complexity of products, processes, and quality controls. The supplier was to be held liable not only for meeting specifications but also for backing up quality assurance with a sound and valid program.

Feigenbaum in his book *Total Quality Control* (see chapter bibliography) had a great impact on the quality control profession during the 1960s. He presented the concept of total quality control and promulgated the principle that everybody in a company shares the responsibility for quality of performance and output. This message was long overdue. At the same time, Juran began to show ways for management to gain improved quality assurance. Juran's *Quality Control Handbook* (see chapter bibliography), still one of the classics in the field, taught management about quality improvement strategies and breakthroughs.

Summarizing the above, all historical evolutionary stages in modern quality assurance blend into integrated and coordinated management programs. All management levels, and not only the middle supervisory staff, become involved in quality planning and control. The program, as a framework and master plan, directs operations, activities, and decisions towards the attainment of predetermined quality goals. Quality assurance more recently has been further linked with management concepts of management by objectives, systems management, project management, and so on.[11,12,13] One can speculate on the further advancement of quality assurance once modern quality management is seen against the historic evolution of and future needs for improved quality assurance management.

Recent Development: Quality Assurance as a Key Management Function

The amalgamation of quality assurance management within the art and science of overall management has just begun. Much research remains to be done before truly effective quality assurance can be established in large and small companies. The current economic recession is at least partly caused by the superior quality and quality assurance of international competitors. Through a strategy for quality improvement, governments and major customers are once again taking leadership in pushing for the broadening of quality assurance into the fields of health services, environmental controls, nuclear power generation and so forth. *Quality of life* and *quality of work life* have become known terms and goals in the public domain that indicate the existence of this movement for quality assurance at the community level.

The question of whether quality assurance is a key management function is not merely an academic matter. In the research and educational institutions much work is left to be done as quality awareness emerges. Whether or not a broad strategy to achieve quality will take hold in the near future depends largely on the success of management and engineering education in this area.

The most promising development of late is that the business community has moved forward by introducing quality assurance programs, partly in response to pressures from competitors, customers, and government regulations. Companies' survival is at stake. These pressures gradually persuade the academic community, as well as those in quality assurance disciplines to create means much greater than statistical quality control and, perhaps, quality circles.

Through cooperation within the academic and business community, advancements can be expected. More and more complexity in products, and advancement in technology, will help propel progress and bring about more quality assurance concepts and practices. Quality control methods have to become a part of modern engineering and managerial technology. We can only hope that the pressure for quality assurance emanating from the market will not diminish in the near future when economic revival sets its pace once again as the norm.

2.3   TOTAL QUALITY ASSURANCE

The concept of total quality assurance seeks commitment of all those connected with a product or service. Management of the producing organization must decide in the free-enterprise economy whether quality is *inspected-in* or *built-in*. Quality assurance policies and systems must range from those where defects are left for the customer to detect, to those where management takes every reasonable step to prevent them. Quality costs at both extremes can be relatively high or low. For instance, in a simple product with defective items that do not cause any harm to the customer (or in which defects occur only occasionally) and that can be rectified by the consumer without much difficulty, quality has extremely low costs. In this case, the consumer would not want to pay for a very costly quality assurance system.

At the other end of the spectrum, a defect can cost irreparable harm to the customer, such as in an air disaster caused by a defective bolt and anchoring device, or a plane carrying passengers running out of gas in mid-air. Both examples may sound impossible, but both have actually happened. Here, the most costly quality assurance system is absolutely warranted and paid for by the consumer. Thus, it is the cost of defects that really decides on a reasonable, affordable, and optimal quality assurance system.

Figure 2.3 shows three different basic kinds of quality assurance systems. The most *comprehensive system* (therefore called *quality planning and control*) *hunts down* defects during the design, production, control, and performance stages. It stretches over the entire product life. The modest *inspection system* on the other hand, provides moderate verifications and testings, either by the worker or the independent inspector, at crucial points in actual production and performance. This simple inspection system attempts to rectify defects when they occur, rather than to also trace these to their causes and clarify prevention. The *quality control system* incorporates more sophisticated inspection and control techniques, such as statistical sampling, and is designed more for prevention rather than only for failure correction. The *quality assurance system* also called quality planning and control system, has to be preventive, and built-in, with audit programs to measure its effectiveness. Obviously, these

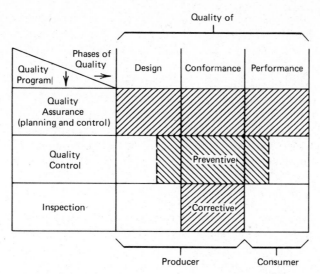

FIGURE 2.3   Three different basic kinds of assurance systems.

are only three basic types and, in each instance, possibly for individual major quality characteristics, special assurance procedures will have to be established that might fall between the basic types.

Total quality assurance or total quality control does not usually mean a cradle-to-grave system with regard only to the product life. The total involvement of all staff in an organization together with suppliers, distributors, and even customers, in bringing about quality and satisfaction has become the more recent understanding of these terms. As has been pointed out, inspection evolved as an independent control function and this led to an undermining of motivation for good workmanship. General technological progress, advancing specialization in the work force, and generally higher levels of training, skills, and workplace expectations suggest a greater delegation for quality from within the organization. Quality circles, job enrichment, and participatory management, are some of the current approaches for a broader and more effective total quality assurance system. The importance of the quality assurance function, as it has evolved from inspection to quality control engineering, now lies in the design, planning, and control of *all* activities related to quality attainment.

2.4   THE SYSTEMS CONCEPT

System is also a term widely used in conjunction with quality assurance. Today we are confronted with such terms as management systems, systems analysis, production systems, marketing systems, information systems, and so on. There is also a theory of systems and practitioners in the field who call themselves systems engineers.

In the most general sense, a system is the phenomenon or idea of an interacting of interdependent elementary components that form a unified whole with a common goal and purpose. Information systems, for instance, can be understood as the regulated, systematic allocation of data or information and systematic processing (arranging, interpreting, transforming) of these inputs. The output of the processed data constitutes a whole, in the sense that it serves its predetermined and predesigned purpose and objective. The input, processing, and output that take place in a system is systematic and systematized.

Wherever an unsystematic interaction of parts prevails, a common purpose or goal can only be achieved in a random fashion, rather than by an orderly and systematic approach. In such a chaotic situation a system is nonexistent, and in the area of information it would mean that data are compiled in a more or less unplanned, irregular, fashion and are processed, communicated, and used in a way that accomplishes its purpose only by chance.

The word *system,* therefore, stands for plan, method, order, and arrangement.

Components of a System

Input, process, and output, are coherent and elementary components of any system. By internal and external forces that are at work within and outside a system, systems or subsystems are created, changed, or dissolved. Any system will show some kind of dynamic inherent equilibrium and operational flow. Its boundaries, when clearly defined, separate a system from other subsystems or parallel systems. Almost all systems are part, in one way or another, of other systems, and are linked to them in many ways. For instance, the concept of a total system describes a complex of interlinked and well-aligned systems.

By its very nature, a system can be cast in a highly abstract language and mathematical model, or it can take the form of a very tangible, physical entity. Often systems in their first phase of existence are conceptual, and are then transformed into an operational and concrete form. Any system must have an inherent logic or rationality which reflects itself in the particular input, output, and objective relation. The objectives for a system normally also determine and influence the underlying principles upon which the system is designed and operates, such as principles of efficiency, economy, or adequacy. Figure 2.4 gives an outline of an education system viewed from a teacher's point of view.

Management Systems and Systems Management

Management is the planning, organizing, initiating, and controlling of operations. As the decision-making force, management is an essential part or function in any organization. Both management and organization can be conceived, described, and established as systems.

Management is one of the subsystems of an organization. Conceived as a system, management is composed of many interacting and interrelated parts—mainly personalities and rules that concertedly perform the function of directing the organization at all levels and phases. The most typical and important output of such a management system are decisions and rules. The input into management systems can be thought of as organizational problems that demand solutions. The intermediate link between input and output in this system, namely the processing unit or *black box,* consists of the various decision-making processes.

A management system is open, in the sense that it has to account for environmental factors and developments, as well as various internal forces and disturbances. It is hierarchical with top, middle, and lower managerial levels. Also, one could distinguish different special areas of management and their respective management systems, such as the functional areas of finance, marketing, and production. In this perspective a management system presents itself

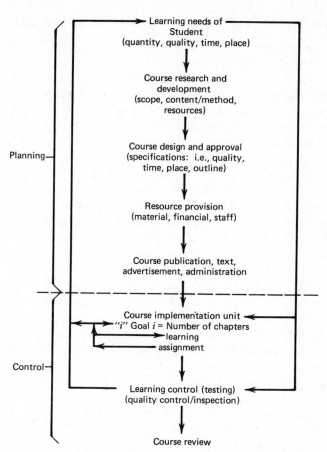

FIGURE 2.4 Systems approach applicable to a course on quality assurance; market: student, employer, professionals, and others.

as being composed of various subsystems. All management subsystems have in common that their main output takes the form of decisions which are correlated within the entire organization.

While we describe management systems in accordance with our general definition of systems, the term management system carries many different meanings within the jungle of managerial semantics. For instance, the term is sometimes used to describe the system of authority and responsibility of members within an organization. Sometimes it is used to typify and classify styles of management such as *conventional* management systems or *modern* management systems. Our concept and explanation of management systems, however, focuses on the common purpose of management within an organization as the decision-making force and helps us to orient our study and practice of management towards the systematic solution of problems. It also provides us with a logical framework of reference and an efficient basis for analysis and design.

Significance and Usefulness of the System Concept in Management

We have seen that systems exist not only as an idea or concept, but in various concrete forms. Any system implies purposefulness, a rational approach to specific goals, and logical, effective coordination of parts or subsystems as a whole towards specific objectives in a smooth flow of operations like a road map. Whether we analyze or design the system or systems of systems, we are bound by logic and rationality in our effort to discover or establish systematic linkages and networks between input and output.

The general significance of the system concept lies basically in the fact that it guides the thinking and, therefore, the action taken towards optimal and effective goal achievement. For the theory (science) or practice of management, this concept must be useful, simply because managing and organizing means to allocate, arrange, direct, and control various resources systematically in a continuously changing competitive environment in such a way that the organization becomes viable, effective, and competitive. The ever-increasing complexity and multitude of problems demand rational decisions and solutions based on a systematic, scientific approach (a systems approach).

Applying the definition of systems to an organization results in a clear definition and a continuous redefinition of specific and general objectives for the organization as a whole and for its integrated parts. These goals, which link any organization with its environment, initiate and direct the various activities or processes within the organization and call for the allocation and input of various resources. Thus the systems concept not only leads managers to recognize their organization as a subsystem within a wider web of all kinds of overriding systems, but also leads them to consider the organization itself as a complex of various kinds of integrated and coordinated subsystems, each having its specific and well-coordinated objectives.

Managers not only become problem oriented, but recognize these problems within the context of the entire organization and its particular objective. While they regard their special area, function, and responsibility in the context of the entire organization, they become open to cooperation across their traditional departmental boundaries for the purpose of solving common problems.

The system concept leads managers to make better decisions for the organization as a whole, rather than simply for the manager's department. Short-term departmental goals receive less emphasis.

The systems concept, in the form of a systems approach to problems, will result in an optimal solution, because the problem is not only recognized in its proper context but carefully and scientifically analyzed to determine the processes that cause the problem. Here the systems concept leads management to apply the various analytical and statistical methods of quality control, operations research, and management science that help discover the root cause of the problem.

The systematic and scientific decision making which is based on the systems concept and systems approach to problems, is the most efficient and often the most simple method for solving problems and complexes of problems. Thus the systems concept not only helps managers to clarify, understand, and master their responsibilities and tasks, but at the same time it also helps to overcome all problems due to the complexity and difficulty of the tasks themselves.

The systems concept allows for the establishment of predetermined decisions or rules and makes it possible for more and more procedures and operations to be better understood in terms of their faults.

Structuring business according to the systems concept does not eliminate the need for the basic functions of planning, organization, control, and communication, yet it provides the basis, the challenge, and the methods for improving these toward the goals of more specific managerial functions and improved performance.

Since all decisions made and actions taken by management ideally revolve around the entire system and its objectives, management efficiency can also be evaluated, measured, and controlled within this wider context. Resource capabilities become better utilized.

Under the systems concept, modern management is transformed to systems management. Managers and their expert consultants analyze existing systems and design, and redesign all subsystems within the context of improving the system for which they are responsible. If faults in the subsystems are corrected for rapid adaption, the atmosphere of blame within the organization diminishes.

The systems concept has particular significance and usefulness, not only for top management or systems experts. Any work that is to be performed in an enterprise constitutes an essential operating part of the whole entity and must contribute towards the smooth effective functioning of the entire system.

Certainly there exist limitations of the theoretical systems concept within practical systems management. While the systems concept stands for utmost rationality, regularity, and order, in performance and goal achievement, management must look after the system itself. As long as management plays an essential part within a system, irrational actions and disturbances must be expected in order to be realistic in problem diagnosis. Also, due to the fact that an organization such as a business enterprise and its individual parts are human-made and based on decisions made by humans, the systems approach can never become the only approach for management. Other approaches, based on a more empirical, historical, psychological, or technical perspective, have their particular value in the study and practice of management. However, the systems concept will, in an essential way, supplement and partly change conventional management into a more scientifically oriented and data-based mode. The idea of systems and management systems has not only helped to clarify prevailing concepts and practices but has also provided the groundwork for more rational and efficient management of organization in the modern environment. The more that managers recognize the extreme usefulness of these concepts and the more they familiarize themselves with the ideas and methods, the sooner a new realization and practice of management will develop that embraces all traditional concepts and approaches.

Application to Quality Assurance

The quality assurance specialists, who have various titles, such as inspector, quality control manager, quality auditor, and so on, need to develop an acceptable quality system in each case that is most effective in minimizing quality costs, maximizing positive quality motivation, and creating highest confidence in product integrity. The underlying concept of a quality system facilitates quality assurance integration and coordination as it envelops the entire organization.

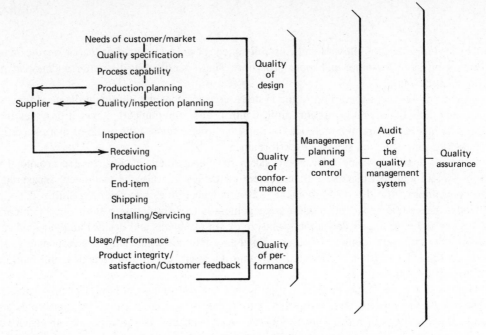

FIGURE 2.5   Essentials of the systems approach to quality assurance.

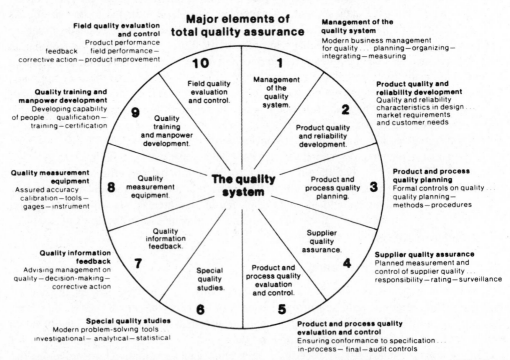

FIGURE 2.6   Major elements of the quality assurance system of Allis-Chalmers. Source: "We Can: Allis-Chalmers Has," *Quality*, May 1981, p. 34, copyright © 1981, Hitchcock Publishing Company. Reprinted by permission.

Figure 2.5 depicts this system approach from which a close relation of quality assurance with production planning and control becomes obvious. While the scheme shows the flow of assurance activities over the whole life cycle of a product, the interdependencies and links between quality assurance, other production functions, and the staff in the individual sections can also be visualized.

This aspect of a productive organization with quality assurance as a subsystem, constructively and powerfully utilizes modern management thought. It complements the conventional line-staff organizational pattern, its hierarchical form, and its reporting lines. In a systems approach, the quality goal directs and motivates all the people working towards achievement. Thus, the management, as the planning and controlling function, not only reorients itself to quality goal attainment; in a well-managed organization, it can and does take on a leadership role. Figure 2.6 is an illustration of how one company views its quality assurance in the framework of a system. Table 2.1 is an organized approach to outlining some of the major units of a quality assurance system.

TABLE 2.1
Various Interpretations of the Major Units of a Quality Assurance System

Division of System	What it Means
Product life-cycle concept	Comprehensive coverage; system starts with marketing research and product or service design and extends over the entire lifespan (i.e., from customers need evaluation to customers need satisfaction).
	Production and distribution phase covered; this system is sometimes called a quality control system.
	Postproduction inspection; inspection is performed on the end-item before distribution, sale, or delivery. This is sometimes called an inspection system.
Product type based	System covers all products or services of an organization, or only selected ones.
Customer based	System is designed for a particular customer or for all customers. In the first case a standard is usually imposed; in the latter case the product is sold in the marketplace.
Organizational unit based	System covers all branches and divisions of the organization or only selected ones. Often these systems are interdependent, such as the corporate and the plant system.
Staff involvement	System covers all jobs and functions, or only selected ones. Degree of staff involvement can also vary.
Marketing based	System adapts to market changes, such as customer demands and supplier capacities.
Technology based	System adapts to innovations and developments in all fields of production, engineering, management, and quality assurance.

TABLE 2.1
Various Interpretations of the Major Units of a Quality Assurance System (*Continued*)

Division of System	What it Means
Community based	System adapts to new laws and socioeconomic changes and developments readily.
	System allows for wide participation in design, implementation, and maintenance phase. People involved include customers, suppliers, company staff, government, employees, and other community representatives.
Information based	System monitors, compiles, communicates, and stores data and information manually or by means of a computer.
Standardization	
External standards	System complies with specific mandatory or optional system standards. These standards can include technical standards, such as for statistical techniques. They may be military standards.
Internal standards	System is standardized within the organization. Documentation exists in the form of a special quality assurance manual.
Design based	System incorporates modern knowledge and experience. For instance, the system might make little use of statistical process control.
Implementation	The system, independent from the quality of design, is effectively adopted by all concerned, including suppliers and customers.
Maintenance	The system is adequately reviewed and audited in order to adapt to changing needs and to continuously improve it.

2.5 KEY TERMS AND CONCEPTS

Quality assurance includes all decisions and actions explicitly purporting the satisfactory attainment of quality in products and services. Form and degree of assurance, as well as its proven effectiveness, influences the confidence of customers and the quality of products and services. The level of quality assurance also depends on quality costs, as it must be cost-effective.

Quality control is an essential element of the wider concept and function of quality assurance, the main purpose of which is to verify conformance to specification. *Inspection* relates to the verification of specific quality characteristics by means of independent tests and predetermined decision criteria.

Statistical quality control accounts for the detection of variations in quality, and applies statistical techniques to control variation and help find its causes.

Quality and reliability engineering is the application of techniques and skills to attain desired and specified quality and reliability of a product.

Quality and reliability management plans and controls programs and procedures designed to attain desired and specified quality and reliability of a product.

Total quality assurance is the composite of all decisions and activities involving the producer, suppliers, distributors, and customers, directed toward attainment of quality.

System is a complex of interacting components that are all directed towards attainment of the same goal and that all share the same environment and resource base. An example is a quality assurance system.

2.6 SUPPLEMENTARY READING

Management Commitment to Quality: Xerox Corp.[14]

I think you will agree that the greatest problems facing American business today are the decline in productivity growth combined with the inability of some industries to meet their customers' quality and reliability expectations. This has affected the ability of many businesses to remain competitive, and we have felt the social and economic impact when this occurs on a large scale.

Let me begin by telling you something about the organization I head at Xerox, which is called the Reprographic Business Group. In this organization there are approximately 21,000 people at sites in the Rochester, New York area; in California; in Canada; and in the United Kingdom and Europe.

The mission of this Group is to deliver technologies and products that provide leadership in the reprographic and electronic printing markets around the world. My responsibilities include strategic business planning, product planning, technology development, engineering, design, manufacturing and procurement. The copier and duplicator products produced by the Group account for somewhere in the neighborhood of three-fourths of the company's total revenue and profit.

In recent years a great many companies have chosen to participate in this rapidly growing industry and have learned to produce high-quality, dependable copiers and duplicators. To carry out its mission successfully in this competitive business environment, my organization must meet several key objectives:

- First, satisfy customers by giving them products that provide better overall value than competitive offerings in terms of quality, performance, cost, and overall responsiveness to their needs.

- Second, improve the quality of life at work for all levels of company employees.
- Third, continue to be responsible corporate citizens in the various countries and communities where Xerox does business.

The objectives I have just described represent no great departure from what Xerox has always tried to achieve. But the criteria for meeting each of those objectives and the level of expectations applied to them are now vastly different.

These new, different, more stringent criteria were imposed on us in a relatively short period of time. They resulted from increased customer expectations—expectations that were enhanced in some cases by competitors who were doing a better job than we were in satisfying customers.

Although it was never easy managing the business to meet traditional objectives under the old criteria, it suddenly became very difficult to meet these same objectives given the new criteria established by competitors and desired by customers. We have asked ourselves more than once how we drifted into such a situation.

You may recall that Xerox owes most of its initial success to the development of an innovative product line. The company was unique in the industry and, in fact, built the industry. In most undertakings, the company went its own way because there was no similar company with which to compare itself. To measure improvement, we compared our performance to our own previous performance. We made progress to be sure, but at our own rate.

In the mid-seventies, competitors began looking at how Xerox managed the business—and in particular market segments they began figuring out how to do it better. But we weren't as curious then about them; we tended still to be much more introspective in defining goals and targets. As a result, in the mid- to late seventies the company began to experience difficulty on a number of fronts:

- In understanding the customers' needs and responding to them as well as the competition.
- In achieving the level of quality and total product value demanded by rising expectations.
- In maintaining profitability in a business environment where our product development and delivery

costs were in many cases higher than competitors' costs.

- And in meeting the needs of employees, who clearly were willing to bring more of their energy and talent to bear on these problems.

By the late seventies—even though the company was still quite profitable—it had become apparent that some fundamental changes had to be made in the way the business was being managed.

It wasn't entirely clear at that time, however, what changes had to be made. We knew the problems of the marketplace had to be faced squarely, but all of the processes needed to determine customers' requirements and assess competitors as the basis for target setting had not yet been put in place. We came to realize that we were not making the most of the ideas of our own employees, and we began to understand that it was management's responsibility to create an environment in which all that talent and energy could be fully tapped.

We knew also that quality and productivity improvements of the magnitude needed would require a continuing effort over a period of years. These improvements would require a cultural change and, most of all, a long-term commitment by Xerox management.

Priorities Redefined

The change process really began on a corporate-wide basis about three years ago when the priorities and direction of the corporation were redefined and widely communicated by Xerox management.

The process of change has occurred in a structured but evolutionary way. There were three major phases— competitive benchmarking, employee involvement, and ultimately a total quality process which at Xerox is called ''Leadership through Quality.'' The first phase was the establishment of competitive benchmarking— the continuous process of measuring our products, services, and practices against the toughest competitors and best functional leaders in the world. These competitors can be in any industry, not just copiers. Our goal is superiority in quality, product reliability, and cost.

Competitive benchmarking gives us a tool we use to establish the standards of excellence which will enable us to achieve leadership. What we have learned

through competitive benchmarking has been used as the basis for establishing quality, cost, and delivery targets, and every organization in the Group now benchmarks its performance in every relevant activity against companies that are the best in each functional area.

These benchmarks are incorporated into our annual operating plan as well as into our five-year business plan. They are tracked continuously and reported on. This process has generated changes in virtually every aspect of my operation. And the benefits have been substantial. One result of benchmarking is a comprehensive survey that has been established as the basis for continually monitoring how well customers like our products and services compared to competitive products. We found that many of the prior measurements we were using simply did not reflect the realities of the marketplace.

Listening carefully to what customers say, and believing it, has enabled us to improve by 21% over the past year, on a composite basis, how our customers feel about us in terms of product function, service, sales, and administration.

Special Projects

We have undertaken a great many projects to achieve the targets we have established:

- The process used to develop and design products in order to reduce cost and schedule has been completely overhauled.
- These design teams are working with suppliers much earlier in the product cycle and much more productively than in the past, enabling suppliers to make a much higher level of contribution to the design and manufacturing process. The number of nonconformance-free parts from suppliers has increased substantially.
- State-of-the-art automated assembly lines have been established for our new ''10 Series'' copiers.
- In our manufacturing and assembly operations, as a result of competitive benchmarking studies, we now utilize only three measurements for quality: For parts brought to the assembly line, quality standards are now set only in terms of nonconformances per million. On the assembly line, we use nonconformances per hundred units. For customers, we measure the

percent of nonconformance-free installations—does the machine work to spec when it's plugged in?

Gap Revealed

Competitive benchmarking revealed a gap between our level of performance and our competitors' performance. The size of that gap showed that parity with our competitors could never be achieved unless employees became involved more fully in the business. Management was just not smart enough to do that job by itself. Employee involvement, then, was the second phase in the evolution toward a total quality program.

Quality Policy

The quality policy we agreed upon reads as follows: "Quality is *the* basic business principle for Xerox. Quality means providing our external and internal customers with innovative products and services that fully satisfy their requirements. Quality improvement is the job of every Xerox employee." Now these words won't go down in history—but we understand them, and all of our employees understand them.

Next we hammered out a set of Quality principles; a set of management actions and behavior; and a set of tools to enable the process.

The five quality principles we agreed upon are these:

1. Quality is the basic business principle for Xerox to continue to be a leadership company.
2. We will understand our customers' existing and latent requirements.
3. We will provide all our external and internal customers with products and services that meet their requirements.
4. Employee involvement, through participative problem solving, is essential to improve quality.
5. Error-free work is the most cost-effective way to improve quality.

The tools we will use to facilitate the "Leadership through Quality" process consist of:

1. The Xerox quality policy.
2. Competitive benchmarking and quality goal setting.
3. Systematic nonconformance- and error-prevention processes.
4. Training for "Leadership through Quality."

5. Communications and recognition programs which reinforce "Leadership through Quality."
6. A method for measuring the cost of nonconformance. . . .

2.7 DISCUSSION AND REVIEW QUESTIONS

(2–1) *Quality assurance management has evolved from quality control and inspection.* Explain, with reference to a particular industry or product.

(2–2) *Concepts of workmanship and quality are closely related.* Explain, using an example.

(2–3) *About 80% of defects are managment controlled, rather than worker controlled and caused.* Explain, and suggest two major remedial actions to be taken by management.

(2–4) *Taylorism inhibits quality control by workers and supervisors.* Why is this?

(2–5) *Modern quality assurance management will further develop in years to come as a mature discipline and profession.* What would some of these developments be?

(2–6) *Without quality of a product or service being specified, quality cannot be assured.* Discuss, and give reasons for acceptance or rejection of this statement.

(2–7) There are three basic types of quality assurance systems or programs. Define and compare these. What program type would you recommend, and for what reason, for the following: airplane, toothbrush, chair, book, bicycle, toy.

(2–8) *Quality assurance is a subsystem.* Explain.

(2–9) According to Peter Drucker, consumerism means that the consumer looks upon the manufacturer as "somebody who is interested, but who doesn't really know what consumer realities are, and who has not made the effort to find out." Perhaps Drucker is right, so far as what many consumers think. However, whether or not this is what consumers think, consumerism seems to have gone further.

Manufacturers are often held liable for injuries even when consumers use products incorrectly. Explain.

(2–10) Most consumers cannot determine the quality of gasoline in any other way than by observing the result of its use when driving the car and possibly by calculating and comparing mileage obtained per gallon. What kind of quality characteristics and indicators are used by consumers to infer the quality of gasoline? Discuss another product where fitness for use is hard to determine before the actual use.

(2–11) An airline pilot had to make an emergency landing because insufficient fuel was tanked. The airport attendants had been confused by the metric measurements and the gauge in the cockpit had been out of order. What should have been done to prevent such a near-fatal accident?

(2–12) For a particular flight, the fare was $164 but only $64 was charged in error. As this error was done for all tickets of the flight and only detected after the takeoff, the pilot was informed about the error and instructed to return. Discuss the error in the system that applies here.

2.8 NOTES

1. For an interesting discussion of management-controllable versus operator-controllable defects, the reader should consult: Juran, J. M., "Quality Problems, Remedies and Nostrums," *Industrial Quality Control,* Vol. 22, No. 12, June 1966, pp. 647–53; Juran, J. M. and Gryna Jr., F. M., *Quality Planning and Analysis,* 2nd ed., McGraw-Hill Book Company, New York, 1980, pp. 106, 139, 314; Deming, Edwards W., *Quality, Productivity and Competitive Positions,* published by Massachusetts Institute of Technology Center for Advancement of Technology, Cambridge, Mass., 1982.

2. For a collection of Taylor's principal writings, see Taylor, F. W., *Scientific Management,* Harper & Brothers, New York, 1947. Published earlier as *The Principles of Scientific Management,* Plimpton Press, Norwood, Mass., 1911, also reprinted by Norton, New York, 1967.

3. "Quality Control Charts," *Bell System Technical Journal,* October 1926, pp. 593–603.

4. Dodge, H. F. and Romig, H. G., "A Method of Sampling Inspection," *Bell System Technical Journal,* October 1929, pp. 613–31.

5. "Reliability of Military Electronic Equipment," report by the Advisory Group on Reliability of Electronic Equipment, Office of the Assistant Secretary of Defense, June 4, 1957.

6. "Military Requirements for Weapon Systems and Sub Systems," MIL–M–26512A, military specification, June 18, 1959.

7. "Maintainability Design Criteria Handbook for Designers of Shipboard Electronic Equipment," U.S. Department of the Navy, NAVSHIP 0967–312–8010, July 1972.

8. Vaill, P. B., "Management and Human Performance." This paper is part of the U.S. Government publication *Zero Defects—The Quest for Quality,* Quality and Reliability Assurance Technical Report TR9, U.S. Government Printing Office, Washington, D.C. (1968), pp. 1–20. See also Gellerman, S. W., "Motivation and Productivity," *American Management Association,* New York, 1963. Newham, D. E., "Zero Defects—Do It Right The First Time," *Journal of Industrial Engineering,* January 1966, pp. 3–6.

9. For a review of motivational programs see Sandholm, L., "Japanese Quality Circles—A Remedy for the West's Quality Problems?" *Quality Progress,* February 1983, pp. 20–3.

10. "Quality Program Requirements," MIL–Q–9858A, military specification, U.S. Department of Defense, 1963.

11. Drucker, P. F., *The Practice of Management,* Harper & Row Publishers, New York, 1954.

12. Cleland, D. I. and King, W. R., *System Analysis and Project Management,* McGraw-Hill Book Company, New York, 1975.

13. Anderson, Douglas N., "Quality Motivation through Management by Objectives," *ASQC Annual Technical Conference Transaction,* 1971, pp. 7–14.

14. Pipp, Frank J., "Management Commitment to Quality: Xerox Corp," *Quality Progress,* August 1983, p. 12. Copyright © American Society for Quality Control Inc., Reprinted by permission.

2.9 SELECTED BIBLIOGRAPHY

ASQC Publications Catalog, American Society for Quality Control, Milwaukee, 1984.

Feigenbaum, A. V., *Total Quality Control*, McGraw-Hill Book Co., New York, 1983.

Fetter, Robert B., *The Quality Control System*, R. D. Irwin, Homewood, Illinois, 1967.

Gilbert, J. O. W., *A Manager's Guide to Quality and Reliability*, John Wiley & Sons, Inc., New York, 1968.

Juran, Joseph M., (editor), *Quality Control Handbook,* 3rd ed., McGraw-Hill Book Co., New York, 1974.

Juran, Joseph M. and Gryna Jr. Frank M., *Quality Planning and Analysis,* 2nd ed., McGraw-Hill Book Company, New York, 1980.

Langevin, Roger G., "What's Ahead for Quality Control in the 1980s?" *Quality Progress,* October 1979, p. 11.

McClain, J. O. and Thomas, L. J., *Operations Management: Production of Goods and Services,* Prentice-Hall Inc., Englewood Cliffs, N.J., 1980.

Rosander, A. C., "Service Industry QC—Is the Challenge Being Met?" *Quality Progress,* September 1980, p. 34.

3 CHAPTER

THE HUMAN ASPECTS

Quality of products and services is created by people, for people. Quality assurance induces confidence in the consumer that the producer will be able and willing to meet quality requirements and specifications. One finds that quality assurance is an interhuman understanding and commitment. However, quality assurance can also mean that the producer has ascertained the expected quality. It is thus a *self-assurance* that we individually and collectively start the production right, keep it on the right track, and finish the work right. This implies that the producer is convinced that the principle of "right the first time" has been observed. Satisfaction from a job well done is also a reassurance that we seek in the work place. Who really can take pride in one's own quality of work when management leaves quality assurance entirely to the workers and to the consumers in the form of a "buyer beware" attitude? The essence of modern quality assurance lies in the fact that the producer must beware as much, if not more than, the buyer.

Quality assurance is a human bond and a contracting arrangement that can make harmonious and mutually beneficial business relationships. The effective motivation of individuals is needed to keep the human bond strong and vibrant. To help achieve company goals, management's fundamental objectives in developing good human relations center around the creation of an environment that makes the satisfaction of human needs compatible with company goals. When we talk of an organization and its people, the human dimension becomes a fundamental aspect, as quality and quality assurance remain deeply rooted in general human existence. The managerial dimension, to be discussed in the next chapter, could perhaps be included under human aspects as well. However, in the context of this book on management it deserves special attention, and thus a separate chapter. The technological dimension, which is often in the foreground and detached from the human as creator and user of technology is also discussed later in the book.

3.1 THE HUMAN AS CREATOR AND USER OF QUALITY

The human relationship in work and its result on quality have become more emphasized, or, better to say, reemphasized in recent years. There is again a great awareness of the need to motivate and support employees, entrepreneurs, teachers, civil servants, and other workers to meet traditional performance standards and to strive for better results. This greater quality effort must take place in individual workplaces and in the community at large.

Quality of work performed at the individual level must be seen as a top priority. However, quality cannot be separated from quantity. To produce more and to sell more is a common company goal. Yet, it is not quantity that will earn the future—and the risk of selling inferior quality becomes higher when quantity becomes a major concern. The concern obviously should not be how much we are producing, but the quality of what we are producing. Productivity must be understood to mean more than the number of items turned out and sold to consumers in a given period. Quality and quantity together make up profit. One thing is clear: poor quality means more and more rework, scrap, and poor productivity, and that in turn means poor business, a bad reputation for the business, diminishing sales, and the rest of the miseries that follow. The economic implications of this are broad and are true for the efforts of a company as much as for a country struggling or striving to stay at the top.

Before we study management and its leadership in quality performance, the rest of this chapter will focus on the individual worker in terms of the organization for quality, the quality assurance team, the motivation for quality, the role of supervisory management, and modern forms of participatory management.

3.2 ORGANIZATION FOR QUALITY

Organization for quality means two things: First, the consideration of quality goals and responsibilities for the entire company, and secondly, the subunits of the organization in charge of quality assurance procedures and programs. The entire company as a team carries the obligation for quality and the individual members share in this responsibility on the job assigned to them. Quality responsibilities thus permeate the organization, ideally from top to bottom.

In the days of homestead economy (with the producer being also the consumer), the interdependence of quality of life and work, and the sharing of common tasks, successes, and failures, was simple, direct, and well understood by all members of the family. In modern industrialized and urbanized society, with its separation of work and home and its specialization distributed throughout large and small companies, these interrelationships have become more impersonal, formal, and complex. To see the responsibility for quality and its many benefits clearly, directly, and convincingly is fairly difficult nowadays, even in the small business settings. Therefore, how can people in large organizations become quality motivated? Tasks are no longer assigned by tradition, practice, mutual understanding, trust, and agreement. In today's organization jobs are interrelated with each other in a formal structure and the jobholder has to meet certain performance standards. Such organizational formality is necessary, because the design of our work systems are no longer as down to earth and simple as before. The products, with their advanced complexity, also dictate complexity in organizations, individual workplaces, and performance standards.

However, ideally speaking, a modern organization of large or small scope with simple or complex products, services, or processes, should function like a successful homestead. Surely, conditions do differ today and organizing for quality and success, particularly in a technologically advanced period is not an easy managerial task. Yet, well-managed companies, with solid positions in the market due to quality assurance, prove again and again that organizing for quality is still possible. In such companies managers usually set an example of good workmanship in the planning and control of operations and work. Thus they *lead,* instead of merely supervising. Under such leadership a harmonious, productive, and cooperative team normally develops. A company and work-team grows and organizes itself well for quality attainment.

Staff Coordination for Quality Attainment

An organization chart, with its work division and job assignment functions, constitutes a formal arrangement of jobs. These jobs become interrelated in a hierarchical order, with an executive on top and the various levels of responsibility subordinated. Normally, only the lowest level jobs do not entail responsibility for the performance of others. Each *job* however, carries with it a defined task, responsibility, respective authorities, reporting lines, qualifications, and remunerations. In a job description performance standards are quality related.

The extent to which the jobholder meets this expectation decides, individually and collectively, the overall attainment of quality in products and services. Obviously, the quality of the job design, the coordination and integration within the wider organization, the workplace, equipment, and the qualification of the jobholder determine actual quality of performance. Quality of supervision, training, work directives, quality systems, and procedures are other factors that set the quality capacity and the motivation for quality assurance. The personality of the jobholder does play a role in quality performance. However, a well-managed company and effective organization would tend to minimize antimotivational personality traits. The manner in which all these contributing factors interplay determines the quality of an organization. All of these are primarily managerial tasks.

Figures 3.1 and 3.2 describe two typical organizations with inclusion of quality assurance jobs. The conventional line-staff organization remains the most conducive arrangement of work. Committees and project teams complement, rather than substitute for, familiar built-in line/staff schemes. However, because of a lack of understanding, quality assurance is not always explicitly considered when organizing a company. Yet, often a mere establishment of a formal organization with task delegation and performance measurement does not suffice in itself to achieve quality. As a result, even highly motivated and qualified staff find themselves handicapped by the formal organizational system, rather than supported by it. Several studies based on actual examination of many products and organization behavior reveal that most quality problems are management-controllable rather than operator-controllable, although many managers prefer to believe otherwise. This attitude regarding worker indifference is discussed later in this chapter.

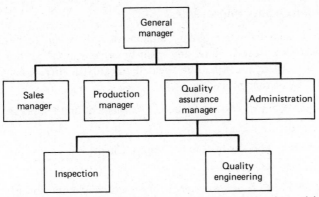

FIGURE 3.1   Organizational structure showing relationships of the product assurance activity (or its equivalent) to other activities and top management.

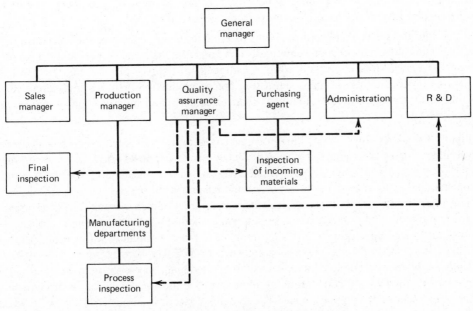

FIGURE 3.2   Quality control as an integral part of the organization.

Reorganization for Quality Improvement

Reorganization can lead to better quality attainment only when reorientation of staff is associated with such measures. Thus, any team or organization can have a unique setup, with sufficient orientation and flexibility geared towards quality. Changes for the better require several forms of reorganization.

1. *Coordination* between various functions (permanent and temporary).
2. *Procedure* and *rule* setting.
3. Concept of *project teams* for quality improvement.
4. *Committees* and *subcommittees.*
5. Continual *training* of all levels of management, including those in marketing, purchasing, design, documentation, inspection, packaging and shipping, and after-sale service in problem solving tools.

3.3 QUALITY SYSTEM WITH CONSUMERS, SUPPLIERS, AND PRODUCERS

Quality goals to be achieved by the entire company and by the individual team members require respective approaches and resource deployment. These goal and approach relationships, when understood as quality systems have several facets.

1. Quality systems and their procedures complement an organization and drive the company forward.

2. Through implementation in the form of quality programs, a quality system mobilizes and energizes an organization and its people. The individual employee becomes directed toward quality performance in the context of the organization.

3. Under the systems viewpoint, people external to the organization, such as consumers, suppliers, and government representatives, also enter the team and movement for quality. Figure 3.3 depicts this interrelationship. Customers, of course, play an important role in specifying quality, either through contracting before production or in test marketing. Contracting customers often impose quality assurance standards and supervise their suppliers and subcontracts.

Figure 3.4 shows three major kinds of contacts between supplier and user/customer or government. There are contacts between producer on the one hand and user or government on the other, during all stages in the product life, including contacts after the production is completed and even after the item is sold to the customer. In the case of *services* we have similar relationships where the customer maintains direct human contact during all stages, as in a restaurant, or contacts after production, when, for instance, repair services are contracted. In fact, any company finds itself embedded in a community with various personal and impersonal relationships, either as a customer or as a supplier.

The systems view of quality assurance, as it is applied in this book, permits the interrelation of all human contacts and cooperation systematically on the individual job level, as well as at the departmental or company level. The systems concept is important because it orients decisions and activities (that is, work structure of technical and managerial procedures)

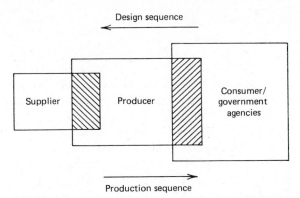

FIGURE 3.3 Quality system with consumers (including government agencies), producers, and suppliers illustrated to depict schematically:
(a) The increasing degree of direct contact (shaded area) starting from the point when raw materials are puchased, processed, and finally sold to the consumers. This typically takes place in the production sequence (lower arrow to the right), where actual operation moves from supply input through production processes to the delivery and servicing.
(b) The increasing degree of value (including price, etc.) of quality (indicated by the sizes of areas drawn). This is feasible when planning based on consumer needs moves to the production capabilities, and only when confirming materials are purchased from the suppliers.

directly towards goal attainment and has a motivating impact. Authority must be given with responsibility. One example is authority given for self-inspection, to operators. When recognizing their personal role, people clearly become more easily motivated to maintain good working relationships. Interpersonal and interdepartment frictions, which often develop in a line-staff organization, tend to diminish when people work together in project teams or work groups with specific goals before them. For instance, a goal of decreasing scrap by a certain percentage within a given period, with some incentives and authority given, will motivate a team under most normal circumstances.

Quality performance ↓	Quality of		
	Design	Conformance	Performance
Required by user (or customer)	User / Producer	User / Producer	User / Producer
Self-imposed	Producer	Producer	User
Government-imposed	Government / Producer	Government / Producer	User

FIGURE 3.4 Major kinds of contact between producer or supplier and user, in situations where quality is either normally expected or imposed and demanded.

3.4 MOTIVATION FOR QUALITY

Machines are designed, constructed, and driven, whereas people act in accordance with their qualifications and motivation. Quality can ultimately be attained only by motivated and committed people. To transform an organization into a team that works together for the achievement of quality assurance, and to establish a dynamic and harmonious movement for quality, describes a major task for management and its quality assurance team.

Quality assurance specialists, such as inspectors, quality and reliability engineers, and quality managers and auditors, prepare procedures and systems that induce and guide management and other staff towards quality goals. Effectiveness and success of this quality assurance team depends largely on the status of the department or function and on the proficiency and personal integrity of the quality assurance specialists involved. The inspector's ability, not only to detect defects and failures, but also to help staff members, suppliers, and other related parties prevent reoccurrence, requires positive personality traits and reinforcing motivation. Quality specialists must be able to establish and maintain positive human contact. This is not easy, particularly when failures are to be detected and prevented and when people are often their main cause.

Quality assurance staff must set a visible example for cooperation and for meeting performance standards. A committed and harmonious quality assurance team exerts a great impact for quality improvement within the company. Profiles and careers in the quality assurance field and profession will be explained later in this book. The design and implementation of a quality assurance system and program has as its prerequisite the proper staffing (by good recruitment and training) of the quality assurance function or department. Those who like to work with people in order to create better quality and improve and enrich jobs will make final quality assurance a most rewarding occupation. For too long inspectors and quality auditors have seen their main task as to inspecting quality into the product. In the future, the quality specialist will have to become a facilitator of better quality who trains and helps other staff members, including senior management.

Theories of Motivation

McGregor[1] distinguished two types of worker attitudes: *X-type* theory assumes that workers are basically uninterested (lazy, untrustworthy, don't care, etc.) in the work they do, and cannot be motivated to better work performance. Hence, managers must take a stick and carrot approach. The *Y-type* theory takes the opposite view that it is the way in which jobs are organized that makes workers unmotivated. There might be some truth and justification for management taking these different attitudes. The first and more pessimistic outlook, although often appearing realistic, leads managers to regulate work strictly, to impose tight quality controls, and to take punitive action in case of inadequate work performance. Much of the "scientific" management of Frederick Taylor reflects a basically X-type attitude. Obviously, this approach will hardly motivate and encourage workers to improve performance, at least not under modern conditions and expectations at the workplace. Today, workers expect a more positive management attitude.

Why should workers be indifferent or be assumed to be indifferent to their work and the results thereof? Particularly in a period of growing outlook and education coupled with economic recession, workers will try harder. Still, this view is fairly negative and pessimistic.

Workers nowadays have had an opportunity to acquire higher education and skill, and, therefore, as a rule, expect more responsibility and challenge on the job. Obviously, everyone at every job is not dissatisfied with what she or he does. Satisfaction also comes from doing things that are challenging, and that provide opportunities for creativity, recognition, and so forth.[2] Peter Drucker and many others point out that in any organization and its people there are dormant talents and special abilities that competent and prudent management can tap for the benefit of all. This potentially valuable asset can best be unlocked in conjunction with quality improvement and quality assurance effects. It is not a coincidence that quality circles have become the vehicle for a modern style of what is called *participative management* (see Sections 3.5 and 3.6 for more details of this). Assuming that workers are potentially motivated, interested, and willing to effectively participate in quality assurance, managers are expected to give leadership and to set an example themselves. This principle of quality starting at the top calls for visible and competent quality planning and programming that concretely involves everyone. Thus, the principle demands a policy of total quality assurance.

William Ouchi[3] has recently formulated a *Theory Z* that essentially describes and critically compares the policies of a number of successful American corporations against a large number of Japanese corporations. Based on a number of traits, characterized by a clannish attitude of American companies which he found similar to their Japanese counterparts, he named those companies Type Z or Theory Z companies. Theory Z organizations are those that are able to develop trust among employees, tend to develop specific images, have management concerned with long-term growth rather than short-term profits, involve a participative style of management with extensive use of a quality circle approaches etcetera. Interestingly enough, a number of similar corporate philosophies were also recognized in the findings of Peters and Waterman.[4]

Role of Senior Management

Quality assurance specialists alone will not be able to achieve quality goals. Senior management has to provide leadership and acknowledgement, along with full support for quality assurance procedures and programs. When management misconceives quality assurance and stresses other goals which may conflict with it, then quality assurance staff soon becomes disregarded. Other staff, supervisors, and operators become unmotivated. However, mainly in response to the Japanese challenge and to more difficult business conditions, management today must forcefully pursue quality assurance and quality goals. This is a most positive sign, and will probably have a compelling impact within and outside American companies.

What motivates people to better work performance?

1. Sufficient and ambiguous performance standards and work procedures.
2. Honest and supportive leadership.
3. Proper recognition of a job well done.

Such responses are not surprising. However, a study of many workplaces still reveals a widespread separation of planning and control on the supervisors side, and prevalence of the mere carrying out of instructions on the operators side. Many supervisors jobs, as well as those of the operators, do not lend themselves to quality motivation. In such cases, designing more challenging and enriched jobs, wherever feasible, would be a starting point on the path for improved quality performance.

After I visited auto plants both in Japan and Europe, my first impressions of those in Detroit were not encouraging. At an engine plant I saw a worker saunter toward his station at the end of a conveyor, where the first of several back-up engine blocks had already fallen several feet to the concrete floor. Never quickening his pace, he picked up the block and sent it on down the line. In another plant I asked a production engineer if they used any statistical control to assure quality.

"You mean the Deming method?" he said. "We used that more than 20 years ago, but we stopped. Why? People would buy the cars anyway. . . ."[5]

The coming of Detroit-Japanese partnerships . . . could eliminate the notion that any Japanese car is inherently better than any American-made car. Then, as more joint-venture cars appear U.S. buyers may start to judge them on their own merits instead of some generalized national reputation. . . ."[6]

3.5   ROLE OF INFORMAL QUALITY ORGANIZATION

One of the most significant contributions to enhancing human motivation to work is the emergence of the concept of informal organization, such as *quality circle*. Quality circles, as developed and implemented in Japan, provide for positive workers–foreman participation in upgrading quality and productivity, and at the same time enhance the quality of work life.

Quality circle is the name given to small voluntary groups of workers that undertake work-related problems as mini-projects, to improve the quality of products and of the work itself. Circles were formed in Japan around foremen, in conjunction with teaching and learning quality control methods. The group selects a theme or problem and then moves through cause analysis towards solving the problem and attaining improvements. Technical specialists assist the group as facilitators.

In North America, where pressure for better quality and higher productivity is strong, the propensity to adopt quality circle as a panacea for economic ills appears to be high.[7] Psychologically oriented motivational programs such as "zero-defects" seem to have petered out. Traditional quality control inspections have not produced very convincing results either. The Taylor method of separating planners and doers on the production floor (which never existed in Japan or Europe to the same extent) has further retarded communication and cooperation for better quality.[8]

Any approach to solving quality and productivity problems must meet existing technical and human conditions. The mere adoption of quality circles in North America has brought many unsatisfactory results. The fact remains that companies and people have developed certain ways of performing their work and are basically reluctant to change.

The following facts reflect on the potential of quality circles as a component of various situations:[9]

1.   Circles do not substitute for effective quality management but are a complementation and expression or development of natural movement towards a less autocratic management style.

2.   Circles with due focus on an individual company's setting can allow development of a broad quality commitment, together with open communication and mutual trust.

3.   Circles do not need to undermine the status of unions; on the contrary, they may be

a vehicle to promote their interests, be they safety, quality of work life, job security, or the like.

4.   Circles alone cannot make a major contribution to quality, as only a small percentage of defects are worker-controllable.

5.   The essence of circles is, first of all, maximum voluntary learning in groups with direct job-related problems, not only to deal with barriers to better quality. Utilization of hidden talents and human resources is the chief benefit.

6.   Quality control methods and techniques must be the tools used in circle activities for discussion of problems. The seven tools to be used are: Pareto Analysis, cause and effect diagrams, stratification, check lists, histograms, scatter diagrams, and the Shewhart control chart. Without a knowledge of these, a tendency of members to think ritualistically or to talk without effect cannot be curtailed. Thus, training of all levels of management in quality disciplines is an important prerequisite.

7.   Circles help reduce Taylorism in most adverse situations, but perhaps not in stubborn cases. Workers and management must have realistic expectations, common goals, and a sense of not infringing upon the rights of one another.

8.   Existing quality programs with standards, procedures, schemes, projects, audits, and so forth, can be strengthened through quality circles.

9.   In stubborn cases, quality circles can prove to be a new and effective training mode, best applied to quality assurance, safety, and related issues, rather than to cost reduction objectives, at least in the beginning stages.

10.   A quality circle's leader is normally the foreman, rather than a quality specialist. The latter can act as advisor if properly trained for the task.

Thus, quality circles, in a way, facilitate positive coordination at the operator level and help overcome other handicaps, such as resistance to change, rejection of more responsibility, information blockage, and general quality assurance apathy. Bocker and Evard,[10] after studying the results of some 450 North American and European companies that had utilized the potential of quality circles, found good return on the investment. Desired prevention of defects, however, needs more than just good coordination and a prevailing spirit of cooperation. Guidance, by management, through the quality assurance program and its individual policies and procedures is needed by the quality circles, so that they remain on track towards quality goals. Future strategy for the actual development of quality systems must, therefore, strengthen both quality circles and quality programs, individually and jointly.

How Big a Circle?

Given the size of the problem, it is not only one area or department that inherits the problems caused by others and vice versa. The fact that a large proportion of perceived quality problems remain management-controllable comes out loud and clear in the results of many independent surveys and studies. Consequently, the challenges and roles of quality circles cannot be limited only to workers. With all the current excitement about this phenomena, it is sad not to recognize conceptually what it is for.

This conceptual lack creates a missed opportunity, for there is not enough talk about management circles. When the seeds of quality problems lie in the higher echelon, one would like to see the radius of circles grow bigger and bigger—reaching to suppliers, subcontractors, consumers, and possibly to the general public. Executives need to readdress the whole quality circle issue, so that a determined effort to manage quality promises to be effective. The proper strategy for the future demands that upper managers not only be versed in financial disciplines, but that they also be able to diagnose the company's position with respect to quality.

Integrating Quality Circles Into Project Teams

Normally, a quality program precedes the formation of quality circles, simply because management initiates and leads. Foremen and operators cannot form quality circles on their own. The point is that both quality programs and quality circles link the various levels in a company, ensuring that the program is prepared with everyone in mind. The real effectiveness of quality circles thus hinges on the suitability of quality programs and vice versa.

While quality circles and respective problem solving efforts remain restricted to small groups, quality programs are company-wide, in all departments and in every function.

What is needed, then, is an integration and coordination of programs, projects, and quality circles. Informality, independence, and voluntary participation in quality circles not only need to be maintained, but need to be reconciled with formality, and with management imposed and enforced quality programs. Quality cannot be attain by the existence of quality circles alone, or, for that matter, only through management-imposed programs.

Figure 3.5 describes the various hierarchical interfaces of a quality assurance organization. While project teams are placed in charge of special contracts, issues, problems, and so on, those that involve direct participation of the foreman and operator in the planning and controlling responsibilities constitute quality circles on a practical level. Project teams are formed at the supervisor level and up, and circles at the work group level. With the scheme depicted here, while both circles and project teams may facilitate communication and decision making differently, they are not mutually exclusive in the hierarchical structure of the organization. The same is true in the lateral and chronological directions. The role and function of steering committees, subcommittees, and other groups is critical here, since they function as approved *decision support systems* designed to allow all levels to perform their responsibilities with authority.

Position (levels)	Management Output (planning/control; decisions)	Organizational Groupings		Audits
		Units	Entities	
Executive	Policies Programs Budgets	Boards/committees	Corporation/ company	Program initiation
Supervisor	Procedures Projects	Project teams	Plant/department/ station	Project
Worker	Job description Instructions Task/performance	Work groups quality circle (voluntary)	Work place/job	Element

FIGURE 3.5   Intracompany interfacial structure.

Establishing Project Teams

Project teams and *committees* have been extensive in American business experience since World War II, when industries were trying to increase military output in the face of shortages of labor, materials, and energy. Senior management has always appointed project teams and the project leader for various purposes and missions, in an effort to analyze, design, and revise chronic problems and to get concrete answers. However, a relatively high degree of specialization has often blocked communication between management and quality specialists. It is the removal of this barrier and the creation of a participatory style that are now sought after.

The real benefit of project teams is that the group work remains task oriented under strong leadership and enjoys adequate managerial and technical support. On a finer scale, both quality circles and project teams encompass project management. Ishikawa[11] describes the difference between the two by explaining, "[w]hile Circles are formed by the people of the same workshop on a permanent basis, teams are organized by people of different workshops or engaged in different functions. They form a team as required to achieve a specific objective and in many cases dissolve the team when the objective has been achieved."

The important factor in achieving satisfactory results is the creation and maintenance of design and development of quality-improvement or project teams. When starting a quality assurance project, management activates the quality movement through project teams, while the grass roots workers participate actively and positively through their circles, both serving the same quality assurance objectives. As need develops, project groups may be further structured to consist of working groups as mini-project teams, committees, or subcommittees, made up of widely fluctuating types of people. Many U.S. companies are already familiar with this approach, known as a *material review board* dealing with major defects, or an *audit* presented as a project.

Table 3.1 compares several different aspects of quality circles and project teams. In spite of the more or less obvious differences, their similarity rests in the grouping of people to solve quality problems and accomplish a common task in which every member has an interest and a stake. Managers more familiar with project management disciplines and less familiar with engineering techniques, or vice versa, can always make their involvement a valuable learning exercise. Quality circles require that managers remain less demanding and aloof, and only become involved when reviewing the results and recommendations. Even when such is the case, managers may not be able to do justice to circle recommendations. However, with the structure of Figure 3.5 in mind, this problem can be alleviated when managers are members of a project team that supervises, among other things, the circles goal setting, training needs, and defect prevention programs.

3.6 PARTICIPATORY MANAGEMENT: NEW APPROACHES

The main concern here is the influx of participation within *all levels of management,* not simply between supervisors and the work force, for a change away from *autocratic* style and towards more of a *participative* nature of management.

In some cases of invalid and poor procedures, workers might have every reason to ignore, or even to overrule, the procedures. What counts is the actual assuring of quality that satisfies

TABLE 3.1
Comparison of the Quality Circles and Project Team
Approaches to Problem Solving

Feature	Quality Circle Approach	Project Team Approach
Main purpose	Quality motivation, worker participation	Attainment of goals, taskforce
Setting of objectives	By group members	By senior management with team's participation
Scope	Depends on group ability/capacity	Determined with objectives by senior management
Need for aids (management)	High	Relatively low
Expertise	Practical; limited	Predetermined for task
Staffing	Volunteer	Appointed, qualifying
Technical approach	Innovative, experimental	Traditional, established
Management involvement	Indirect, relatively little	Direct, delegation of authority and responsibility
Organizational status	Ad hoc, without formal line function	Ad hoc, within formal organizational structure
Size of group	Small	Varying, task-dependent
Performance standard	None	Normally preceding task assignment
Performance control	Little	Usually undertaken
Qualification requirements	Little	Qualification required
Accountability	Little	Qualification required
Termination	Not without group approval	Decision by senior management

the customer. It is not only doing the job right the first time, but the second time, third time, and every time, which saves money and creates new customers.

At the risk of seeming to take a cookbook approach to an inevitably complex set of issues, we offer a *10 step approach* based upon actual field tests with many small firms. This approach incorporates formation of quality circles and project teams, integrated together as a permanent feature, for any management committed to either reviewing its current quality status or studying to establish a better one. The closing of the quality gap is looked at by considering each step on a project-by-project basis to get the most out of this integration. If followed, this 10 step approach should result in an effective integrated quality assurance program, such as is needed by any company today.

STEP 1: *Determine existing practices for assuring quality.* No matter at what stage a company is in its quality improvement program, you have to know where you stand with regard to *practices* as opposed to *procedures*. Remember, the two differ from one another in that procedures are documented while practices are not. Such examinations need to be done for all departments, not just for the one officially assigned with quality assurance responsibilities. If the chain of the quality system is broken in any one area, the effect will be felt in the end result.

STEP 2: *Formulate the practices in writing and verify these with the persons concerned.* Many managers discover, often the hard way, that a quality system will not function if too rigidly enforced. This is mainly because they do not dedicate themselves to a vision of the future. Many things can and do change. Judging by a company's instability, conditions such as new product design, new markets, new customers, changed supply sources, technological changes, inexperienced staff, and other similar factors always contribute to the flexibility that a system is required to have.

People in an organization usually hold many conflicting views about the best method of reaching a goal. Verifying with concerned personnel in early stages makes it easier to discover the facts, whatever the basis of people's conflicting perspectives may be.

STEP 3: *Form a quality circle (or call a meeting) for review, and finalize the procedures.* Everything is accomplished through people. Building a long-term relationship with the people in quality control, in other departments, is important to restore confidence in quality. Before a project is approved, early involvement of workers avoids conflicts and delays. The best rule, in our experience, is to choose a project that is virtually certain to be approved by the employer and to choose one that can be experimented with and has all chances of success in the end.

If one project in one department is successful and people are simultaneously trained, and become receptive to quality circles, an attempt should be made to repeat the exercise in every department, one after another. At first, however, efforts should be focused on the ability to do well only those limited things which are of utmost importance to quality attainment.

STEP 4: *Document generally approved practices as formal quality assurance procedures.* Remember that the project began by gathering information, looking for facts, building up a coalition, building confidence, and, most of all, agreeing upon what each party sought as an objective behind the objective. Now it is important to document the approved procedures.

STEP 5: *Compare procedures with applicable quality program standards; determine voids and major deviations, make corrections.* Quality programs, as described in the respective program standards, have a broad effect on an organization. The way a company's system measures up and reports performances must be understood and guided by the blueprint of standards. If it is not consistent with those standards, it should go back to the drawing board.

STEP 6: *Bring together the formalized company-wide quality assurance program in the form of a quality manual.* Many companies have their quality policies and procedures scattered among an amorphous collection of individuals, groups, and departments. The lack of a clear-cut, unified document, is an impediment to progress. Responsible managers must remember that a quality manual is, in effect, a user's manual. If imposed from high up, without full knowledge and participation of all, it is almost sure to be unworkable. We hear about defective product recalls very often, yet nobody has heard of a recall of manuals that consistently fail to serve their purpose.

STEP 7: *Prepare special quality circle principles and operating procedures in conjunction with Step 3.* There is no substitute for assured continuity of participation. This is a positive and challenging way of looking at the success of all quality programs. If management fails to recognize circle activities, management's responsibility becomes suspect. Put simply, no dream of "zero-defect" or "right the first time" will ever be realized if management's real intentions conflict with the goal of never-ending improvement. When a company prepares plans for circles' operation, the potential of quality circles is put in proper perspective.

STEP 8: *Establish a project team in support of and in cooperation with quality circles.* Even though the principle of quality circles is to retain voluntary participation, its success or effectiveness depends upon whether it is built-in with or loose from the overall system. The idea of a project team at this stage is to provide guidance, help, and necessary training for all circle activities.

Earlier, in Step 3, quality circles were being formed and gaining experience. Now the circles become positively ready for useful recommendations to feed the project teams. This cross-fertilization is for both groups, and, in the long run, accountable towards a company's quality goals.

The basic thrust is to keep reinforcing effective participation. The scope of involvement of any number of workers in a given project team should depend on the nature of the problem, from mere consultation to participation, then moving on towards self-control qualifications. Recommendations and solutions should ooze out of the circles almost unnoticed, without any need to extract them.

STEP 9: *Audit planning.* Audits are particularly important for examining ongoing operations and assessing necessary improvements. In short, an audit's purpose is to review progress against goals. Executives will want to review a system so that they know one exists. Nevertheless preparation of an audit program is a task by itself. During this phase, like the others, consultations with quality circles and other project teams is vital to maintain momentum and continuity.

STEP 10: *Conducting the audit with a mandate to follow up with the audit reports.* Beginning audits should be conducted as projects in conjunction with other quality programs, but in a hands-off mode with quality circles. Having a meaningful statement of objectives and a long-range plan is essential to this function. The audit is best done annually and should be conducted first by an internal team and then ensured through some external third party for objectivity.

By integrating internal quality audits into a comprehensive management audit program that extends from production to supplier quality control and customer services, the entire quality system becomes linked, as if it constituted one large unit.

Motivation for Ongoing Improvement

Maintaining control over each of the products and processes is one thing, but struggling to get everyone motivated for continual improvement is something else. This is where the real challenge of management lies. Convincing everyone to agree upon and accept *never-ending quality improvement goals* should be done by making each program flexible enough to fit these features. Frequent auditing helps to define where the efforts should be directed most. When the status is known, all persons take their initiative from knowledge of the needs.

3.7 KEY TERMS AND CONCEPTS

Quality organization is the composite sum of all functions responsible for planning, controlling, and performing quality assurance activities. It can consist of one function or job, or of a multilevel responsibility and authority scheme. The organization is responsible for the quality assurance program and procedures, while actual quality assurance is also the responsibility of all other staff, as part of the assigned job and function of each.

A **quality circle** is a small group of workers that together with their supervisor voluntarily analyze work-related problems, recommend solutions, and seek improvements in work performance, quality, and productivity. Facilitators provide expertise in quality assurance methods. Quality circles originated in Japan, where they were initiated subsequent to workshops in statistical quality control. They have been adopted in Western countries with mixed success. Quality circles seem to function best, and motivate staff to improve performance most, when coupled with formal quality assurance programs and project teams.

Project teams are different from quality circles in that they are management imposed, directly controlled taskforces. Breakthroughs for improved quality and quality assurance can be attained by assigning respective tasks or projects to such teams, possibly chaired by a quality assurance expert. Members are appointed from various departments and functions.

Motivation at the workplace normally refers to a positive attitude towards performance and workmanship. McGregor's *X-Type* worker is considered unmotivated and unmotivatable, in contrast to the *Y-Type;* in very general terms, the Taylor concept assumes an *X-Type,* while quality circles assume a *Y-Type.* Ouchi's *Z-Type* worker or company develops a family-like work environment that positively motivates, and thus improves the quality of products and work life.

Participatory management expresses a management style under which all staff members are given an opportunity to share traditional managerial responsibilities of planning and control within formally assigned responsibilities. It is designed to enhance quality assurance, productivity, and general effectiveness.

The **10 step approach** incorporates principles of participatory management, in that it builds upon actual practices in order to develop and establish an effective quality assurance program that complies with applicable standards. It also incorporates audits in order to assure direct involvement of supervisory and senior management.

3.8 SUPPLEMENTARY READING

An Inquiry into Values[12]

. . . I took this machine into a shop because I thought it wasn't important enough to justify getting into myself, having to learn all the complicated details and maybe having to order parts and special tools and all the time-dragging stuff when I could get someone else to do it in less time—sort of John's attitude.

The shop was a different scene from the ones I remembered. The mechanics, who had once all seemed like ancient veterans, now looked like children. A radio was going full blast and they were clowning around and talking and seemed not to notice me. When one of them finally came over he barely listened to the piston slap before saying, "Oh yeah. Tappets."

Tappets? I should have known then what was coming.

Two weeks later I paid their bill for 140 dollars, rode the cycle carefully at varying low speeds to wear it in and then after one thousand miles opened it up. At about seventy-five it seized again and freed at thirty, the same as before. When I brought it back they accused me of not breaking it in properly, but after much argument agreed to look into it. They overhauled it again and this time took it out themselves for a high-speed road test.

It seized on *them* this time.

After the third overhaul two months later they replaced the cylinders, put in oversize main carburetor jets, retarded the timing to make it run as coolly as possible and told me, "Don't run it fast."

It was covered with grease and did not start. I found the plugs were disconnected, connected them and started it, and now there really *was* a tappet noise. They hadn't adjusted them. I pointed this out and the kid came with an open-end adjustable wrench, set wrong, and swiftly rounded both of the sheet-aluminum tappet covers, ruining both of them.

"I hope we've got some more of those in stock," he said.

I nodded.

He brought out a hammer and cold chisel and started to pound them loose. The chisel punched through the aluminum cover and I could see he was pounding the chisel right into the engine head. On the next blow he missed the chisel completely and struck the head with the hammer, breaking off a portion of two of the cooling fins.

"Just stop," I said politely, feeling this was a bad

dream. "Just give me some new covers and I'll take it the way it is."

I got out of there as fast as possible, noisy tappets, shot tappet covers, greasy machine, down the road, and then felt a bad vibration at speeds over twenty. At the curb I discovered two of the four engine-mounting bolts were missing and a nut was missing from the third. The whole engine was hanging on by only one bolt. The overhead-cam chain-tensioner bolt was also missing, meaning it would have been hopeless to try to adjust the tappets anyway. Nightmare.

The thought of John putting his BMW into the hands of one of those people is something I have never brought up with him. Maybe I should.

I found the cause of the seizures a few weeks later, waiting to happen again. It was a little twenty-five-cent pin in the internal oil-delivery system that had been sheared and was preventing oil from reaching the head at high speeds.

The question *why* comes back again and again and has become a major reason for wanting to deliver this Chautauqua. Why did they butcher it so? These were not people running away from technology, like John and Sylvia. These were the technologists themselves. They sat down to do a job and they performed it like chimpanzees. Nothing personal in it. There was no obvious reason for it. And I tried to think back into that shop, that nightmare place, to try to remember anything that could have been the cause.

The radio was a clue. You can't really think hard about what you're doing and listen to the radio at the same time. Maybe they didn't see their job as having anything to do with hard thought, just wrench twiddling. If you can twiddle wrenches while listening to the radio that's more enjoyable.

Their speed was another clue. They were really slopping things around in a hurry and not looking where they slopped them. More money that way—if you don't stop to think that it usually takes longer or comes out worse.

But the biggest clue seemed to be their expressions. They were hard to explain. Good-natured, friendly, easy-going—and uninvolved. They were like spectators. You had the feeling they had just wandered in there themselves and somebody had handed them a wrench. There was no identification with the job. No saying, "I am a mechanic." At 5 P.M. or whenever their eight hours were in, you knew they would cut it off and not have another thought about their work. They were already trying not to have any thoughts about their work *on* the job. In their own way they were achieving the same thing John and Sylvia were, living with technology without really having anything to do with it. Or rather, they had something to do with it, but their own selves were outside of it, detached, removed. They were involved in it but not in such a way as to care.

Not only did these mechanics not find that sheared pin, but it was clearly a mechanic who had sheared it in the first place, by assembling the side cover plate improperly. I remembered the previous owner had said a mechanic had told him the plate was hard to get on. That was why. The shop manual had warned about this, but like the others he was probably in too much of a hurry or he didn't care.

While at work I was thinking about this same lack of care in the digital computer manuals I was editing. Writing and editing technical manuals is what I do for a living the other eleven months of the year and I knew they were full of errors, ambiguities, omissions and information so completely screwed up you had to read them six times to make any sense out of them. But what struck me for the first time was the agreement of these manuals with the spectator attitude I had seen in the shop. These were spectator manuals. It was built into the format of them. Implicit in every line is the idea that "Here is the machine, isolated in time and in space from everything else in the universe. It has no relationship to you, you have no relationship to it, other than to turn certain switches, maintain voltage levels, check for error conditions . . ." and so on. That's it. The mechanics in their attitude toward the machine were really taking no different attitude from the manual's toward the machine, or from the attitude I had when I brought it in there. We were all spectators. And it occurred to me there *is* no manual that deals with the *real* business of motorcycle maintenance, the most important aspect of all. Caring about what you are doing is considered either unimportant or taken for granted. . . .

3.9 DISCUSSION AND REVIEW QUESTIONS

(3–1) What is meant by "motivating for quality" and how can it be achieved?

(3–2) *The quality system concept and view promotes cooperation between customer and supplier.* Explain.

(3–3) *Good human relationships between inspectors and workers is important for quality assurance achievement.* Explain.

(3–4) *Senior management must play a leadership role in quality assurance.* Discuss this need and ways to fulfill it.

(3–5) *Much of senior management's involvement in quality assurance depends on the ability of quality assurance specialists to communicate.* Explain.

(3–6) *In a well-managed company there is no need for formal quality assurance groupings, such as quality circles.* Discuss.

(3–7) *For management, conventional task forces or project teams are more suitable to achieve quality assurance goals than quality circles.* Discuss.

(3–8) A somewhat frustrated chief inspector in a plant put a giant heap of scrap at the employee entrance complete with a sign reading "Last year's scrap cost us $800,000! How much did YOU make?" and the sign on each machine read "Scrap here costs $3.50." Discuss.

(3–9) *Workers seek self-improvement opportunities. Quality circles respond to this desire, as do various training courses.* Discuss.

(3–10) A union representative for workers in a plant making small copiers states, "the design department of our company developed a new kind of copier usable as a desk copier. It was already pilot tested and demonstrated at major trade shows and conventions. But now headquarters has decided to close our plant and gave the production of the innovation to a company in a foreign country." Discuss this statement with reference to quality assurance aspects and policies. Include in your discussion the question you would ask for further clarification of the situation.

(3–11) In order to create more jobs for people, various part-time and job sharing schemes have been suggested. For instance, to reduce general work time from 40 to 35 hours per week. The wage in all these systems is to be per hours worked. For each of the following arrangements list the positive and negative impact on quality of performance and product or service quality.

a. Reduction of work week from 40 to 35 hours
b. Flexible work time within total of 35 hours per week
c. Sharing of one job by two persons
d. One year leave for every six years of full service
e. Job offered with varying time requirements per month (or per day) between certain limits of maximum and minimum time
f. Jobs with completely flexible time below a maximum

(3–12) *Quality assurance management makes all the difference.* This fact has apparently been demonstrated by the case of the Motorola plant for TV manufacturing in Illinois. When the Matsubishi Electric Industrial Company bought the plant in 1974 the same workforce continued under the new management. Before the takover, records show that inspectors found 140 defects per 100 TV sets on the average. Within the next five to six years the figure had dropped to about 5 per 100 sets. How could this have been achieved? What is the message concerning quality assurance management?

(3–13) *About 80% of defects arise from poor management and faults in the system.* Take your own company situation as an example and discuss.

3.10 NOTES

1. McGregor, D., *Human Side of Enterprise,* McGraw-Hill Book Company, New York, 1960. The theory was originated by Maslow, A. H., in *Motivation and Personality,* Harper & Bros., New York, 1954, and popularized by Douglas McGregor.

2. This theory is attributable to Frederick Herzberg. See, for example, Herzberg, F., Mausman, B., and Synderman, D., *The Motivation to Work,* 2nd ed., John Wiley & Sons, Inc., New York, 1959.

3. Ouchi, William, *Theory Z: How American Business Can Meet the Japanese Challenge,* Addison-Wesley Publishing Company, Reading, Mass., 1981.

4. Peters, Thomas J. and Waterman Jr., Robert H., *In Search of Excellence: Lessons from America's Best-Run Companies,* Harper & Row Publishers, New York, 1982.

5. Grove, Noel, "The Automobile And The American Way, Swing Low, Sweet Chariot!" *National Geographic,* Vol. 164, No. 1, July 1983, p. 17.

6. "The All-American Small Car Is Fading," *Business Week,* March 12, 1984, p. 95.

7. For an annotated bibliography of some 42 research papers on quality circles, the reader should consult an article by Konz, Stephan, "Quality Circles: An Annotated Bibliography," *Quality Progress,* April 1981, pp. 30–5.

8. The influence of the Taylor System on quality control is the topic discussed in a series of eight papers by Juran, J. M., "The Taylor System and Quality Control," *Quality Progress,* May to December issues, 1973.

9. See, for example, Goodfellow, Matthew, "Quality Control Circle Programs–What Works and What Does Not," *Quality Progress,* August 1981, p. 30. This article describes reasons for failure of circles in many organizations. Other informative sources on quality circles in the North American environment are: Gryna Jr., Frank M., *Quality Circles: A Team Approach to Problem Solving,* published by AMACOM, a division of American Management Associations, New York, 1981; Cole, Robert E., "Will QC Circle Work in the U.S.?" *Quality Progress,* July 1980, p. 30.

10. Bocker, Hans and Evard, Klaus, "The Quality Circle Concept—A Challenge to Management to Meet the Productivity Crisis" *Zeitschrift fuer Betriebswirtschaft* (in English), November–December, 1982, pp. 1053–78.

11. *Quality Control Circles Koryo,* (in English), Japanese

Union of Scientists and Engineers, 1980. A related article is, Yamamoto, M., "The Japanese Homogeneity Promotes Ikigai," *Quality Progress,* September 1980, p. 18.

12. Pirsig, Robert M., *Zen and The Art of Motorcycle Maintenance,* copyright © R. M. Pirsig, 1974, pp. 30–5. Reprinted by permission of William Morrow & Company.

3.11 SELECTED BIBLIOGRAPHY

Drury, Colin G., "The Human Factor in Industrial Inspection," *Quality Progress,* December 1974, p. 14.

Gilmore, Harold L., "Quality of Employee Performance," *Quality Progress,* May 1980, p. 14.

Harris, Douglas H. and Chaney F. B., *Human Factors in Quality Assurance,* John Wiley & Sons, Inc., New York, 1969.

Ishikawa, Kaoru, *Guide to Quality Control,* Asian Productivity Organization, 1976 Unipub, New York, also available through the American Society for Quality Control (ASQC), 230 West Wells Street, Milwaukee, WI. 53203.

Johnson, Ross H. and Stone, Raymond N., "Supervisors and Quality Control Supervisors," *Quality Progress,* September 1982, p. 16.

Konz, Stephan, "Quality Circles: An Annotated Bibliography," *Quality Progress,* April 1981, p. 30.

Yamamoto, Mititaka, "The Japanese—Homogeneity Promotes *Ikigai*," *Quality Progress,* September 1980, p. 18.

4 CHAPTER

MANAGERIAL DIMENSIONS

There is now a theory of management for improvement of quality, productivity, and competitive position. No one can ever again claim that there is nothing in management to teach. Students in a school of business now have a yardstick by which to judge the curriculum that is open to them.

W. Edwards Deming
Quality, Productivity and Competitive Position

In this book we consider that anybody who plans and controls work acts as a manager. Most people holding a job in an organization plan the work to a varying degree, and, therefore, participate in overall management. Even at the lowest level in the organizational structure where work procedures must be followed, some work preparation and independent decisions need to be made by the operator. Given the above understanding of management, even a low level job allows the person to manage and to self-control performance and achievement. Everyone, in one way or another, participates in management. This is also true when the job does not allow supervisory power and care for the work performance of others. This broad concept of management that envelops the entire team and company has become a prerequisite for total quality assurance.

4.1 THE QUALITY MANAGEMENT PROCESS

A strict separation of management and nonmanagement or labor, as implemented under Frederick Taylor's ''scientific management,'' has largely become outdated under modern technological and social conditions. Factories with mass production, fabrication, and assembly lines normally employing low-skilled workers, allow them some participation in planning and controlling, perhaps recognizing that the extreme separation of mind and hand has never been possible.

Management means planning and controlling of work. This role implies decision making, leading, supervising, administering, risk taking, and so on. In terms of quality, management concerns one's own work and the work of others. A manager has to manage his or her own work well first before it can be expected that others on the team and in the company will accept that manager's leadership and follow his or her example. For instance, if management establishes poor, unworkable, policies and procedures on quality assurance, which are supposed to direct all activities relating to quality attainment, then even highly motivated staff cannot function effectively.

In man-made productive systems, management designs, operates, and maintains production processes and creation of quality in products and services. Quality assurance, as an inherent element of production processes, is an essential component in each job. After all, processes divide into individual jobs; a job is a small integrated part of a production system and thus a subsystem, that is, a micro production system.

Quality Starts at the Top

We pointed out before the importance of supervisors setting an example of good workmanship. This means that quality assurance *must come from the top* in the form of procedures, policies, directives, and leadership. When quality planning and control responsibilities are shared by everyone in a company, quality awareness and motivation must permeate down from the top in the organization.

All work being planned and controlled for quality attainment must also meet deadlines and quantity standards.* A sound work plan strikes a balance between quality and quantity standards. Quality attainment requires sufficient time for the worker. Often tight time standards cause poor quality of workmanship. However, quality of work depends also on the quality of the workplace, of teamwork, and on the quality of others work. A proficient manager designs production processes, plant layouts, workplaces, and work methods with more than the required quality of the product in mind. Proper consideration must be given to the worker's qualifications and personality. It stands to reason that prudent management seeks wide-ranging participation of operators, and their collective representation in the planning and controlling of work.

Quality can also emerge from the bottom of the organizational hierarchy. Defects may be controllable by management, but poor workmanship and disregard of work procedures do cause many failures. Policies and procedures by themselves do not suffice to bring about the proper quality of products and services. For instance, many special processes need skill, talent, and high motivation of the operator to assure the desired quality. Thus, with the help of the operator's influence on procedures, and possibly even on policies, communication should work in both directions, from the top to the operator level, and in the reverse.

4.2 GROWING AWARENESS FOR QUALITY ASSURANCE

In the recent past major breakthroughs in the management of quality have occurred in the form of greater quality awareness, liability and risk protection, and in the development of dormant benefits resting in the system of quality assurance. This has come about mainly through strong international competition, generally demanding improved business conditions and pressure for improved quality from consumers and governments. The following are some statements that indicate an evolution in top managements commitment to quality assurance.

Quality is, in its essence, a way of managing the organization. It does not depend upon geography nor upon particular social culture nor solely upon special techniques . . . . [e]xperience demonstrates that strong modern quality programs have become major business resources and business assets in themselves for companies today.[1]

If U.S. Corporations are to become basically quality-oriented, they will need, perhaps . . . incentives that motivate managers.[2]

The realization is finally taking hold that the design and production of high quality goods and services is not just a quality manager's technical problem on the factory floor but a general manager's problem throughout the entire corporation. It leads everywhere and touches everything. . . . What is needed, then, is revolution in the way managers think about the continuum of product development activities. By this we do not mean a shift only in their conventional approaches to quality problems but also in their readiness to make the long term investments in people and equipment necessary to make better products less expensive.[3]

High quality, after all, is not achieved by a few random management decisions but by a complex, all-encompassing, interactive management system that has the uncompromising long-term support of top management. . . . Quality, to the Japanese, means error-free operation . . . in most of the Japanese factories, the quality chart on the walls measure the defect rate not in percentages but in parts per million: Their defect rate between 300 to 500 ppm, and their ''near-term goal''—100 to 200 ppm. And the long term? ''Zero, of course.''[4]

Crosby[5] describes various stages of quality maturity in management. These range from implied quality assurance in sound business practice such as organized quality control and inspection functions, to formal quality management programs. These programs currently constitute the highest level and most advanced stage in modern quality assurance management. Quality assurance through such programs ranks with other managerial responsibilities, such as securing return on capital, proper employment opportunities, a solid market position, and cost-effective operations.

Greater maturity in management regarding quality assurance means more than just awareness of such obligations cognizance of the many benefits. Education and training of supervisors and senior echelon employees in quality assurance concepts and methods is needed, before major breakthroughs in companies can occur and suitable and effective quality assurance procedures and integrated programs can emerge.

4.3 MANAGEMENT PRINCIPLES IN QUALITY ASSURANCE

In order to attain quality goals, operations and performances must be planned, monitored, and controlled, so that deviations are recognized in time and corrections made to prevent their recurrence.

This fundamental planning and control activity, as it is to be carried out by management, implies the use of numerous principles.

1. Determining goals on a long, medium, and short-term basis that are strong in depth and commitment.

2. Correlating quality assurance objectives, in addition to the time perspective, with other goals that might possibly conflict, be neutral, or even aid in quality assurance. Assigning a budget.

3. Ensuring that goals are valid regarding the need and realistic regarding the companies' immediate and long-term capacities for quality assurance.

4. Assigning the proper people with sufficient training and experience tasks in the planning and implementing of quality assurance procedures and programs.

5. Seeking the broadest possible participation of staff in the design and implementation of quality assurance programs.

6. Designing systems and programs that are to create better quality in products and services, in the form of small projects that then become assignable to project teams.

7. Communicating objectives and plans as widely and clearly as possible, so that people are informed.

These rules, and many others, have been and will be followed in well-managed companies as a matter of normal practice. In fact, a company that has been successful in both good and difficult times does have a quality assurance program. However, it might not always be formulated in writing through procedures, but nevertheless it exists in and through the daily routine and work practices. Probably one of the most fundamental managerial principles is to know where one stands and then where one wants to go. If the quality assurance goal appears promising and worthwhile in light of what is needed and what one can do, then plans and actions will be forthcoming from an energetic individual and a strong team.

In this book we shall describe modern management thought on and practice of quality assurance. In principle this means that we apply both standard and some innovative management ideas to a relatively new field of management activity. For the many quality professionals, these innovative ideas and practices belong, for instance, to:

1. Quality program standards, as a guide to management.

2. Quality projects, as task and action oriented packages for quality attainment.

3. Auditing of quality assurance systems as a tool to review progress and achievement.

4. Quality circles, as a form of participatory management.

5. Quality systems management, where a total approach in quality assurance is envisioned and understood, for the actual designing and implementing of quality programs.

Quality assurance management should, therefore, be considered as a unified whole, consisting of the above items, together with many others, including quality concepts for those in purchasing, marketing, and so on.

Thus management appears as a subsystem within the wider-ranging system, such as the plant, company or department. The quality system was introduced in Figure 2.5 and Table 2.1 in Chapter 2. Here, management is described as a subsystem of such a quality system, that, as such, includes planning, control, initiation, and directing functions.

Management of Quality Assurance Projects and Programs

Quality assurance projects and programs both involve planning and control. Projects have a particular well-defined task and objective, while programs are more for general procedures and activities. Projects are carried out by teams; programs are administered by functions of departments.

Figure 4.1 relates the quality assurance management system to various programs. An overview of a quality management system structure is shown in Figure 4.1(*a*). Proceedings

General	Quality Assurance (company level)	Quality Plan (product/service)
Objectives	Program standards, strategic requirements of quality planning and analysis (Chapters 14, 18)	Marketing survey (customer needs), technical/nontechnical specifications, quantification, uniformity in purpose and definitions (Chapters 8, 10, 13)
Approaches	Quality policy, organizations, program manual, management commitment (Chapters 15, 18)	Production and inspection/test plans, resource evaluation (Chapters 9, 10)
Decision making	Policies for risks and uncertainties, performance evaluation programs (Chapters 7, 11)	Inspection/test procedures, process control techniques, vendor–vendee relations (Chapters 9, 11, 13)
Implementation, supervision	Job description, staffing, quality improvement projects and programs, motivation (Chapters 4, 14, 15)	Inspection and process control standards, evaluation techniques, contacts/communication (Chapter 13)
Control, corrective action	Statistical control techniques, self-inspection, training in quality (Chapters 11, 12)	Use of charts, graphs, and product/process evaluation, improvements in information system (Chapters 5, 6, 16)
Review, audit	Internal/external audit, corrective action (Chapters 18, 19)	Audit standards, reports (Chapters 18, 19)

(Arrows connect General stages downward: Objectives → Approaches → Decision making → Implementation, supervision → Control, corrective action → Review, audit)

(b)

Typical Ingredients	Main Theme
Objective(s), policies	Written, understood, implemented, and maintained (at all levels)
Quality management system	A planned and systematic action (with a road map) to achieve the stated objectives (to include the life cycle approach)
Quality improvement programs and projects	Subject matter of improvement with built-in audit system to measure and interpret the improvement

FIGURE 4.1 (a) An overview of a management system structure. (b) Ingredients of an effective management commitment for quality assurance.

from setting objectives to verifying attainment of quality program components in a company-wide environment are described. Quality policies and programs, nevertheless, remain the essential ingredients for a successful management commitment for quality (Figure 4.1[b]). Planning and control of a project is directed towards the attainment of the project goal and purpose. The project plan's key events can be called milestones. Figure 4.2 demonstrates that a major project can be divided into smaller ones, each with a particular task. Project teams, when organized in conjunction with quality circles, produce unexpected results as the following examples taken from real companies demonstrate.

Example 1: A chemical company wanting to diagnose quality costs occurring in each of its departments started a program by establishing project teams with steering committees as its diagnostic arm. However, it was the input from quality circles that suggested higher than normal shakedown inspection, cutting the cost by 35% in six months.

Example 2: In a ductile iron foundry, when the project teams and circle members realized that some one person no longer owned the problem, they cut down the sheer waste of granular

Objectives	Improvement of quality of supplies, for major items within the period, from _____ to _____
In-charge	Mr. A
Supervisor	Mr. B
Project	Quality Manual, section 3.2.6 under "quality action projects"
Procedure/references	
Measures	*Determination of questionable supply items. Compilation of all available pertinent data
	*Listing of major defects and other problems with quality impact
	Survey and evaluation of suppliers
	*Comparative analysis of price–quality relationships
	Review of current purchasing practices regarding quality assurance, by an ad hoc committee
	Search for available standards and guidelines, including those of competitors
	Establishment of a supplier liaison function.
	*Assessment and possible introduction of statistical acceptance sampling and other pertinent decision-making techniques
	Submission of a project report to management
Members of project team	

*Asterisks indicate where quality circles would play the role. Where the problem (or project) is interdepartmental, two circles may meet on it, or a temporary circle may be formed by members of each unit until the problem is solved.

FIGURE 4.2   A typical example of the role played by quality circle members in a quality improvement project.

nickel by weighing each charge exactly in paper bags, rather than the accustomed way of throwing the material with scoops into the pouring ladle. This amounted to thousands of dollars of savings each year, plus consistency in metal microstructure.

Example 3: A pressure vessel manufacturing company decided to register its name with the National Board of Boiler and Pressure Vessels Inspectorate in Ohio, in order to be able to build vessels conforming to codes established by the American Society of Mechanical Engineers (ASME), and thus, expand its market. A project was instituted with the expressed goal of establishing a quality control program and producing a manual that complied with the code requirements.

This goal was operationalized by defining subgoals, milestones, and tasks. Resulting work packages were then assigned to the team members and the interrelationships shown in a PERT network[6] that allowed interaction with foremen and workers through their circle members for procedure writing. Once the project plan was approved by management, employee awareness meetings were held to inform everyone about the company's mission. Smaller projects were given to circles, while the project team worked on manual design. After a target date, the full program was put into operation, the success of which was monitored by an internal quality program audit team. The company passed the final audit by the ASME and now holds "U" and "UM" code symbol stamps as a certified manufacturer.

On one other occasion, the project team was given the goal of cutting down in-plant rework and scrap to 2% in one year. Figure 4.2 highlights the plan and the task assignments of phase one of the scheme entitled "improvement of quality of supplies," involving both the project team and quality circles approaches to problem solving.

Projects for Development of a Quality Assurance Program

Figure 4.3 depicts the major quality management system activities with regard to quality programs considered through project management steps. At each stage of a specific activity, such as setting objectives or determining an approach, a coordinate task for the quality program of quality assurance activity can be defined. For instance, the quality system or program evolves after the need has been analyzed and the requirements have been established. Quality program standards guide management in this phase (see Chapter 18). Once the program has been documented and implemented, auditing (Chapter 19) helps to maintain program effectiveness.

4.4 THE QUALITY MANAGER: FUNCTIONS AND ACCOUNTABILITY

Effective quality management depends on the ability of those who manage an understanding of the essentials of modern management concepts and methods. The tasks—quality performance, keeping customers satisfied and making work productive by maintaining a balance between the attainable limits and available resources—are what basically make up the profile of a quality manager. Appendix A of this book contains a detailed outline of a job description which the manager will find a useful example, either in making his or her own job description, or for drafting one for a subordinate, or both.

Preparation Phase	Determination of quality policy Establishment of quality assurance program Formulation of procedures Establishment of quality assurance information system
Organizational Phase	Management presentation of need, goals, and approaches Formation of quality improvement projects or teams Execution of approved projects on a pilot basis Establishment of training and advisory services
Implementation Phase	Management presentation of results of pilot projects Conducting of quality assurance training Compilation and priorization of projects and other plans Initiation, supervision, and coordination of projects Preparation and dissemination of progress reports Submission of final reports Presentation of total project results, acknowledgement and rewards; establishment and announcement of new objectives and projects
Control Phase	Establishment of performance control procedures Establishment of quality audit teams Planning and execution of audits Analysis of audit reports, followed by corrective action and performance control Review and report on total quality program (annually)

FIGURE 4.3   Quality assurance through project management, showing four general categories of steps essential to the preparation of a master plan.

The quality assurance function is in charge of the planning and control of all procedures and activities aimed at assuring quality of the products and services. The people who hold quality assurance jobs are held responsible for an effective quality assurance program which might entail no more than end-item inspection, or even less than that.

Functions and jobs designed to assure quality through respective planning and programming constitute an integral part in the wider concept of the organization. The quality assurance function can be arranged and assigned as part of a job in a small enterprise or as a hierarchy of many subfunctions and jobs in a large organization.

The move from staff to line helps to further develop quality assurance into subfunctions and a more independent force. With the growth of the company and/or the quality assurance program, an expansion of the organization becomes necessary.

One major organizational question is whether quality assurance at the inspector level should come under the supervision of the foreman or another supervisor, such as the production manager. Both direct supervision by the local area supervisor or by the independent quality assurance manager have advantages and disadvantages. Program standards suggest there should be *independence* of the inspector from foreman supervision, and direct reporting to the quality manager. A recent trend promotes *self-inspection* of the operators, and quality audits through the quality assurance function as inspection control. Figure 3.2 has already

Function	Quality Control Manager	Chief Inspector	Deputy Chief Inspector	Inspectors	Statistician
Quality control manual	xx	x	x	x	
Inspection and test plan	xx	x	x	x	
Procedures	xx	x	x	x	
Document control	xx	x			
Test equipment		xx	x		
Purchasing	x	xx			
Incoming inspection			x	xx	x
In-process inspection			x	xx	x
Final inspection			x	xx	x
Identification traceability			xx	x	
Handling and storing		x	x	xx	
Special processes		xx	x		
Shipping		x	x	xx	
Nonconforming items		xx	x		
Customer-supplied items		x	xx		
Quality records	x	xx			x
Corrective action	xx	x	x		
Statistics quality costs	x	x			xx
Consumer feedback	xx	x			x

FIGURE 4.4 Organization charts including the functions and accountability of various quality control personnel.

Function	Quality Control Manager	Chief Inspector	Deputy Chief Inspector	Inspectors	Statistician
Liaison with design and engineering	xx	x			
Training	x	xx			
Auditing	xx	x			
Budgeting	xx	x			

xx–Primary responsibility.
x–Secondary responsibility.

FIGURE 4.4 (*continued*) Organization charts including the functions and accountability of various quality control personnel.

demonstrated one of the many hybrid solutions in which the inspector reports to both the foreman and the quality manager. Another important trend is keeping the quality manager on an equal level in the company hierarchy with the production manager or the manager of purchasing and similar other executives. A quality manager, or the quality department that reports directly to top management, is clearly in a better position to remain unbiased and, therefore, is able to fulfill its quality assurance function more effectively.

An often more informative description of an organization is accomplished by using the *matrix* form, in which responsibilities and positions are interrelated. An example is given in Figure 4.4.

Any organization chart must be complemented with adequate job descriptions or position guides. These provide information about specific duties, responsibilities, authorities, supervisory and lateral relationships, accountability, performance standards, and required qualifications.

4.5 QUALITY ASSURANCE MANAGEMENT DURING ECONOMIC RECESSION

When conditions of economic recession prevail, new tasks and opportunities arise for quality assurance managers. Benefits from effective quality assurance become more visible and pronounced during these hard times, when many companies fight for survival in the market. Some companies and industries are more affected by the downturn of demand, saturation in the sales market, loss of markets to foreign competition, and so forth, than others. The automobile industry is an example of foreign competitors having nibbled away demand for cars from large domestic car producers. The computer industry in North America stands for a growth industry in spite of an all-around economic recession. This indicates that structural and technological changes favor some and disfavor others.

Top management can help their companies not only to cope with complex forces in the marketplace, but to use these for the company's advantage and benefit. It is no coincidence that quality assurance is fairly neglected in good times and moves into the limelight of managerial attention in bad times. Under the conditions of a sellers' market, investments into ever-larger plants can be made, because the output can be sold fairly easily. Quality assurance means to comply with minimum specifications that still seem to satisfy the ever-changing needs and demands of the customers. Once the recession sets in, several changes occur that have impact on quality assurance.

1. Demand diminishes in most sections and remaining customers become more quality insistent. Product liability claims increase; the public becomes more restless and militant in cases of poor quality of products and services.

2. Lack of demand and a surfeit of liability claims lead to plant closures. Public opinion begins to favor the "victims," against the "powerful corporations," particularly in times of general economic hardship.

3. New plants with innovative modern technology and improved all-around quality assurance conditions cannot be financed because of relatively inflated costs, major government safety regulations, and other regulations that are only mandatory for new plants, and because of not enough increase in demand.

4. Older plants offer many advantages during periods of cost increases and expensive financing. Fixed costs are embedded and many reserves and opportunities for rationalization and cost reduction exist from previous years of relatively easy growth. Workers and management, afraid of losing their jobs due to a plant shutdown, are now motivated to achieve quality assurance, if they were not already motivated before. This leads to better utilization of productive capacity and thus creates conditions for financing new quality assurance programs.

5. Unemployment, in the form of sporadic layoffs, offers the company a chance to retrain staff internally and externally. Returning and often selected staff is usually not only more productive, but superior in quality of work performance. Management, in the meantime, often has used the shutdown interval to regroup and to establish new quality assurance procedures and programs.

6. Large corporations strengthen their ties with suppliers which, as a rule, are small enterprises. Quality assurance in small enterprises, if it has not been neglected in the formal sense, can much be much improved when required. In shifting work towards suppliers, large corporations impose and induce quality assurance programs for mutual interest and benefit.

7. As customers and governments become more demanding concerning quality of products and services in periods of recession, so do industrial customers, become more demanding of their suppliers. This pressure moves through the entire product or service life cycle.

8. Under the leadership and quality insistence of large industrial customers, small enterprises further strengthen their position via quality assurance programs.

9. Large corporations can convincingly prove to their own staff, owners, community, unions, customers, and other concerned parties that defects have become extremely costly; more so during periods of inflation than during growth periods. Therefore, a strategy of prevention of defects and the stressing of integrity of design and workmanship may set new goals with which everyone involved can readily identify. The expertise required for quality control and for the application of methods such as statistical quality control grows, not only among managers, but also among supervisors and workers. The quality assurance specialist

now becomes a facilitator and resource person, instead of the former watchdog, informer, and enforcer.

10. With the all-around revival of quality assurance in companies under the strain of inflation and unemployment, quality assurance involvement and careers become newly defined and attractive. Young employees see their opportunities in this special area growing, and also recognize quality assurance on their current job, both individually and as a group, to be a steppingstone to important future benefits.

11. Governments also come under pressure to give leadership in quality improvements in many different ways. As a provider of government services, the public demands not necessarily more, but better services. Creation of new employment opportunities, public works projects, and social services for those affected by the economic recession are some areas where public funds must be spent and where quality must be assured.

12. Government regulatory bodies, such as the Food and Drug Administration have made important strides, not only under the pressures of greater quality and safety demands from the public, but in following the trend towards defect prevention via quality assurance programs. Ripple effects have emanated from regulatory bodies as these demands have been received by them.

In summary, quality assurance brings attractive and beneficial leverage to the efforts of economic revival and survival. Economic recession, with its inflation and unemployment, itself inherently creates the challenge and opportunity for new quality assurance efforts and movement. Quality assurance benefits can be achieved with little investment, public resistance, and organizational change. What is required, however, is that management in private and public organizations recognize its quality assurance role and be able to learn and adapt to the situation in time and ''while the iron is hot.'' This requires a more direct assessment of the situation, and requires that managers and all those participating in management learn about modern quality assurance, or at least review respective knowledge and experience.

4.6 MANAGEMENT TOOLS AND TECHNIQUES

Quantitative methods developed in the general discipline and practice of management facilitate rational decision making. These methods are applicable and important in quality assurance management as well.

Quantitative methods and models interrelate numerical data and variables pertaining to a problem and inform managers about factual conditions. As not every determinant and element in a problem, decision, and action can be quantified, these models and techniques help abstract facts from fuzzy reality. In quality assurance, customer dissatisfaction can be measured in terms of complaints and by the severity and details of such feedback, but psychological and emotional factors often escape quantification. The frequency of such complaints, when analyzed with statistical techniques, bear important messages for management. The compilation, analysis, presentation, and communication of these statistics and pieces of information, help to monitor, observe, and detect problems and to take remedial actions. Without data there cannot be a real basis for problem solving.

Statistical Quality Control

These methods and models of quality control are traditionally of the statistical type in the quality assurance field. In fact, quality control has introduced modern quantitative methodology of decision making in many companies, and in particular in the production and operation function. A process control chart, for example, exemplifies quantitative planning and control, because control limits, usually based on performance data, serve as a standard for future process behavior, and thus can predict future process performance. The control chart abstracts the truth from the factory floor production process, in that it selects from numerous, important, predefined quality characteristics.

If the data from an ongoing process is properly and continually retrieved and analyzed, conclusions on process behavior can be readily drawn by operators and nonstatistical experts alike. Thus, the method allows its user to *predict* whether the process will remain in control for a given condition. Many, if not most quantitative methods, such as the statistical charts, and decision making models, have this *predictive* value. They do not find the *causes* of problems however. These and other techniques of statistical quality control are described in subsequent chapters where appropriate.

There are other kinds of models that demonstrate relationships such as histograms, checklists, graphs, Pareto Analysis, and cause-and-effect diagrams. These too are abstract, and reveal fact. Use of many of these models of various kinds can become an integrated element in any complex decision-making process. However, statistical quality control is by no means the only applicable quantitative method for quality assurance. Nevertheless, their simplicity makes them very attractive for use by operators and nonspecialists alike.

Other Relevant Analytical Methods

Most of the other methods and models used in management science serve equally well in quality assurance management. These are listed in the context of a general decision-making process in Figure 4.5. The individual quantitative tools for managers vary with regard to, for instance, their predictive power; that is, whether they use fixed (deterministic) or random (statistical) variable inputs, whether they produce an optimum alternative (i.e., linear programming), or a feasible one (i.e., heuristic programs). Most of these techniques can make use of modern computer facilities and in this fact rests their particular attractiveness to managers. A computer, even in the form of a pocket calculator, allows wide use of quantitative statistical methods and helps even those users who lack technical knowledge and expertise.

Why Not Use All Management Tools?

In past teaching and practice of quality assurance management, many methods other than the familiar statistical ones have been largely neglected. One reason for this may be the awareness of the risks for misapplication of the techniques. In Part Two of this book, we include such important quantitative methods and models. Their potential use for answering questions regarding quality assurance matters will be shown. Wherever data that seem pertinent can be quantified, these tools, as listed in Figure 4.5, will be assessed for their practical usefulness, (selected mainly for demonstration purposes) and incorporated with the particular theme of that chapter. Application of these methods is not restricted to specialists and higher ranking managers. Emphasis will remain on application in the total environment of the many people that participate in decision making: operators, supervisors, specialists, and executives.

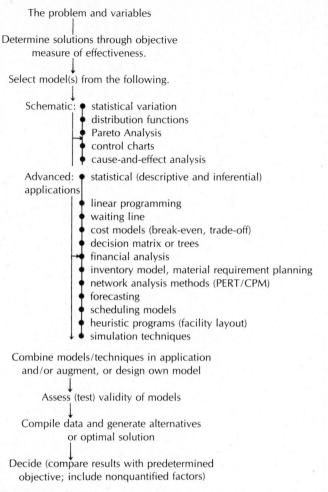

The problem and variables

Determine solutions through objective
measure of effectiveness.

Select model(s) from the following.

Schematic: ● statistical variation
● distribution functions
● Pareto Analysis
● control charts
● cause-and-effect analysis

Advanced: ● statistical (descriptive and inferential)
applications
● linear programming
● waiting line
● cost models (break-even, trade-off)
● decision matrix or trees
● financial analysis
● inventory model, material requirement planning
● network analysis methods (PERT/CPM)
● forecasting
● scheduling models
● heuristic programs (facility layout)
● simulation techniques

Combine models/techniques in application
and/or augment, or design own model

Assess (test) validity of models

Compile data and generate alternatives
or optimal solution

Decide (compare results with predetermined
objective; include nonquantified factors)

FIGURE 4.5   Summary of management tools and techniques available to
handle quality problems.

Modern management of quality assurance can only gain in strength through the complementation of its decision making by use of these techniques. Here there must come a thrust
for a new breakthrough and technological advancement. In Figure 4.6 we outline the curriculum for learning the use of various tools and techniques. Good management should give
everyone a chance to learn those techniques appropriate to improving quality, be it through
self-study, internal seminars, or formal courses in universities.

4.7   QUALITY AND PRODUCTIVITY IMPROVEMENT: A CONTEMPORARY MANAGEMENT ISSUE

A position often taken is that the revival of productivity (meaning higher output per worker)
must be fueled by technological advances, a new era of labor–management cooperation, and
a more experienced work force. During the 1970s, U.S. productivity grew by 20%, while

FIGURE 4.6 — A guide to the use and application of quality control tools by category of management.

(a) Basic statistical tools

Tools and Techniques	All Employees (From top down)	Process Oriented Staff	Project Management and Supervisory Staff	Training, Safety, and R&D Oriented Personnel	Office Management, Vendor Relations, and Inventory Control	Engineering and Quality Control Staff	Operators and Technicians	Comments
Variation	×	×	×	×	×	×	×	Basic mathematics and the use of the statistical function on a calculator; the theory-to-application transition should be made by selecting real process case examples and engineering problems for data gathering, making sure that everyone understands and applies these techniques.
Distributions	×	×	×	×	×	×	×	
Cause-and-effect diagrams	×	×	×	×	×	×	×	
Pareto chart	×	×	×	×	×	×	×	
Control charts	×	×	×	×	×	×	×	

(b) Advanced applications

Tools and Techniques	All Employees (From top down)	Process Oriented Staff	Project Management and Supervisory Staff	Training, Safety, and R&D Oriented Personnel	Office Management, Vendor Relations, and Inventory Control	Engineering and Quality Control Staff	Operators and Technicians	Comments
Reliability		×	×			×		The important point is not mathematical elegance but rather, quick and effective application of simple yet appropriate statistical tools.
PERT/CPM			×			×		
System charting			×		×			
Design assurance		×				×		
Factorials/EVOP						×		
Regression		×	×	×		×		
Computer operations		×			×	×	×	

72

Japan's increased output per worker peaked at 145%, Germany's advanced by 75%, and France's by 77%. According to the data from the United States Bureau of Labor Statistics, if an comparison is made of U.S. productivity growth against its foreign rivals, the standing remains in the same order from greatest to least, Japan, France, Germany, and the United States.

Workplaces can be redesigned, new tools and automated processing can be introduced, the worker can be retrained, and more skilled workers selected, but what about the fact that less defective items produced in output per period (or per unit of resources) will increase real productivity?

The following information, extracted from recent publications, relates quality and productivity improvements. Organizations and their management can use quality assurance strategically and tactically, in long and short run operations, to attract and satisfy customers, increase market share, build favorable images, and become more forceful and successful competitors. It is suggested that higher quality and resulting enhanced customer satisfaction and sales normally also increase productivity.

Deming's answer to the question of why productivity increases with improvement of quality is very practical: "less rework."[7] However, there exists an enormous lack of knowledge. A survey of 197 corporate executives undertaken in 1981 showed that "they do not generally understand the relationship between productivity and quality." When asked what critical issues they were facing, 82% considered productivity extremely or very critical and ranked it as the top issue; other critical issues were rising costs (76%), government regulations (74%), and quality control (68%).[8]

If nothing else, the data reported above should compel management to rethink their approaches to quality improvement, because it is management who has the authority to modify, improve, or change the system. However, quality gains are not just the result of a new robot here and a quality circle there; rather, they are achieved by sound quality management practices that are committed, and deliberately and systematically applied.

4.8 KEY TERMS AND CONCEPTS

Management in conjunction with quality assurance means to plan and control decisions and actions in an organization, in order to attain optimal quality of products and services in a cost-effective manner.

Management is the discipline (art and science), the function or institution in an organization (practice), and the career (profession) of organizational planning and controlling. This responsibility and activity implies supervising, directing, decision making, and risk taking. Management is not identical to, although it is related to, entrepreneurship and administration. In a hierarchically structured organization, senior managers supervise middle managers, who in turn supervise operators. Under a style of participatory management all staff members share in the planning and

controlling of operations, and thus share in management as part of their assigned job functions. This style, as opposed to autocratic senior management, appears to be more conducive for effective quality assurance and motivation.

Quality assurance management is the planning and controlling of decisions and activities directly concerned with the quality of products and services during the design, production, and performance stages. The quality assurance function and its management, plans and controls respective procedures and programs and, as such, is an essential carrier of an organization's quality assurance.

Quality assurance system management is the planning and controlling of all decisions and activities for the design and maintenance of an effective program and the individual procedures purporting to provide expected and satisfactory quality of products and services.

Management methods relate to the art and science of management. In analytical and predictive management or decision models, variables that have a bearing on the decision are quantified, and alternative paths of action are then selected on the basis of predetermined criteria or desired outcomes. Statistical methods, along with other quantitative methods, can be used for making rational decisions.

4.9 SUPPLEMENTARY READING

What Top Management Must Do[9]

Ways of doing business with vendors and with customers that were good enough in the past must now be revised to meet today's requirements of quality and productivity. Drastic revision is required.

It is not enough for everyone to do his best. Everyone is already doing his best. Efforts, to be effective, require direction. Statistical methods furnish direction.

It is not enough that top management commit themselves for life to quality and productivity. They must know what it is that they are committed to—i.e., what they must do. These obligations can not be delegated. Mere approval is not enough, nor New Year's resolutions. "High level committees" accomplish nothing without knowledge of what to do.

Only top management can bring about the changes required. Failure of top management to act on any one of the 14 points listed ahead will bring failure to reach the maximum quality and productivity that could otherwise be achieved. For example, only the top management of a company can reconstruct the terms of reference for managers of purchasing. It will no longer suffice to purchase materials and parts on price alone. Consideration of quality is necessary. The requisite quality can be produced only by vendors that learn and apply process control by statistical methods. Moreover, the only useful proof of quality and of cost is statistical evidence of process control. How many purchasing managers know this? They must learn a new profession. Quality by inspection is outmoded.

The 14 points shown below provide an outline of the obligations of top management. No one in management need again ask, "What must we do?" They also provide a yardstick by which anyone in the company, stockholders, and the bank, may measure the performance of the management. Never again may anyone claim that there is nothing about management to teach in a school of business.

It is unfortunate that there may be no way to assemble top management in America as there was in Japan.

New equipment and automation are not the answer. It is, instead, necessary to learn to use effectively the equipment on hand. This can be done only by statistical methods.

The 14 Points in Outline

1a. Innovate. Allocate resources to fulfill the long-range needs of the Company and of the customer. The next quarterly dividend is not as important as existence of the company 5, 10, or 20 years from now. One requirement for innovation is faith that there will be a future.

b. Put resources into plans for product and service for the future, taking into account:
   - Possible materials, adaptability, probable cost.
   - Method of production; possible changes in equipment.
   - New skills required, and in what number?
   - Training and retaining of personnel.
   - Training of supervisors.
   - Cost of production.
   - Performance in the hands of the user.
   - Satisfaction of the user.

c. Put resources into education (points 12 and 13).

2. Learn the new philosophy. We can no longer accept defective material, material unsuited to the job, defective workmanship, defective product, equipment out of order.

3. Eliminate dependence on mass inspection for quality. Instead, depend on vendors that use process control through statistical techniques. The purchaser is entitled to the control charts for critical characteristics of purchased material, as evidence of quality, uniformity, and cost.

4. Reduce the number of suppliers for the same item. You will be lucky to find for any item one vendor that can furnish evidence of repeatable, dependable quality, and that knows what his costs will be. Price has no meaning without evidence of quality. Demand and expect suppliers to use statistical process control, and to furnish evidence thereof.

5. Use statistical techniques to identify the 2 sources of waste: faults of the system, or common causes (85%), and local faults (15%): strive constantly to reduce this waste. (Dr. Joseph M. Juran said this years ago.)

6. Institute better training on the job with the help of statistical methods.

7. Provide supervision with use of statistical methods; encourage use of these methods to identify which defects should be investigated for solution. The aim of supervision should be to help people to do a better job.

8. Drive out fear throughout the organization. The economic loss resulting from fear to ask questions or report trouble is appalling.

9. Help reduce waste by putting together as a team the people that work on design, research, sales, and production.

10. Eliminate use of goals and slogans posted for the work-force in an attempt to increase productivity. Zero defects is an example. Such slogans, in the absence of quality control, will be interpreted correctly by the work-force as management's hope for a lazy way out, and as indication that the management has abandoned the job, acknowledging their total inadequacy.

11. Examine closely the impact of work standards in production. Work standards are exacting a heavy toll on the economy. There is a better way.

12. Institute elementary statistical training on a broad scale. Thousands of people must learn simple but powerful statistical methods.

13. Institute a vigorous program for retraining people in new skills, to keep up with changes in materials, methods, design of product, and machinery.

14. Make maximum use of statistical knowledge and talent in your company.

Statistical quality technology is a method, transferable to different problems and circumstances. It does not consist of procedures on file ready for specific application to this or that kind of product.

One obstacle to recovery is the supposition that the benefits of the statistical control of quality are accomplished suddenly by affirmation of faith. It is not so simple: it will be necessary to study and to go to work.

Another obstacle that lies in the road to improvement of quality and productivity in America is that so many people in executive saddles have not the faintest idea what to do, nor any idea that there is anything to do. An example of depths of despair is the following quotation, which I copied from a bulletin put out by a well known company to call people together to a conference to study productivity.

   . . .business managers must direct their heavy weapons at that difficult to define, that often elusive target called productivity.

This quotation can only be described as abandonment of all hope for the sinking ship.

Business concerns must look ahead and plan for the future, if they plan to stay in business. Schools of business teach this maxim, but do not practice it themselves. They have failed to perceive opportunities and challenges that lie ahead. In fact, some schools of business have moved toward the teaching of skills, displacing the teaching of the creation of knowledge.

4.10 DISCUSSION AND REVIEW QUESTIONS

(4–1) *The nail that sticks up is hammered down* (Japanese proverb). Discuss with reference to controlling the quality of work.

(4–2) *Quality assurance means planning and control, and, as such, is an element in each job. It also means that each jobholder participates in management within a predetermined area of responsibility.* Explain.

(4–3) *Quality starts at the top.* Why is this important for effective quality assurance, and in what ways can this principle be carried out?

(4–4)   *Quality also starts at the bottom of the organizational hierarchy. But management must take action to influence and direct this quality assurance in order to make it effective.* Explain.

(4–5)   *Management must be committed to quality assurance.* What are the reasons and how can this committment be put into practice?

(4–6)   *Quality assurance can be viewed as a system.* Describe the various features of this system, and compare it with other systems in an organization, such as the marketing, procurement, and finance systems.

(4–7)   *An effective quality assurance system is closely related to other systems in an organization and is thus an integral part of a larger system.* Describe such a relationship with reference to a particular company.

(4–8)   *Japanese success, to a large extent, is the result of application of quantitative methods in quality assurance.* Explain.

(4–9)   *American managers have more enthusiasm for excellence than skill to attain it. Hundreds of thousands of Japanese have been trained in quality assurance, but only a handful of U.S. colleges offer a degree in the subject.* Discuss.

(4–10) *Inflation and unemployment create relatively favorable conditions for promoting quality assurance management.* List some of the major reasons for this and give an example demonstrating this fact.

(4–11) *A relative neglect of quality assurance during growth periods pays off during subsequent recession, if the time for a quality assurance breakthrough is not already too late.* Explain.

(4–12) *In recessionary times, large manufacturing corporations, such as automakers, tend to increase their reliance on suppliers, rather than investing in large new plants.* Explain this strategy from a quality assurance point of view.

(4–13) List for each of the quantitative methods mentioned in this chapter, (such as linear programming, and waiting line systems), one example from quality assurance where the respective method appears applicable.

(4–14) What can explain the fact that many quantitative methods were not applied in quality assurance management in the past?

(4–15) Consider the six principle features of quality control in terms of:
a.   Total quality control involving suppliers, workers, and distributors.
b.   Quality control audits.
c.   Education and training.
d.   Quality control circles.
e.   Application of statistical methods.
f.   Nation-wide quality control activities.
Assume you are a quality assurance manager in a small furniture plant with 100 employees. What action would you recommend in all six areas?

(4–16) Using the sample in Appendix A, prepare a job description for a quality assurance manager in a manufacturing plant, a hospital, and an airline company.

(4–17) *Quality assurance managers must cooperate closely with other functional managers.* Describe needs for and ways of such cooperation, using a particular example.

(4–18) *The same conditions that promote defect-free operations and production also increase productivity.* Explain why productivity increases with an increase in quality.

4.11 NOTES

1. Feigenbaum, A. V., "Quality and Business Growth Today," *Quality Progress,* November, 1982, p. 22.

2. Main, Jeremy, "The Battle for Quality Begins," *Fortune Magazine,* December 1980, p. 28.

3. Leonard, Frank S. and Sasser, W. Earl, "The Incline of Quality," *Harvard Business Review,* September–October 1982, p. 163.

4. Hayes, Robert H., "Why Japanese Factories Work," *Harvard Business Review,* July–August, 1981, p. 57.

5. Crosby, Phillip B., *Quality Is Free,* McGraw-Hill Book Company, New York, 1979.

6. Program Evaluation and Review Technique (PERT) is a potent management tool in planning and executing almost any major project, from developing and defining of project objectives, to identifying the most effective chronology of the steps required for their attainment, required resources, major problem areas and uncertainties, pointing out alternative approaches and evaluating progress. See Chapter 13 for examples of the use of this technique.

7. Deming, W. Edwards, *Quality, Productivity, and Competitive Positions,* published by Massachusetts Institute of Technology Center for Advanced Engineering Study, Cambridge, MA., 1982.

8. Special Report, "Survey Shows CEOs Uninformed About Quality," *Quality Progress,* May 1981, p. 14.

9. "Japan: Quality Control & Innovation," special advertising section, *Business Week,* July 20, 1981, the section "What Top Management Must Do," by Deming, W. Edwards, pp. 19–21. Reprinted by special permission, © 1981 by McGraw-Hill, Inc., New York.

4.12 SELECTED BIBLIOGRAPHY

Barjaria, Hans J. (editor), *Quality Assurance: Methods, Management, and Motivation,* Society of Manufacturing Engineers/American Society for Quality Control, 1981.

Bingham, R. S., "Management Expectations for Quality Controls," *Quality Progress,* June 1973, p. 18.

Chase R. and Aquilano, N., *Production and Operations Management,* 3rd ed., Richard D. Irwin, Homewood, Ill., 1981.

Feigenbaum, A. V., "Business Quality Systems—New Key to Profitable Growth," *Quality Progress,* January 1981, p. 21.

Juran, Joseph M., *Managerial Breakthrough,* McGraw-Hill Book Co., New York, 1964.

Juran, Joseph M., "That Uninterested Top Management," *Quality Progress,* December 1977, p. 18.

Lessig, Harry J., "A True Test of Management," *Quality Progress,* April 1979, p. 20.

Nixon, Frank, *Managing to Achieve Quality and Reliability,* McGraw-Hill Book Company, London, 1971.

Schroeder, Richard E., "How Quality Control Relates to Insurability," *Quality Progress,* April 1975, p. 14.

Smith, Martin R., "Making Quality Management Effective," *Quality Progress,* December 1974, p. 20.

PART II

PLANNING AND CONTROLLING OF QUALITY

In Part II, planning and control of product and service quality is presented against the background that was described in Part One. Quality has emerged as a major obligation in business, the mainspring for success, and a major objective for management. In order that specified and satisfactory quality can be attained and assured, all activities and processes in an organization must be planned and controlled accordingly. The product and service, with their life-cycle phases of design, production, and performance, and the respective quality assurance phases, is the focus of this part.

In the first few chapters we introduce statistical concepts and methods as a major planning and control device. For many readers this will offer a timely and useful review; for others it might lead to further reading and study. The interface of production and quality assurance management deals mainly with the important interrelationship of quantitative and qualitative aspects in production and operations management. Concern for the consumer, quality goals, and production efficiency are to become inherent elements in design, production, and actual consumption or application.

Chapters 8 through 13 follow the logic of product or service design production, conformance, and performance. The original quality of various resources becomes changed in a systematic and creative manner, so that demands for products and services in the market are effectively and efficiently satisfied. We shall show the interactions of consumers, producers, and suppliers in their drive for quality. Managers play an essential role as planners and controllers, using statistical and other scientific methods for rational decisions and actions. Total quality assurance accompanies a product or service from cradle to grave.

Part II will set the stage for planning and controlling a company-wide quality assurance program that embraces all individual products, lines of products, contracts, and so on. ''Planning and Controlling of Quality,'' the title of this part, focuses on the quality of the item for consumption. The next part, ''Managing Quality Assurance,'' analyzes and describes quality oriented decisions and activities with reference to the organization, rather than to the individual consumer.

CHAPTER 5

PLANNING AND CONTROL THROUGH STATISTICAL DATA

The objective of this chapter is to present the basic concepts of statistics that are widely used and adopted as one of many managerial tools in solving quality problems. Industries, businesses, and governments all generate and use a mass of data to make decisions. Handling of data, or use of various mathematical and statistical techniques has, for many good reasons, become such a vast field in itself that in the context of quality assurance management it deserves special coverage in this book.

The reader is cautioned however, that statistics is just one of many tools available for solving quality problems. One must always keep in mind the holistic view of the management of quality assurance. The collection, analysis, and use of data for information gathering is not only for use of the quality control department, but is equally beneficial in sales, marketing, product planning, customer services, and other functions of a company. The message here is to quantify all information, whether obvious or hidden, as much as possible, and analyze it scientifically to get a good fact base, be it a good quantitative picture of a company's sales, of customers' problems, or be it a market analysis, or a survey of competitors. As a matter of fact, few of us can think of anything in connection with our own jobs that is not associated in some way with numbers or data. For effectiveness however, data analysis must remain simple enough to be useful by most of the personnel.

If used, understood, and applied by all levels of management, there are many outstanding benefits of planning and control through statistical data. The statistical methods help:

1. To analyze data properly and draw correct conclusions, taking into account the existence of variation, and thus minimizing wrong decisions.

2. To make sure that production and operation methods for quality attainments are sound, and that they meet the demands of warrantees and consumerism.

3. In refining requirements to make the production and operation methods as good as possible.

4. In meeting the needs for increased reliability of products and services. Such aspects begin with analyzing data to quantify process variation and then further comparing the variation to engineering tolerance limits (called a process capability study).

5. To save costs and to avoid company-wide waste of resources and supplies. If there are no measurements of any other resources (besides money), how can it become evident whether management is moving in the right direction?

6. To wisely apply various sampling techniques based upon recognized standards and acceptance sampling plans.

7. To establish appropriate quality control procedures and inspection plans in order to meet corporate goals for productivity.

5.1 OVERVIEW OF THE CONCEPTS AND ANALYSIS OF DATA

The scope of the uses and application of statistical data is, in fact, unlimited. To try to limit its use only to the quality control areas is to defeat its purpose and ignore its scope. These methods can be applied to almost any problem facing management, whether the problems lie in engineering, manufacturing, inspection, purchasing, or in general quality control. However, where and how to use the specific techniques depends a great deal on knowledge and practice. Just for the sake of its applicability, think of a typical mini-sheet of problems facing a hypothetical company.

1. Too many defects leaving inspection.
2. Bad parts or materials coming into the plant.
3. Dissatisfied customers, loss in sales.
4. Trouble in meeting delivery schedules.
5. Too much cost in repair, rework, or scrap.
6. Too much inspection.
7. Inability to trace the causes of troubles.
8. Trouble in finding assignable causes.
9. Atmosphere of blame.
10. Too many adjustments in engineering specifications.
11. Too much variations from machine to machine.

The above list could be made many pages long. What appears the most crucial in many business settings is the need to translate the common language of a company's problems into plans that are communicable and based upon statistical reasoning and facts. For efficient planning and control, the data must make sense and help improve problem situations by tracing them to their real causes.

Consider for a moment the following itemized translations of the above mini-sheet of a company's problems in relevant statistically oriented questions, such as the following.

1.   How many and what kind of defects pass through inspection, and why?

2.   How many and what kind of bad parts or materials are coming in? Why?

3.   Customer dissatisfaction; why and how? How much is it costing the company in direct sales for each of the product lines?

4.   How many times were deliveries delayed? What were the reasons? Are all instances recorded?

5.   How much were the repair and rework costs? Are they all itemized, explaining the reasons?

6.   Why is there so much dependence on inspection?

7.   Why aren't causes of the problems being traced? Is everyone busy fire-fighting?

8.   Is there a lack of a system or corrective action plan?

9.   Why the atmosphere of blame? Is it perhaps a lack of well-defined authorities and responsibilities? Is it because there are no training or motivation programs in existence?

10.   How many times were adjustments or changes in specifications needed, and why? How could they have been avoided or minimized?

11.   How much variation is there and on which particular machine?

It is clear that unless the questions are obvious and direct enough, no answer given or obtained will be helpful.

As the managerial process has increased in complexity, a need has grown for the use of devices that clarify significant relationships and pinpoint the most important ones. To meet this need, various tools, including pictorial tools (such as diagrams, graphs, charts, etc.) and other quantitative methods must be utilized as aids in decision making. Reporting by means of charts, diagrams, and graphs is an essential first step in the analysis of this large volume of data. Likewise, it is necessary to organize the facts in ways that will help make their meaning understandable.

This chapter is concerned with the importance of such schematic analysis and problem solving techniques. It is important that the reader understand the fundamentals of these techniques so that they can think through possible applications in areas other than those illustrated.

We begin with basic definitions and terminology before proceeding to data analysis techniques.

Statistics mean the collection, organization, analysis, interpretation, use, and meaningful presentation of data aimed at solving a given problem.

Data is factual information used as a basis for reasoning, discussion, or calculation. Thus, any data is useful only to the extent that it accurately and readily promotes reasoning. The reasoning will be as good as the data upon which it is based. If the data is inaccurate, it is certain that the resultant reasoning can be no better.

The concept of variation is one of the vital facts of life, which states that no two objects will be perfectly identical. This is true in nature as well as in every manufacturing operation, no matter how precise and delicate the constants are. Thus, whenever we have variation we have a statistical problem. Sizes of holes drilled from the same machine in exactly identical conditions vary, as do apparently identical twin babies at birth. Even in numerically controlled lathe machines, the overall diameter of machined parts varies slightly from one to another, and from batch to batch, for the same material characteristics.

Thus, in such problems, we should be using methods designed for analysis of data, that is, statistical methods. In fact, what the examples given illustrate, is that the closer we look, the more we are bothered about variation. It is sufficient at this stage to say that variation is a way of life.

Patterns of variation can be found in virtually all measurements or readings of data, provided we have a fairly large collection of data. A tabulation gives us a picture of how the data looks, that is, a *pattern of variation*. The pattern, or distribution, may or may not present any clue to the problem, and needs further subjective analysis.

Accuracy and precision of data relate to patterns of variation. However, the difference between the two terms must be understood. For example, the accuracy of a reading or data is the extent to which the average of a long series of observed measurements differs from the known true value. In other words, accuracy is a measure of the closeness of the measured values. On the other hand, a given set of data or readings is said to be precise if the extent to which they repeat, one after another, is close and agrees with each other, within the limits of measurement.[1] Figure 5.1 illustrates the distinction by showing the results of a series of rifle shots at a target.

Accuracy errors are commonly known as *systematic errors*, that is, when the readings are either plus or minus. As an example, measuring equipment, such as a caliper, must be corrected by equipment adjustments to make it read accurately (calibration). Precision errors

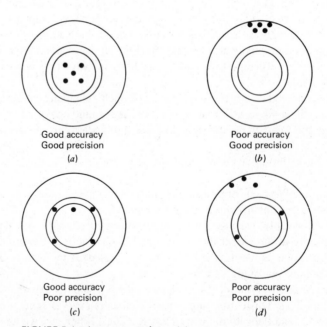

FIGURE 5.1 Accuracy and precision.

are often called *random* or accidental errors, and their causes must be investigated using statistical concepts. In this case, the readings can be both plus and minus.

The *units of measurement* are called the language of measurement. The three systems in use are the English, the metric, and the Systeme International d'Unites (or SI). Most of the countries around the world use or are getting ready to use the metric system and SI system, both of which are decimal based (related to each other by factors of 10). We assume that readers are already familiar with these systems and the conversion factors from one system to another.

Types of Data may be classified as either discrete or continuous. *Discrete* data can take only certain fixed values (countable), such as the number of defective screws in a box. One has either 5 screws defective or 6 screws defective, but not 5.3 screws defective. *Continuous* varying data can take any value within specified (or realistic) limits. Newly born human babies can weigh anywhere between 3 pounds and 15 pounds; the head thicknesses of a given lot of machined bolts, measured to the nearest 0.001 millimeter, would be recorded as varying say, for example, 0.213, 0.212, 0.211, 0.210, or 0.209, millimeter, and so on. Discrete variables can be listed individually in tables, whereas continuous variables are better presented in terms of groups, intervals or cells. However, both types of data become easier to visualize and analyze when presented in *grouped* or *tabular* form.

The term *average,* with which everyone is familiar, will be called the arithmetic mean. The average of a set of *n* numbers (or observation), say, $X_1, X_2, \ldots, X_n$ is given by $\overline{X}$ (irrespective of whether *n* observations are made on one thing or one observation each on *n* things); thus:

$$\overline{X} = \frac{\text{Sum of } X\text{'s}}{\text{Number of } X\text{'s}} = \frac{X_1 + X_2 + \ldots + X_n}{n} = \frac{\sum\limits_{i=1}^{n} X_i}{n}$$

(Sigma i.e., the symbol Σ is the Greek letter notation of summation, $\sum\limits_{i=1}^{n}$ meaning the sum of all values of X, from X_1 to X_n inclusive.)

The *Range, R* of a set of numbers is the difference between the highest observed value (number) and the lowest. The formula is thus: range, $R = X_{(max)} - X_{(min)}$

5.2 ORGANIZING DATA IN TABULAR FORM: FREQUENCY DISTRIBUTION

Consider the following raw data set consisting of a series of measurements or observations, presented in Table 5.1.

These numbers could represent almost anything, such as:

1. The ordered quantity of materials in different boxes.
2. The values of tensile strengths of steel samples.
3. Marks obtained in an examination.
4. Voltage readings across similarly wound electrical coils.

There are 30 observations in Table 5.1. Thus, the so-called sample size is $n = 30$. Each individual number is called X_i, where the subscript $i = 1$ to 30 represents each individual

TABLE 5.1
A Set of Raw Data

100	48	42	68	75
78	65	64	74	51
98	83	90	70	75
103	53	89	86	97
80	58	60	91	80
75	65	85	109	77

TABLE 5.2
Ordered Array of Data

42	64	75	80	91
48	65	75	83	97
51	65	75	85	98
53	68	77	86	100
58	70	78	89	103
60	74	80	90	109

reading. The maximum is $i_{max} = 30$. The presentation of any raw data in the fashion of Table 5.1 is nonetheless very confusing, especially when looked at with the naked eye. Many questions arise: What is the highest and the lowest number? What is the average? What is the spread of the data? Is the data clustered tightly or loosely?

Table 5.2 shows the next step of ordering the data in ascending numerical order, from which we can now easily see, for example, the highest and lowest readings.

To make things clearer and simpler, making a frequency table (or distribution) is the next logical step in the analysis of the data. The data may be grouped or categorized into intervals of their occurrence, or groups or cells to provide a better summary. Table 5.3 shows seven groupings of data, their tally of values, frequency, and cumulative frequencies. It can be seen that condensing data like those in Table 5.3 somewhat hides full details, but brings out other information.

Graphical Presentation

Using a convenient horizontal scale for data values and a vertical scale for cell frequencies, frequency distributions may be plotted graphically in several ways as shown in Figure 5.2. The frequency bar chart is obtained by erecting a series of bars, centered on the cell midpoints,

TABLE 5.3
Tally of Data and Frequency Table[a]

Data Intervals	Tally	Frequency (Number of Observations)	Cumulative Frequency	Relative Frequency	Percent Frequency	Cumulative Percent Frequency
40–49	//	2	2	0.067	6.7	6.7
50–59	///	3	5	0.100	10.0	16.7
60–69	//////	5	10	0.167	16.7	33.3
70–79	////// //	7	17	0.233	23.3	56.6
80–89	////// /	6	23	0.200	20.0	76.6
90–99	////	4	27	0.133	13.3	90.0
100–109	///	3	30	0.100	10.0	100.0
	Total $n = 30$			1.000		

[a]A frequency table such as this is better known in the production organizations when it is simply called a tally card.

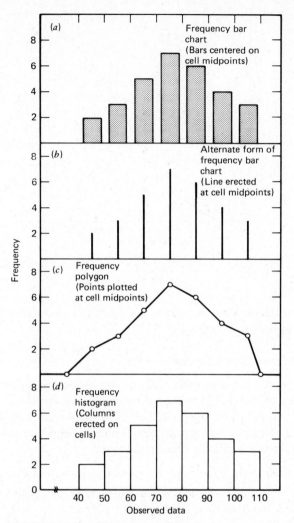

FIGURE 5.2 Graphical presentation of a frequency distribution.

with each bar having a height equal to the cell frequency. Alternatively, a bar chart may also be constructed by using only lines rather than bars. Another common form of presentation is obtained by placing a series of vertical columns along the graduated horizontal scale, each having a width equal to the cell width and a height equal to the cell frequency. Such a graph is called a *histogram*. A variation of the frequency histogram is the frequency polygon. This is a single-line graph constructed by connecting the midpoints of the bars in the histogram.

Frequency tables or distributions can be converted into relative (percentage) frequencies. This distribution is sometimes called the *empirical* probability distribution. Frequency histograms and polygons can be prepared for the relative values in the same way as for the original data (Figure 5.3). In quality control applications, the relative frequency has its greatest usefulness as a measure of fraction defective or fraction nonconforming, in which

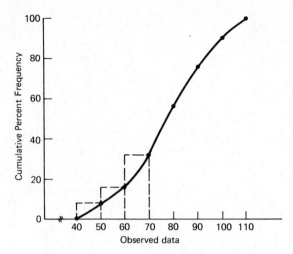

FIGURE 5.3   Cumulative frequency distribution of the data of Figure 5.2.

case it is the fraction representing the ratio of the number of observations outside a specified limit to the total number of observations.

Characteristics of Frequency Distributions

Frequency distribution curves have certain identifiable characteristics that can be put to use in making many day to day decisions once the data are condensed and summarized in the form of histograms or bar chart distributions. Simply, with a sketchy knowledge, the graphical presentation of a distribution makes it possible to visualize the nature and extent of observed variation. Two important identifiable characteristics are:

  1. *Pattern of distribution,* concerning the symmetry or lack of symmetry of the data.

  2. *Peakedness,* concerning whether the curve is quite peaked (referred to as *leptokurtic*) or whether it is flatter (referred to as *platykurtic*).

Figure 5.4 shows different drawings that illustrate the above characteristics. Several representative statistical measures (Section 5.3) are available for describing these characteristics, but by far the most useful are the arithmetic mean, $\overline{X}$, the standard deviation, σ, the skewness factor, g_1 and the kurtosis factor, g_2—all being some form of algebraic function of the observed values.

The coefficient of skewness g_1, of a sample of n numbers, $X_1, X_2, \ldots , X_n$, is defined by the expression:

$$g_1 = \frac{\sum_{i=1}^{n} (X_i - \overline{X})^3}{n\sigma^3}$$

Note that for a symmetrical distribution, $g_1 = 0$.

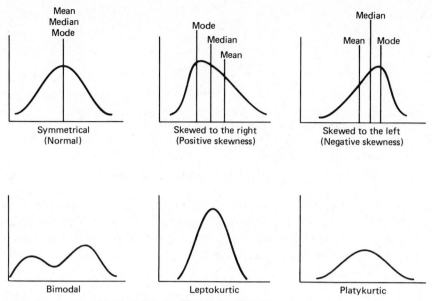

FIGURE 5.4   Characteristics of frequency distribution.

The coefficient of kurtosis, g_2, for a sample of n numbers, $X_1, X_2, \ldots, X_n$ is defined by the expression:

$$g_2 = \frac{\sum\limits_{i=1}^{n} (X_i - \overline{X})^4}{n\sigma^2} - 3, \text{ where } \sigma = \text{standard deviation.}$$

Both g_1, and g_2 are pure numbers and can be either positive or negative. In general, g_1 is negative if the long tail extends to the left, and positive, if the long tail extends to the right; g_1 is negative; when a distribution is flat-topped with small tails (relative to the normal distribution), and positive when a distribution has a sharp peak, thin shoulders, and fat tails (relative to the normal distribution).

5.3   MEASURES OF CENTRAL TENDENCY

The central position of the data distribution or the tendency of data to build up in the center is conveniently described by a measure of central tendency. There are three techniques in common use: the mean, the median, and the mode (see Figure 5.4).

The Mean.   The most simple mean, called the arithmetic mean, was described earlier, represented by the symbol, $\overline{X}$. Other less frequently used mean values are, for example, weighted mean, when a number of means are combined with different frequencies. The formula is given by:

$$\overline{X}_w = \frac{\sum\limits_{i=1}^{n} w_i \overline{x}_i}{\sum\limits_{i=1}^{n} w_i}$$

where $\overline{X}_w$ = weighted mean, and w_i = frequency of the ith mean.

Geometric mean, of a sample of n numbers, $X_1, X_2, \ldots, X_n$, is the nth root of their product. The formula is:

$$\text{geometric mean} = \sqrt[n]{X_1 X_2, \ldots, X_n}$$

The Median.   The median is the value of the middle item in an array. If there are an even number of items, there is no middle value and any figure between the two middle items of the array might be considered the median, since there are an equal number of items on each side. This is exemplified in the following table for an assumed distribution of five items.

| 50.5 | 74.5 | 76.5 | 89.5 | 93.5 |
| 80% | 89% | 89% | 91% | 93% |

When data are grouped into a frequency distribution, the median is obtained by finding the cell that has the middle number, and then interpolating within the cell. The interpolation formula is given by:

$$M_d = L_m + \left(\frac{\frac{n}{2} - cfm}{fm} \right) i$$

where:

M_d = Median.
L_m = Lower boundary of the cell with the median.
n = Total number of observations.
cfm = Cumulative frequency of all cells below L_m.
fm = Frequency of median cell.
i = Cell interval.

The median is an effective measure of central tendency if the distribution curve is skewed (positive or negative), and where an exact midpoint is desired (see Figure 5.4).

To illustrate the use of the formula, data from Table 5.3 will be used in conjunction with the frequency distribution shown in Figure 5.2(*d*). The required values are: $L_m = 70$, $n = 30$, $cfm = 17$, $fm = 7$, and $i = 10$. Substitution in the above formula gives:

$$M_d = L_m \left(\frac{\frac{n}{2} - cfm}{fm} \right) i = 70 + \left(\frac{\frac{30}{2} - 17}{7} \right) \times 10 = 67.15$$

The Mode.   The mode, M_o, of a set of numbers is that value which occurs with the greatest frequency. For example, a series of numbers, say, 10, 13, 15, 15, 15, and 21 has a mode of 15, the series of numbers 1, 2, 3, 4, 5, and 6 does not have a mode, and the series of numbers 8, 9, 9, 9, 12, 18, 20, 20, 20, 22, 23 has two modes, 9 and 20. A given series can thus be unimodal (having one mode), bimodal (having two modes) or multimodal (having more than two modes).

The mode is a quick measure of the central tendency and for most practical purposes can be used to describe the most typical value of a distribution. From a histogram, the mode is easily found as the intersection point of diagonal lines drawn from the upper corners of the rectangle representing the modal class to the upper corners of the adjacent rectangles.

When a distribution is only moderately skewed, it can be shown that there is an approximate relationship between the three measures of location expressed as: (mean − mode) = 3(mean − median).

In many cases, when the measurements are carried out for a data, the central tendency from a production lot or for incoming materials may not exhibit the desired value. Such single data with questionable occurrence is, in fact, common in industries. These should also be analyzed as an opportunity for improvement, and taken seriously as evidence that some important but unrecognized effect has taken place. The idea here is to follow through and check whether the occurrence is spurious or genuine. It may be an important signal of unsuspected but critical factors.[2]

Measures of Dispersion

The measures of dispersion describe how the data is spread out or scattered on each side of the central value. In most applications, measures of dispersion and measures of central tendency are both needed to describe the data.

Standard Deviation

The standard deviation is one of the most useful measures of dispersion for the subject matter covered in statistical methods of quality control. It is understood as the root-mean-square deviation of the observed values about their mean. The formula is:

$$\sigma = \sqrt{\frac{\sum_{i=1}^{n} (X_i - \overline{X})^2}{n - 1}}$$

where:

σ = Standard deviation.
X_i = Value of ith number in a series.
$\overline{X}$ = Mean of the series.
n = Number of observation in the series.

Example: To illustrate the computation of standard deviation, we will use the raw data of Table 5.1, and show the simplified calculation when data are grouped.

From the raw data:

$$\text{mean } \overline{X} = \frac{X_1 + X_2 + \ldots + X_i}{n}$$

$$= \frac{100 + 48 + \ldots + 77}{30} = \frac{2289}{30} = 76.30$$

$$\Sigma X_i = 2289, \; X_i^2 = 182{,}991$$

$$\sigma = \sqrt{\frac{182{,}991 - \dfrac{(2289)^2}{30}}{30 - 1}} = \frac{8340.3}{29} = 16.96$$

Sigma squared (σ^2) is called the variance. An example of the computation of standard deviation from a grouped data is shown in Table 5.4. Note that in either examples, the unit of standard deviation is straightforward (dollars, pounds per square inch, etc.) but the unit of variance is difficult to interpret.

Quartile Deviation

The median divides the distribution in half and the quartile divides the distribution into quarters. Three points must be located: the first quartile (Q_1) is the point that has one-fourth

TABLE 5.4
Computation of the Standard Deviation from a Frequency Distribution (Fracture load from tensile tests of ceramic fibre of 100 lots)

Fracture Load in Pounds (cell interval)	Number of Lots, f	d'	fd'	f(d')²
70–74	1	−4	−4	16
75–79	6	−3	−18	54
80–84	17	−2	−34	68
85–89	29	−1	−29	29
90–94	20	0	0	0
95–99	17	1	17	17
100–104	13	2	26	52
105–109	10	3	30	90
110–114	6	4	24	96
115–119	3	5	15	75
120–124	2	6	12	72
125–129	1	7	7	49
Total	125		46	618

Calculation: $\sigma = i\sqrt{\dfrac{\Sigma f(d')^2}{N} - \left(\dfrac{\Sigma fd'}{N}\right)^2} = 5\sqrt{\dfrac{618}{100} - \left(\dfrac{46}{100}\right)^2} = 12.25$

in which: N = lot size = 100; i = cell interval.

of the frequencies smaller and three-fourths larger; the second quartile (Q_2) is the median; the third quartile (Q_3) has one-fourth of the frequencies larger and three-fourths smaller.

The measure of dispersion based on the interquartile range is called the quartile deviation, and is given by the formula:

$$Q = \frac{Q_3 - Q_1}{2}$$

The quartile deviation is approximately 67% of the standard deviation. If the frequency distribution is skewed *positively*, the distance between Q_1 and Q_2 will be less than the distance between Q_2 and Q_3.

5.4   THE USES OF A HISTOGRAM IN DECISION MAKING

There are many uses of histograms that can help day to day decision making on all levels of management, including the operator level. After the data are condensed and summarized in the form of a histogram or bar chart distribution, a look at the graph itself makes it possible to visualize the nature and extent of deviations observed or computed. In this section we give a few examples of practical problems that were solved using histograms as a tool with the following particular features in mind

1.   The shape of the histogram.
2.   The centering of the histogram.
3.   The width of the histogram.

Examples: The main question is whether the histogram appears symmetrical. When a normal curve (or bell-shaped curve as in Figure 5.5[*a*]) is expected, then any deviation can be suspected of being caused by manufacturing conditions or other conditions. This calls for investigation of the reason(s) for the variability. The two histograms of figure 5.5(*b*) (drawn in solid and broken lines) illustrate a situation where a quality control inspector took up the problem of reducing the amount of scrap of aluminum base plates because holes were being drilled too large. After an improvement action (in this case it was simply a matter of changing over to a different drill machine), the new histogram showed considerable change in terms of average hold diameter (lowering of $\overline{X}$ and standard deviation, σ, both in favor of needed improved quality.

In the example illustrated by Figure 5.5(*c*), the large step-like appearance on the left edge compared to the gradual and normal behavior on the right edge was taken up as a curious problem worth looking into. It was found that inspector *A*, who kept rejecting certain parts, was never informed by the design review of some relaxation in the lower specification limit that was allowed by the vendor and had been enforced in the company three months previously.

The histogram shown in Figure 5.5(*d*) warns about potential trouble when the measurements are beyond the tolerance limits.

One customer was puzzled to see the appearance of a histogram (Figure 5.5[*e*]) on hardness tests of ductile iron hub castings received in a lot from one of its supplier's foundry during summer months. Although the sample taken indicated that the lot met hardness tolerance

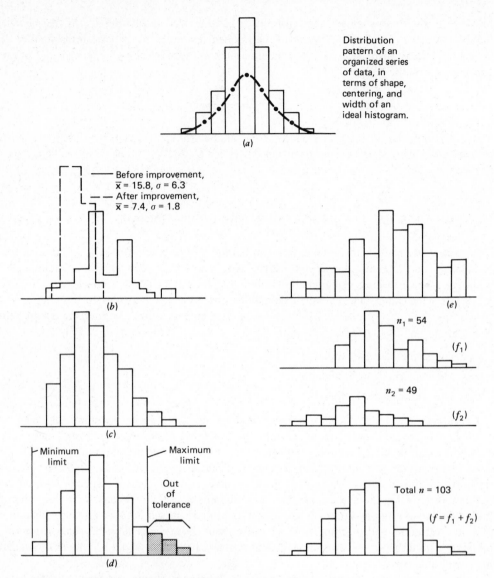

FIGURE 5.5 Typical examples of insight provided by abnormal histograms. See text for explanations.

limits, the customer, realizing the chance of some mix-up, checked further. Through microstructural examination of the castings on an optical microscope, it was revealed that the foundry had poured some of them in a different grade of metal than the rest. Eventually, the foundry accepted responsibility for this mistake (the melting foreman had been on summer holiday) and ended up paying for this scrap.

The effect noticeable from the width of the histogram is equally interesting. A subcontracting company specializing in welding did all the welding jobs for a pressure pipe manufacturing organization that supplied the steel pipes, elbows, and other parts, including welding rods. However, on one occasion several batches of piping components showed cracks during hydrostatic pressure testing. Subsequently weld failure analysis was carried out by taking hardness measurements. The results are shown by the histogram distribution of Figure 5.5(*f*), indicating a broad distribution for the hardness of the steel pipes. It was discovered that the parent company had purchased steel pipes from two different mills, and the hardness distributions of each were different (Figures 5.5[f_1] and [f_2]). Notice how a lack of quality checks of the materials received in the first place can render the production wasteful. Knowing the complete history of the data and keeping a record is important in improving and controlling the manufacturing process. In this case, a stratified histogram helped discover the cause.

If the relative size of the variables or components of the total are to be emphasized, a *pie chart* is also an appropriate tool.

Limitations of Histograms

Histograms based on too few measurements or observations can lead to wrong decisions. A good rule of thumb is that at least 50 to 60 measurements must be made to reveal conclusive evidence, either for or against an assumed hypothesis. In addition, one must interpret all three aspects of a histogram—the shape, the width, and the centering—with some knowledge of the particular manufacturing process in question.

A symmetrical (well-defined bell-shaped, Figure 5.5[*a*]) histogram is not always free from hazardous conclusion. The reason is that one cannot see the trend in the data with the progress of a manufacturing operation. The data in a histogram may appear symmetrical, but only when plotted individually in order of operational progress, will they show an increasing or decreasing trend to tell us whether the process has really been stable or not.

5.5   HANDLING DATA BY USING PARETO DIAGRAMS

To a quality manager or chief executive of a company the word "problem" has multiple meanings that must be resolved: loss in sales, market share, defects, product recalls, cost savings, quality control, return on investment, and so forth. While each of these topics is a good deal more complex than the discussion in this section, the fact remains that each problem consists of so many smaller problems that it is difficult to know just how to *prioritize* them. A definite need exists to know the magnitude of a problem before solving it.

To find the area that needs improvement, or the problem of most concern, use is made of an indispensable tool, called Pareto Analysis.[3] Let us review, for instance, the data on defectives, collected in Table 5.5, from inspection reports submitted by a group of inspectors in a manufacturing company making pressed metal automotive components. Data from this table have been plotted into a bar chart (Figure 5.6) in order of occurrence of major defects. The horizontal axis lists the defective items from the major item on the left to the minor item on the right, with others arranged in descending order of magnitude. This type of graphical presentation is called a Pareto diagram.

A major use of the Pareto concept is that such studies can furnish an important basic direction where attention and efforts for quality improvements are needed most with the least

TABLE 5.5
Record of Defectives

Date_____   Number Inspected_____

Total Number Inspected 2112

Defect Type	Number of Defectives	Percent	
		Defectives	Defectives (cumulative)
Crack	210	51.5	51.5
Deformation	35	8.5	60.0
Flash	108	26.5	86.5
Stain	37	9.1	95.6
Rough edges	18	4.4	100.0
	408	100	

amount of analytical study. It can thus help in designing quality improvement programs, by identifying the *vital few* from the *trivial many,* since it is the two or three tallest bars which account for the majority of the problem, the smaller bars being lesser causes. Thus, one can definitely say that a pareto diagram is the first step to making improvements. In fact, the subject of study taken up by one quality-conscious company found this out, when they compared the techniques used by their personnel in case reports filed over a five year period (Figure 5.7).

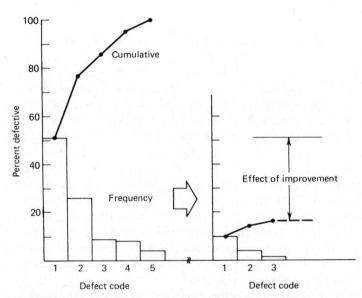

FIGURE 5.6   Pareto diagrams for defective components (before and after improvement).

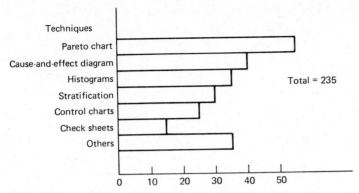

FIGURE 5.7   Quality control techniques most frequently used by a company in case reports.

Problems that Can Be Solved Using the Pareto Concept

In quality assurance it is not one area or one department that inherits the problems caused by others and vice versa. Improvement in the organization is not only a question of improvement of quality—there are also many other problems: improvement in productivity and efficiency, safety of workers, cost reduction, energy conservation, marketability, and so on. The following example illustrates the application of Pareto Analysis. The student is encouraged to think of problems that they face or know of at various workplaces, in terms of the use of a Pareto diagram, by asking:

1. What are the *outstanding* (i.e., vital few) *causes* that need immediate attention for improvement?

2. What *action,* among all available, is the most appropriate for remedy?

Example: The diagram shown in Figure 5.8 was used to lower the rate of customer complaints at a bank. The branch manager started his effort by finding out and determining the type of errors that were getting through the system and the source of these errors. A simple categorical listing of the type of errors were made differentiating between dollar encoding errors and high speed processing errors. On the basis of this analysis, the goal for improvement was determined to be debits and credits information maintenance.

Generally, if measures taken are proven effective, then first of all, the length of the bar for each item will change. Experience has shown that it is easier to reduce the taller bars by half than to reduce a short bar to zero. Even if the amount of effort required to reduce a tall bar and a short bar were the same, there is no doubt that the effort should be concentrated towards reducing the tall bars. The causes of all errors are intermingled, and if the vital few errors are removed, the rest are taken care of as a consequence. Secondly, the order in which the bars appear may or may not change, depending upon how the roots of problems are tied to one another, that is, whether the manufacturing process is product oriented, process oriented, or machine or operator oriented.

One well known use of the Pareto technique is *analysis of vendors.* This can take several forms, for example:

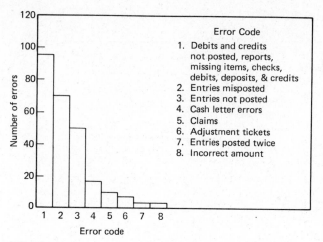

FIGURE 5.8   Pareto diagram of cumulative errors applied to a bank branch problem.

1. Analysis of defects, lot rejection, monetary losses, and downtime, according to defect type, seriousness classification, material type, loss of reliability, and other factors.

2. Analysis based on product types aimed at deciding whether the company's procurement policy should be based on few vendors and large quantity or many vendors and small quantity.

3. Analysis based on process types aimed at receiving better quality from superior processes used by one vendor over another in, for example, coil winding, heat-treated metal parts, electroplating, and so forth.

4. Analysis based on costs incurred.

5. Analysis based on workmanship, reliability, and factors of similar importance.

Several other pertinent analyses can be made in some instances for comparing workers, machines, shift-to-shift operation, efficiency, controllability studies and other matters.

Other Uses

As previously stated, a Pareto diagram is the simplest and most effective management tool which defines priorities for quality improvement projects by way of dramatizing the subject by showing graphically where the effort of management should be stressed.

5.6 CONDUCTING ANALYSIS TO DETERMINE THE CAUSE: SYMPTOMS AND DIAGNOSIS

Excuses for scrap or defective products made by companies are many and varied. However, the fact that one company can excel compared to another, when making the same product or providing the same service, is surely a matter for consideration. In fact, no product is ever made with zero defects, and there are no companies consistently making zero scrap.

One reason is the influence of a large number of variables that are due to humans, machines, materials, methods, or systems. Even when there is only slight contribution from each of these factors, the differences add up to a great deal in the end product.

The purpose of this section is to stimulate further logical thinking and provide more effective tools while looking for causes of defects. Collection of empirical data is one thing, but actual knowledge comes from interpretation (understood as an act of thought). Interpretation consists of testing various hypotheses or theories based upon prior knowledge and other available collateral information—qualitative, quantitative, or experimental.

A diagnostic study is properly carried out in the sequence of: knowing what *is* happening wrong; and, knowing what *should* correctly be happening. The observed difference between the two is the lack in proper or desired quality, simply stated as a *problem*. Figure 5.9 shows an organized approach for handling defect diagnosis in a step-by-step sequence. The scheme shown makes no distinction between a small problem or a larger or more difficult problem, since they all need to be understood fully to make the effort preventive. A suggested scheme for the collection, organization, and use of data is given in Figure 5.10.

Cause-and-Effect Diagram

A cause-and-effect diagram (also known as a fish bone diagram or Ishikawa diagram[4] is a powerful tool useful in helping to sort out the real cause of a problem. The causes giving rise to a quality problem are not only always interlinked and dependent on one another, but unfortunately, they vary from one diagnostician (or analyst) to another.

On many occasions an analyst either may not be creative when collecting evidential data or may be weary about the atmosphere of blame. It is important that one must think and speak frankly when discussing a problem's causes in a group (brain storming). It is equally important to develop respect for other members' suggestions.

Example: A ductile iron foundry producing heavy wheel hubs (weighing approximately 80 lbs each) for a local farm equipment machinery company was plagued by blow holes appearing on flat surfaces of the castings. Obviously, the question was, why did blow holes occur suddenly. When the foremen, lead hands, and plant manager got together to investigate, they first made up a list of all operational factors (Table 5.6) since any one factor, or perhaps a combination of factors could have given rise to the defect. Next, a cause-and-effect diagram was made, to enumerate cause types (Figure 5.11; the inset describes a skeleton in major categories). The advantage here was that all causes were listed and no major ones were left out. Since listing the causes was done with a purpose in mind, the diagram served as a focus for discussion. By going through recorded data, it was possible to pinpoint the high moisture content in the molding sand as the cause of the blow holes in the castings.

Other Uses

A cause-and-effect diagram, being educational in itself, can be used for many problems, provided that the relationships between causes and effects are studied in a rational manner. Think for example, of its applications in improving safety at the workplace, building up a good team of workers, improving the quality control program, increasing sales, reducing scrap, and other similar projects.

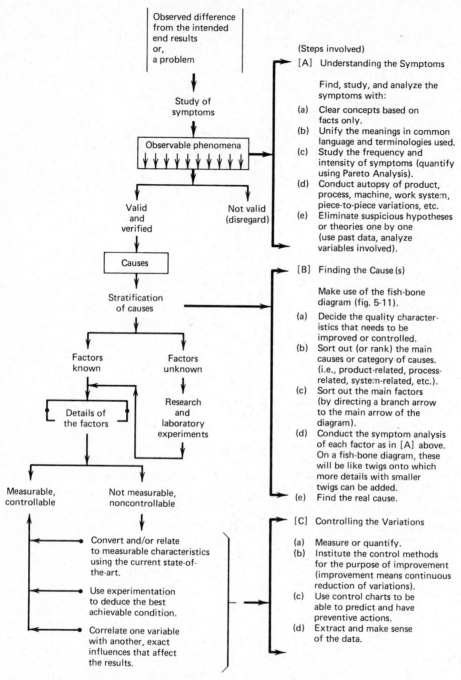

Observed difference
from the intended
end results
or,
a problem

↓

Study of
symptoms

↓

Observable phenomena

Valid and verified / Not valid (disregard)

Causes

↓

Stratification
of causes

Factors known / Factors unknown

Details of the factors / Research and laboratory experiments

Measurable, controllable / Not measurable, noncontrollable

Convert and/or relate
to measurable characteristics
using the current state-of-
the-art.

Use experimentation
to deduce the best
achievable condition.

Correlate one variable
with another, exact
influences that affect
the results.

(Steps involved)

[A]   Understanding the Symptoms

Find, study, and analyze the
symptoms with:

(a)   Clear concepts based on
facts only.
(b)   Unify the meanings in common
language and terminologies used.
(c)   Study the frequency and
intensity of symptoms (quantify
using Pareto Analysis).
(d)   Conduct autopsy of product,
process, machine, work system,
piece-to-piece variations, etc.
(e)   Eliminate suspicious hypotheses
or theories one by one
(use past data, analyze
variables involved).

[B]   Finding the Cause(s)

Make use of the fish-bone
diagram (fig. 5-11).

(a)   Decide the quality characteristics that needs to be
improved or controlled.
(b)   Sort out (or rank) the main
causes or category of causes.
(i.e., product-related, process-
related, system-related, etc.).
(c)   Sort out the main factors
(by directing a branch arrow
to the main arrow of the
diagram).
(d)   Conduct the symptom analysis
of each factor as in [A] above.
On a fish-bone diagram, these
will be like twigs onto which
more details with smaller
twigs can be added.
(e)   Find the real cause.

[C]   Controlling the Variations

(a)   Measure or quantify.
(b)   Institute the control methods
for the purpose of improvement
(improvement means continuous
reduction of variations).
(c)   Use control charts to be
able to predict and have
preventive actions.
(d)   Extract and make sense
of the data.

FIGURE 5.9   A scheme of organized approach to quality problem solving.

Gather
background
information
that
relates to the
problem
|
Start
data
collection
|
Analyze
the data
|
Review
and
eliminate
unwanted data
|
Present the
result
|
Relate your
conclusion to
the problem

Planning for Data Collection

- Make sure that the data have been produced in a state of statistical control.
- Decide on the type of data needed (e.g., variables, discrete), its impact on decisions, cost, and other consequences.
- Sort out assumptions in the use of data, including risk, uncertainty, variability, precision, and accuracy.
- Ensure suitability of data for use; avoid listing redundant data.
- Determine and note the important characteristics or circumstances under which data have been collected.
- Record variability conditions.
- Collect the data.
- Apply statistical techniques to analyze the data.
- Plot the data. Charts and graphs are better than tables that contain numbers. Only analyzed data are meaningful.

FIGURE 5.10   A suggested scheme for the collection, organization, and use of data.

5.7   LINEAR REGRESSION AND CORRELATION: THE SCATTER DIAGRAM

When there is a well-established relationship between two or more causes (or data groups) it is possible to make estimates of one variable from known values of the others. For example, in the above case study, varying the percentage of moisture content in the molding sand might be used to discover the optimum level against the size of vent holes for blow hole defects in the foundry.

Using the relationship between a known variable and an unknown variable to estimate the unknown one is called *regression analysis*. The uses of regression analysis include forecasting (see Chapter 7), determining the important variable influencing a quality characteristic, and locating optimum operating conditions. The measurement of the degree of relationship between two or more variables is called *correlation analysis*. A *scatter diagram* is a graph between two variables: the known variable, normally plotted on the *x*-axis, and the variable to be estimated, plotted on the *y*-axis.

Figure 5.12 is an example of a scatter diagram for the relationship between three different raw material ingredient compositions (*x*) and material strengths (*y*). The device used for estimating the value of one variable from the value of another consists of a line (or curve) through the data points on a scatter diagram, drawn to represent the average relationship between the two. This is called a line of regression.

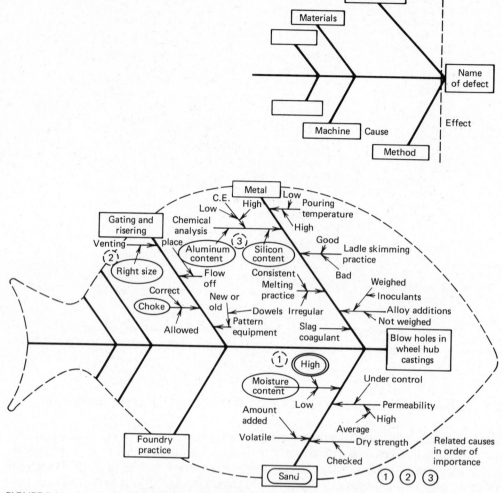

FIGURE 5.11   A typical cause-and-effect diagram (or fish-bone diagram) for discovering defect causes as applied to blow hole defects in a foundry.

The straight line is represented by the equation:

$$Y = a + bX$$

where:

a = A constant value of Y that represents the point where the regression line intercepts the Y axis.

b = A constant value that describes the slope of the regression line (that is, the amount by which Y changes every time X is changed by one unit).

Y = The dependent variable that management wants to forecast.

X = The independent variable being used to forecast the dependent variable.

TABLE 5.6
Operational Conditions and Defect Cause Check List

Iron type	Grade: 80 (ductile iron) (wheel hub)	Date:_____
Foreman	_____	Shift:_____
Inspector	_____	

Metal	Pouring temperature
	Chemical analysis
	Aluminum content
	Ladle skimming practice
	Furnace operator
	Inoculant/alloy additions
	Melting practice
	Slag coagulant
Sand	Moisture content
	Strength
	Permeability
	Volatiles
	Dry strength
Gating and risering	Venting
	Flow-offs
	Condition of bottom boards
	Chills used or not
	Core points capped or not
	Pouring speed
	Chocked gating system
	Pattern equipment

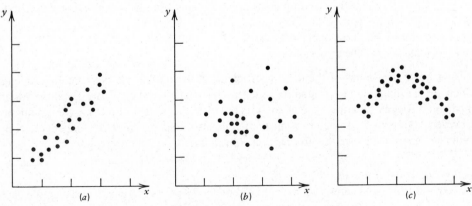

FIGURE 5.12   Scatter diagrams indicating (a) positive correlation, (b) no correlation, and (c) a trend.

There are many possible curved regression equations that might be fit to the data. The one used quite commonly is a second degree parabola shown by the equation $Y = a + bX + cX^2$, where c is a constant and other terms are as previously defined.

Regression Lines

A regression line or line of regression is one that represents the slope of correlated data as accurately as possible. With a given equation of the straight line regression, such as $Y = a + bX$, the slope, b, and the intercept, a, are calculated from:

$$b = \frac{\Sigma XY - \dfrac{(\Sigma X)(\Sigma Y)}{n}}{\Sigma X^2 - \dfrac{(\Sigma X)^2}{n}}, \text{ and}$$

$$a = \frac{(\Sigma X)(\Sigma XY) - (\Sigma Y)(\Sigma X^2)}{(\Sigma X)^2 - n\,(\Sigma X^2)}$$

where $n =$ number of pairs of X, Y values.

Multiple regression analysis is a method of taking into account simultaneously the relationship between all the variables, when two or more independent variables are to be used in making estimates of the dependent variable.

For example, in a steel manufacturing company it is always required to control the chemical analysis of more than two elements. For a typical low-carbon alloy steel, it is not only the percentage of carbon that has to be controlled in the melting operation for desired tensile strength, but that of other elements as well, such as silicon, manganese, sulpher, phosphorus, nickel, and a few others. Thus, we have here more than one input variable. The regression equation for two variables would be in the form:

$$Y_{X_1 X_2} = a + b_1 X_1 + b_2 X_2$$

where:

$$Y_{X_1 X_2} = \text{The predicted value of the response variable (say, tensile strength of}$$
steel, based upon variables X_1 and X_2 (any two alloying elements).
$$a = \text{The } Y\text{-axis intercept.}$$
$$b_1 \text{ and } b_2 = \text{The slopes with respect to } X_1 \text{ and } X_2.$$
$$X_1 \text{ and } X_2 = \text{The two variables.}$$

To estimate the values of the three coefficients in the model, we perform experiments in which we set the levels of X_1 and X_2 at the corresponding values of the response variable. The main difficulty with this type of analysis has been the burden of making the calculations. However, the use of computers is now of immense help, as there are excellent, simple, nonlinear, and multiple regression programs available.

Method of Least Squares

The method of least squares is a widely used method of fitting a curve to data, to establish the trend. The basic problem is selecting a method of fitting a trend line with the criterion of a good fit. The most common criterion is to decide whether (for the best selected line) the sum of the deviations of the actual values from the trend values is a minimum. If so, the line drawn can be regarded as the line of best fit. Another criterion is to decide if the sum of the squared deviations is a minimum. The name *least squares* is derived from the use of this criterion.

Logarithmic Trend Lines: Probability Paper Graphing

Even when the actual values of the trends from a given set of data seem entirely different and not comparable, use can be made of logarithmic plots to present the data. A nonlinear trend on an arithmetic chart may become a straight line on a semilogarithmic plot. Instead of computing the trend of the data (Y), the logarithms of the data are found, and a straight line is fitted to the series log Y, instead of the series Y. This is called a semilogarithmic plot. Probability paper is a special form of logarithmic graph paper with scales that are adjusted so that a perfectly normal distribution will plot as a straight line. Semilogarithmic paper has other uses, including plotting a geometric series where, for example, a statistician wants to emphasize *rate of growth*. If the *amount* of change is more important, simple arithmetic paper will serve the purpose. Figure 5.13 shows schematics of other general plots that are most effectively observed by the use of probability paper graphing. Nelson[5] provides details of probability plots with practical examples.

The size of the logarithmic paper is designated by the number of cycles. Note that in Figure 5.13, curve 1 stretches to three cycles whereas curves 2 and 3 were sufficiently drawn with only two cycles. The actual data of Figure 5.13 represent three different properties tested for computer software reliability. Since the ruling in all cycles is exactly the same, the value of any line in any cycle is always exactly 10 times the value of the corresponding line in the cycle immediately below.

The above brief treatment of the subject of regression and correlation analysis is merely an introduction. The reader is advised to consult other statistical texts mentioned at the end of this chapter and the next.

5.8 KEY TERMS AND CONCEPTS

Statistical data are numerical observations on groups of items (statistical populations) that measure and express variation.

Statistical quality control applies statistical inferential methods to measure quality characteristics (variables and attributes) and to make decisions concerning lot quality or process performance.

Patterns of variation can be shown in form of frequency distributions.

Accuracy measures systematic errors from a known true value.

Precision relates to random errors expressed in statistical laws.

A **Pareto diagram or analysis** interrelates cumulative proportional or percentage distribution, such as for defectives, over various classes or types. The inequality of the distribution allows for concentration on the vital few types.

A **cause-and-effect diagram** is a simple, fish bone like chart, used to trace actual or possible causes of a particular defect. This cause analysis permits wide application by all staff and is an important aid for prevention.

Other terms and concepts in this chapter should be checked either in this text or in other statistics texts.

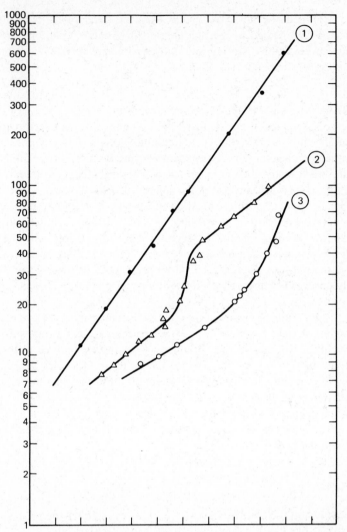

FIGURE 5.13 Smoothing out effect of probability plots; (1) represents perfectly normal distribution, (2) represents data from two different populations, and (3) signifies skewed distribution with a positive skewness.

5.9 SUPPLEMENTARY READING[6]

. . . Here is an example taken from a Japanese book, issued by the Asian Productivity Organization to teach statistics to workers at the factory floor. Suppose we wish to make a perfect plate of rice every time. How should we go about it? To make a perfect plate of rice requires certain ingredients: The rice, the pot to cook it in, the water, the stove, the method of delivery and presentation. The workers are taught to draw an influence diagram, called a "fish-bone diagram," which shows a plate of rice as the output and indicates lines from the inputs to the outputs to show schematically how things flow together and in what order. Then, as part of the perfect plate of rice example, they discuss how to control each input. What kind of rice? How old? How stored? How clean? What kind of pot? How clean? How much water? What temperature? How clean?

How measured? What kind of stove? What temperature? How long? What kind of plate? Served at what temperature? How long a delay?

The point is made that if a perfect plate of rice is desired, each and every one of the inputs must be under control. The workers need to understand how much variation can be allowed in each of the inputs and still produce a "perfect plate of rice." They need to define what it means to say a plate of rice is "perfect." If the variations are under control, a perfect plate of rice will occur every time. It is that simple.

The workers then learn to make the same analysis of the product they are making. With this kind of education, they are in a position to make observations about the work they do. And they are ready to participate in joint efforts with the management, to find ways to do the work better.

Some people advise omitting the statistical training and just introduce quality circles in the factory. They may win some early approval, but unless they give the workers the tools with which to observe and report their results statistically, the movement will eventually die, as the workers and the bosses run out of useful things to do. You can obtain some gains without statistics. But you will never reach 97 percent up time for your production systems without statistics.

Quality circles are important elements in any strategy to improve quality and productivity, but they are not the most important sources of improvement. According to the experiences of the Nashua Corp., only 10 to 15 percent of the gains in productivity arise from the initiatives in Quality Circles. At Nashua, the most important gains have been made through managerially directed programs in which engineers, statisticians, physicists and workers went to work as a team, systematically studying a process. Management involvement is a key element for success for often-times, in fact, most of the time, the problem transcends departmental boundaries and may even lead to changed vendor relations. Quality circles help prepare workers to introduce and understand innovations, but are not expected to be the major source of such innovations.

We cannot overstate the importance of everyone learning statistical methods.

You will appreciate the importance of statistics after you have seen a Japanese factory using the "just in time" method of manufacture. In this method the seats for an automobile arrive from the vendor an hour or less before they go into the car. Not the day before. Not ten hours before. But "just in time" as the name says. This means that the material in inventory is reduced by an order of magnitude.

In these systems, the assembly line moves as a single unit, without buffer zones between stations. Everything works right. The gains in productivity are enormous.

Yes, you can do some things without statistics. But you cannot reach the standards of the new competition. It would be like entering the Olympics and deciding you did not have to train. You will never design a system in which everything works right the first time if you do not understand its statistical variations.

Of course it would be utterly foolish to take the workers away from their jobs and put them through a three credit hour college course in statistics. The teaching should always be made part of the work. Techniques should be introduced as they are needed, as the workers are trying to improve the process. The work itself should be the basis for the education, and the need-to-know the motivator for learning.

There are some people who believe that these methods are intended only for the factory floor. These statistical methods may be used anywhere that systems of people and machines are used to get the work done. This goes for the office as well as the factory, the service industry as well as manufacturing industry.

Consider a trucking firm managed by a man educated according to current management methods taught in many schools of business management. If you ask him to define his job he will say it is to run the company as profitably as he can and to expand its business. To do so, he may call upon the best consultants he can get to help him design the best possible systems for dispatch, routing, and maintenance. He may set up work standards for the drivers and institute computer-based procedures to keep track of the performance of the drivers, trucks, and dispatchers. He will institute the best methods of dispatching he can devise, either on his own or as gleaned from trade magazines or meetings of his association. He will study his markets and their opportunities. He will keep extensive records of income and expense, ever on the alert for opportunities to increase his profit.

Of course he will not be able to manage his empire

alone. As the organization grows, he will institute methods to see that his desires for efficiency and performance are carried out. Perhaps he will adopt "Management by Objectives" and teach the methods to those who report directly to him, and to others. He may assign as much as 5 percent of his work-force to data gathering and performance evaluation, helping him to search for possible profit opportunities.

He may hire outside lecturers to give "motivation" lectures to his workers. He asks his division managers to nominate a "driver of the month" who will be suitably recognized and rewarded for superior performance.

In short, his idea of a good manager is one who sets up a system, directs the work through subordinates and by making crisp and unambiguous assignments, develops a basis to set standards of performance for his employees. He sets goals and production targets. He rates employee performance objectively against these targets. . . .

He understands that a trucking system is subject to great variability. Traffic conditions change, trucks break down, shipping docks are not always ready to receive or load, mistakes are made in routing and addressing. Packages are not always ready to go. People are not always ready to sign receipts. There are countless ways for the system to go wrong and out of control, decreasing quality and increasing cost. The manager knows that these things occur randomly. To observe and report on such a system requires a knowledge of elementary statistics. The manager arranges for everyone in his organization to learn elementary statistics, on company time, and to apply what they learn by keeping their own records. Drivers keep their own control charts to describe how long they have to wait at dock to load or unload. They look for trends, for correlation with other events. They meet regularly, with the encouragement of the management, with the dispatchers and compare the results when the dispatchers make different arrangements with customers. They keep data on the performance of their trucks and discuss their observations with the buyers, so that when trucks are purchased they have the best characteristics. Their discussion is not based upon casual or prejudiced observations, but upon statistical evidence. When the manager tells his purchasing agent to buy on "quality"

and not just first cost, the data are available to do just that. Everyone in the system is involved in studying it and suggesting ways to improve it. It is a way of life. The employees will see the setting of work standards as a dumb idea, one that inhibits them from doing better. They will not understand anyone who tries to "manage by objectives" because they will be constantly engaged in refining their objectives themselves and recording the performance of the system.

That last phrase is important: "the performance of the system." It is delusionary to regard the performance of the people in the system as due entirely to their performance and not a function of how the system works. Data by Juran suggest that 85 percent of the time the problem is with the system, not the worker. A manager who acts as though whenever anything goes wrong it is the worker's fault deserves to have a militant union. Indeed, the workers will need one to protect themselves from such bad management.

Because the problem is generally with the system and not the workers, the manager spends considerable time making sure that the workers feel secure in their jobs. They know, for example, that if the productivity is improved they will share in the profits and will not be laid off. They know that if they complain about poor maintenance, and have the statistics to back it up, the management will not regard them as troublemakers, but will welcome their contributions and act upon them.

No one will have to teach this manager to be nice to people. He will see them as partners in his desire to improve the output and reduce costs. He will not think of them as robots who happen to be made of flesh and bone. He will regard them as human observers, capable of using their intellects to help him improve things. He will not plaster the walls with empty slogans such as "Do it right!" or "Zero defects" for he knows that such slogans mean nothing to people who are constantly trying to eliminate the causes of defects. . . .

5.10 DISCUSSION AND REVIEW QUESTIONS

(5–1) From your own experience, list six areas in which statistics can be useful in making quality control decisions.

(5–2) Differentiate the meaning of the words "statistic" (singular) and "statistics" (plural).

(5–3) What common techniques are important in analyzing raw data. Describe the kind of information needed to analyze data in the following types of operations.
- a. A small, one owner, watch repair shop
- b. A service facility at a drive-in bank
- c. A local theatre

(5–4) Explain the difference between a continuous variable and a discrete variable. Give examples of each.

(5–5) What are the advantages of presenting statistical data in graphical forms rather than in tabular forms?

(5–6) Why is it important that something be known about the shape of the distribution?

(5–7) Why is the standard deviation considered a better measure of dispersion than the average deviation?

5.11 PROBLEMS

(5–1) The following data consists of 60 measurements for hardness values of welded parts of a special grade of steel.

35	41	33	32	34	41
30	40	40	34	29	28
37	29	29	28	37	33
41	35	37	37	41	31
34	42	40	33	32	42
29	32	34	41	30	33
30	37	32	36	36	29
34	30	31	29	32	28
38	29	38	32	32	31
31	41	28	40	28	30

- a. Summarize the data in tabular form in the ascending order.
- b. Draw a histogram using the above data.

(5–2) A foundry that makes ductile iron castings tries to meet railway specifications concerning tensile strengths by a strict control of chemical composition. The table below gives 50 measurements of the percentages of silicon content that were sampled at random intervals from the drillings taken out of finished castings.

2.1	1.8	2.5	1.9	2.2
3.0	2.5	2.2	2.0	2.2
2.4	2.4	2.5	2.5	2.8
3.1	2.8	2.0	2.2	3.1
2.0	1.9	1.9	2.4	2.8
2.4	2.0	2.6	2.0	2.2
2.2	2.2	2.8	3.1	2.5
2.5	2.5	2.2	2.8	2.6
2.0	2.6	2.5	2.0	3.0
1.9	3.0	2.0	1.9	2.5

- a. Summarize the data in tabular form.
- b. Summarize the data in graphical form.
- c. If the specification calls for (2.5 ± 0.3) as the percentage of silicon, how many lots would have (i) too high, and (ii) too low silicon contents? Assume that each of the readings corresponds to one lot of shipment.
- d. If a lot is made up of 1000 castings, determine (i) the total number of castings containing high silicon; and, (ii) the total number of castings containing low silicon.

(5–3) An electrical inspector who checks the resistance between two terminals of finished circuit boards at random finds ten readings (in ohms), as follows: 2.59, 2.57, 2.60, 2.55, 2.49, 2.54, 2.57, 2.58, 2.50, and 2.60.
- a. Compute the arithmetic mean.
- b. Determine the range.
- c. Compute the geometric mean.

(5–4) Determine the median of the following groups of numbers.
- a. 32, 21, 25, 18, 28
- b. 45, 38, 43, 45, 53

(5–5) Determine the mode for data, as given in problem 5–4.

(5–6) a. Define "standard deviation."

b. For the following observations, compute the standard deviation, variance, and average deviation about the median: 25, 33, 25, 50, 21, 69, 55, 39, 45, and 63.

(5–7) The following frequency distribution shows the time required to process 101 drivers' licenses. Compute the variance and standard deviation.

Time (in minutes)	Number of Licenses
1.0 to 1.5	7
1.5 to 2.0	11
2.0 to 2.5	28
2.5 to 3.0	35
3.0 to 3.5	15
3.5 to 4.0	5

(5–8) a. Describe the usefulness of a Pareto diagram.

b. Discuss the interpretation of a histogram based upon its appearance in terms of (i) the shape, (ii) the width, and (iii) the centering.

b. Asymmetrical with more concentration of data to the left.

c. Bimodal distribution

d. Relationship with specification limits

(5–10) The management team in a company is establishing priorities to attack a serious internal quality problem. You, as a quality manager, have been requested to establish a data analysis system to find and direct this attack. Describe your approach in brief.

(5–11) The following table gives data on defective items taken from the assembly line of automatic clothes-washing machines over a two-week period.

a. Prepare a Pareto diagram to help decide where improvement must begin.

b. With which items do you suggest that improvement should begin, and why?

c. Consider the manner in which defectives appear over the two week period and prepare corresponding Pareto diagrams for each week. Draw cumulative lines on each Pareto diagram.

	Week 1						Week 2				
Defect Day	1	2	3	4	5	—	8	9	10	11	12
Faulty switch	2	3	10	12	5		2	3	6	9	7
Defective motor	1	0	2	1	3		2	4	1	2	3
Off-centering	9	7	5	3	2		0	3	7	8	5
Poor angling	3	0	0	2	0		5	0	0	3	2
Loose screw	5	10	3	8	7		6	0	11	12	7
Washer	4	7	3	8	9		10	11	13	15	8
Scratches	3	4	0	3	2		1	0	3	5	0
Flexible hose	6	5	4	0	7		0	0	0	3	4
Panel nuts	3	5	2	2	4		1	3	4	2	4
Others	1 (forgotten washer)						3 (forgotten clamp)				

(5–9) Plot sample histograms showing the following effects.

a. Symmetrical (bell-shaped) distribution

(5–12) Preparing coffee is very similar to a production process in a factory. The coffee beans (raw material) are ground (processing treatment), and a measured amount of coffee and water is percolated in a coffee-making machine (another process). Make a cause-and-effect diagram showing the steps necessary to make a good pot of coffee.

(5–13) Prepare a cause-and-effect diagram for improving employee morale towards quality of workmanship suitable to your company.

(5–14) Prepare a cause-and-effect diagram for the following.
   a. Quality control in your company
   b. Getting top management of your company interest in the company's quality mission

(5–15) The relationship between wheat yield and summer temperature in Canadian prairie provinces is calculated by $Y = 140 - 1.50 X$, where Y = Average yield per acre (bushels), and X = Average summer temperature (°F).
   a. What would the average wheat yield per acre be for a year in which summer temperatures averaged 75°F?
   b. What would the average yield be if summer temperatures averaged 80°F.

(5–16) During the production process of a ductile iron hub casting, it was found necessary to decrease the moisture content of molding sand to reduce the appearance of blow holes. The data shown below represent the percentages (by weight) of water content added to the sand mixer (X) and the amount of moisture found in the laboratory tests (Y). A pair of corresponding data is formed by X and Y.
   a. Plot this data on a scatter diagram. Which observation should you use as the X variable? Why?
   b. Explain the relationship between X and Y. How can this information be used in reducing defects?
   c. Compute the straight line regression equation for the data and plot the regression line on the scatter diagram.

obs.[a]	X	Y	obs.	X	Y	obs.	X	Y	obs.	X	Y
1.	4.0	3.1	9.	5.1	4.1	17.	3.8	2.9	25.	5.0	4.1
2.	4.5	3.6	10.	4.3	3.5	18.	4.3	3.4	26.	3.9	2.8
3.	4.3	3.4	11.	4.4	3.4	19.	4.1	3.0	27.	4.1	3.0
4.	4.6	3.5	12.	4.3	3.1	20.	4.7	3.6	28.	4.2	3.3
5.	4.1	3.9	13.	4.8	3.6	21.	4.8	3.8	29.	4.4	3.5
6.	4.8	3.8	14.	4.9	3.8	22.	5.1	3.8	30.	4.8	3.7
7.	4.9	3.7	15.	5.1	4.0	23.	5.3	4.1	31.	4.9	3.8
8.	5.0	4.0	16.	4.4	3.5	24.	3.9	2.8	32.	5.2	4.0

[a]obs = observation.

5.12 NOTES

1. See *ASTM Standards on Precision and Accuracy for Various Applications,* American Society for Testing and Materials (ASTM), E177–71, Philadelphia, 1977.

2. The concept of handling outliers is described in Nair, K. R., *Biometrika,* Vol. XXXV, 1948, p. 118, and Vol. XXXIX, 1952, p. 189.

3. Named after Vilfredo Pareto, an Italian economist (1848–1923), who studied the distribution of wealth and quantified the extent of wealth inequality. M. O. Lorenz later used Pareto Analysis to plot cumulative curve to depict concentration of wealth in graphic form.

4. Named after its developer, Professor Kaoru Ishikawa, of Tokyo University, Japan. Because of the appearance of this diagram, the common Japanese name is Sakana No Hone (fish bone). For details see, for example, Ishikawa, K., "Cause and Effect Diagram," *Proceedings, International Conference on Quality Control,* Japanese Union of Scientists and Engineers (JUSE), Tokyo 1969, p. 607–10; *Guide to Quality Control,* published by Asian Productivity Organization, Tokyo, 1981; Inoue, M. S. and Riggs, J. L., "De-

scribe Your System with Cause and Effect Diagrams,'' *Industrial Engineering,* April 1971, p. 26–31.

5. Nelson, Wayne, *How to Analyze Data with Simple Plots,* Vol. 1, published by the American Society for Quality Control, 230 W. Wells Street, Milwaukee, Wisconsin, 53203, 1979.

6. Tribus, Myron and Hollomon, J. Herbert, ''Productivity . . . Who Is Responsible for Improving It,'' *Agricultural Engineering,* Vol. 63, No. 7, July 1982, pp. 10–20. Reprinted by permission of the authors.

5.13 SELECTED BIBLIOGRAPHY

Baker, E. M., ''Why Plan Your Inspection?'' *Quality Progress,* July 1975, p. 22.

Batterby, Albert, *Mathematics in Management,* Penguin Books, New York, 1966.

Charbonneau, H. C. and Webster, G. L., *Industrial Quality Control,* Prentice-Hall, Englewood Cliffs, N.J., 1978.

Gedye, R., *A Manager's Guide to Quality Control,* John Wiley & Sons, Inc.; New York, 1968.

Ishikawa, K., *Guide to Quality Control,* Asian Productivity Organization, 1981.

Juran, Joseph M. and Gryna, Jr, Frank M., *Quality Planning and Analysis,* 2nd ed., McGraw-Hill Book Company, New York, 1980.

Nixon, Frank, *Managing to Achieve Quality and Reliability,* McGraw-Hill, London, 1971.

CHAPTER 6

PROBABILITY AND STATISTICS IN MANAGEMENT DECISIONS

In the preceding chapter we discussed the need and applicability of some basic tools and techniques as an important part of an overall quality improvement strategy. Excluding much of the jargon of mathematical symbols and formula derivations, this chapter likewise deals with the treatment of probability and statistical inferences that have many applications in quality control decisions. All that is sought for, in a way, are estimates of probabilities in order to assign consequences, and, in the end, the selection of an appropriate decision criterion for choosing a course of action. Quantification of items by use of such techniques offers a more precise way of conveying meaning than other methods, which may many times remain inexact, no matter how precise the words they employ may be.

6.1 CONCEPT OF A SAMPLE AND A POPULATION

In most cases we get data through sampling, since one cannot simply measure, inspect, or check every single unit. For example, a cook who tastes a spoonful of curry in order to form a conclusion regarding the whole kettleful is using a sampling technique. A *sample* is thus a limited number of items taken from a larger source called a *population*. The statistical methods with which observations made on a random sample are used to derive inferences about a population are found very useful. Random samples mean that each possible sample of say, *n* items (sample size, *n*) has an equal chance of being selected. A population is considered to be *finite* when we know the total number of units contained in it; otherwise it is *infinite*.

113

Sampling has a number of advantages.

1. As an alternative to the prohibitive cost of inspecting the entire population.
2. As an alternative to the physical impossibility of examining all of the population.
3. As an alternative to the destructive nature of some tests.
4. Less time is needed than for other methods.
5. Adequacy and authenticity, as they are based on sound mathematical theories.

The selection of a sample is a matter of intelligent choice made by considering the stages and exact location in the manufacturing operation where estimates can best be applied and serve the particular need. This depends a great deal on prior knowledge of the characteristics of the materials, products, processes, or service criticality. The variability of sample averages decreases with an increase in the sample size, n. For example, averages calculated from eight measurements will have only half the variation of averages calculated from two measurements.

For all possible samples from an infinite population, the following rules hold.

1. The average of the sample averages is equal to the population mean.
2. The standard deviation of the sample averages is equal to the standard deviation of the population divided by the square root of the sample size, n (i.e., $\sigma_{\bar{x}} = \sigma/\sqrt{n}$).
3. If the population has a finite variance and a finite mean, then as the sample size increases the distribution of the sample mean approaches a normal distribution with mean, μ and variance σ^2/n. This rule is generally referred to as a form of the *central limit theorem*.

In order to distinguish between sample and population characteristics, the use of two different symbols should be noted as shown in the following table.

	Sample	Population
Average	$\bar{X}$	μ, (A Greek letter, *mu*)
Standard deviation	s	σ, (A Greek letter, *sigma*)

The relation between sample and population, expressed by the second rule stated above, suggests that as the sample size, n increases, the standard error of the means decreases, and the sample means will shrink more closely around the true population mean, as shown in Figure 6.1.

Sampling Bias

It is to be noted that the methods of probability in sampling suggest that for the sampling to be *unbiased* every item in the population must have a chance of being chosen in the sample. One commonly used method of assuring this is called simple *random* sampling. If a process is operating in a random manner, then any part of its output may be viewed as a random sample from the output as a whole. The random sampling can be facilitated by using an unbiased method of selection and avoiding any method that can influence in any way the

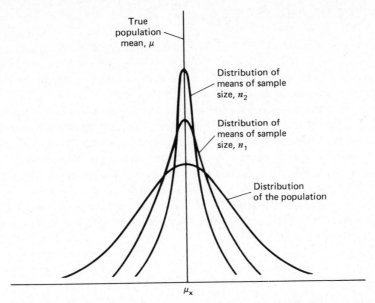

FIGURE 6.1 Distribution of population items and sample means with varying sample sizes, $n_2 > n_1$. ·

selection procedure, such as if, for example, a predetermined pattern were obvious to the population. Some examples of biased selection are:

1. Selecting samples from only the poorly run machines most of the time.
2. Selecting only those products that appear to be defective.
3. Ignoring other lots that meet specifications most of the time.
4. Sampling based upon convenience (or excuses).

If a systematic pattern is already known for the population, *stratified* sampling may be used. Each strata (boxes on the top of the pile, scrap made during the night shift work hours, etc.) should then be sampled independently. The number of items choosen in the strata need not be related to the proportion of items in the various strata. This is called *nonproportional stratified sampling.*

If the population is finite (such as in a lot or batch of items) and if there are no physical difficulties in selecting items from the universe, *random numbers* provide an excellent way to pick a random sample. Random digits are generated in any way that gives each digit from zero to nine an equal chance to be selected. The random number tables have been prepared in this way even using fast computers where the computer itself is given the task of deriving the random numbers.[1]

In quality control practice, if it is feasible to assign a different number to each item in a lot and to draw an item from any place in the lot, a formal scheme for drawing a random sample may be adopted with the help of a random numbers table. If we cannot assign numbers to the members of a population, it should be insured by methods like thorough mixing (helpful in liquids), stirring, grinding, and other mechanical means.

6.2 CONCEPT OF PROBABILITY

The concept of probability has many applications in everyday life. Will it rain or snow? Will my business succeed at this location? Will this product be a hit in the market? In fact, there are many practical applications in management decision making where the outcome of a particular event is known by the use of probability theories.

An *event* is the outcome of some experiment. Events are said to be *mutually exclusive* if they cannot happen simultaneously. For example, if a coin is tossed we cannot get a head and a tail at the same time. When more than one event can occur simultaneously, they are called *not mutually exclusive* events. When the occurrence or nonoccurrence of one event does not in any way affect the likelihood of the occurrence or nonoccurrence of another event, the two events are said to be *independent* events.

A convenient definition of probability is one based on a frequency interpretation. If an event E can happen in A different ways and cannot happen in A' different ways, then the probability that event E will occur is:

$$P(E) = \frac{A}{A + A'}$$

and the probability that event E will not occur is:

$$P(E') = \frac{A'}{A + A'}$$

Since total events = number of ways something can happen + number of ways something cannot happen we have: $n = A + A'$.

Thus $P(E) = \dfrac{A}{n}$ = Probability of occurrence, and $P(E') = \dfrac{A'}{n}$ = Probability of nonoccurrence, where $P(E) + P(E') = 1$; or $P(E_1) + P(E_2) + P(E_3) + \ldots + P(E_k) = 1$, where k represents the number of possible outcomes.

The above indicates that a probability can be expressed as a number which lies between 1.0 (or, 100% certainty that an event will occur) and 0.0 (i.e., impossibility of occurrence). Thus, the rule is that *the probability of an event is equal to, or lies between, zero and unity.*

The estimate of the probability of E as n approaches infinity is given by:

$$P(E) = \lim_{n \to \infty} \left(\frac{A}{n} \right)$$

Basic Theorems of Probability

The following theorems are useful in securing unknown probabilities that involve the mathematics or calculus of probability based on four rules—two addition rules and two multiplication rules. In the discussion that follows, the reader will note that the probability theorems have been stated in terms of two events only but can be expanded for any number of events.

Addition of Probabilities

Theorem 1: If A and B are two events, then the probability that either A or B occurs is $P(A \text{ or } B) = P(A) + P(B) - P(A \text{ and } B)$.

Example: Suppose that a newly designed military parachute is being tested for success of design and operation. The probability that the trigger device will open it is 0.55, the probability that it will get jammed half-way is 0.30, and the probability that the parachute will open but get jammed is 0.10. Therefore, the probability that either of the events occur is $0.55 + 0.30 - 0.10 = 0.75$ or 75%.

Theorem 2: If the two events A and B are mutually exclusive (i.e., cannot occur simultaneously), then the probability that either A or B occurs is $P(A \text{ or } B) = P(A) + P(B)$.

Multiplication of Probabilities

Theorem 3: If A and B are two events, then the probability that events A and B occur together is $P(A \text{ and } B) = P(A) \times P(B/A)$ where $P(B/A)$ = probability that B will occur assuming A has already occurred.

Example: Assume that a steel rolling mill buys 80% of a given steel used in the plant from supplier A, who has been able to supply with an average of 4% defective. Supplier B, selling the rest needed, that is, 20% of the steel, has an average of 6% defective. The probability that a given item selected at random from inventory stock of A will be defective is 0.04. The probability of defectives being supplied by supplier A is $P(A) = 0.80$. The probability of a defect, $P(B/A) = 0.04$. Thus, the probability that the item is defective and was supplied by A is $= 0.80 \times 0.04 = 0.032$ or 3.2%. Similarly, the probability of defective item supplied by $B = 0.20 \times 0.06 = 0.012$ or 1.2%. Total defectives $= 3.2 + 1.2 = 4.4\%$.

Theorem 4: When the occurrence of one event A, has no influence on the probability of the other event B, then the probability of both occurring is $P(A \text{ and } B) = P(A) \times P(B)$.

Example: A circuit board has two critical components that operate independently of one another. The probability of survival of one is 0.95; the corresponding probability for the second component is 0.90. According to design reliability estimates, both components must operate successfully for the total system to be passed by the inspection department. In this case, the probability of the successful operation of the circuit board is $0.95 \times 0.90 = 0.855$.

Bayes' Theorem: If E_1, E_2, E_3, . . . , E_n are mutually exclusive events, one of which must occur, and if these events can be associated with another event E, then:

$$P(E_i/E) = \frac{P(E_i) \times P(E/E_i)}{\sum_i [P(E_i)P(E/E_i)]}$$

where:

E_i = The possible outcomes.

$P(E_i)$ = Prior probability that is, the probability of each possible outcome prior to consideration of any other information.

$P(E/E_i)$ = Conditional probability that the event E will occur under each possible outcome, E_i.

$P(E_i) \times P(E/E_i)$ = Joint probability, determined by Theorem 3.

Example: Suppose that in an annual supplier quality survey (audit) conducted by a large automotive manufacturing company, 60% of the suppliers in the listing received a good rating and 40% of the suppliers received a poor rating. It is known to the company that, on the average, the good suppliers have late delivery only 5% of the time, while the suppliers rated poor have been late about 15% of the time. If a particular supplier selected at random from the company's list is checked late on a particular day, what is the probability that this supplier is not rated overall as poor?

The probability that the supplier is one with a good rating is:

$$P(E_1/E) = \frac{(0.60) \times (0.05)}{(0.60)(0.05) + (0.40)(0.15)} = \frac{0.03}{0.09} = 0.33 \text{ or } 33\%$$

The probability that the supplier under question has a poor overall quality rating is:

$$P(E_2/E) = \frac{(0.40 \times (0.15)}{(0.60)(0.05) + (0.40)(0.15)} = \frac{0.06}{0.09} = 0.67 \text{ or } 67\%$$

Note that $P(E_1/E) + P(E_2/E) = 0.33 + 0.67 = 1.0 \text{ or } 100\%$

Using Probability in Decision Making: Tree Diagram

A *tree diagram* is one of the convenient ways of graphically presenting probability numbers when various events are occurring simultaneously. A simple example to illustrate the concept and application of tree diagrams in decision making is described in the following example.

Example: Suppose a large shipment of machined parts contains 4% defective. A random sample of 25 parts is selected and inspected. If there are four or more defectives, the shipment is rejected. If there are one, two or three defective parts, an additional sample of 50 parts is selected and inspected. At this time, if there are four or more defectives in the total sample of 75, the lot is rejected. Otherwise, the shipment is accepted. The probabilities of defective parts in the two samples of 25 and 50 items are shown in Table 6.1.

The problem was to test the given sampling plan to determine the probability that it would reject a shipment of items that was 4% defective.

The tree diagram helpful in deciding acceptance or rejection of a lot based upon probability calculations is shown in Figure 6.2. Each of the five branches of the tree represents one of the events that can occur with the first sample. The probability of having a different number

TABLE 6.1
Probability of Defective Parts in a Given Sample

Number of Defective Items in the Sample	Probability[a]	
	Sample of 25	Sample of 50
0	0.368	0.135
1	0.368	0.271
2	0.184	0.271
3	0.061	0.180
4	0.019	0.143
	Total = 1.000	= 1.000

[a]The method of determination of these probabilities is based upon Poisson distribution, discussed in later sections.

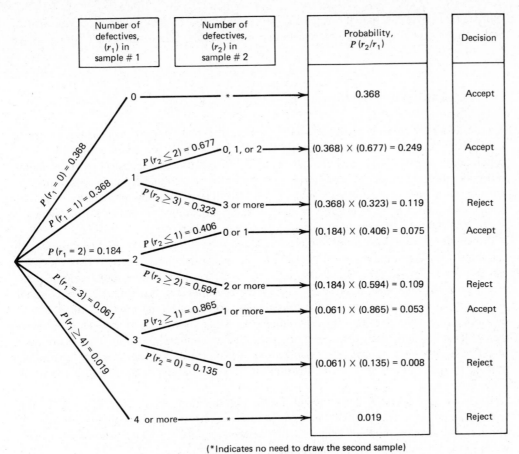

Number of defectives, (r_1) in sample # 1	Number of defectives, (r_2) in sample # 2	Probability, $P(r_2/r_1)$	Decision
0	*	0.368	Accept
1	0, 1, or 2	$(0.368) \times (0.677) = 0.249$	Accept
	3 or more	$(0.368) \times (0.323) = 0.119$	Reject
2	0 or 1	$(0.184) \times (0.406) = 0.075$	Accept
	2 or more	$(0.184) \times (0.594) = 0.109$	Reject
3	1 or more	$(0.061) \times (0.865) = 0.053$	Accept
	0	$(0.061) \times (0.135) = 0.008$	Reject
4 or more	*	0.019	Reject

$P(r_1 = 0) = 0.368$
$P(r_1 = 1) = 0.368$
$P(r_1 = 2) = 0.184$
$P(r_1 = 3) = 0.061$
$P(r_1 \geq 4) = 0.019$

$P(r_2 \leq 2) = 0.677$
$P(r_2 \geq 3) = 0.323$
$P(r_2 \leq 1) = 0.406$
$P(r_2 \geq 2) = 0.594$
$P(r_2 \geq 1) = 0.865$
$P(r_2 = 0) = 0.135$

(*Indicates no need to draw the second sample)

FIGURE 6.2 Use of a tree diagram in accepting or rejecting a lot. Adapted by permission of the publisher from an example in Stockton, John R. and Clark, Charles T., *Introduction to Business and Economic Statistics*, 5th ed., South-Western Publishing Co., Cincinnati, 1975, pp. 116–18.

of defectives for the second sample consisting of 50 parts is entered on the second set of branches. Clearly, a knowledge of the probability of accepting or rejecting a lot that is 4% defective can be used to decide if the sampling plan is good.

Figure 6.2 shows that with this sampling plan there are four situations in which rejection can occur and four situations where acceptance is possible. Therefore, if the buyer decides to take larger samples, he or she will have a smaller probability of accepting a lot that is 4% defective. Note that this method can be utilized similarly for deciding on other sampling plans with any other known defective percentages.

One major weakness behind the slow development of the Bayesian approach in statistical decision making lies in the assigned prior probabilities. There is a considerable amount of discussion in the literature regarding the role of subjective judgment which it is arguable, may have built-in errors. Furthermore, assigned prior probabilities may depend on the forecaster's intuitive judgments, rather than being based on actual data. The advantage of this approach however, is that the rules of Bayes allow the initial forecasts in the form of probabilities to be revised when more data become available, so as to arrive at a compromised form of action.[2]

6.3 PROBABILITY DISTRIBUTIONS

A *probability distribution* is defined as a theoretical distribution of all possible values of some characteristics and the probabilities associated with the occurrence of each in the population. For example, a shipment consisting of 200 electrical switches is known to contain 100 good ones, 80 repairable ones (with loose screws), and 20 that are damaged. Thus, the probability of finding a damaged switch is $P_d = 20/200 = 0.10$; the probability of finding a repairable switch is $P_r = 80/200 = 0.40$; and the probability of finding a good switch is $P_g = 100/200 = 0.50$. Figure 6.3 is a plot of these probabilities.

Probability distributions are of two types: *discrete,* for attributes data; and *continuous,* for variables data. When the events are discrete, (i.e., when the characteristics being measured can take only certain specific values, e.g., 1, 2, 3, etc.) as in the above example, the probability distribution is referred to as a *discrete probability distribution.* The common discrete types of distributions are the Poisson, hypergeometric and binomial (or *Bernoulli* distribution). When the events are continuous (i.e., when they can take any value), or when the events are equally likely to occur in any part of the continuum, the probability distribution is called a *continuous probability distribution.* Examples of known forms of continuous probability distributions are: normal distribution, exponential distribution, and Weibull distribution.

The main application of these several distribution functions is as an aid that can be used with a sample of observations to make predictions about the large population (finite and infinite). A common problem with the users however, is the difficulty of knowing exactly when to use each one of these distributions.

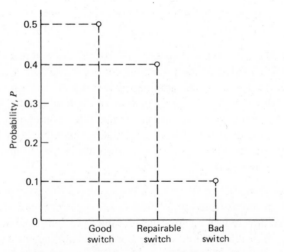

FIGURE 6.3   Simple discrete probability distribution.

Consider a most popular example involving the flipping of coins (one, two, or more) to determine the chance or likelihood (probability) of obtaining a single head or tail or combinations of at least a head and a tail, and so on. In such examples, one is always dealing with a sample of definite size to determine the potential number of times a certain event (obtaining a head or a tail or the number of defects) may be observed in the sample. This approach encompasses all the possible outcomes within the given sample space (binomial distribution). Consider now a problem, for instance, the number of flaws in a roll of steel wire. Here we are dealing with isolated events that occur in a continuum of time, that is, a problem in which the number of occurrences of an event can be determined but not the number of nonoccurrences (Poisson distribution).

In computing desired probabilities, there are many instances where we assume a finite population, because simpler calculations assuming a finite population gives results close enough for practical purposes.

Quantification of Events

For solving many problems in probability, it is necessary to know the trial number of possible outcomes. The rules of *permutations and combinations* facilitate counting of the occurrences and nonoccurrences of an event. The symbol P_r^n represents the number of *permutations* of n things taken r at a time. For example, $P_4^{20} = 20 \times 19 \times 18 \times 17 = 116,280$.

In general,

$$P_r^n = (n)(n - 1)(n - 2) \ldots (n - r + 1) = \frac{n!}{(n - r)!}$$

The written notation ! is called factorial. By definition, $0! = 1! = 1$. For a set of n things, a subset of r, chosen without regard to their order of selection, is called a *combination*. The number of different combinations is denoted by C_r^n, and calculated by the following formula:

$$C_r^n = \frac{n!}{(n - r)! \, r!}$$

where:

> C_r^n = Number of combinations of n objects taken r at a time.
> n = Total number of objects.
> r = Number of objects selected out of the total number.

For example, if we consider n repeated trials, each with two possible outcomes (i.e., defective or not defective), then the possible number of different arrangements each having x defectives and $(n - x)$ nondefective is C_x^n. After a sample has been drawn, each new drawing changes the proportion of defective and nondefective articles in the remaining portion of the lot; the probability changes from draw to draw. However, if the population is large enough (compared to the sample), the changes in probability from one draw to the other can be considered negligible.

Hypergeometric Probability Distribution

Hypergeometric distribution is correct to use when the probabilities of an event change as successive events occur; that is, when the population is finite and the random sample is taken without replacement. The formula for the hypergeometric probability distribution is given by:

$$P_{(r)} = \frac{C_r^R \times C_{n-r}^{N-R}}{C_n^N}$$

where:

$P_{(r)}$ = Probability of r defectives in a sample of size n.

R = Number of defectives in the lot.

N = Total number of items in the lot.

$N - R$ = Number of good items in the lot.

$n - r$ = Number of good items in the sample.

C_n^N = Combination of all items.

C_r^R = Combination of defective items.

C_{n-r}^{N-R} = Combination of good items.

Example: A lot of 10 digital watches purchased has 4 with defective alarm buttons. What is the probability of drawing one defective in a random sample of 4?

From the given statement of the problem, $N = 10$, $R = 4$, $n = 4$ and $r = 1$. Using the formula:

$$P_{(r)} = \frac{C_r^R \times C_{n-r}^{N-R}}{C_n^N}$$

and substituting appropriate values, we get:

$$P_{(1)} = \frac{C_1^4 \times C_{4-1}^{10-4}}{C_4^{10}}$$

$$= \frac{\dfrac{4!}{1!\,(4-1)!} \times \dfrac{6!}{3!\,(6-3)!}}{\dfrac{10!}{4!\,(10-4)!}}$$

$$= 0.3809$$

The Binomial Probability Distribution

The binomial probability distribution is useful in calculating discrete probabilities with an infinite number of items that have attributable characteristics, such as defective or not defective, pass or fail, or success or failure. Assume that a finite population contains a fixed proportion of articles in good and bad categories, represented respectively by p and q. The general binomial expansion is given by:

$$(p+q)^n = p^n + np^{n-1} \times q + \frac{n(n-1)}{2} p^{n-2}q^2 + \ldots + q^n$$

which reduces to:

for $n = 2$, $(p+q)^2 = p^2 + 2pq + q^2$

for $n = 3$, $(p+q)^3 = p^3 + 3p^2q + 3pq^2 + q^3$, and so on.

where n = number of trials or the sample size.

It should be noted that the general usefulness of binomial distribution applies where:

1. Success and failures (i.e., p and q) are essentially unchanged by the drawing of the sample.

2. The number of successes is small when used in experiments.

3. Repeated trials of the same events are identical.

Example: A lot of 20,000 electrical motors is known to be 10% defective. What are the probabilities that a random sample of four motors drawn from this lot will contain (a) all four defective motors, (b) three defective motors, (c) two defective motors, (d) one defective motor, and (e) no defective motors?

With four samples drawn, n = 4 and the pertinent binomial expansion becomes:

$$(p + q)^4 = p^4 + 4p^3q + 6p^2q^2 + 4pq^3 + q^4$$

where:

p = Probability of no defect = 90% = 0.90.
q = Probability of defect = 10% = 0.10.

Solving the equation with p = 0.90 and q = 0.10, we find the solutions as (a) probability of all 4 defectives = q^4 = $(0.1)^4$ = 0.0001 or 0.01% (b) probability of 3 defective motors = $4pq^3$ = $4(0.9)(0.1)^3$ = 0.0036 or 0.36% (c) probability of 2 defective motors = $6p^2q^2$ = $6(0.9)^2(0.1)^2$ = 0.0486, that is, 4.86% (d) probability of 1 defective motor = $4p^3q$ = $4(0.9)^3(0.1)$ = 0.2916 or 29.16%, and (e) probability of no defective motor = p^4 = $(0.9)^4$ = 0.6561 or 65.61%. (Note that total probability = 0.0001 + 0.0036 + 0.0486 + 0.2916 + 0.6561 = 1.0). A plot of this probability distribution will be found similar to Figure 6.3, but if p = q (as in the case of tossing of a coin with equal chances of getting a head and a tail p = ½, q = ½) the distribution will be symmetrical, regardless of the value of n.

The expansion of the binomial terms when n increases beyond three or four becomes hard to remember. An easy way to obtain the numerical coefficients (i.e., $n(n-1)/2$ etc.) of the terms is to use *Pascal's triangle*, which tabulates the coefficients for varying values of n, as shown here.

Number in Sample, n	Coefficient in Expansion, $(p + q)^n$
1	1 1
2	1 2 1
3	1 3 3 1
4	1 4 6 4 1
5	1 5 10 10 5 1
6	1 6 15 20 15 6 1
7	1 7 21 35 35 21 7 1

The other method of obtaining the powers of values of p and q is from a *lattice diagram*. This can be used in small lots of known quantity for rapid inspection. The lattice diagrams

Number of defectives	0	1	2	3	4	5	6	7
7								p^7
6							p^6	p^6q
5						p^5	p^5q	p^5q^2
4					p^4	p^4q	p^4q^2	p^4q^3
3				p^3	p^3q	p^3q^2	p^3q^3	p^3q^4
2			p^2	p^2q	p^2q^2	p^2q^3	p^2q^4	p^2q^5
1		p	pq	pq^2	pq^3	pq^4	pq^5	pq^6
0	1	q	q^2	q^3	q^4	q^5	q^6	q^7

Number inspected → 0 1 2 3 4 5 6 7

FIGURE 6.4a   Lattice diagram for p and q values.

enable the user to find the probability of getting a certain number of defectives quickly. Figures 6.4a and 6.4b show lattice diagrams completed for seven items inspected. In Figure 6.4a the values of p and q are given that correspond to their numerical coefficients shown in Figure 6.4b. When the two diagrams are superimposed, one can determine the probabilities for any particular number of defectives in a batch of any size. Note that these diagrams can be extended beyond the seven numbers shown.

Tables of binomial probabilities are available (U.S. Ordnance Corps, Pamphlet No. ORDP 20–1) which save a great deal of time in calculations, and give probabilities for every sample size up to 150.

For many practical applications in sorting out defectives, one is not interested in the entire binomial expansion. If the way in which the number of defectives are drawn is not important,

Number of defectives	0	1	2	3	4	5	6	7
7								1
6							1	7
5						1	6	21
4					1	5	15	35
3				1	4	10	20	35
2			1	3	6	10	15	21
1		1	2	3	4	5	6	7
0	1	1	1	1	1	1	1	1

Number inspected → 0 1 2 3 4 5 6 7

FIGURE 6.4b   Lattice diagram with numerical values.

one can use a formula for the number of combinations:

$$C_r^n = \frac{n!}{(n-r)!r!}$$

where:

C_r^n = Number of combinations of n objects taken r at a time.
n = Sample size.
r = Number of occurrences.
! = Factorial (example: 5! = 5 × 4 × 3 × 2 × 1; note: 0! = 1.).

The binomial expansion may now be expressed by the following probability equation:

$$P_{(r,n,p)} = \frac{n!}{(n-r)!r!} \cdot p^r \cdot q^{n-r}$$

where $q = 1 - p$.

A few general forms of binomial expression are shown in Figure 6.5.

In the defective electrical motor example discussed earlier, the probability of finding defectives can be calculated by using this formula as well. It is left to students as an exercise.

Values of the individual terms of the binomial expression are available through Tables A1 and A2 (Appendix B). The binomial expansion, however, would not be practical without computer assistance when n is more than 10 or 15. However, when n is large and p is small, one can take advantage of another distribution that was developed by the French mathematician S. D. Poisson (1781–1840). The characteristics of the binomial distribution can now be summarized as follows.

1. The distribution being discrete gives the possible values of the variable increases in steps of size $1/n$; where the minimum possible value of sample proportion is 0, and the maximum is 1.

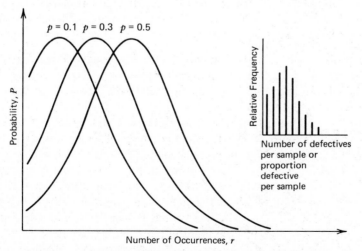

FIGURE 6.5 Representation of binomial forms having different probability numbers. Inset shows a plot for relative frequency of occurrences of defectives versus the number of defectives that has a binomial form with $p = 0.3$.

2. The highest relative frequency of occurrence is associated with the sample proportion equal to its population proportion. However, the symmetry of the distribution depends on the population proportion, such that when the population proportion < 0.5, distribution is skewed to the right; when the population proportion = 0.5, distribution is symmetrical; and when population proportion > 0.5, distribution is skewed to the left.

3. The binomial distribution has a mean and a dispersion like other distribution functions. If n = size of sample, p = fraction defective, then np = number of defectives. The standards deviation of the binomial distribution for the case using number of defectives is: $\sigma = \sqrt{n\bar{p}(1-\bar{p})}$ where $n\bar{p}$ = average number of defectives. This important characteristic forms the basis of np charts in statistical quality control technique, as described in Chapter 11.

The Poisson Distribution

The formula for the Poisson distribution is:

$$P_{(r,n,p)} = \frac{(np)^r \times e^{-np}}{r!}$$

where:

n = Number of trials or occurrences per unit of time or volume, such as number of defects.

p = Probability of occurrence.

r = Number of occurrences.

np = Expected number of occurrences of an event.

Applying the Poisson distribution requires a prior knowledge of the expected number of occurrences of an event. Analogous to the binomial distribution, it also consists of a number of probability terms which add up to 1.0. Most common applications include the basis of the defect control chart for attributes, and the evaluations of acceptance sampling procedures.

Example: Suppose that a production line for fuses is normally known to be 10% defective. A random sample of 50 fuses is selected from the lot for analyzing a particular problem of the day. What are the probabilities of finding r defectives (say, r = 0, 1, 5, 10, 15) during the random sampling?

Table B in Appendix B provides the probability of r defectives in a given sample of n units. The expected number of defectives in the sample of 50 fuses is np = (0.1)(50) = 5. In actuality, the number of defectives in the sample will not always be exactly 5, but will vary from 0 to 50. However, it is known that on the average, the production line is 10% defective (i.e., on the average only 1 in 10 fuses manufactured is defective), and we can be confident that a sample of 50 will not contain more than 5 defectives. From Appendix B, Table B, taking np = 5, we can find out various probabilities of r. The result is shown in Table 6.2. This table reveals that the probability of finding five defectives is only 17.55% or 1755 out of 10,000 times. The probability of finding 10 or 15 defectives further decreases drastically. A tabulation of daily defectives, if carried out, will help decide if the defectives are due to chance causes or to assignable causes, and when to stop the production line for investigation of the causes.

TABLE 6.2
Tabulation of Poisson Probabilities

r	Probability
0	0.0067
1	0.0337
5	0.1755
10	0.0181
15	0.0002

Table C in Appendix B gives cumulative Poisson probability terms. Knowing the np, given at the left column with a line under the column r, (which means acceptance number equaling, say, 0, 1, 2, . . . , events), cumulative probabilities can be read. A *Thorndike chart* showing cumulative probability curves for a Poisson distribution can be found in published *Sampling Inspection Tables* by Dodge and Romig.[3]

Equivalence of Poisson and Binomial Expressions

Earlier, an expression was derived for binomial expansion that involved p and q, such that $(p + q)_n = 1$. Now, introducing another important quantity, e^x (where e is a constant called *natural* logarithm with a value of 2.7183 . . .) and expanding e^x into an infinite series, we get:

$$e^x = 1 + x + \frac{x^2}{2!} + \frac{x^3}{3!} + \frac{x^4}{4!} + \dots$$

Multiplying both sides by e^{-x},

$$e^x \times e^{-x} = \left(1 + x + \frac{x^2}{2!} + \frac{x^3}{3!} + \frac{x^4}{4!} + \cdots\right)(e^{-x})$$

or,

$$1 = e^{-x} + x \times e^{-x} + \frac{x^2}{2} \times e^{-x} + \frac{x^3}{6} \times e^{-x} + \frac{x^4}{24} \times e^{-x} + \dots$$

Substituting x with np (a quantity which is the expected number of occurrences of an event), we get:

$$e^{-np} + np \times e^{-np} + \frac{(np)^2}{2} \times e^{-np} + \frac{(np)^3}{6} \times e^{-np} + \cdots = 1$$

Comparing this result with the general formula for binomial expansion:

$$(p + q)^n = p^n + \frac{np}{1!} \times q^{n-1} + \frac{p^2 n (n - 1)}{2} \times q^{n-2} + \cdots + q^n = 1$$

one finds that the Poisson approximation of the binomial is good for small values of p (and large values of n) over the practical range that is of interest in most quality control applications. When the mean of the binomial is small, the Poisson distribution gives a fairly close estimate, since the binomial approaches the Poisson distribution as a limit, as the value of fraction defective decreases and n increases in size.

6.4 CONTINUOUS PROBABILITY DISTRIBUTIONS

There are three types of most common continuous probability distributions: normal, exponential, and Weibull.

Normal Probability Distribution

The normal distribution curve has the well-known bell shape and is symmetrical about the center (Figure 6.6). The normal density function is given by:

$$Y = \frac{1}{\sigma\sqrt{2\pi}} \times e^{-(X-\mu)^2/2\sigma^2}$$

where:

σ = Standard deviation of the population.
μ = Population mean.
π = 3.141.59.
e = 2.71828.

The curve for normal distribution is related to a frequency distribution and its histogram. The shape of the histogram itself is an indication of the probability distribution of the population (see Figure 6.6). It is not necessary, however, that an observed histogram always look like a symmetrical one. The midpoint μ (mu), the mean of the population, occurs under the highest point on the curve. The total area under the curve represents all possible events and equals 1. Any portion of the area under the curve equals the relative frequency of events falling within that portion.

To solve any problem, one needs to know only the average (μ) and the standard deviation (σ) of the population. The corresponding parameters for a sample of known quantity drawn

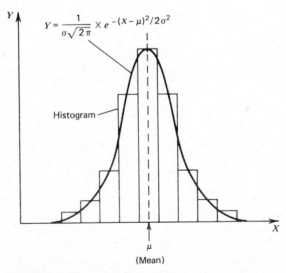

FIGURE 6.6 The normal curve.

from a population are the statistics $\overline{X}$ and s described earlier, which remain the best estimates of the true values of μ and σ. When the size of the sample is small, the changes that sample statistics will be in error are considerably greater than for sampling cases when the sample size is large. For most engineering applications, a sample size of 30 or more gives good enough prediction accuracy, since when the sample size approaches the population size, the sampling error approaches zero.

Typical Properties of the Normal Curve

Area under the normal curve and probability: Figure 6.7 is a plot of a normal curve identified with a mean of zero and standard deviation in multiples of $\pm 1\sigma$, $\pm 2\sigma$, and $\pm 3\sigma$. The total area under the normal curve is 1, meaning that it encompasses 100% of the events described by the distribution. For example, the area under the normal curve bounded by $\mu \pm 1\sigma$ would encompass 68.27% of the items described by the distribution. Similarly, 95.45% of the area is contained by vertical lines measured $\pm 2\sigma$ from the mean and 99.73% of area has $\mu \pm 3\sigma$ limits. $\mu \pm 4\sigma$ limits encompass 99.94% of the area.

Inferences using the curve: If a process or observation is known to follow a bell-shaped (normal) curve, it is possible to estimate the proportion that will be below or beyond a given value (tolerance or specification limit). This may relate to, for example, process capability studies, tolerance, or improving capability. It is important to emphasize that a process must

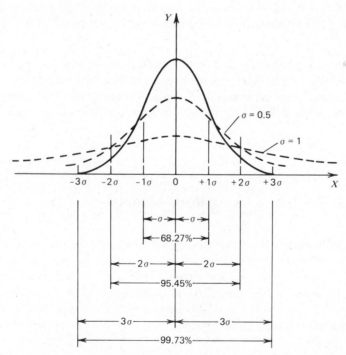

FIGURE 6.7 Plot of normal curve with positions to indicate a mean of zero and a standard deviation of $\pm 1\sigma$, $\pm 2\sigma$, and $\pm 3\sigma$ (solid curve). The smaller σ is, the higher the peak at $X = 0$, and the steeper are the descents on both sides (broken curves).

be in a *state of statistical control* in order for any prediction of probabilities to be valid. There should be no further attempts made to go ahead, either with the process or with the results of it, unless examination of the data is done according to the question of the state of statistical control that produced the data. However, the fact remains that a process may be in a state of statistical control, yet could produce defectives. Inferences drawn simply from curve behavior can, at times, be misleading, since every practical problem is different in its causes. However, if a given characteristic is normally distributed, and if the estimates of the average and standard deviation of the population are known a technique called *normalization* (see discussion on Z scale) can be used to estimate the total percentage of a lot that will fall within the engineering specification limits.

Z scale: In Figure 6.7, several specific σ values are identified on the horizontal axis. Most practical data however, do not necessarily fall within these simple multiple limits. Therefore, some form of conversion factor is needed so that any fraction of the horizontal scale can be read in terms of standard deviation multiples. This conversion from X to Z scale (called normalizing) is done by the following formula, which has a standard normal distribution with $\mu = 0$ and $\sigma = 1$.

$$Z = \frac{X - \mu}{\sigma}$$

In case of sample lots drawn from the population, $Z = X - \overline{X}/s$. Table D given in Appendix B provides a tabulation of Z values with the corresponding area under the normal curve. The probability that Z lies between any two numbers a and b (assuming that $a < b$) is read from this table.

Example: An airline keeps a record on the length of life of an electronic module used on its entire line of similar jets. The population parameters were found as $\mu = 10,000$ hours and $\sigma = 225$ hours, based on actual flying time. The acceptable time range is from 9500 to 10,600 hours. What is the likely number of burnt-out modules for each 5000 of them used?

Here:

$$Z_1 = \frac{\mu - X_1}{\sigma} = \frac{10,000 - 9500}{225} = 2.22$$

$$Z_2 = \frac{\mu - X_2}{\sigma} = \frac{10,000 - 10,600}{225} = -2.67$$

From Table D of Appendix B we find the percent area contained between the mean (μ) and a Z value of $+2.22$ and -2.67. For $Z = +2.22$, a value for the area is $= 0.4868$. For $Z = -2.67$, a value for the area is $= 0.4962$ (ignoring the minus sign). Therefore, the total area between μ and $+2.22$ and $-2.67 = 0.4868 + 0.4962 = 0.9830$. Hence, 98.3% of the 5000 modules or, $983 \times 5 = 4915$ modules may be expected to fall within the acceptable range of 9500 and 10,600 hours. The likely number of burnt-out modules will be 1.7% or 85 modules.

Variation of means: For various samples (sample sizes of $n \geq 30$), drawn one after another from the same lot or population, the mean of each sample varies from one another. However, since every normal distribution is defined by its mean and by its standard deviation, logically this would apply equally to the distribution of sample means (compare, central limit theorem). The resulting curve of the distribution of sample means will therefore resemble the normal curve of Figure 6.7 (compare it with Figure 6.1). A standard deviation of sample means is referred to as standard error of the mean given by:

$$\sigma_{\bar{x}} = \frac{\sigma}{\sqrt{n}}$$

where $\sigma_{\bar{x}}$ is standard deviation of the distribution of sample means (or the standard error of the means) and σ and μ are as previously defined.

In practice, however, neither is a population infinite nor are the samples, once checked, inspected, and tested put back in the population. For these cases, the above equation for the standard error is modified as follows:

$$\sigma_{\bar{x}} = \frac{\sigma}{\sqrt{n}} \cdot \sqrt{\frac{N-n}{N-1}}$$

in which N is the size of the population and other symbols have the same meaning as before.

When the population (N) is large, $\sigma_{\bar{x}}$ is approximately equal to $\sigma/\sqrt{n}$, and $\sigma_{\bar{x}}$ is known as the *standard error*. Assuming the distribution of $\bar{X}$ being normal with mean μ and standard deviation $\sigma/\sqrt{n}$, one can normalize the variation by $Z = \bar{X} - \mu/\sigma/\sqrt{n}$.

Confidence limits and confidence intervals: The *confidence limits* are the limits within which the true value of the characteristic is said to have a given probability of the statement being correct. The *confidence interval* is the interval between the upper and lower confidence limits and aims at bracketing the true value of a population parameter, such as its mean or its standard deviation. For example, from Figure 6.7 the probability that the mean of any randomly chosen sample will not differ from the true mean of the population by more than 1, 2, or 3 standard errors is 68.27%, 95.45%, and 99.73% respectively. Expressed differently in terms of a confidence interval, we can also say that a $\bar{X} \pm 3\sigma_{\bar{x}}$ confidence interval will contain the true population mean 99.73% of the time. In terms of probability, the meaning is that if $3\sigma_{\bar{x}}$ is added to and subtracted from $\bar{X}$, the probability of any sample mean falling within this interval will be 0.9973. This value of probability is then equivalent to certainty in measurement or decisions. The illustration of Figure 6.8 shows the 4σ confidence interval for two samples having individual means. Some of the most commonly used standard deviation values and their corresponding area under the normal curve are given in Table 6.3.

A confidence interval for the population mean, μ can be estimated from the relation: $\mu_{\bar{x}} \pm k.\sigma_{\bar{x}}$ where, $\mu_{\bar{x}}$ = mean of the distribution of sample means, k = measure of standard errors from $\mu_{\bar{x}}$ ($k = 2.58$ for 99% confidence interval, $k = 1.96$ for a 95% confidence interval and $k = 1.64$ for a 90% confidence interval) and $\sigma_{\bar{x}}$ = standard error of the means as previously defined (see Table 6.4 for more common values of k).

If only one sample is drawn from a population, σ and $\mu_{\bar{x}}$ are replaced by s and $\bar{X}$ respectively, and $\sigma_{\bar{x}}$ equals $s/\sqrt{n}$, n being the sample size.

Example: A pharmaceutical company continuously produces a famous brand of its headache pills in mass quantities. Every hour a random sample of 150 capsules are drawn (without replacement) from a production lot of 2000. Tests are done to determine a critical chemical ingredient. The mean weight and the standard deviation were found to be 0.425 mg and 0.05

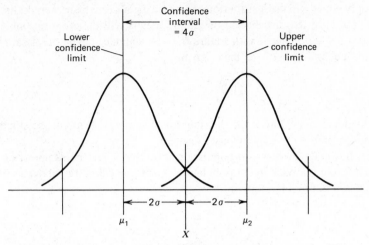

FIGURE 6.8   The 95.45% confidence limits and confidence interval illustrated for two separate sample means having separate mean values μ_1 and μ_2.

mg, respectively. For the production lot under review, find the 95% and 99% confidence intervals for the true mean weight of that chemical ingredient.

Since the samples are drawn without replacement,

$$\sigma_{\bar{x}} = \frac{\sigma}{\sqrt{n}} \cdot \sqrt{\frac{N - n}{N - 1}}$$

in which, $\sigma = 0.05$, $n = 150$ and $N = 2000$. Substituting these values in the above equation and taking the appropriate k-factor from Table 6.4, we get the 95% confidence limits as:

$$\frac{0.05}{\sqrt{150}} \cdot \sqrt{\frac{2000 - 150}{2000 - 1}} = \frac{0.05 \times 0.96}{12.25} = 0.004 \text{ mg}$$

TABLE 6.3
Most Common Standard Deviation Values Corresponding To Area Under the Normal Curve

Limits	Percent of Area under Normal Curve	Percent of Area Not under Normal Curve
$\mu \pm 1\sigma$	68.27	31.73
$\mu \pm 1.64\sigma$	90.00	10.00
$\mu \pm 1.96\sigma$	95.00	5.00
$\mu \pm 2\sigma$	95.45	4.54
$\mu \pm 1.58\sigma$	99.00	1.00
$\mu \pm 3\sigma$	99.73	0.27
$\mu \pm 4\sigma$	99.94	0.06

TABLE 6.4
Values of k For Some Common Probabilities

k	Probability, P (in percent)
0.67	50
1.15	75
1.28	80
1.64	90
1.96	95
2.33	98
2.58	99
3.09	99.8
3.29	99.9

Using the k-factor of 1.96, the 95% confidence limits are $0.425 \pm 1.96 \times (0.004)$ or, (0.425 ± 0.00784) mg. The 99% confidence limits (using $k = 2.58$) are $0.425 \pm 2.58(0.004)$ or, (0.425 ± 0.01032) mg.

Confidence Intervals for Sample Size Less Than Thirty

For sample sizes less than 30, approximation to normal curve is less valid, and in that case another distribution, called *student t distribution,* is used to determine confidence limits. The t distribution is also symmetrical like the normal curve, but it is slightly flatter and spreads more for smaller values of sample size. Using this distribution, confidence limits are expressed by:

$$\overline{X} \pm t_c \cdot \frac{s}{\sqrt{n}}$$

where t_c is the confidence coefficient, taken from student t tables, that varies for each different number of *degrees of freedom* and for the *level of significance* desired.

Number of degrees of freedom, $dF = n - 1$, and level of significance $= 1 -$ respective probabilities (e.g., a level of significance of 0.01, 0.05, and 0.10 correspond to 99.0%, 95.0%, and 90.0% confidence intervals).

Table E in Appendix B gives values of probability for different degrees of freedom. Use of the table of t values first requires finding the line for the correct number of degrees of freedom and then reading across the table to the column giving the desired level of significance. The table gives areas values of one tail only. To apply to an actual example, it is necessary to double the area values given in the table. Figure 6.9 shows a student t distribution for 90% confidence limits. The test of statistical data such as the one using the student t test, however, is based on several assumptions about the parameters of the population from which samples are drawn.

1. The samples are drawn from a normally distributed population.

2. Each of the observations in the sample are independent of one another.

3. The populations have the same variances as the samples. This type of test is referred to as a *parametric statistical test.* In tests based on the above assumptions, all related

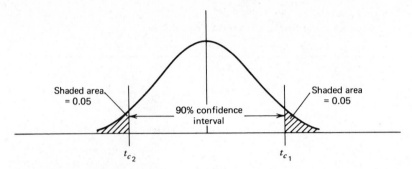

FIGURE 6.9 Student t-distribution for 90% confidence limits.

parameters, such as control limits, process spread, distribution of average, and distribution of individual values, are interdependent, as shown in Figure 6.10. In contrast, a *nonparametric statistical test* is one in which the model does not specify the parameters of the population from which samples are drawn and is frequently used with very small samples without any assumption for the nature of the population. One such distribution function is the well known *Chi-square distribution*.

Chi-square Distribution

The Chi-square distribution is used for many nonparametric tests. A quality control manager might wish to know, for example, whether a test conducted for mean time between failures (MTBF) on new circuit boards would actually fall within the prescribed interval 90% of the time or, regardless of whether failure has occurred or not; and, what would the true failure rate of a lot of circuit board assemblies be with a 90% confidence limit given by knowing the failure history of only a few circuit boards? Such decisions can be made by using chi-square distribution.

From previous sections we have seen that for each of the possible combinations of n sample items out of a population of N, there is a mean $\overline{X}$, a standard deviation s and a

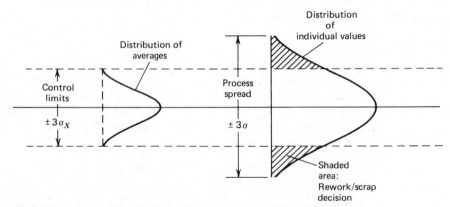

FIGURE 6.10 Relationships of distributions for individual measurements, averages, and control limits.

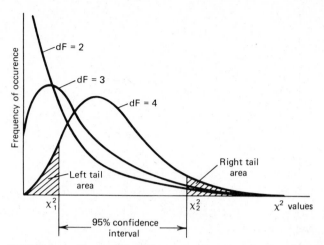

FIGURE 6.11 Illustration of chi-square distribution with various degrees of freedom. Left and right tail areas of the distribution with dF = 4 is shown for a 95% confidence interval. The values χ_1^2 and χ_2^2 for which 2.5% of the area under the curve lies in each tail are determined from χ^2-table for $\chi_{.025}^2$ and $\chi_{.975}^2$.

variance (s^2). Just as $\overline{X}$ has a normal distribution so does s^2 have a near-normal distribution. The distribution of a value $(n - 1)\, s^2/\sigma^2$ is, in this instance, usually called a chi-square distribution. The probability values are calculated by finding the areas under the χ^2 distribution curve (Table F in Appendix B). The family of χ^2 distributions are distinguished by values of degrees of freedom, dF like the t-distribution. As an example of the use of the table, suppose we have to find the two values on the χ^2 distribution which have 95% of the distribution between them. The lower limit will be called $\chi_{0.025}^2$ and the upper limit as $\chi_{0.975}^2$. If the degree of freedom, dF = 4, (dF = $n - 1$), $n = 5$, then the lower and upper values from Table F (Appendix B) are obtained as 0.484 and 11.14 respectively (see Figure 6.11).

Exponential and Weibull Distributions

The exponential probability function is given by:

$$Y = \frac{1}{\mu} \times e^{-X/\mu}$$

The shape of an exponential distribution curve is shown in Figure 6.12. This type of distribution is of special interest in reliability and life testing (see Chapter 8). Making predictions based on an exponentially distributed population requires an estimate of the population mean. Suppose that in *a fatique failure test* the resulting histogram is found to resemble an exponential type of probability plot, as in Figure 6.12. For a MTBF value of 679 hours, the probability that the time between two successive failures of test bars does not exceed 40 hours can be simply calculated by finding the area under the (exponential) curve beyond 40 hours. Such standardized tables of negative exponential curves in terms of

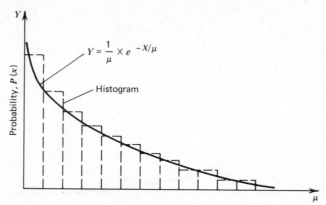

FIGURE 6.12 The exponential curve representing the data of a histogram.

$e^{-X/\mu}$ for various X/μ values are given in Table G of Appendix B. In our present problem, $X = 40$, $\mu = 679$, so that $X/\mu = 40/679 = 0.06$. From exponential tables, the area under the curve for $X/\mu = 0.06$ is 0.9418. Therefore, the probability that the time between two successive failures is greater than 40 hours is 0.9418 or 94.18%.

The Weibull distribution function has similar applications in reliability and life testing, for which the reader should refer to the appropriate discussion section on product reliability in Chapter 8.

6.5 STATISTICAL INFERENCES: HYPOTHESIS TESTS

The statistical inferences drawn from the results of a sample not only help us in many ways to predict the quality of a population (because of time and cost limitations); but, in addition, lend a degree of certainty as to how good our inferences are (i.e., the probability of being right or wrong).

The Two Types of Sampling Error

The use of $\overline{X}$ as an estimate of μ gives rise to two possible errors, called Type I and Type II.

1. *Type I errors* occur when we decide from the sample data that the process is not operating right, when, in fact, it is. In other words, we *reject* the hypothesis when it is *true*. The probability (or the risk) associated with this type of error is denoted by α.

2. *Type II errors* occur when we decide from the sample data that the process is operating right, when, in fact, it is not. In other words, we *accept* the hypothesis when it is *false*. The probability (or the risk) associated with this type of error is denoted by β.

Note that, practically speaking, one is never free of risks of one type or the other. All that can be desired is to reduce one type of risk at the expense of the other type. For example, reducing the Type I error will mean that we have to make very sure that the process is behaving falsely, and vice versa. In theory, the only way to reduce both types of errors is to increase the sample size. In practice, however, this may not be feasible for many reasons.

The *test of a hypothesis* is a rule by which a hypothesis is either accepted or rejected. A hypothesis is an assumption regarding a population being sampled. The major objective behind testing a hypothesis is to design a test so as to minimize both α and β.

The assumption of primary interest is called the *null* hypothesis, H_0. Consideration of a Type II error, β, in terms of probability, requires the formulation of another hypothesis, called *alternative* hypothesis, denoted by H_1.

Two simple examples will be used for illustration.

Example 1: Suppose that we take samples out of a truck load of microwave ovens for quick quality checks (using a check list), and from five readings of critical defects compute an average warm-up period at maximum load to be 30 seconds per unit and the standard deviation to be 10 seconds. Let us assume that a previous method of 100% inspection showed the average time to be 35 seconds. How do we know whether the time observed with the present sampling method is not a chance variation? Is this a possible value, given the information that we have? If the specifications call for an average of 30 seconds, we might be willing to accept the new results as more valid. However, how do we ascertain this by formulization? The steps for testing the hypothesis and drawing conclusions from it are as follows.

STEP 1: *State the hypothesis.* Hypothesis is used here as an assertion made about a population. For example, the warm-up period at maximum capacity load for the microwave ovens could be 30 as well as 35 seconds. This is called the hypothesis to be tested, or null hypothesis, denoted by H_0; written as $H_0: \mu = 30$ (i.e., the hypothesis to be tested is that $\mu = 30$ seconds). The alternative hypothesis is, $H_1: \mu = 35$ seconds.

STEP 2: *Choose a level of significance.* This is the probability level (common values of α are 0.01, 0.05, or 0.10, see Section 6.4) for which we have to determine whether the hypothesis is true ($\mu = 30$) or false ($\mu \neq 30$). (That is, if there is a statistically significant difference between the sample result and the value of the parameter stated in the hypothesis.)

STEP 3: *Choose a test statistic.* Let us assume that the mean of the population from which the sample was taken has a similar normal distribution curve, and we adopt $\alpha = 0.05$. If $\overline{X} = 30$, and $s = 10$, (say for a sample of 4), choosing a normal distribution function (test statistic given by $Z = \overline{X} - \mu_0/\sigma/\sqrt{n}$), we would now like to know whether in the sampling distribution of means of sample size 4, a mean of 35 would be probable.

STEP 4: *Determine the range of values of the test statistic resulting in acceptance or rejection of the hypothesis.* First we calculate the standard deviation of the mean, s_n. Since $s_n = s/\sqrt{n}$, we get $s_n = 10/\sqrt{4} = 5$ (seconds).

When a hypothesis is rejected, one normally concludes that the value specified in the hypothesis is wrong for the confidence level of $(1 - \alpha)$ percent.

In the above example, let us suppose that the Type I error does not exceed 5% ($\alpha = 0.05$). The distribution models are summarized in Figure 6.13. If H_0 is true, the acceptance region can be obtained by locating values of mean time that have only a 5% chance of being exceeded when the true mean time is 30 seconds.

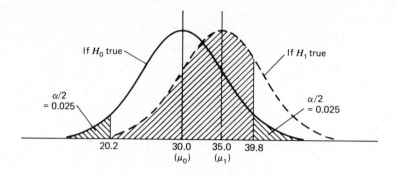

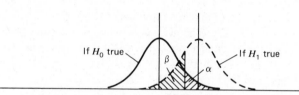

FIGURE 6.13   Test of the hypotheses. For $H_0 : \mu_0 = 30.0$ the acceptance range is defined as between 20.2 and 39.8 ($\alpha = 0.05$). For an alternative hypothesis, $H_1 : \mu = 35.0$, the hatched area on the top drawing shows an 83% chance of incorrectly accepting the hypothesis.

Table 6.3 shows that for 95% of the area under the normal curve, $k = \pm 1.96$. Then, under the original hypothesis that $\mu = 30$, 95% of the sample means will fall within ± 1.96 $\sigma_{\bar{x}}$ of 30, such that:

$$\text{Upper limit} = 30.00 + 1.96 \times \frac{10}{\sqrt{4}} = 39.8$$

$$\text{Lower limit} = 30.00 - 1.96 \times \frac{10}{\sqrt{4}} = 20.2$$

Since the mean of a random sample of 4 is within the acceptance range, the hypothesis is accepted.

If H_1 is true, the population is characterized by a mean of 35 seconds. The type II error, or β, is the probability of accepting the original hypothesis ($H_0 : \mu_0 = 30$ seconds) when it is false. To know the β error, one must find the probability that a sample mean will fall between 39.8 and 20.2 when the true mean of the population is 35. This alternative hypothesis can be written as $H_1 : \mu = 35$.

The probability can be calculated by finding the areas bound between the upper and the lower limits and using: $Z = \bar{X} - \mu / \sigma_{\bar{x}}$. With $X_1 = 39.8$, $X_2 = 20.2$ and $\sigma_{\bar{x}} = 10/\sqrt{4}$; total area $= 0.1685 + 0.0015 = 0.1700$ (Table D of Appendix B). The probability of acceptance region is therefore, $1 - 0.1700$ or 0.8300 (83%). This is the chance of incorrectly accepting the original hypothesis $H_0 : \mu = 30$.

Notice from Fgiure 6.13, that the acceptance region was determined by dividing $\alpha = 5\%$ into two equal parts. This is called a *two-tail test*. It is also likely that the entire error could be located at either tail of the distribution curve. These are *one-tail tests*. This is illustrated by another example in the following section, by choosing the Chi-square distribution. If the probabilities of accepting H_0 for various alternative H_1 values (for a given α and n) are plotted for population parameter (μ), the resulting curve is known as the *operating characteristic curve* (OC). Several standard statistical tests usually present operating characteristic curves for the common hypothesis[4].

Generality of the above method for testing a statistical hypothesis based on student t distribution can be similarly developed. For testing hypotheses concerning variances, distributions such as χ^2 (discussed in Section 6.4), or the F distribution can be used. The mechanics differ but the basic idea remains the same.

Use of Chi-square Distribution for a Test of Significance

To test whether or not a sample variance fits the distribution expected for a statistic of chi-square distribution, we proceed in a manner similar to our use of normal distribution; but in this case we will have to hypothesize a specific value of σ^2. Since the χ^2 distribution is skewed (not symmetrical), a two-tailed test of hypothesis that equally divides the α risk requires that the critical regions for the test be unequally spaced from the mean of the distribution.

For example, consider the χ^2 distribution shown in Figure 6.14 for sample size, $n = 6$ and a degree of freedom, dF = 5. As in the tests of hypothesis for the mean, we specify two critical regions; corresponding to χ_1^2 and χ_2^2 (shaded area in Figure 6.14a values. If the calculated value of the test statistic lies in either of these regions, we reject the hypothesis. Normally the two critical values will not lie at equal distances from the mean ($\bar{\chi}^2 = 5$) because of the assymmetry. Thus, the critical regions define how far from a mean value of $\chi^2 = 5.0$ we can expect a computed value of χ^2 to fall.

For a two-tailed test, say, if we choose a risk of $\alpha = 0.05$, each tail (shaded area) will contain an area of 0.025. The χ_1^2 and χ_2^2 values will be obtained with the help of Table F of Appendix B, and marked off accordingly on the graph paper.

For a one-tail test, where, for example, one is concerned only with increase (or upper limit) in the variability of the process, the entire risk is represented by one area only (Figure 6.14b placed under the right-hand tail of the χ^2 distribution curve.

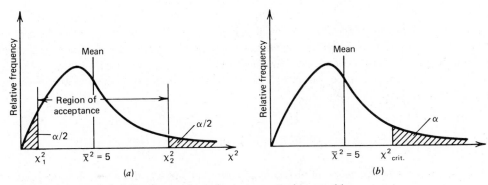

FIGURE 6.14 Use of χ^2-distribution (sample size, $n = 6$, degree of freedom, dF = 5) for a test of significance of variance. (a) Two-tailed critical regions. (b One-tailed critical region.

Example 2: Manitoba Aircraft Systems has long been producing turbine rotor blades out of specially designed new alloy castings. After all the machining operations and final heat treatment, the finished blades had a variance in hardness of 4.0 RC (Rockwell Hardness Number). A group of metallurgists in the company's quality control department came up with a new heat treatment procedure that would give the same average hardness but that is much more economical. The manager of quality assurance wanted to know from the metallurgists whether the new procedure variance was consistent with a population variance of 4.0, or if it was representative of a higher variance?

The metallurgists proceeded by taking hardness readings on six random sample blades. According to the statement of the hypothesis, this was a one-tail test. The quality assurance manager was interested only in knowing whether or not the sample variance exceeded what would otherwise be obtained from the conventional process. The steps in the test of significance used are:

STEP 1: *Statement of hypothesis* $H_0: \sigma^2 = \sigma_0^2$; H_1: $\sigma^2 \neq \sigma_0^2$, where σ^2 and σ_0^2 signify the variance of the new process and conventional process, respectively.

STEP 2: *Choosing a level of significance.* $\alpha = 0.05$, degree of freedom, dF = 5.

STEP 3: *The test statistic.* χ^2 distribution, $\chi^2 = (n - 1)\left(\dfrac{\sigma^2}{\sigma_0^2}\right)$

STEP 4: *Determine the range of value of the test statistic.* Since dF = 5, sample size $n = 6$, and for these six samples, if the hardness readings found were 48, 53, 49, 55, 54, and 57, a variance of 12.0 is obtained.

To test the hypothesis, we calculate the value χ^2 for the sample size of six and compare the calculated value with the Table F, (Appendix B) values for the assumed dF = 5 and $\alpha = 0.05$. If the value of the calculated statistic falls in the critical region (the shaded area of Figure 6.14b), we take this as evidence that the hypothesis was incorrect. To find the critical value of χ^2 from Table F we look in the column headed $\alpha = 0.05$ opposite dF = 5 and read 11.07 as the critical value.

Calculating the χ^2 value from the equation, we get,

$$\chi^2_{cal.} = (6 - 1) \times \frac{12}{4} = 5 \times 3 = 15$$

STEP 5: *Conclusion.* Since the calculated value ($\chi^2_{cal.} = 15.0$) is greater than the critical value ($\chi^2_{crit.} = 11.07$), we reject the hypothesis. The rejection simply means that even though the new heat treatment procedure is economical, with a risk of error implied by the selected $\alpha(0.05)$ it will produce more variability in the hardness (quality characteristic) than the conventional process.

6.6 DESIGNING AND PLANNING OF EXPERIMENTS

The theoretical basis for analyzing problems affecting quality was described in the preceding chapter. In this section, particular emphasis will be placed on the tracing of those causes that are hidden to an extent that warrants special investigation through experimentation. The *experiment* consists of running special tests (or doing something different) to collect data or observations from the process, product, or system in a preplanned manner and arriving at a decision to prove or disprove a hypothesis until the real cause is identified. Depending upon what the results show, it may be necessary to collect more data, revise the technique, or, perhaps, perform further experiments.

The designing and planning of such experiments is a crucial step. It means deciding on:

1. How and what kinds of data or observations should be made within the limits of available resources (i.e., defining the problem).

2. What techniques should be used.

3. The related method of analysis: simple or complex, statistical or nonstatistical; their pattern, cycle, freaks, tendency, grouping, stratification, and introduction of randomness into the experiment.

4. What factors are to be considered (dependent or response variables), and their prioritization.

5. Ascertaining the reliability of measurements, particularly when the data are small and all conclusions must be based upon only a few measurements.

6. The types of experiment (trial and error method, special runs, pilot runs, theoretical versus practical, operations research methods, etc.).

7. Reporting the results of the experiments.

Some Techniques of Experimentation

There are many tools available to researchers and management groups in handling problems that confront them in their quality planning and analysis. Several of these have already been mentioned in earlier chapters. Still more of those will be discussed in their appropriate places throughout the text. For the purpose of this section we highlight the ones that are helpful in the designing of experiments. It is not intended to study the statistical methodology in great detail, but to indicate the basic principles behind them.

1. *Randomization* is essential in any quality control experiment to minimize bias in the results and to turn systematic errors into random errors, so that the effect of one factor will not be masked or confused by that of another factor. It is like shuffling cards. It is important, however, to recognize, with respect to the variables, whether randomization is effective for the purpose in mind. The methods of using random numbers were already explained.

2. *Replication* is the repetition of an observation or process that can give almost an identical observation of the process as the original one. In scientific studies performed by metallurgists and material scientists to examine the microscopic details of fractured metal surfaces, replicas (made out of plastic or carbon using special techniques) permit detailed examination of the tiny surfaces under electron microscopes. In this kind of experiment the metal pieces cannot be used for many reasons, so all scientific observations must be made using replication. Again, errors of measurement are to be detected and analyzed from one replica to another. In quality control experiments, this can mean designing a product and putting it to the test exactly the way a customer will use it, or at least close to it, even under laboratory conditions.

3. In many situations laboratory experimentation to evaluate the effect of one variable upon the other does not always yield conclusions that are applicable to factory or field conditions. *Evolutionary operation* (EVOP) (originated by G.E.P. Box and K. W. Wilson) is a technique for optimizing the quality of the output by varying and evaluating factors experimentally in the actual production line. When the need arises, a pilot run may be set up to evaluate the process variables. However, all conditions must be left close to the real situation.

The experimental steps involve first selecting only two or three independent variables (such as temperature, time) and varying the levels of these factors sufficiently, in planned cycles, to create evidence of change (deterioration or improvement). When the designated effects become significant, the cycles are performed again for different ranges or average points. The procedure is repeated until the optimum condition is reached.

4. The field of *production and operations management* (POM) is rich with many useful techniques that can be applied to quality control problems. We illustrate such applications throughout this text in appropriate chapters. Readers should refer to the index for selected topics under "simulation experiments," "forecasting," "waiting line analysis," "inventory control methods," "materials requirement planning," "planning evaluation and review technique (PERT)," and other techniques.

5. In handling complex problems when one is not definite about the number of major variables and their interplay, it is useful to *dissect* major variables into their components. The normal attempts for *dissection* should concentrate on elements like product-to-product variation, process-to-process variation, lot-to-lot analysis, operator-to-operator variation, time-to-time variation, defect-to-defect variation, and the like.

When an analyst starts to study factors of variation in each of these categories, any system or randomness in the data becomes identifiable with a definite cause. The statistical methods of analyzing variations are discussed in detail in Chapter 11.

6.7 KEY TERMS AND CONCEPTS

Random sampling is selecting units from a statistical population in such a manner that all units have an equal chance of being selected. Stratified Sampling is random sampling from various predetermined strata, subunits in a lot or batch, or process phases.

Probability, simply put, is a measure of the likelihood of occurrence or nonoccurrence of an event. *Probability distributions* are either based on statistical laws, such as the normal, binomial, and Poisson distributions, or on empirical observation and data. Important terms and concepts are: Bayes' Theorem, tree diagram with probabilities, binomial probability distribution, hypergeometric probability distribution, continuous probability distributions (normal, exponential, and Weibull), and Chi-square distribution.

Hypothesis testing is a systematic assertion after

drawing inferences on a statistical population based on the observations from a sample.

Statistical estimation is the process of analyzing a sample result in order to predict the corresponding value of the population parameter. The *confidence interval* is a range of values that includes (with a preassigned probability called a confidence level) the true value of a population parameter. Confidence limits are the upper and lower boundaries of the confidence interval.

6.8 DISCUSSION AND REVIEW QUESTIONS

(6–1) Define the following terms.
    a. Probability
    b. Event
    c. Mutually exclusive events
    d. Independent event

(6–2) By means of a suitable example, explain the meaning of each of the terms in problem (6–1).

(6–3) What is a tree diagram? Discuss the role it plays in making decisions based upon probabilities.

(6–4) What is a probability distribution?

(6–5) What is the appropriate distribution to use, when:
  a. the probability of an event is constant?
  b. the probability that an event will occur is not constant from one trial to another?

(6–6) a. What relationships exist between the binomial distribution and the normal distribution?
  b. When can the Poisson distribution be used to get a good approximation of binomial probabilities?

6.9 PROBLEMS

(6–1) What is the probability of an event that is certain to occur, and for an event that will not occur?

(6–2) The probability that it will not snow on Christmas eve is 90%. What is the probability that it will snow on Christmas eve?

(6–3) When rolling two dice, what is the probability of rolling a 6 or an 11?

(6–4) A certain product is assembled out of four components, A, B, C, and D. If the probabilities of each of these being defective are found respectively to be 0.2, 0.3, 0.4, and 0.5, what is the probability that the assembled product

  a. will be defective?
  b. will not be defective?

(6–5) The monthly report of an appliance assembly department indicates a defect classification as follows.

Major defects	2175
Minor defects	6010
Both major and minor defects	250
No defects	50,400
Total =	58,835

  a. What is the probability that an item chosen at random from this output will have (i) no defect? (ii) a minor defect? (iii) a major defect?
  b. What is the probability that an item chosen at random will have a defect (either major or minor)?
  c. In the next production month, scheduled for 2000 items, how many would you expect to be without defects?

(6–6) Suppose that your company, as a buyer, has received a large shipment of valves that is claimed to have 3% defective. A random sample of 30 valves is selected for inspection and tested. If there are no defective valves in the sample, the shipment is accepted. If there are three or more defective valves, the shipment is rejected. However, if there are one or two defective valves, a second sample of 60 valves is drawn and tested. If there are three or more defective valves in the total of 90 in the sample, the shipment is rejected. Otherwise, the shipment is accepted.

The probabilities of defective valves in the two samples are given in the following table.

Number of Defective Valves	Probability	
	Sample of 30	Sample of 60
0	0.1851	0.0138
1	0.3561	0.2478
2	0.2672	0.3015
3	0.1015	0.1123
4 or more	0.0901	0.3246
	Total = 1.0000	= 1.0000

a. Show all possible sample combinations that could lead to the acceptance or rejection of the shipment with the illustration of a tree diagram.

b. What is the probability that the shipment will be (i) accepted? (ii) rejected?

c. Supposing that the manufacturer of the valves believes that 95% of the valves will pass buyers' tests. If this is true, what is the probability that, in a sample of 10 valves, 1 or fewer will be defective?

(6–7) A lot of 20,000 light bulbs received is known to be 10% defective. Determine the probability that a random sample of four bulbs drawn from this lot will contain:
a. Zero defective bulbs.
b. One defective bulb.
c. Three defective bulbs.
d. Four defective bulbs.
Use binomial expansion to find the answers.

(6–8) If the probability is 0.86 that a single motor coil is defective, what is the probability that a sample of 25 will contain:
a. Zero defective motor coils.
b. One or less defective motor coil.
c. Two or less defective motor coils.
d. Three or less defective motor coils.
Use the Poisson method as an approximation of the binomial distribution.

(6–9) A manufacturer of television picture tubes claims that his product does not have more than 2.15% defective items. A random sample of 1000 tubes were tested and four were found defective. Do you believe his claim is correct? How many samples do you need to test to verify the manufacturer's claim? What about attaining zero defect?

(6–10) From a continuous production line filling soda pop bottles, a sample of 30 bottles is randomly drawn for inspection of liquid weights on each work shift. The population has a mean value of 4 ounces (in weight) and a standard deviation of 0.5 ounces.

a. What is the mean of the distribution of sample means?
b. Determine the standard error of the means.

(6–11) To determine the quality of the molten metal bath in a large electric furnace, a metallurgical inspector draws 3 scoops of sample for carbon analysis. Assuming a normal distribution, determine the 90%, 95%, and 99% confidence limits for good quality bath, based upon percentage of carbon content. The mean value of carbon content was found to be 0.20 (weight percent) and the standard deviation 0.02.

(6–12) The following data were reported by the receiving inspection department of a computer manufacturing firm based on the life testing of two reliable suppliers of electronic modules.

	Supplier A	Supplier B
Number of units tested	6	6
Mean life (hours)	1850	1500
Standard deviation (hours)	30	45

It appears that supplier A is more quality-minded than supplier B. Is the difference in experimental data significant? Using the student t-distribution, show your computation and explain your reasoning.

(6–13) The following distribution shows repair costs for solar circuits of a new model of electronic calculator after a random check of 160 warranty claim cards.

Repair Costs ($ in dollars)	Number of Claims
Between $40–45	8
45–50	30
50–55	48
55–60	50
60–65	21
65–70	2
70–75	1
Total = 160	

Using a Chi-square test and a level of significance of 0.10, test the hypothesis that the universe distribution is normal.

(6–14) A city transit facility defines quality of service for its citizens by claiming that bus arrivals at any bus stop in the metropolitan area are within two minutes (late or early) of the specified time of running every ten minutes. Interruptions of this service may occur, however, due to many unforeseen reasons. Extensive spot checks made by transit inspectors indicate that there were 385 unscheduled interruptions in 1982 and 495 in 1981. Assuming an exponential distribution:

a. Calculate the mean time between unscheduled interruptions, assuming the transit service to be available continuously.

b. What is the chance that buses will be available to all users without interruption for at least 24 hours during Christmas day?

6.10 NOTES

1. Some of the better known sources of such tables are: Table XXXIII of R. A. Fisher and F. Yates, *Statistical Tables for Biological, Agricultural and Medical Research;* Interstate Commerce Commission, Bureau of Transport Economics and Statistics, *Table of 105,000 Random Decimal Digits.*; and Rand Corporation's, *A Million Random Digits,* published by The Free Press, Glencoe, Illinois, 1955.

2. Bonis, A. J., "Why Bayes Is Better," *Proceedings, Annual Reliability and Maintainability Symposium,* January 1975, p. 340.

3. Dodge, H. F. and Romig, H. G., *Sampling Inspection Tables,* 2nd ed., John Wiley & Sons, Inc., New York, 1959.

4. Bowker, A. H. and Lieberman, G. J., *Engineering Statistics,* Prentice-Hall, Englewood Cliffs, N.J., (1959); Natrella, Mary G., *Experimental Statistics,* National Bureau of Standards Handbook 91, Government Printing Office, Washington, D.C.,

6.11 SELECTED BIBLIOGRAPHY

Burr, I. W., *Engineering Statistics and Quality Control,* McGraw-Hill Book Company, New York, 1953.

Dixon, W. J. and Massey, F. J., *Introduction to Statistical Analysis,* McGraw-Hill Book Company, New York, 1957.

Duncan, A. J., *Quality Control and Industrial Statistics,* Richard D. Irwin, Inc., Homewood, Ill., 1974.

Ehrenberg, A. S. C., *Data Reduction,* John Wiley & Sons, Inc. Publishers, New York, 1975.

Grant, E. L. and Leavenworth, R. S., *Statistical Quality Control,* McGraw-Hill Book Company, New York, 1974.

Juran, J. M. and Gryna Jr., F. M., *Quality Planning and Analysis,* 2nd ed., McGraw-Hill Book Co., New York, 1980.

Kurnow, E., Glasser, G. L. and Ottman, F. R., *Statistics for Business Decisions,* Richard D. Irwin, Homewood, Ill., 1959.

Ott, E. R., *Process Quality Control,* McGraw-Hill Book Company, New York, 1975.

Shewhart, W. A., *The Economic Control of Quality of Manufactured Product,* Van Nostrand, 1931, American Society for Quality Control, 1980.

7 CHAPTER

THE PLANNING INTERFACE: QUALITY AND PRODUCTION

Planning is required to attain satisfactory quality of products and services. Without planning, the design of a product or service will fail to meet the expectations of the customer, the production will, in all likelihood, not conform to the design, and the customer will not be able to gain proper use and performance of the item. Planning is thus the essential mental managerial activity associated with any work.

Production is organized work and utilization of resources for the creation of products and services. Resources are people, machines, materials, knowledge, organization, and so on; that is, everything of value for the purpose of a particular production and the resulting product. Effective production means maximizing productivity of available resources, benefits, and also quality. In what follows, we shall focus on the planning of production in general terms and then relate the need of and environment for specific quality planning and control.

Planning involves predicting and analyzing quality needs for products and services. Once these are assessed and established with sufficient certainty factors, planning in the form of forecasting becomes designing, specifying, testing, and other activities leading to the final product design. The product plan can then be translated into a production plan. Both product and production plans are decisions and the preceding planning is decision making. Plans are usually the selected alternative paths of action that appear most feasible and promising under prevailing conditions. Planning continues in the actual production in the form of start-up plans, control and inspection plans, training and supervisory plans, employment plans, and so on.

Logically, product quality requires a large degree of sound planning. Without planning, production is dictated by rule-of-thumb, tradition, individual taste and talent, and the conditions and mood of the day. In a highly diversified and advanced economy and technology with complex products and services, unplanned production has become unfeasible. Certainly, quality assurance cannot be thought of without considering underlying production planning.

7.1 THE PLANNING CYCLE

Planning is a continuous task that is performed in repeating steps in a hierarchical fashion. Planning by senior management influences that of middle and operational management. Figure 7.1 demonstrates these intimate relationships. Policies provide general guidelines for subordinated levels in the organization. This corporate planning must comply and seek compliance and must consider many laws and external planning constraints. In addition to policies, senior management directs and approves programs, projects, and systems.

Middle management translates policies into more detailed work directives, called procedures. Plant and branch managers, for instance, know the prevailing conditions in their operational realm better than those above them. With communications up and down the hierarchy, planning becomes interactive and the plan cycles itself between drafting, approving, and finalizing. The delegation of planning thus becomes based on a sound principle. Planning also describes a cycle with regard to determination of the objectives, approaches, resources, and other elements of the plan. Once the plan is finalized, implementation follows and control procedures monitor execution of the plan. A plan can also mean a product design, project, problem resolution, or action program. The interplay between controlling and re-planning continues until either the objectives have been achieved or a decision is made to discontinue execution of the plan.

Controlling is an essential part of planning and, as such, will not be dealt with separately. Any planning results in action, and performance and goal achievement. During the planning and implementation activities, uncertainty of developments disturbing smooth and effective execution of plans prevails. These conditions, though very familiar to quality assurance personnel and to management in general, require continuous monitoring of performances. Controlling then serves the purpose of anticipating and detecting deviations from the plan.

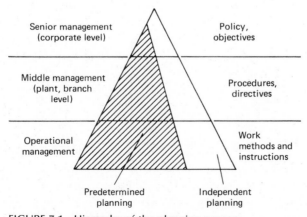

FIGURE 7.1 Hierarchy of the planning process.

This occurrence triggers cause analysis and possible remedial action. The replanning can include review and adjustment of original objectives, performance standards, and available resources. Such a review could also lead to a termination of plans in cases where even a goal adjustment would not make attainment possible. Figure 7.2 shows such a control system that involves replanning and adjustments. The analogy of a ship and its crew comes to mind, where in spite of weather conditions the vessel is to keep on course unless the destination is changed.

Advantages of the Hierarchical Quality Planning Process

Top management cannot do everything by itself. To illustrate this character of the hierarchical planning system that is both simple and effective, Figure 4.1 shows a breakdown of various general management activities. When this is combined with the perspective of Figure 7.1, several advantages of hierarchical decision making and quality planning become apparent.

1.   It puts quality decisions where they belong. Managers on all levels do not feel pressured to get involved in everything that goes on in the organization. This simplifies their thinking and plan of action.

2.   Conventional planning and analysis become decentralized, and problem- and priority-driven. Managers at each level know exactly which problem they are responsible for in production and quality, what reports they have to produce, and the consequences and results thereof.

3.   Managers can identify themselves with the number of detailed variables that are to be controlled for quality improvements.

4.   Lines of authority, responsibility, and command become clearer, and performance appraisals or reviews become much more meaningful.

5.   Time schedules become easier to meet.

6.   Decision making becomes independent, which makes job performance more challenging within everyone's constraints (elements of motivation).

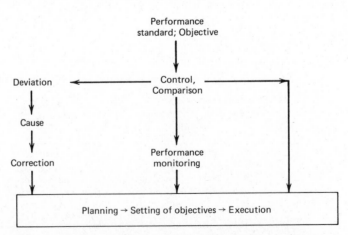

FIGURE 7.2   Control system involving replanning and adjustments.

7.2 PRODUCTION PLANNING AND QUALITY ASSURANCE

Production planning and quality assurance are closely related and interdependent. Only sound and effective production planning has the best chance of resulting in the respective quality of production and products. The better the plan and the more thorough the planning, the less failures and malperformances, the less fire-fighting, less replanning, and less corrective action will be necessary. The strategy of getting it right the first time describes the most reasonable and convincing production strategy for both the long and the short run.

Quality assurance concepts and methods can and should be applied to production planning, and, for that matter, to all planning. In fact, any organization attempting to rationalize and improve its operations might well start with surveying and planning its quality assurance function, whether this involves the quality of products or services. In the following discussion the relationship between production and quality assurance planning will be shown. A production plan is usually the prerequisite for planning of quality assurance. Quality systems and programs can not be established in isolation. A close integration and coordination with production is essential.

Production means the creation of products and services that are in demand. This definition implies that the output of productive systems has a price in the market and, therefore, allows the use of scarce resources and the coverage of costs. It further implies that, in the long run, customers are satisfied with the quality and the quality assurance.

7.3 THE PRODUCTIVE SYSTEM CONCEPT

Production is carried out by a multitude of productive systems. Figure 7.3 schematically describes such a system with inflowing resources, transformation processes, and outflowing products and services. Any productive system is embedded in an economic and social environment with which it constantly interacts. In this context quality assurance has to be named as one of the main subsystems of a productive system. Other subsystems are the capital and cashflow system, manpower planning, management information, and decision-making systems. These subsystems are conceptually and practically interdependent and interactive. One can also distinguish productive systems with reference to the managerial levels of responsibility in a corporate production system in which the plants, branches, and individual jobs and operational systems are subordinated and integrated.

Productive systems are not restricted to manufacturing industries, where materials, parts, and supplies are transformed into higher-value goods. Practically any business or enterprise uses various resources in order to sell its products and services in the market. Therefore, retail stores, theaters, insurance agencies, and so on, are all to be considered as productive

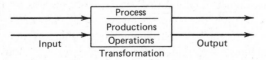

FIGURE 7.3 Schematic of the input, transformation, and output processes.

systems. This then also logically means that these business entities, private or public, have a subsystem for quality assurance. In many, if not most cases, this subsystem is not, as yet, explicitly conceived, formalized, or organized.

"Production" and "productive system" are generally applicable terms and concepts. However, what makes each productive system unique and distinct from others are many factors, such as the type of output, applied technology, resources, scope, ownership, maturity, productivity, and so forth.

Categories for Productive Systems

Categories for productive systems that aid their conceptualization and design are, for instance:

1. One in which customers place their order with the producer or supplier before actual production. This enables both partners to specify quality and other conditions, to meet the customer's particular requirements.

2. One in which customers place an order for an item that is already produced or fully designed and standardized.

3. Mass production with standardized products differing from job shop production or production of custom-made items that use intermittent production processes.

4. Projects that describe production of large items with considerable complexity and uniqueness.

5. Production (provision) of services, normally with direct customer contact before and during production. Such service industries include transportation, public services, insurance, professional services, and the like.

These are the main types of productive systems as they occur in reality. Other differentiations can be made, such as between small and large businesses. Each type of system has certain aspects in common, and these help to plan products and production more adequately. Managers can orient themselves by common aspects of relevant types, and thus simplify their planning. Once the overall production has been properly established, the design of a suitable quality assurance system is also greatly enhanced. Table 7.1a gives some examples of productive systems quite familiar to readers. Table 7.1b relates quality expected by customers with the type of productive system.

Quality Assurance and Productive Systems

Quality assurance is closely related to any production of goods and services. Production means utilizing human and material resources for creating higher valued products and services for which customers will pay a price. These tangible and intanginble outputs from productive systems must have satisfactory quality. Moreover, the quality of a productive system's design, implementation (start-up), and operation (control, workmanship) determine the quality of its output. The quality assurance function and system concentrates on this dimension of productive systems.

Productive systems, in reality, differ from each other uniquely and to certain extents. They can be categorized and typified in regard to the product or service, industry, scale of operation, market, type of customers (contract, project, commercial), location, ownership, and so on. Each type has some features in common, for instance, product characteristics, technology

TABLE 7.1a
Productive Systems Examples

Productive Systems	Major Inputs	Major Process	Major Output
Electronic assembly television set	Components	Assembly	Circuit board
Test laboratory	Sample specimen	Test procedure, equipment	Test result, report
Book publishing	Manuscript	Review, editing final manuscript	Book
Printing	Original copy	Editing, corrections, printing	Book, pamphlet, etcetera
Management, in general	Object, objective	Decision making, supervising recording, analyzing	Directives, decisions, plans reports, information
Quality assurance	Quality specification; standards	Designing, inspecting, training	Satisfactory quality image

and skills in processes, suppliers, and other resource bases. In some productive systems customers (users), producers (suppliers), and government (public), work more closely together in the design, operation, and performance phases than in others.

A major customer or procurement agency might stipulate details of a product (service, project, etc.). In this situation, the productive system must be accordingly amendable because the expected output (custom-made item) determines the processes and inputs. Both customer and producer must cooperate and harmonize. For other products and services, so-called "commercial items," distribution enterprises get involved and standardization with mass production becomes feasible. Standardized products allow standardization of processes, performances and inputs. Customers buy after, not before, production. In spite of the different

TABLE 7.1b
Productive Systems Examples Related to Quality Goals of Customer Expectations

Productive System	Customer Expectations
Car assembly plant	Reliable and comfortable transportation
Farm	High grade in crop, cattle
Theatre	Entertainment, education
Restaurant	Pleasant surrounding, good food, drinks, service
Service station	Fast service, easy self-service, gasoline to specification, reliable mechanical service
Food processing	Healthy, tasty, well-wrapped, and hygenically distributed
Office	Reliable and timely preparation of records, reports, correspondence, handling of applications, information, telephone messages

level of involvement of customers, the nature of the productive system and of the quality assurance of output remains unchanged in principle.

The effectiveness of a productive system—a plant, an office, shop, and so on—can be measured as (output/input) ratios, usually in mere quantitative terms, as productivity. This is how many of the present management generation are conditioned to think of business growth. However, what is not considered is the loss of productivity due to poor quality work—so well known to management in terms of scrap, rework, replacement of warranty items, reinspection of the rejected units, maintenance of unduly high stocks of spare parts, and so forth. Thus, the real effectiveness of a product system hinges on quality production: (output/input) $\times$ quality factor.[1] This is how a more accurate and more realistic market oriented productivity concept ought to be described. Relating the quality of output to input and transformation processes could and should assess the real system effectiveness.

Productive System as a Part of the Overall System

A productive system is a system of systems that is, an aggregate of interdependent subsystems. It is usually also itself a subsystem of a higher ranking system, such as a corporate entity. Quality assurance, as has been pointed out, is one of these subsystems of a productive system. Design and operation of a quality assurance system must be done in close relationship to the overriding productive system. Parallel to the quality assurance system are those systems of physical production, scheduling, inventory control, distribution, procurement, and so forth. Each of these has a unique relationship to quality assurance. All are integrated into the productive system (plant, branch, etc.).

Productive systems, and each of their subsystems, need a particular planning and control force. These are the management, or management-information/decision-making systems. Quality assurance does have its own management system. Tangible outputs are procedures, manuals, policies, and decisions. Such outputs inform people, and thus direct their decisions and actions towards the ultimate achievement of the desired product or service quality.

7.4 PLANNING AND CONTROL OF PRODUCTIVE SYSTEMS

The planning and control of productive systems move through the phases similar to those described for planning cycles. The main phases are the system design, the startup, the maintenance, and the termination. These phases describe the *life cycle* of a productive system. If the productive system is associated with one particular product, as is usually the case in project type productions, then the product life directly determines the productive system's existence. If, for instance, the quality of a product builds a sound quality image in the market, then the supporting productive system and its quality assurance subsystem are strengthened as well. Effectiveness of planning and control enhances growth and length of life.

In order to plan and control a productive system, each case must be clearly defined and delineated. A useful starting point is to determine the output (current or expected), because the purpose and objective define the productive system. For instance, if the purpose is to assure the quality of a computer chip, the product would have to be specified with regard to application, design criteria, and so forth, along with quantity, delivery mode and timing, location of market, and production and resource bases. By clarifying the output in terms of the material, time, and place dimensions, process capacities and inputs can then also be determined.

Once the productive system is conceptualized in general aggregate terms, the various subsystems, such as the quality assurance system can be designed. Various aspects, such as the management system, subsystem, plants, departments, and specific markets or customers, each having a direct relationship to the expected output, help to define the quality assurance system under consideration. Here one can see that the systems view is a powerful management aid and basic for any systematic planning and control of production and output quality.

Each phase in the system life demands specific planning and control activities and management involvements. Forecasts of developments and control information induce frequent review and correction in design, startup and other aspects of production. These phases can have a multitude of complex detailed planning objects and problems. When phrased as questions, problems are more easily understood. Actually, asking the right question at the right time to the right person, makes a manager and planner proficient. How do you test an airplane to know how far it can glide from 30,000 feet in the air if found without fuel? The only answer can be that it is too late. If such emergencies are not planned for ahead of time, the chances of anyone surviving will be nil.

Following are some pertinent sample questions regarding product quality and its assurance, related to each phase of a production system's life cycle. Such questions could be further developed into checklists and schematics helpful to management in developing quality strategy.

Conceptualization (birth of system): What are customers requirements, their actual or potential demands in the market, in the short- and long-term? What is the expected volume, timing, place or region? What kind of product or service would meet this demand? How do you know? What technological, human, and production capacity would be required? What about training and motivational needs?

Design of productive system: What are the major and critical quality characteristics of the product or service? What are the major defects, hazards, risks, and so on? What research have you done to convince yourself? What are the variations in scope or quality, timing, and place or location? Do you know? What are the current respective organizational, technical, and human capacities? Is the product requiring fundamentally new systems designs, or is an adaption of a current system feasible? Would the resulting product live up to customer's expectations? What inspection and quality control system is required? How should the inspection stations be arranged? What is the required qualification of inspectors? Is it only numerical quotas that must be met? What about senior executives, engineers, foremen, and supervisors? What is the layout of facilities and what is its impact on workmanship and product quality? What are the production procedures and their translation into individual performance standards? What impact do these standards have on quality? Do they specify quality that everyone in the company can understand as meaning the same thing. What is the general arrangement of production with regard to individual orders, lots, and so on? Is the production of individual (customer-determined) design required or is mass production more adequate? What is the degree of standardization of product and production? What is the general arrangement for the inventory control system? How can inventory be minimized? How can quality be maintained in stockkeeping, handling, and shipping?

Startup of system: Are all production plans and procedures sufficiently finalized and communicated? Is everybody involved in the production and distribution or servicing properly informed and trained? If not, why not? Can you explain? Why aren't the payroll department and finance people trained in quality? Are special plans and implementation aids finalized so that necessary adjustments can be made quickly and correctly? What are the control devices for monitoring defects and deficiencies? Are these considered in the production plan, so that unpleasant surprises can be avoided and so that prevention is possible? What cost system are there to account for defects?

Maintenance of system effectiveness: Has a steady state of performance been achieved and measured by leveling out of the number of deficiencies, adjustments, downtime, failure costs, and other interruptions? What about obtaining and maintaining a steady state of performance of operators, machines, and supervision? What is being done about these? Does the control system provide sufficient information for monitoring performance and prevention of deficiencies? What is your quality image in the market? What is the degree of satisfaction of new and old customers? How do you know for sure? What is the company's market share or position and its development? What are the audit reports indicating with regard to quality assurance effectiveness? How are the improved effects measured? What have been the problems in maintaining effectiveness of the productive system in the past? What system do you have to correct vendor problems? Does the quality department participate with the purchasing department? Do problems indicate the need for a basic review and redesign, or is termination to be considered?

Termination of system: What are the costs and benefits of discontinuing the system (product, processes, etc.) in whole or in part? How much does each of the defective products or processes cost? Should the termination be temporary or final? On what criteria is this decision to be made? How should the termination proceed? Should the production be phased-out gradually or more quickly? What is the least costly phase-out schedule? What can be safeguarded for revival? How can the quality image and supporting assurance system be utilized in conjunction with another existing or newly created system?

Production planning produces plans that show to a certain degree the product's quality characteristics. Assembly charts, production flow diagrams, and other major planning documents will be discussed in the following chapters. Product and production design are practically inconceivable without quality consideration. However, the emphasis of assuring quality with the customer in mind will vary. It is often unduly overshadowed by other considerations, such as maximizing quality in terms of productivity, meeting deadlines, maximizing short-term benefits, and so forth. The question arises as to how management prepares itself to meet these goals, or measures up to continued self-appraisal.

7.5 FORECASTING

Management involves making decisions under conditions of risk and uncertainty. For evaluating risks we have to obtain information on probabilities, which is not always feasible. Under uncertainty, without prior informationn on the status of a situation, risks and uncertainties present quite a scenario for management. To minimize this uncertainty, pertinent information is required regarding, for instance:

1. Customer demands and expectations for quality.
2. Frequency of customer complaints, defects, and product failures.

3. Competitors in the marketplace and their impact on quality specification, and sales.

4. Government policies and regulations relevant to current quality specifications.

5. Technology and process design and development and its impact on customer demand and quality requirements and specifications.

6. Workmanship and competence of people in the production process to create required and new quality.

7. Availability and quality of supplies, including capital, equipment, and human resources.

8. Management competence, aids, and methods facilitating quality assurance and quality of design, production, and performance.

These and many other, more specific, uncertain future conditions vary by degree, risk involved, time, place, and the competence of the decision maker. The more accurately developments and future conditions can be predicted, the more valid and reliable will be the decision made and thus, the more effective subsequent actions and goal achievements will be.

Forecasting of quality demands and the impacts of various quality determinants are fairly difficult, because the definition of quality is very complex. Not all customer expectations and requirements concerning products already available can be fully quantified and specified. How much more difficult is it then, to predict what kind of car, computer, food, garment, leisure products, and so on a customer will buy and competitors will sell next year, or next season. The more uncertain the future, the more variable and dynamic the consumer needs and changes of taste become. The stiffer the competition, the more likely intervention from government, social, and economic forces, and, consequently, forecasting of quality becomes complex.

Forecasting of quality has been undertaken mainly in the form of *market research*. Consumer behavior analysis, new product management and design assurance have helped to predict the product quality that would be required at some future point in time.

Quality assurance management does not start with the receiving of quality specifications, either directly from the customer or from internal company functions, such as marketing or engineering. Assurance implies knowing what the customer expects of a product and service in both the near and more distant future. Making of good quality takes time, assuring satisfactory quality takes even longer. The reason is that quality assurance starts with the design planning phase and not when actual production begins. The lead time before customers receive products and services of a *new* design is much longer than for an established product and design reviews, changes, improvements, field testing, and so on, are necessary because customer expectations, competitor's products and services, and other determinants for a company's success change continuously.

Forecasting Methods

Three basic types of forecasting methods that can be used by quality assurance managers are *qualitative techniques, time series analysis* and *cause/effect models*. The *qualitative technique* uses judgement of people concerning the future. Mainly through questionnaires, it brings to bear the knowledge, experience and wisdom of a group, deemed to be able to predict accurately. The quality of the forecast itself can be well measured when prediction and actual outcome are compared. People change their judgement and thus, as time proceeds, qualitative techniques allow for reviews and reforecasting (in the *Delphi* technique this is done through repeated use of questionnaires). As one can imagine, qualitative techniques can be used to predict customer requirements, developing technology, availability of resources, ecological outcomes, and many other forces that have an impact on quality requirements and respective production potentials.

Predicting defects and defectives, number of cars per family, size, and other quantitative indicators for consumers can be done by a quantitative and statistical approach. *Time series analysis* extrapolates the parameters that, for instance, measure quality characteristics. The information obtained can be traced regarding its accuracy, and thus the quality of the information can be controlled to a certain degree. Time series analysis and statistical control charts are very closely related, with the forecasting being based on the historical data. The decomposition of time series into *trend, cyclical, seasonal,* and *random* components, is similar to the tracing of random and nonrandom shifts of entries in the control chart to assignable causes. Control charts can also be used to monitor the quality of forecasts, as will be shown later. Control chart techniques are discussed separately in Chapter 11.

Cause–effect models combine quantitative and qualitative techniques of forecasting. Regression and correlation analysis (discussed in Chapter 5) extends time series analysis, particularly with regard to the trend. Obviously, defects normally have several known and unknown causes. Delineating major causes and applying mathematical models, such as regression and correlation, not only explains defects as the effect, but at the same time allows us to use parameters, such as the slope, to predict cause–effect relationships in quantitative terms. What about qualitative cause–effect relationships? As an example, accidents and user injuries with an agricultural implement were traced to ignorance, rather than carelessness of the user; remedial action in the form of providing more proper information to the users will most likely have some beneficial effect in the future. In other words, cause–effect analysis followed up by remedial action are not only closely related logical steps in looking at problems, but also allow prediction of outcomes.

Purpose and Objective of Forecast

1. Determination of quality assurance policy; long term customer quality requirements in the intended target market, including shifts and changes.

2. Defining major quality characteristics of product or service in terms of aggregate (product/service line), specific lines or items, parts, components, type of use, levels, degrees (tolerance of defects), and services required with product.

3. Assessing available technology and its potential for new products or services, quality characteristics, quality assurance.

4. Assessing new quality assurance management and engineering methodology, particularly in regard to breakthroughs and strengthening for competing in the market with quality, including standard writing.

5. Assessing development, challenges and opportunities of current capacity to compete with quality of workmanship successfully.

6. Determining development of supply and subcontracting resources and community services for strengthening quality and quality assurance performance.

Each specific objective can be achieved by determining the following: forecast information required, timing and planning horizon, forecasting methods applicable, data required, form of communication, use of forecasts, and maintenance of quality of forecast information.

7.6 ORGANIZING FOR KNOWLEDGE IN FORECASTING

In the following two examples (one from a service industry and the other from manufacturing), we explain in more practical details the assessment of quality assurance forecasting situations and the use of the available methods.

Example 1: Quality assurance policy creating a favorable quality image for a fast food restaurant that allows branching out is the main consideration in this example.

Major forecasts required in this case by management are:

1. Demand development and pattern in a specific region or community.

2. Projection of competition in regard to type of food, service, delivery, premises, customers, timing, income, and employment development.

3. Availability of employees, suppliers, and managers.

4. Opportunities for field testing of various alternatives.

5. Financial resources and services.

Let us assume that the above information is to be provided for each of the next five years. The main method of forecasting here is market research with particular emphasis on quality aspects. Price-cost relationships and quantitative growth are forecasts that can be combined with the quality forecast. Because the branch operation is the very competitive fast food industry is to be started off in a suitable community that offers field testing and development opportunities, forecasting must not create obstacles and countermeasures by competitors. Causal methods could be used for projection of number of competing outlets, price cost developments, relationships of various types of restaurants, and so on. The above information would allow one not only to assess the opportunity for business success, but also to assess the kind of quality level that offers a niche in the market.

Data required for policy related forcasts are dependent on the information required. *Questionnaires* can be used for questions concerning the development of customer expectations and preferences. Is the tendency towards more expensive food, better services and surroundings, or towards faster service, more standardized food, more alcoholic beverages, and so on? Judgement by customers and experts must be analyzed separately.

Once the first fast food restaurant goes into operation, information from customers will become very important for forecasts of policy changes and their impact on the restaurant's quality image and on satisfaction among customers. Such quality assurance policy changes would more clearly delineate the clientele because customer suggestions and complaints can be explained by means of systematic cause–effect analyses.

Now, let us suppose that the fast food restaurant has arrived at the stage for branching out. The original quality assurance policy has either been proven effective or has been modified. Now, after the start-up phase for the first restaurant is accomplished, forecasts turn to more specific questions concerning the quality and quality assurance of foods, services, suppliers, staff, and so on, in all new locations. As more and more factual data become available from its total operation, time series analysis and casual methods can be used. For instance, quality cost reports and customer complaints reveal the strengths and weaknesses of individual products, services, branches, and operations. Only when major customer dissatisfaction develops or competition encroaches, would the quality assurance policy have to be reviewed and forecasts relevant to quality assurance policy be repeated. Again, at the mature development stage of this fast food restaurant, assuming that it has, in fact, been launched successfully, forecasts are relatively more reliable concerning quality, because of the available historical data.

Example 2: A manufacturer of buses wants to prepare its overall quality organization for improved quality assurance during the next five years. Currently, major customer complaints have affected the quality image of the company. In order to be able to satisfy customers more adequately and to rebuild a positive image, the manufacturer needs to know customer expectations concerning bus quality in the upcoming period and, in particular, after a five year adaption period. The following information is desired from respective forecasting.

1. Quality characteristics expressed in number required in different categories of demand and bus use, such as school busing, urban transportation, travel, or privately owned. Other quality attributes and variables are size, seating, on-board services, comfort, safety, reliability testing, technical innovation.

2. Defects and nonconformances that are associated with major quality characteristics and perceived as unacceptable for buyers and users of buses; for the current year and for five years later. What might still be currently acceptable might not be so in the future (because of new safety regulations or a competitor's stronger product and services.)

3. Quality assurance requirements of purchaser: standards imposed, procedures and evidence demanded, competitors and government regulations on safety, etcetera.

4. Services required by purchaser and also by the bus user in addition to the inherent transportation service, such as financing, bus maintenance, driver training, travel guide services, fleet management services, and operational guides.

5. Technology changes within the next five years and the expected technological state after five years with regard to bus design and design assurance, production technology, servicing and maintenance, concerning body, style, power track, electronics, fitting, safety engineering and so on.

6. Methodology for the various quality assurance management functions: concerning not only quality assurance procedures and programs, but also computerized inspection and information systems, auditing, staff motivation and training, automation, dealers, customer relationships, parts availability, and maintainability.

7. Resource availability for creating the required quality and quality assurance in terms of manpower (including management), finances, management aids, quality assurance specialists, dealer and supplier relations and quality indoctrination, and government regulation changes.

8. General developments and status in five years in the market, competition, economy, income, employment, infrastructure, consumer behavior concerning bus transportation, and so on.

9. Life phases of the bus production from the time of design, to field testing, rapid growth period, turning point to steady state, and major modification of design or termination of design and production.

A planning horizon of five years was assumed in this example. Once the information is further assimilated, it will be disaggregated and defined concerning the type, target market, type of user, and so on. The next step calls for the selection of the appropriate forecasting method. For instance, if the manufacturer enters the market with an entirely new design the early life phase would not allow use of time series analyses because adequate historical data would not yet have been available. However, is the manufacturer already experiencing the rapid growth period, or in the steady state? This question for the forecast concentrates on the shifts and turning points in the quantitative as well as qualitative demand. Usually, when the market becomes saturated a more forceful competition with quality and associated services (availability, maintainability) sets in.

As shown in the first example for determining a quality assurance policy for fast food service, forecasting quality expectations and requirements do not always give scope for quantification. The exception is when the quality that preferences are to be inferred from shift in certain types, defects, customer complaints, or usages. This necessitates change in the statistical parameters. Casual models can also be used in this example when, for instance, the indicators for urban sprawl, price for gasoline or diesel, income, parking costs and facilities, and so forth, can be assessed with additional impact on bus transportation demands. Regression and correlation models and factor analyses can also be selected if data are available.

While this information and forecasts were to assist quality assurance management in preparing a future quality assurance program, much of this information can also be used for other functions and decision making. In a well-structured forecasting system, as an integrated part of the overriding decision-making and management information system, information requests induce forecasting; thus, data compilation, processing, and information communication to decision points is coordinated.

A quality assurance forecasting report would give, for the information requested, a clear definition of the question, forecasting method selected, time horizon applied, data used, assumption made, and functions involved and informed. Quality related forecasts, once they are valid and reliable, offer competitive advantage and should be kept confidential.

7.7 QUANTITATIVE METHODS OF FORECASTING

Any series of historical data taken from a defined and delineated statistical population reveals certain properties through statistical analysis. The number or proportion of defectives compiled over a period by product line, department, region, and so on, vary and indicate trends (regular and irregular fluctuations), and cycles from which assignable causes can normally be detected. Moreover, through *extrapolation technique* future outcomes can be predicted.[2]

Statistics compiled for using the forecasting are the *averages* and *indexes*. In the following examples we show a time series analysis by means of the *moving average* model, followed by the *weighted moving average* and the *exponential smoothing* model. If a trend is observed, a regression analysis can further describe the effect of long-term determinants.

Analysis of time series and other similar methodology should be conducted when, for example, effects of measures for reducing defects are under study, comparisons are to be made between plants, shops, or regions, concerning quality and quality improvements, and when nonrandom causes such as substandard performances, poor supplies, and ineffective quality control are to be studied.

Example 1: Joran Torres is a Madrid luthier whose production of guitars for the past several years has been as follows:

Year	Number of Guitars
1977	14
1978	18
1979	17
1980	24
1981	22
1982	26

He is concerned about the growth in demand that will affect quality of workmanship and wants to take precautionary steps.

1. Using a four year moving average, determine expected demand for the years 1981, 1982, and 1983.

2. Do the same using a weighted moving average method.

3. Do the same using an exponential smoothing method.

Solutions

1. Four Year moving average: Forecast for year five = average of previous four years.

Year	Number of Guitars	Simple Four Year Moving Average Forecast
1977	14	
1978	18	
1979	17	
1980	44	
1981	22	18.25
1982	26	20.25
1983		22.25

2. Weighted moving average method: This method is used to make the forecast depend more heavily on a given particular (for example, most recent) year.

The weighted moving average is calculated as follows

$$\text{Forecast} = \frac{(\text{Weight} \times \text{Amount})}{(\text{Weight})}$$

If weights are assigned as follows:

> Most recent year, weight = 50
> Second most recent year = 25
> Third most recent year = 15
> Oldest year = 10

then the forecast for 1981 is:

$$\frac{(50 \times 24) + (25 \times 17) + (15 \times 18) + (10 \times 14)}{100} = 21$$

Forecast for 1982 is:

$$\frac{(50 \times 22) + (25 \times 24) + (15 \times 17) + (10 \times 18)}{100} = 21.35$$

Forecast for 1983 is:

$$\frac{(50 \times 26) + (25 \times 22) + (15 \times 24) + (10 \times 17)}{100} = 23.8$$

3. Exponential smoothing: Let the smoothing constant, $\alpha = 0.2$; $0 \leq \alpha \leq 1$; weight assigned for forecasting error.

The current forecast = the previous forecast, plus α times the difference between the previous period actual and the previous forecast.

To start the forecast procedure, let the forecast for 1977 = 14.

Year	Actual Production	Forecast ($\alpha = 0.2$)
1977	14	
1978	18	$14 + 0.2(14 - 14) = 14$
1979	17	$14 + 0.2(18 - 14) = 14.8$
1980	24	$14.8 + 0.2(17 - 14.8) = 15.24$
1981	22	$15.24 + 0.2(24 - 15.24) = 17$
1982	26	$17 + 0.2(22 - 17) = 18$
1983		$18 + 0.2(26 - 18) = 19.6$

Example 2: The Bricklin Motor Company in Canada recorded the following sales of passenger cars over an eight year period.

Year	Cars Sold
1972	207,692
1973	231,598
1974	258,980
1975	225,293
1976	210,049
1977	247,427
1978	248,285
1979	236,437

1. Compute a simple four year moving average forecast for car sales. What is the forecast of sales for 1980?

2. What is the demand estimate for 1980, derived by using least squares regression analysis? Illustrate with a graph.

Solutions

1.

Year	Sales	Four Year Moving Average
1972	207,692	
1973	231,598	
1974	258,980	
1975	225,293	
1976	210,049	230,891
1977	247,427	231,480
1978	248,285	235,437
1979	236,437	232,763
1980		235,550

2. See Figure 7.4.

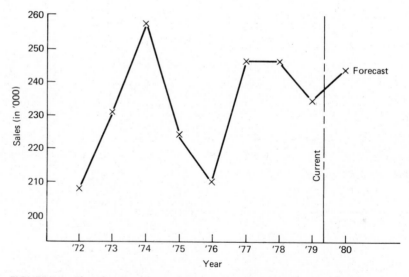

FIGURE 7.4 The plots in the graph indicate a wide spread and poor fit to a line; the R-square value is 0.14 (very low). The model will be a poor predictor. The linear equation for $Y_{year} = -5,287,526 + (2,794)_{year}$; $Y_{1980} = -5,287,526 + (2,794)_{1980} = 244,594$ cars.

Example 3: An automobile manufacturer has gathered the following history of defective units of one particular make of car.

Year	Number of Defectives
1970	42,025
1971	46,021
1972	44,127
1973	49,504
1974	51,226
1975	50,895
1976	52,041
1977	57,121
1978	55,142
1979	59,985

1. In order to plan their warranty programs, the company must forecast the number of defective vehicles expected for 1980. Do so.

2. If a new quality assurance program is implemented for 1980, defectives are expected to be reduced by 25%. In this case what is the forecast of number of defectives?

Solutions

Select a forecast method for a six year moving average.

Year	Number of Defectives	Six Year Moving Average Forecast
1970	42,025	
1971	46,021	
1972	44,127	
1973	49,504	
1974	51,226	
1975	50,895	
1976	52,041	47,300
1977	57,121	48,969
1978	55,142	50,819
1979	56,985	52,655
1980		53,902

1. Therefore, the manufacturer can expect 53,902 defective units in 1980.

2. If the quality assurance program is implemented, expected defects drop by 25%, and the expected number is $53,902 \times 0.75 = 40,427$

Example 4: Studebaker Corporation passenger car sales from 1957–63 were as follows.

Year	Passenger Cars
1957	79,011
1958	61,434
1959	162,143
1960	111,575
1961	84,891
1962	94,922
1963	76,108

Use a four year moving average to predict sales for 1964.

Solution

$$\text{Forecast for 1964} = \frac{111,575 + 84,891 + 94,922 + 76,108}{4} = \frac{367,496}{4} = 91,874$$

In the following examples the effect of self-inspection by the operator, measured by number of defectives taken over a certain period of time, is to be determined. The study is to analyze the data by means of forecasting models as described in previous examples. To make the data comparable to a fixed base, production volume for the number of defectives must be taken. For varying production output the correlation factor and other related statistics give insight into the relationship of defects and output. A positive correlation would normally be expected. However, with the introduction of more reliable processes at high output level, a negative correlation, that is, less defectives with increasing output, is possible. In the following examples the initial regression analysis will be further followed up with a second related data series, the output. This then leads to a correlation analysis.

Example 5: Assuming that the production volume is constant over the year, monthly results on the number of defectives detected are tabulated as follows.

Month	Number of Defectives Detected by Self-inspection[a]	Number of Defectives Detected by External Inspection
January	24	10
February	30	7
March	10	15
April	15	12
May	22	11
June	7	20
July	8	18
August	13	15
September	10	12
October	15	8
November	18	12
December	12	6

[a]Self-inspection applies when the workers do their own inspection.

Do the following:

1. Regression analyses for self-inspection and external inspection separately (time series analysis).

2. Correlation analysis for self-inspection and external inspection.

Solutions
See Table 7.2.

Example 6: From the monthly production data given in the following table:

1. Prepare a scatter diagram.

2. Conduct correlation analysis of defects and output.

Month	Production	Defects
January	345	30
February	420	35
March	450	40
April	407	32
May	391	30
June	305	21
July	380	28
August	420	21
September	480	18
October	520	27
November	505	22
December	510	19

Solutions

1. See Figure 7.5.

2. The correlation coefficient: production $= 1$, defects $= -0.20$. Defects appear to be stochastically independent from the level of production. Figure 7.5 gives a clearer picture.

Other useful applications of the regression/correlation method can be:

1. Assessment of supplier performance.

2. Results of series of tests.

3. Maintenance of quality during storing.

4. Analysis of customer complaints and warranty statistics.

Example 7: An example for a regression and correlation analysis in compiling data on defectives monitored through self-inspection and through external inspection is shown here. In order to assure the reliability and validity of data from self-inspection, special controls are needed to avoid distortion and falsification. The slopes of the linear function and the standard error in such cases express the relative effectiveness and reliance of self-inspection versus external inspection. A reasonable hypothesis would be to state that where the number of defects detected through self-inspection is high, subsequent external inspection is less

TABLE 7.2
Regression and Correlation

Analyses for the Example 5 Problem		

Defects$_{month}$ = Intercept + Slope$_{months}$

	Self-inspection	*External inspection*
Intercept	21.15	13.03
Standard error	4	2.7
Slope	−0.89	−0.13
Standard error	0.54	0.37

Defects per month decrease during the year for self-inspection more strongly than for external inspections. Assuming that the production remained constant, one can conclude that either quality improved or quality of inspection was lost in effectiveness. The relative scatter as measured by the standard error of estimate, is larger for self-inspection, possibly indicating less consistency than the additional external inspection.

Correlation analysis

	Self-inspection	*External inspection*
	1	−0.64
	−0.64	1

The fairly strong negative correlation indicates a possibility that high defect detection by self-inspection leads to low results by the complementary external inspection; not a surprising observation.

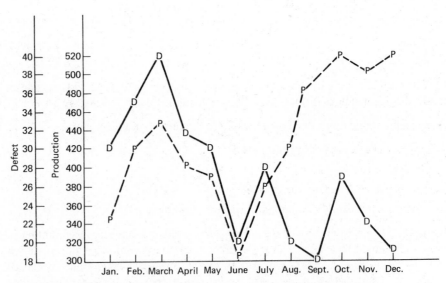

FIGURE 7.5 The scatter diagram indicates a break in the pattern with declining defects and increased production during the second half of the year. The weight of individual defects is not considered. D = defect, P = production.

successful and vice versa. The following is a table for the number of defectives detected in self-inspection versus external inspection, using the regression/correlation model.

Production	Defects	
(units)	Self-inspection	External Inspection
2500	145	44
3000	160	67
2800	120	41
2340	105	45
1800	56	20
2400	87	56
2650	95	60
3220	103	71
3400	110	68
4120	126	75

The results are analyzed in Table 7.3 and illustrated graphically in Figure 7.6.

Superficial analysis of scatter diagrams and data series can hardly be justified. However, realizing the difficulty and costs for attaining valid and reliable data, superior information is generated with the methods and models described and exemplified in previous examples.

TABLE 7.3
Summary of Data Compiled Using Regression and Correlation Analysis

Regression Analysis	Self-inspection	External Inspection
Intercept	47.31	−8.59
Standard error	40.3	14.14
Slope	0.02	0.02
Standard error	0.01	0
Correlation analysis	Self-inspection	Production
	1	0.49
	External inspection	Production
	1	0.85
	Self-inspection	External inspection
	1	0.47

Regression for self-inspection and output shows a poor linear fit, as already indicated in the scatter diagram (Figure 7.6). The analysis for external inspection results is more conclusive, as the fit is closer, showing more defects as output increases.

The correlation between self-inspection and production is expectedly very weak, while external inspection results are much more correlated positively with the production output. Consequently, correlating results from self-inspection and external inspection is positive, but not very strong. One conclusion to be drawn is that operators, when performing inspection, work under greater pressure with increasing production, than external inspectors.

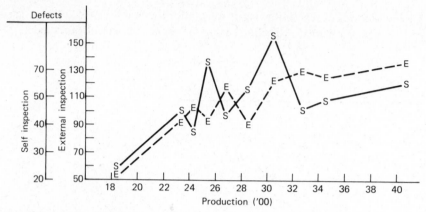

FIGURE 7.6 Plots for self-inspection (S) and external inspection (E) results, in terms of the defects detected, show an increase with increased production. In the middle range, self-inspection is relatively erratic. The higher the output the less effective self-inspection becomes relative to external inspection.

They are also not to be used as part of a permanent reporting system but mainly for special studies and projects in quality policy and program planning. The average statistic and time series analysis, combined with a statistical control chart, can give many informative values for control of ongoing operations.

7.8 QUALITY OF A FORECAST

Many crucial decisions must be based on forecasts and as a result of such forecasts investments are made, staff is hired, and new products and processes are designed. If the forecast is negative, opposite reactions are effected. Reliability and validity of forecasts is an important objective in forecasting, although perfect accuracy can seldom be achieved. The more conditions change and the more complex they become, the more difficult is the assurance of the forecast's quality.

Quality of a forecast is expressed in the reliability, validity, and relative accuracy of the information generated. The difference between the forecast and the subsequently observed and determined actual value can be used as a quality determinant. In qualitative forecasts no quantitative measurement can be made, because only a judgement about the future state of affairs was made. In this case the quality characteristic of the forecast is the attribute of whether or not the prediction has actually come true at the stated time. It is like the *go/no go* test for product quality control.

Forecast Reviews

Forecasts based on qualitative techniques, such as Delphi, can be controlled for quality through frequent or sporadic revisions. This intermediary appraisal of the already revised forecast allows for adaption to new developments. It is not unlike the course correction of an airliner, when having strayed off the flight path or when given an entirely new direction and destination. This appraisal of a forecast in the light of new developments or even for the matter of routine analysis should always apply the techniques discussed above. In the example of bus manufacturing (Section 7.6), the checklist of question asked, information provided, specific method used, data sources used, persons involved, and so on, can serve as a main document for the appraisal. In most cases, repeating the original forecast operation is too time consuming and costly.

A sample will probably allow updating and the sample size could be increased in case major events affecting the original forecast quality have occurred. In the extreme case in which no confidence in the forecast has remained in the mind of decision makers, the forecast has to be repeated. The reason for the forecast being invalidated, once carefully analysed, will induce remedial actions, as in any other quality control matter and nonconformance situation. Such corrective action can mean further clarification of the information sought, of the forecasting method and expertise used, and of the data sources applied. The effort of forecasters to substitute for or complement judgements of persons with factual and quantitative data will be very strong when judgements seemed to have failed.

Forecast Error Control

Error is the numerical difference between the forecasted and actual values. Using quantitative forecasting methods such errors cannot be considered defects, because perfect accuracy of forecasts remains practically impossible. Nevertheless, the error can be minimized and controlled in a way similar to controlling defects in a product or service. Consequently, conventional control methods and, in particular, control charts are applicable.

Measurements for forecasting errors are the mean absolute deviation (MAD) and the mean squared error (MSE). The definitions are as follows.

$$\text{MAD} = \frac{|\text{Actual} - \text{Forecast}|}{n}, \text{ and}$$

$$\text{MSE} = \frac{\sum(\text{Actual} - \text{Forecast})^2}{n - 1}$$

where, n = number of forecast periods.

If errors approximate the normal distribution and have a mean of or near zero, the following relationship between MAD and MSE exists:

$$1.25 \times \text{MAD} \cong \sqrt{\text{MSE}}$$

For controlling the error from forecasting period to period, a tracking signal (T) can be computed:

$$T = \frac{\sum(\text{Actual} - \text{Forecast})}{\text{MAD}}$$

MAD updating is performed using the exponential smoothing model.

$$\text{MAD}_t = \text{MAD}_{t-1} + \alpha \left[|\text{Actual} - \text{Forecast}|_t - \text{MAD}_{t-1}\right]$$

The value of alpha (α) ranges between zero and one and expresses as a proportion the adjustment for the last MAD. Finally, when the normal distribution (ND) of errors can be confirmed, the tracking signal (T) and thus the quality of the forecasts can be controlled statistically. The ND can be assumed when the average error is equal or near zero. The second parameter of the normal distribution, the standard deviation is approximated by

$$\text{Standard deviation, } s = \sqrt{\text{MSE}}$$

Example: In a company producing automobile parts the monthly output is about 5000 items. Final inspection at the end of each month counts defectives recorded during the month. A forecast of defects to be expected in the next month is then made using exponential smoothing and the production quota is raised accordingly. The quality of the forecast is to be controlled, because errors seemed to be too high and unnecessary costs have been incurred.

For the control chart of the tracking signal, ±3 standard deviations are to be applied for setting the upper and lower control limits, assuming that a normal distribution can be confirmed. Data on defects have been compiled for the last year and it is intended to start using the new control at the beginning of next year. The first six months are to be used for the initial MAD.

Month	A[a]	F	Error	|e|	CUM (e)	CUM |e|	e²	MAD$_t$	Tracking Signal, T
1	54	57	−3	3	−3	3	9		
2	62	60	2	2	−1	5	4		
3	63	62	1	1	0	6	1		
4	69	71	−2	2	−2	8	4		
5	70	68	2	2	0	10	4		
6	67	70	−3	3	−3	13	9		
7	65	66	−1	1	−4	14	1	$2.17 + .2(1 − 2.17) = 1.94;$	$−4/1.94 = −2.06$
8	66	62	4	4	0	18	16	$1.94 + .2(4 − 1.94) = 2.35;$	0
9	57	59	−2	2	−2	20	4	$2.35 + .2(2 − 2.35) = 2.28;$	$−2/2.28 = −0.88$
10	53	52	1	1	−1	21	1	$2.28 + .2(1 − 2.28) = 2.02;$	$−1/2.02 = −0.49$
11	57	54	3	3	2	24	9	$2.02 + .2(3 − 2.02) = 2.22;$	$2/2.22 = 0.90$
12	48	49	−1	1	1	25	1	$2.22 + .2(1 − 2.22) = 1.98;$	$1/1.98 = 0.50$
				25			63		

[a]A = actual, F = forecast, CUM = cumulative.

$$\text{MAD} = \frac{25}{12} = 2.08 \qquad \text{MSE} = \frac{63}{12 − 1} = 5.73$$

$$1.25(2.08) \cong 5.73$$
$$2.6 = 2.4$$

or,

$$\text{MAD}_t = \text{MAD}_{t−1} + \alpha (|e|_t − \text{MAD}_{t−1}); \; \alpha = 0.2$$

Initial MAD = 13/6 = 2.17; Then

$$\text{average error} = \frac{\sum e}{n} = 1/12 \cong 0 \qquad \text{(normal distribution)}$$

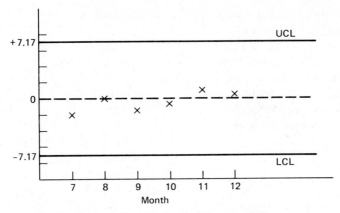

FIGURE 7.7   Control chart for the tracking signal, *T*.

For control chart:

$$\text{UCL} = \overline{X} + 3s; \quad \text{LCL}, = \overline{X} - 3s; \quad s = \sqrt{\text{MSE}} = 2.39$$
$$= 7.17; \qquad\qquad = \quad -7.17$$

$$\overline{X} = \text{Average error}$$
$$\text{UCL} = \text{Upper control limit}$$
$$\text{LCL} = \text{Lower control limit}$$

UCL and LCL $= \pm 7.17$, Mean $= 0$; (ND); Figure 7.7 illustrates the control chart.

The tracking signal for the error remains well within the limits and is randomly distributed very close to the center line. The forecast of defects appears to be reliable and valid.

7.9   KEY TERMS AND CONCEPTS

Planning, as a management responsibility and activity, means determining an objective and desired outcome using optimal approach and required means, and the employment and deployment of available resources. The resulting *plan* provides guidance and a basis for performance and goal achievement. Need for changes of the plan, such as different quality specifications, is monitored by control of the plan implementation and leads to review and possibly replanning.

Controlling is a verification of performance according to plan and the initiation of corrective action in case of deviation, with built-in opportunities for improvement of the plan and respective operations and performances. Both planning and controlling are interdependent and integrated in a cycle of repeating decisions and activities.

Production (also called *manufacturing*) is the creation of products and services in accordance with market needs and opportunities and by means of plans and various human, technological, and material resources. Quality is an essential factor and aspect of any production. Production is often identified only with physical transformation processes, such as manufacturing. *Operation* is used as a more generic term for explicitly including service systems under production. Customers demand quality in both. Products and services have to be produced, that is, created, in terms of quality. The *productive system* is the complex of all interdependent and interrelated components that create the desired products and services in a given environment and resource availability.

Forecasting is a systematic and analytical prediction of future events, developments, and other outcomes under conditions of risk and uncertainty. Quality assurance, particularly in conjunction with production and production planning, requires many reliable and valid forecasts, such as information on quality trends and consumer expectations, costs of defects and defect prevention in upcoming periods, quality assurance strategies and measures of the major competitors. Forecasting methods facilitate quality and reliability of the information to aid the needs of decision makers.

7.10 DISCUSSION AND REVIEW QUESTIONS

(7–1) List main reasons for planning quality of products and services.

(7–2) *Planning for quality and for production go "hand in glove."* Explain.

(7–3) List information and data required for planning quality.

(7–4) Define the quality and quality assurance of a plan prepared by quality assurance management.

(7–5) List examples of productive systems and the kind of quality to be achieved in terms of products and services.

(7–6) *Quality expectations of the customer permeates the entire productive system.* Explain.

(7–7) *Quality assurance itself can be considered a productive system, with certain outputs and inputs.* Explain.

(7–8) *Quality planning and production planning are always closely related and interdependent in goal achievement.* Explain.

(7–9) *Under the systems view quality assurance planning can be described as a subsystem of the production system.* Explain.

(7–10) *Forecasting of quality is an essential prerequisite of quality assurance planning.* Discuss.

(7–11) *Quality of demand is not the only subject of forecasting in quality assurance and under the productive system view.* Explain.

(7–12) *Usually qualitative and quantitative forecasting methods must be applied jointly in determining quality assurance policies and actions.* Explain.

7.11 PROBLEMS

(7–1) T. E. Turner[3], a quality control professional, suggested the following "quick and dirty" checks for assessing the need for improved quality management. The checklist included the following questions.

1. "Does your department ever reject anything and, if it does, does it stay rejected—at least most of the time?"
2. "How frequently are you unpleasantly surprised or even dismayed that 'such a defect' got through?"
3. "Do you learn anything from rejects? If so, does corrective action follow throughout the entire plant organization?"
4. "Do your reports ever change action or cause anything to happen?"
5. "Can you name one customer who was lost last year for quality reasons or would you even know if a customer was lost for quality reasons?"
6. "Are you catching defects early enough?"

The questionnaire for a survey into the state of quality control in a local manufacturing community can also be used for providing basic information for quality planning. Give your comments and elaborate on the above questionnaire sheet.

(7–2) It has been pointed out in this chapter that in the study of a productive system, the necessary first step is the development of a description of that system. Once the salient features of the system have been described, one is able to determine why the quality of that system is good or poor. For the following examples of service operations, try to organize your description of the productive system into the well known categories; namely input, production (or transformation) process, and output. Illustrate your answers with the help of flow charts.

a. A restaurant
b. City transit
c. Hospital emergency service
d. Newspaper home delivery system
e. Postal service

(7–3) A quality assurance manager wants to assess the forecasting errors for rejected parts at the receiving inspection point. The actual and predicted values are shown below.

Period	Rejects	Predicted
1	30	25
2	95	105
3	57	51
4	20	21
5	15	12
6	32	40
7	86	82
8	53	50
9	94	85
10	62	68

a. Compute the mean absolute deviation (MAD) for the fifth period, then update it period-by-period. Use the exponential smoothing method with a constant of 0.3.
b. Compute a tracking signal for periods 5 through 10 using the initial and updated MADs.
c. Plot the tracking signal on a graph using action limits ± 3 standard deviation. Comment on the quality of the forecast.

(7–4) A manufacturer of home computers uses a great number of test gages. These must be replaced quite frequently and, because of considerable leadtime, the demand is to be forecasted. Actual demand over the last 12 month period is shown below.

Month	Demand	Month	Demand
January	48	July	70
February	64	August	83
March	86	September	96
April	92	October	90
May	74	November	85
June	68	December	70

a. Predict the demand for January of next year using a three-month moving average.
b. Plot the demand on a graph and estimate the demand for next January through to next December.
c. Compute the 2-sigma control limits for the actual demand above, and enter your estimates from b. onto the control chart. Comment on the results.

(7–5) An electric razor company sells a new design and expects that customer complaints and frequency of repairs will gradually be reduced. Complaint statistics for the new electric razor are shown below.

Month	Complaints	Month	Complaints
January	560	July	320
February	437	August	306
March	470	September	349
April	380	October	302
May	350	November	270
June	375	December	260

a. Smooth this data using a three-month moving average.
b. Plot the data of the moving average and comment.
c. Forecast the complaints for next January.
d. Use an exponential smoothing model with a smoothing constant of 0.4 and predict the complaints for next January. Use 500 as the base.

(7–6) Prepare a statistical control chart for the complaint statistics in the above example with the mean at the center line and control limits set at ± 3 standard deviations. Comment on the result and the application of the control chart for forecasting.

(7–7) A producer of chocolate bars has developed a linear trend equation that can be used to predict returns from customers and also from dealers. Because such returns will be replaced immediately without cost to the consumer, these are to be predicted using the equation.
Given the equation $y_t = 48 - 5t$, where y_t = monthly return, and t_0 = January
a. Are the monthly sales increasing or decreasing, and by how much?
b. Predict the returns for next October.

7.12 NOTES

1. Feigenbaum, A. V., "Quality and Productivity," *Quality Progress,* November, 1977 p. 16; "Quality and Business Growth Today", *Quality Progress,* November, 1982, p. 22; "Business Quality Systems—New Key to Profitable Growth," *Quality Progress,* January, 1981, p. 16.

2. The readers interested in more examples of these analyses and techniques should consult Meredith, Jack R. and Gibbs, Thomas E., *The Management of Operations,* John Wiley & Sons Inc., New York, 1980, pp. 84–100.

3. Turner, T. E., *Quality Progress,* May 1973.

7.13 SELECTED BIBLIOGRAPHY

Abernathy, W. J., "The Limits of the Learning Curve," *Harvard Business Review,* January–February 1974, pp. 87–95.

Clark, Charles T. and Stockton, John R., *Introduction to Business and Economic Statistics,* South-Western Publishing Co., Cincinnati, Ohio, 1975.

Deming, W. E., *Some Theory of Sampling,* John Wiley & Sons, New York, 1966.

Feller, William, *An Introduction to Probability Theory and Its Applications,* 3rd ed., John Wiley & Sons Inc., New York, 1968.

Moroney, M. H., *Facts from Figures,* Penguin Books, New York, 1956.

Smith, David Eugene, *Qualitative Business Analysis,* John Wiley & Sons, New York 1977.

CHAPTER 8

QUALITY OF DESIGN: PRODUCT AND SERVICES

When paying a price for a commodity or service the consumer expects certain quality from the producer. To satisfy the consumer demand for an adequate design, the product must be designed either with the consumer directly involved or with one in mind. For instance, a pencil must be useable for writing notes, an airplane for transporting people. Both products, with vastly different complexity, must be designed to create ultimate satisfaction, when used for the intended purpose.

The complexity of products and services determines technology of production and the kind of productive system. Products and services are outputs of productive systems. Under the system view these have to be designed before they can be produced. Such planning of commodities, items, projects, and services can be called product design.

Three distinct phases of product design are:

1. Determination of customers need and demand.
2. Assessment of respective technology and production capacity.
3. Assessment of required human, material, and organizational resources.

Product design and production design are closely related. Accordingly, *quality of design* will be described in three chapters respectively covering product and services, production and operations, and resources and supplies. After a producer has completed these three design phases, production planning is completed and production, understood as the creation of products and services, can start.

In this chapter we are concerned with the design phase and the product reliability. Product is anything useful for a consumer that has a value and a respective price. Such a product, as the creation of production, is said to have utility for the consumer. Whatever this utility is in the mind of the consumer, and the producer, is difficult to define and, in a practical sense, not completely determinable. Expectations and satisfaction are psychological reflections and expressions of individual persons. A product is as much a concept in peoples' minds as it is a tangible or intangible carrier of utility. A pencil is still largely used for writing. But why are there so many varieties? There are simple wooden ones and those in a golden casing. Some even have watches and calendars built-in. In fact, any product at closer view is a very complex phenomenon.

What the consumer sees in the product and expects from it and what the producer thinks the consumer sees in the product are not necessarily the same. The more the views of the consumer and the producer match, the more successful is the "meeting of minds," that is, the contract between both business partners. In our context of quality assurance, psychological, sociological, economical, and other more general aspects of product design cannot be pursued, although these would be interesting and relevant.

Quality of product design can be measured by the degree to which it meets the expectations of consumers. Quality expectations of consumers relate basically to three familiar dimensions: the material quantity and quality, the time, and the place. For instance, a particular consumer wants a certain number of pencils of specified quality and price, at a particular time and place. These consumer demands aggregate into markets or demands for custom-made items; for example, special orders. Existing or potential consumer demand communicated through various channels induces producers to consider a product design. Often a product idea from the producer sprouts new products and consumer demands.

8.1 THE DESIGN PHASE

However, product quality, as one major dimension in product design, remains intimately intertwined with mere quantitative, timing, and spatial (locational) product design problems. When large volumes of an item can be produced and sold, quality will be standardized and then marketed through various distribution channels. Of course, the timing is important as well, as the products do describe life cycles. The most suitable quality product can only be sold at a particular time and place. If one design does not please the customer, competitors might serve the customer better. Money may as well be spent or saved on substitutes or entirely different items.

Quality assurance planning accompanies product design activities and decisions through all phases (Figure 8.1). New product ideas, market opportunities, new technologies, materials, skills, and competitors successes in the market are among the main factors that trigger product design procedures. Directly or indirectly, customers are initiating products and services. Careful vetting of ideas and product research leads to preliminary and final designs, before senior management can approve actual production.

Quality assurance remains involved with the design process. This helps to ensure that quality will be adequately specified from the consumers' point of view. Design engineers and other design specialists often tend to emphasize mere technological features, rather than actual consumer needs and perceptions. Moreover, quality assurance ensures that the final

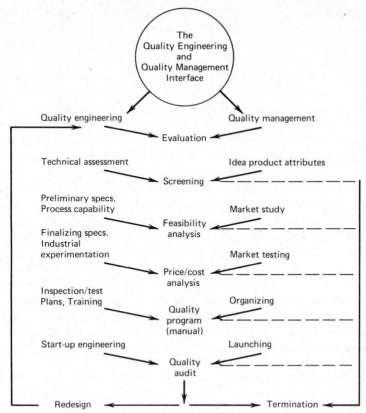

FIGURE 8.1   Quality assurance planning process that accompanies new product design activities and decision making through all phases in the life-cycle approach.

design allows proper inspection and verification of major quality characteristics. With the modern trend toward preventing defects as early as possible, participation of quality assurance experts in product design has become increasingly useful and important.

Quality assurance of design can take many forms, ranging from simple practices in order processing to elaborate design assurance and reliability programs. Product complexity and the quality impact of defects on performance dictate the degree of quality assurance during the design phase. However, in each case of product design, major quality characteristics should be carefully specified. This is the obvious prerequisite for conforming to such standards during production. Therefore, we shall describe quality specification first, before any other principles and steps in design assurance.

Specification of Quality

Quality starts conceptually and practically with consumers. The designer translates consumers needs and desires into a product design. Quality thus becomes defined, clarified, characterized, and adequately specified.

Specifying quality means to establish the desired, measureable or assessable attributes of a product. This is a complex task as not only consumer requirements must be observed. Many legal, technical, and economical constraints must play an ever-increasing role. Specifying quality has become a difficult decision-making task in which many specialists must participate. The outcome and decision should be a specification which is realistic and valid with regard to customer needs.

A useful quality specification entails all attributes and measurements that together define and describe the product in clear and unambiguous terms. A good product specification informs the consumer and the producer sufficiently about the product. The producer, of course, needs possibly more detailed documentations in the form of blueprints, procedures, recipes, and so on, for planning subsequent production and procurement.

Specifications vary in formality and technical content. A specification is always a technical, determinable, and required attribute or measurement of a quality characteristic. Any specification entails at least three data: the quality characteristic description or measurement; possible hazards and conditions of application; and, tests for verification.

Other relevant data in a specification can be costs, standards, tolerances, defects, and failures or hazards. Information given for specifications can be too detailed or too general. In order to strike the optimal balance the particular purpose of the information should be kept in mind. Often consumers require less technical data than the producer. Whenever possible specifications should be standardized, as this simplifies definition, compliance, and verification.

Specifications tend to fall into three basic types:

Type A Individually specified, regardless of specifications of the other units in the same product.

Type B Specified in terms of distribution, which the product as a whole must meet.

Type C Specified in terms of allowances for a certain percentage of units or lot of products submitted. For example, the tensile strength of steel rods shall not exceed 70 p.s.i. However, steel rods will be acceptable if say, only 5% of the samples tested exceeds this limit.

Depending upon the expected reliability of the end item, or the complexity in manufacturing process control, the trend is either to individually specify each and every measurement (Type A) or to go along with a distribution in mind (Type B). The normal distribution of quality characteristics is important for two main reasons: it makes it easier to compute the percentages of the product that are likely to fall outside the specification limits; and, it can be assumed for behavior of the averages without the necessity of knowing individual variation. Both of these reasons have important uses in statistical theory. One of the special uses is in the control chart techniques where specifications are controlled, adjusted, and readjusted to suit the design specifications. The reader will be familiarized with such terms as "upper" and "lower" control (specification) limits in a different chapter.

In *process capability studies* where processes are to be compared with specifications, the process itself first must be in a state of control. If the process is verified not to be in control, engineers can avoid unnecessary costs by changing the specifications, the tolerances, or the process performance. Sorting and repairing is an expensive way to handle specification trouble. With the advantage of Type B specification, a process capability analysis (described later in Chapter 11) can lead to much easier detection of out-of-specification conditions.

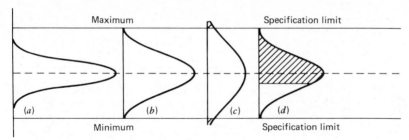

FIGURE 8.2   Illustration of four basic relationships between process and specification (type B).

Assuming normal distribution, four cases are described in Figure 8.2: (a) process spread narrower than the specified limits; (b) process spread equal to the specified limits; (c) process wider than the specified limits; and (d) process being off-center.

Standards

A standard is a generally *approved and accepted* specification that guides product design and saves time and cost. Standards are based on experience, experimentation, and consensus of specialists and the consumers.

Tolerance

The intended and specified attributes, measurements, and other characteristics in the design usually differ somewhat from the actual product or service. An acceptable variation around a nominal value is called a *tolerance*. When parts with tolerances become assembled in a product, the joint maximum allowable variation must be carefully determined and specified. Tolerance limits may at times be exceeded during production, but not if proper controls are used. Process capability analysis helps to determine whether specifications, and in particular tolerances, are tighter than the current production capability allows. Process capability studies apply statistical methods that will be discussed in a later chapter.

When assessing customer-specified tolerances, the producer should determine the natural variation in actual process by statistical control chart techniques. Thus tolerances, as desired control limits, would have to be validated. Moreover, tolerances should not be set tighter than required, because resulting higher number of rejects, production stoppage, process improvements, and so on lead to unwarranted costs.

Figure 8.3a shows a number of important factors that must be considered during both the design and the manufacturing phase, keeping in mind a similarly large number of possible variation sources (Figure 8.3b).

Statistical Meaning of Tolerances, Clearances, and Fits

Whenever parts are made with allowable tolerances, their fits in an assembly create new dimensions. For the proper functioning of the assembled unit, an understanding of the distribution of tolerances is essential. One can use the basic statistical laws in order to arrive at economical and workable situations.

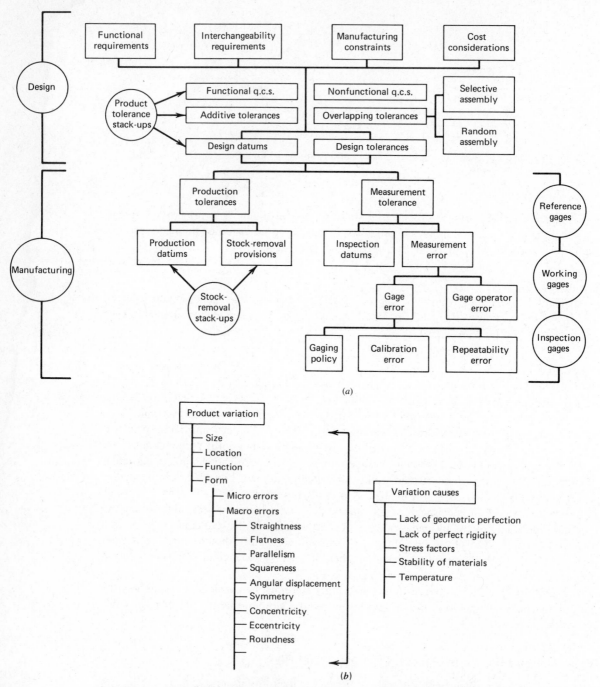

FIGURE 8.3 Major elements to consider for analyzing tolerances. (*a*) Product and manufacturing tolerances structure. (*b*) Product variation and variation sources. Source: Kirkpatrick, E. G., *Quality Control for Managers and Engineers*, John Wiley & Sons, Inc., Publishers, New York, 1970. Reprinted by permission of publisher and the author.

1. *Law of addition:* If two or more parts are assembled in such a way that one dimension is added to the other, the average dimension of the assembly $= X_1 + X_2 + X_3 \ldots$ If the parts are assembled with the knowledge of their standard deviations, the standard deviation of the assembly will be given by:

$$\sqrt{(\sigma_1)^2 + (\sigma_2)^2 + \ldots}$$

Using standard deviation formula, one would get narrower tolerances, since the square root of the sum of the squares is less than the actual numbers merely totaled.

2. *Law of differences:* If the parts are assembled such that one dimension is subtracted, the average dimension of the assembly will be $X_1 - X_2$.

3. *Law of sums and differences:* If the parts are assembled in such a way that some of the dimensions are added while others are subtracted, the average dimension of the assembly will be the algebraic sum of the average dimensions, that is,

$$X_1 + X_2 - X_3 + X_4 - \ldots$$

Statistically, whenever dealing with standard deviation, the problem is to predict the limits of variation of individual measurements in the total population based on a sample of data. However, the problem is simplified, at least to a first approximation, by setting the tolerance limits at $X \pm 3s$, where the average X and standard deviation s of the sample can be used directly as estimates of the population values. For preciseness, the relation $X \pm Ks$ is sometimes used where K is a function of the confidence level desired. The design determinants, through required tolerances that can be stated as Ks, must fall within $\pm$ tolerance. Statistical tolerance limits can also be computed using the range instead of the standard deviation.[1]

Specifications Define Defects

From the quality assurance viewpoint, any major attribute or measurement must be verifiable during production and in the end product. Specifications define acceptance criteria and indirectly define defects as nonconformances. A defect can take many different forms and appearances. Often mere symptoms lead to detection of defects and deviations. Defects can be minimized or entirely avoided once they are anticipated and clearly defined during the design phase. Faulty designs create the most costly defects.

For defect prevention all major specifications should list hazards and major causes of negative quality impacts. This comprehensive analysis and information facilitates avoidance of defects and generates control procedures. When products and services are designed with the consumer in mind the obvious source for defect information is the consumer. While the consumer, as a layperson, might not know the technical causes of defects, symptoms are usually fairly easily detected during testing of design. Figure 8.4 is an example of a consumer information request. Ways to communicate with consumers effectively for quality assurance will be shown later.

Figure 8.5 is a simple example of a product specification that was prepared by a layperson. Consumers quite frequently, through inquiries and complaints, aid producers in arriving at a realistic and sound product specification. When ordering a custom-made item, consumers will stipulate and negotiate quality characteristics that are desirable and feasible.

Customer Feedback*

Name of Restaurant: _____ Date _____

The following items did not live up to my standards.

1. _____ 2. _____ 3. _____

I believe the following to be the cause. _____

_____

I would recommend the following corrective action. _____

_____

***What's at Steak?**

Quality Characteristic	Consequence	Assurance Action
Thickness of meat	Uneven amount of heat available to product.	Measurement of meat thickness.
Weight of meat	Variations in cooking time; customer complaints.	Weighing of meat before or after cooking.
Thawed age of meat	Too much exposure to tenderizer, high bacteria counts, poor appearance.	Tagging of meat trays with dates; reporting of poor looking meat.

FIGURE 8.4 Typical consumer information request form. Customer feedback may provide a direct clue to the assurance action required.

How to Specify

Certain major rules underly a sound specification that informs the producer as well as the consumers. The following principles help accomplish this objective.

1. Specifications must meet customers' needs.

2. Producers' capabilities must allow conformance to specifications.

3. Specifications form a bench mark for decisions that require proper production procedures.

4. The formulation must be brief, concise, unambiguous, numerical where possible, augmented by models, photo, drawings, and other visual aids, and understandable by the person to whom it is directed.

5. Specification should be consistent with general company policies, laws, and mandatory standards, and with other related specifications.

6. It should be categorized and typified, as to degree of importance, mandatory or optional, maximum or minimum.

Item No.	Component	Quality Characteristics			Hazard	Assurance Element
		Parameter	Spec.	Tolerance		
1.	Handle dimensions 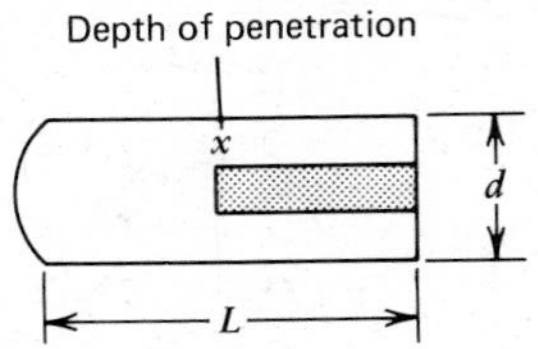 Depth of penetration 	1.1 Length = L 1.2 Diameter = d 1.3 Penetration depth = x	$1\frac{1}{2}$ in. $\frac{1}{2}$ in. $\frac{1}{2}$ in.	$\pm\frac{1}{8}$ in. $\pm\frac{1}{16}$ in. $\pm\frac{1}{16}$ in.	1.1 Undersize: could split on assembly 1.2 Inadequate: would not withstand rated torque	1.1 Measurement 1.2 Ditto 1.3 Ditto
2.	Handle material	2.1 Unbreakable, clear plastic (Ref: Standard A 1234) 2.2. Color: yellow			2.1 Safety hazard if breaks during use	2.1 Measurement and drop test
3.	Shaft dimensions 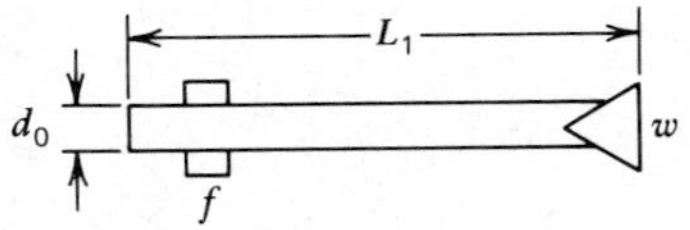 	3.1 Length = L_1 3.2 Diameter = d_0 3.3 Flange = f 3.4 Width = w	3 ins. $\frac{1}{8}$ in. $\frac{1}{4}$ in. $\frac{1}{8}$ in.	$\pm\frac{1}{8}$ in. $\pm\frac{1}{32}$ in. $\pm\frac{1}{16}$ in. $\pm\frac{1}{32}$ in.	3.2 Undersized: breaks, shatters, or bends 3.2 & 3.3 Oversized: could split handle 3.4 Exceeding tolerance: may damage tool or screw	3.1 Measurement 3.2 Ditto 3.3 Ditto 3.4 Ditto
4.	Shaft material	4.1 4040 stainless steel			4.1 Shaft could twist before rated torque reached	4.1 Metal testing
5.	Overall assembled length	5.1 Length	4 ins.	$\pm\frac{1}{8}$ in.	5.1 Too long: improper penetration of handle and subsequent failure	5.1 Measurement
5.	Torque test of completed assembly	6.1 75 in. per pound		± 5	6.1 If torque under specification: failure of product	6.1 Measurement

FIGURE 8.5   A simple example of a product (pocket screwdriver) specification prepared by a layperson.

7. Specifications should be identified regarding product, order number, date, persons in charge, approvals, and so on.

8. Possible hazards, defects, and causes of nonconformance should be analyzed and listed.

8.2 DESIGN ASSURANCE

Design assurance is the quality assurance of the design. Building a product with the customer in mind is, of course, basically the task and responsibility of those in charge of specifying quality. The verification of adequacy through testing and design reviews, undertaken by independent specialists, complements, rather than substitutes for, the careful and continuous observation of high design standards. The separate examination of specifications and design activities serves the sole purpose of affirming and possibly further improving the product or service design.

To a certain extent and with some objectivity, the review can be undertaken by the designer or by the design team. This is the case, for instance, when quality assurance specialists participate during the entire design project or at important milestones. However, independence and proficiency of the reviewer must also be visible and explicit. An audit or review, planned and conducted according to special procedures and standards, not only strengthens the design, but builds product liability protection. A formal design review is visible and convincing evidence that the producer has observed all necessary care in ensuring the quality and integrity of the product. Reviews and design audits should therefore comply with quality audit standards, which will be discussed later in the book as a separate chapter.

The scope and objective of the design review varies and depends largely on design assurance activities preceding such a review. For instance, defects and possible causes of failures are usually not sufficiently analyzed because emphasis is on positive, rather than on negative, specifications. The review will then have to conduct or extend such an analysis, in order to prevent occurrences entirely or partially.

Principles of Design Assurance

The design reviewer first establishes scope and objective of the design assessment. Only after the clarification and communication of the tasks checklist can other working papers be prepared. Questions raised and evidence sought involve, for instance, the following.

1. Have customer's needs, requirements, and stipulations been adequately researched?

2. Has the required production capacity been assessed, and are human and material resources and facilities available?

3. Does the current state of the art in technology allow the production? Can required technology be developed?

4. Have all principles, standards, and procedures been complied with? If not, what were the reasons?

5. Are specifications communicated, understood, and approved by customer and producer?

6. Do specifications meet required standards of form and content?

7. Have possible defects and failures been analyzed and recorded?

8. Can major specifications and related defects and failures be inspected and controlled during production and performance of the product?

The reviewer will seek evidence for such questions and other checks through direct interviews or control of documents. Design engineering should participate as much as possible without undue interference, and mainly for the purpose of speedy corrective action and design improvement. Any design review should lead to a formal recording and report of the results. Senior and supervisory management receiving such reports, can become properly informed and then decide on further action. As has been pointed out, the reviewer only examines the design and does not adopt design responsibilities.

When a special defect and failure analysis has not yet been undertaken during the primary design project, the reviewer will normally recommend this necessary extension and complementation. Reviewers conducting such an analysis now join the design team, rather than remaining as independent reviewers. Therefore, the term ''review'' is used, rather than ''audit,'' because an audit would not allow actual involvement in the design activity. Figure 8.6 is a flowchart illustrating a system for evaluating product hazards.

Prediction of Design Faults

For virtually every product it is possible to conceive of a variety of ways in which a flawed performance may be manifested. Defect and failure mode and effect analysis are obviously important design assurance functions, particularly for critical specifications and hazardous products.

A *defect* is an unacceptable deviation from a specification of an attribute or measurement of quality. Such deviations are called nonconformances, and a defect is sometimes any undesirable feature of a product or service. For quality assurance, defects, by definition, relate to specification, and not to the perceptions of a person.

A *failure* is a defect that relates to reliability in performance.

Reliability is the probability of failure-free performance of an item during a specified period and under required conditions.

An analysis of defects and performance failures during the design and experimentation stage of the product is to determine cause and effect relationships. The more these are known and understood, the more likely is the prevention of detrimental causes and the detection of occurrences later.

Defects and failures have different impacts on quality of conformance and performance. For each major and critical specification, at least one major defect or failure can be stated, namely, that production has not conformed; or that the user has either not complied with instructions or the item itself does not perform as intended and expected. *Nonconformances* are all deviations from specifications.

Defects and failures should be directly related to and recorded in reference to the relevant specification. This has been demonstrated earlier when specifications were discussed. In this further analysis for critical products and specifications, all possible (negative) cause–effect relationships should be determined. As a first step, defects and failures should be classified by type and description of features and symptoms, component, part, and place, relative to

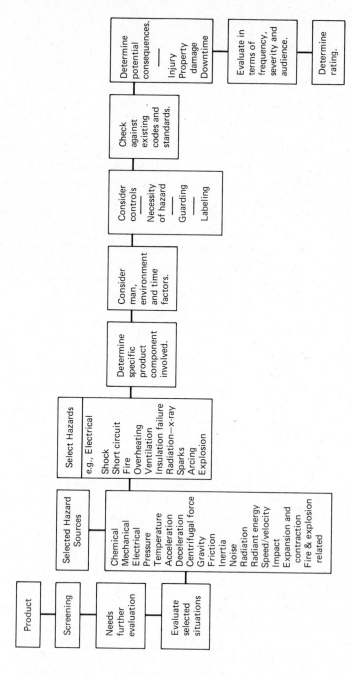

FIGURE 8.6 A typical system for the identification of product hazards. This system involves the identification of hazard sources and their possible interaction with product users and the environment. From the U.S. Department of Commerce Bulletin, *Consumer Product Safety—Responsive Business Approaches to Consumers' Needs*, Superintendent of Documents, U.S. Government Printing Office, Washington, D.C., April 1981.

frequency, relative importance and costs, and required tests and controls for detection. Such classifications help in monitoring the defects and failures, if they can not be eliminated through redesign or other countermeasures. Moreover, full description of defects allows the traceability, determination, and possible prevention of all major causes. When this can be done during the design phase, then the degree of design assurance will be raised, and defect and failures costs will be considerably reduced.

8.3 TECHNIQUES OF FMEA AND FTA

Failure Mode and Effect Analysis (FMEA) has been developed as a special technique for design assurance. With this approach, defects and failures are systematically studied as effects of certain causes. Probabilities and conditions of occurrence allow application of statistical techniques as well. The design specialist should apply such an analysis as an integral element of the design procedure, because adequate technical and analytical expertise is required. Reliability engineering applies such failure mode and effect analysis. Techniques of the analysis range from simple systematic listing of defect- and failure-related causes to comprehensive and complex *fault tree analysis,* (FTA). Fault tree analysis was developed at Bell Laboratories and became widely used and refined in the U.S. space program and defense safety systems in the 1960s. Risk studies of nuclear and chemical plants are now required in many countries by law and FTA technique becomes very helpful in analyzing data for such studies.

Fault tree analysis defines relationships between functional effects and product components, looking from the top downward in order to find the combination of effects that causes failures. On the other hand, a FMEA is looked at by considering the failure of individual components of the system (one at a time) and noting its consequences from the bottom upward. In other words, in a fault tree analysis, each hazard (or event) on the list is tackled in the following order: listing of events in priority; connecting individual events with possible causes; and, looking for ways to avoid the origin of causes. The approach is the reverse in the case of FMEA, which starts with origins and causes and looks for any negative effects. Fault tree analysis is a well accepted technique for finding failure relationships, and is fault oriented where one looks for all possible ways in which a system (or product) can fail. For a simple product with relatively few parts that could fail, FTA provides a fairly rapid analysis. However, for complex systems large fault trees become difficult to understand, bear no resemblance to the system flow sheet, and mathematical logic tends to become complex.

Failure mode and effect analysis examines all possible failures of every component. Its main advantages are that it is easy to understand, based on a standardized approach, noncontroversial, and nonmathematical. However, depending upon the complexity of the system, it may be very time consuming, and may often be misleading where combinations of failures are involved.

Procedure for Construction

1. Define the flaws giving rise to functional defects, that is, the specific malfunctions responsible for the effect, events, faults, or failures. For example, the car won't start, the light does not go on, and so forth. (The real challenge would be to define the flaws of the system and lead in steps to identifying the failed part.)

2. Identify the most direct or possible causes, list them with their direct and indirect effects which relate to the malfunction. When a car does not start, it may be related to more than one cause (battery low, no gas, starter motor defective and so on), but when the house light does not go on, it may only be the fuse, the switch, or the bulb.

3. Quantify the event's probability so far as is possible by securing functional data, knowing prominent defects, or conducting laboratory tests on the simulated environmental conditions (consumer's uses and abuses etc.).

4. Determine the time dependencies of an event's probability to find out which particular fault of a particular kind is more likely to occur (a) soon after the product is manufactured (b) soon after the consumer's use, and (c) after a given period of use by the consumer.

Figure 8.7(b) is drawn in the form of a fault tree considering the events described in the block diagrams of Figure 8.7(a). The rectangular blocks are always identified for events

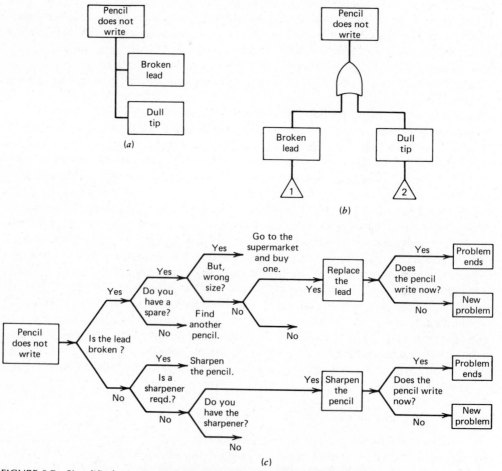

FIGURE 8.7 Simplified representation of a fault tree analysis. (a) Functional events. (b) Functional events in fault tree form. (c) Tree diagram for a given problem.

which are connected by an arrowhead-shaped logical symbol (called an *OR gate*). The blocks underneath an OR gate further subdivide the main event into two or more events. The triangular symbols merely signify reference keys, allowing continuation of the tree branches over on a separate page. Figure 8.7(*c*) is another simplified version of a tree diagram deduced from simple and logical yes or no answers.

Many times the analysis performed is not as simple as the one described above. One event breeds another and the fault propagates to higher levels. Until the analyst comes to a definite conclusion, he may use "conditional" events or an *AND gate* to depict his findings. These and other symbols commonly used are described in Figure 8.8. When fully developed, the results obtained by FTA and FMEA should be equivalent, provided both analyses are complete and correct. Figure 8.9 gives an example of a fully developed fault tree for a clutch problem in a car.

Knowledge of Failure Probabilities

A knowledge of failure probability acquired by using realistic failure rates for the parts in question, is an important step in discovering a failure mechanism. It helps analysts to make theoretical estimates of achievable reliability and points out the areas where high failure rates (low probability of survival) may lie concentrated for a particular design. To compute failure rates or the probability of survival, requires that experimental tests be run on the product. This aspect, with other techniques of predicting equipment reliability, will be discussed at length in the latter part of this chapter. However, in terms of simple probability theorems

Symbol	Meaning
	An event, usually a fault
	Basic component fault
	A fault not developed further as to its cause
	A conditional event
	An event expected to occur in normal operation
	AND gate
	OR gate
	Reference key to another part of the fault tree
	INHIBIT gate, used for conditional event
	Reference key to another fault tree

FIGURE 8.8   Illustration of basic fault tree symbols.

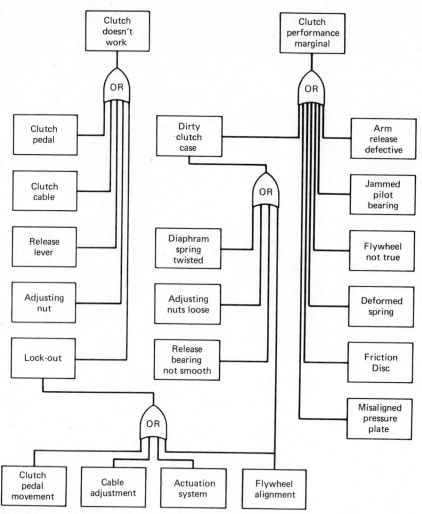

FIGURE 8.9 An example of a complex fault tree analysis.

(described in Chapter 6) it is interesting to see how the algebraic equivalents of probabilities in fault trees are developed, whenever *OR/AND* gates are used; (see Figure 8.10).

Results of failure mode and effect analyses highlight potential problem areas and thus direct further design activities. Checklists and fault trees, make valuable information accessible and are instrumental in effective design assurance. More such an analysis of defects and failures brings about a deeper grasp of underlying causes; thus the easier it is to take preventive and corrective actions. For instance, fault trees can show if causes interact or have an isolated impact on the ultimate product quality. Once the cause has been firmly established through experimentation and testing, remedies can be determined and prescribed.

Design review not only verifies that failure mode and effect analysis have been properly conducted, but a special final test of the design confirms whether or not a validity review

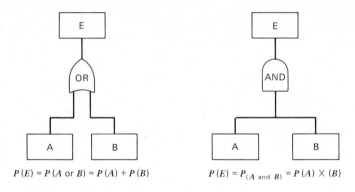

$$P(E) = P(A \text{ or } B) = P(A) + P(B) \qquad P(E) = P_{(A \text{ and } B)} = P(A) \times (B)$$

FIGURE 8.10   Fault tree algebraic equivalents.

might be undertaken. A design is valid and reliable when the actual performance of the item under the intended conditions satisfies customers' expectations and other imposed requirements. Such tests that finalize a design require the production of a prototype. The flight test of a new aircraft model is a typical example.

Once the review of specifications of the complete design, product, and underlying design process has affirmed the adequacy of the design and its compliance with appropriate design standards, preparation for planning the production can commence. All documentations describing the design in necessary detail must now be complete and approved.

8.4   DESIGN APPROVAL AND CONTRACT REVIEW

Management must decide, partly on the basis of design review reports and possible customer approval, when the design is ready for production planning. A formal design approval is to screen the integrity of the design by senior management in addition to the preceding design review. Such careful approval procedure demonstrates actual involvement of senior management in finalizing the design.

As senior management usually also finalizes and signs formal contractual documents with the customer, the design approval should be preceded by a special contract review. All obligations of the contracting parties for quality and quality assurance have to be documented and validated. Obviously, any lack of clarity and understanding at this crucial point can have costly repercussions.

Naturally, not all causes and factors that create defects and failures can be recognized and anticipated during the early design phase. This is particularly true for relatively innovative new products and services where little practical experience exists. Design assurance must therefore also review the monitoring of production and performance in order that the validity of the design can be maintained. Simple trouble reports (Figure 8.11) and other communications from customers, external and internal nonconformance analysts, and so forth can indicate faulty designs and new causes for defects and failures. Periodic design reviews should follow up original designs and seek modifications and adaptations to changed conditions. Therefore, design is more or less a continuous process as long as the product line is not terminated or entirely new models are not introduced. At a certain point in time and

Trouble Report

From: Foreman/Inspector

To: Manager

Date:

Hour:

Place:

I report the following trouble.

I know/think (scratch out one) that the trouble is due to the following cause.

I suggest the following corrective action.

I would like you to keep this report confidential/not confidential/confidential for the time being (circle one).

FIGURE 8.11   An example of a simple trouble report form.

in the life of a product, repair, rather than redesign, become the reasonable and economically wise action to take. However, design changes made in the heat of a production crisis without the necessary design assurance can come back to haunt management years later. In Appendix C is a list of questions which, if used by management for evaluation, could aid in avoiding many of the problems discussed.

8.5   DESIGN DOCUMENTATION AND INFORMATION SYSTEM

Product and service design creates much paperwork, because many people need to participate, proceedings have to be documented, and results formally recorded and communicated. Quality assurance of the design hinges considerably on the quality of the associated recording and reporting systems. There are, for instance, the design principles and procedures, codes and standards, guides and design aids. The product idea must be formulated and then researched, feasibility studies distributed and approved, proposals formulated, test results recorded and analyzed, until intermediate and final specifications can be documented.

Design assurance and reviews include documentations, communications, controls, and examinations. They seek evidence that is more often in manuals and records. Results of these reviews are then reported and communicated to those concerned.

Design documents of design proceedings, reviews, and descriptions of the design itself are an integral part of the overriding management information system. Quality assurance has its own subordinated information system that includes design assurance. These information systems will be described in a later chapter. At this point the importance of proper documentation and information planning and control presents itself in conjunction with design assurance.

The importance of design and specification *documentation* rests on the following.

1. Design activities and decisions must be traceable in case of future failures and defects.

2. Customer and producer must have clear information concerning quality characteristics; contract reviews are especially conducted to affirm the understanding of the mutual agreement.

3. Production planning requires final product descriptions and drawings.

4. Documentation facilitates timely data compilation, data processing, and communication of information. Special form and records allow the standardization of documents.

5. Design reviews are not feasible without documentary evidence and formal communications.

Whenever design changes occur, the documentation and information system must allow quick and reliable responses. A major failure and modifications due to events that invalidate the design, trigger remedial and corrective redesign. Design change control must be accompanied with documentary control. Moreover:

1. Reason for the change must be documented, analyzed, and accepted by authorized persons in accordance with predetermined procedures.

2. Changes must be arrived at under predetermined and authorized procedures.

3. Modified and new documents, such as specifications, should be released only after review, and with authorization.

4. Obsolete documentation should be withdrawn simultaneously with its replacement. This requires that for important documents that the distribution is controlled. Obsolete documents should be destroyed.

5. Information on the change must be communicated without delay to those concerned.

Design changes can be controlled through special change forms. These forms serve as evidence for design change reviews.

Probably the most important change, that of contracted specifications, needs particularly careful treatment. Changes can render a design unfeasible regarding current production capacities and can violate codes and standards. A special contract review must ascertain that changes are properly studied and cleared. Changes internally initiated usually arise from inspection and nonconformance analysis. Quality of design is only valid when design activities are closely coordinated with production. In fact, under the view of productive systems, quality of design leads from the specification of the product and service into derived specification of production and the required inputs of resources.

Design assurance within the quality assurance system is a parallel system to production and performance assurance. The following two chapters will pursue quality of design in the production and supply phase.

8.6   PRODUCT RELIABILITY

A commonly used definition of reliability is as follows. "Product reliability is the probability of a product performing its intended function over its intended life and under the operating conditions encountered."[2] This definition is from the famous AGREE Report.[3] There are a few other formal definitions that have been proposed; they differ in exact phrasing but are very similar in their general intent. The common point in all definitions is that reliability is quality, stated in terms of time. However, since no product lasts forever, reliability is a function of probability.

The above definition of reliability encompasses four associated basic factors: probability (a numerical value); performance; life (time of operation); and operating conditions (environment).

The term probability is quantitative. For example, in terms of life expectancy, when we say that the probability of the survival of a car is 80% for five years of normal driving, it means that, on the average, 80 out of 100 cars (of a given make and model) will last for five years. In the real world, however, we live with substantial deviations from the ideal design and road test conditions. In the best interest of all possible intentions, we can and do look for a probability of functioning well for the intended use. As will be shown later, different probability distributions can be used to describe the failure rate of a given product. Obviously, when a manufacturer states the product to be guaranteed reliable for two years, he would have to make it absolutely sure (again through probability shifts) that those products will function for a minimum of two years at the hands of consumers.

The second factor, quality of performance, refers to the intended function of the product. A car is not to be used for towing another car! The intended life of the product (third factor) is the time over which the product will function without failure. This in turn forms the basis of the definition of reliability. The last factor simply involves the operating conditions necessary for a product to function reliably. This is understandable, since even the best designed car will not function properly if abused by customers. On the other hand, products such as foods or pharmaceuticals must contain the prescribed or guaranteed amounts of the contents desired and must contain none or at least not more than a permissible amount of undesirable species. Correct labeling also comes into the picture as a part of reliability in food and drug administration.

It becomes obvious that to achieve high reliability in products, manufacturers need to define their reliability tasks in the form of what are known as reliability programs. Many of us will rightly argue that there is usually no way to guarantee (for the world we live in) in absolute terms the functioning of any product. However, considering all circumstances, what we can do is to make the probability of functioning sufficiently high for consumers.

A typical reliability program includes the following activities.

1. Choosing, specifying, and procuring reliable parts.
2. Selection of suppliers.
3. Reliability testing.
4. Consideration of environmental factors.
5. Worst-case analysis.
6. Institution of failure analysis programs.

7. Corrective action system.
8. Design review.
9. Quality control of manufacturing.
10. Failure mode and effect analysis.
11. Prediction and life testing.

8.7 TERMINOLOGY AND CONCEPTS IN RELIABILITY STUDIES

Since the definition of reliability includes quantitative probability determinations, we start by defining some critical terms often used as reliability indexes.

Failure Rate

The failure rate, λ, is defined as the number of failures in a given time interval.

$$\lambda = \frac{\text{number of failures}}{\text{total unit operating hours}}$$

Failure rates are normally expressed in three ways: failures per hour; percent failures per 1000 hours; and, failures per 10^6 hours. Alternatively, failure rate may also be in failures per million miles, failures per reversal, and so on. Irrespective of whichever way the failure rate is expressed, it is easier to convert it from one form to the other by using the factors given below.

To Obtain Failure Rate in	Multiply by	If Failure Rate Is Given as
Failures/h	10^{-5}	% failures/1000h
% failures/1000h	10^5	Failures/h
Failures/$10^6 h$	10^6	Failures/h

Failure rate depends on factors that are directly or indirectly related to engineering properties (mechanical, electrical, metallurgical, thermal, etc.) of materials or parts out of which the product is made up. Realiability engineers are guided by various handbooks and government–industry data banks available for use in prediction of reliability.[4] There are many more considerations which the designers must take care of in reliability predictions, such as safety factors, past history data, industrial experience of the identical setups and so on. Because of the environmental sensitivity including sensitivity to thermal, electrical, corrosion effects, and mechanical vibration, the reliability of functioning is of great importance. The reliability standard for the electronic equipments is the well known MIL-HDBK-271C[5] that deals with the reliability problem and limitations of reliability predictions. Another military standard, MIL-STD-781B, outlines test levels and test plans for reliability qualification (demonstration), reliability acceptance, sampling tests, and longevity tests.

Mean Time Between Failures (MTBF)

The mean time between failures (MTBF) is defined as the mean (or average) time between failures of a repairable product. For many products that can be repaired after each failure, MTBF is a useful measure of reliability, but it is not correct for all applications. Generally speaking, MTBF is the reciprocal of the failure rate with an appropriate conversion factor.

$$MTBF = \frac{1}{failure\ rate,\ \lambda} \times (conversion\ factor)$$

The following table summarizes three of the most frequently used conversions.

Failure Rate, λ	MTBF (in hours)
Failures/h	1/failures per h
% failures/1000h	10^5/% failures per 1000h
Failures/$10^6 h$	10^6/failures per $10^6 h$

Mean Time to Failure (MTTF)

The mean time to failure (MTTF) is the term defined as the mean or average time to failure of a nonrepairable product, or the mean time to first failure of a repairable product. Another term, MTFF, is also common. It designates the mean time to first failure of a repairable product.

Failure Pattern: The Three Periods of Equipment Life

Suppose that a piece of equipment is put to a given test and is run until it fails. The time to failure is recorded, the equipment is repaired, run again, and the next time to failure is recorded. When this procedure of repair, test, and record is repeated numerous times, a plot of collected data between failure rate and time, often behaves in a familiar pattern of failure known as the *bath tub curve* (Figure 8.12).

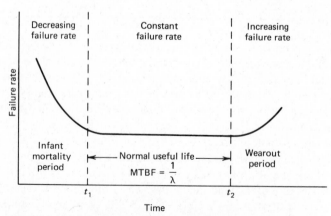

FIGURE 8.12 Failure rate versus time curve.

Three major periods are distinguishable. The first of these is *the infant mortality period.* During this period, the failure rate starts out as high because of the presence of weak or substandard components or owing to poor quality design, misuse, and so on. After one repair and another, the failure rate keeps dropping until a relatively low constant level is obtained at time, t_1. In practice, it it possible to simulate this range in test (debug) or overstress the equipment (in electronic equipment this is called "burn-in") to find the beginning of a useful life at t_1.

The second period distinguishable is *the useful life period.* During this period, only random (unpredictable) failures occur because of the absence of weak and/or defective components that were replaced in the period of infant mortality. It is this period (between t_1 and t_2) during which the concept of MTBF applies.

During the third period, called *the wear-out period,* as more and more components wear out or become used up, the failure rate increases rapidly.

If the number of failures that occur during the wearout period is plotted against time, a normal distribution type curve is obtained. The failures of all components during this wear out phase are assumed to be caused by wear out. The so-called mean-life of a component is taken as the mean of the normal curve for the wear out period. *Longevity* then, is simply the wear out time for the part or component. When the failure rate is constant the distribution of time between failures is similar to that given by the exponential distribution. The properties of equipment failures using parameters such as MTBF, MTTF or, TBF (time between failures) can thus be studied using the characteristics of these probability distribution functions.

8.8 METHODS OF PREDICTING RELIABILITY

The Exponential Distribution for Reliability

When the number of failures that occur in equal time intervals remains constant, the probability of survival, P_s (or the reliability, R) may be predicted by the exponential distribution. If the frequencies are expressed in relative frequencies, they become estimates of the probability of survival. The probability of survival (that is, when no equipment failure occurs) which gives a measure of reliability, R, is expressed by:

$$P_s = R = e^{-\lambda t} = e^{-t/\mu}$$

where:

e = Base of the natural logarithm (= 2.718).

t = Specified period of failure-free operation.

λ = Failure rate.

μ = Mean time between failures (MTBF) = $\dfrac{1}{\lambda}$.

In terms of MTBF, the above equation can be rewritten as:

$$P_s = R = e^{-t/\text{MTBF}}$$

Figure 8.13 describes the probability distribution, reliability curve, and the failure rate curve for the exponential distribution.

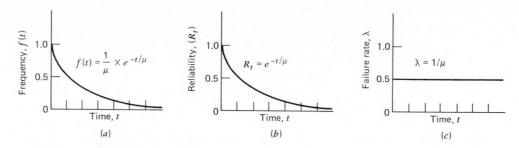

FIGURE 8.13 Probability distribution, reliability curve, and failure rate curve pertaining to the exponential distribution. (a) Relative frequency distribution. (b) Reliability as a function of time. (c) Failure rate as a funtion of time.

Probability of Survival of Series Systems

If there are two or more components (or subsystems) connected to each other in series and we have a situation where all of them are critical to system survival, the total probability of survival of the whole system can be found from: $P_{\text{system}} = P_1 \times P_2 \times P_3 \times, \ldots, \times P_n$ where, $P_1, P_2, P_3, \ldots, P_n$ are the probabilities of survival of individual items (or components).

For example, if a stereo system has four speakers with reliabilities of 0.986, 0.990, 0.950, and 0.970, then the total speaker system reliability is 0.899. Notice that the reliability of the system as a whole has decreased. This fact is well illustrated by the diagrams in Figure 8.14 which show that the fewer the components (n = number of components), the better the reliability. Calculating the system reliability in this case, however, is based on certain assumptions: that the reliability of any one speaker does not depend on others; and, that the failure of any one speaker will cause the system to fail. Although these assumptions may not be valid in actual practice, they give a very convenient approximation for the interre-

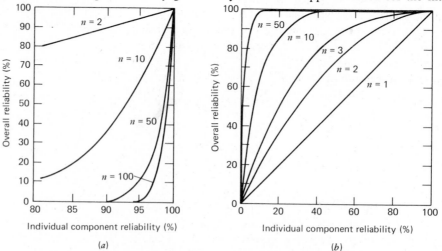

FIGURE 8.14 Relation of part and system reliability. (a) Series system reliability. (b) Series parallel system reliability. Source: Meredith, Jack R. and Gibbs, Thomas E. *The Management of Operations*, 2nd ed., John Wiley & Sons, Inc., New York, 1984, Reprinted by permission.

lationship behavior when many components are involved in a system. Furthermore, since $R = e^{-t/\text{MTBF}}$, it will also be noted that an increase in MTBF does not result in a proportional increase in reliability. Students should test this for themselves as an exercise, say, by assuming $t = 1h$ and MTBF of 5, 10, 20, and 100 hrs.

Probability of Survival of Redundant Systems

When many parts or subsystems are connected in a series, all of them must function together for the overall system to function. Sometimes, however, designs can be made so that the failure of one part will not cause the whole system to fail. This is done by the use of redundancy. With the help of appropriate probability calculations, the designer can predict in quantitative terms the effect of redundancy. Using redundant parts (or subsystems) either in parallel or series-parallel combinations, it is possible to obtain a configuration where all elements must fail before there is an overall failure of the system.

Parallel Systems

Figure 8.15(a) depicts a parallel redundant system. For the overall system to function, either A or B, or both, must function. Recalling the probability theorems outlined in Chapter 6, the reliability of this system is:

$$P_{A \text{ and/or } B} = P_A + P_B - P_{A \text{ and } B}$$

or $= P_A + P_B - P_A P_B$

If A and B are identical subsystems, $P_A = P_B = P$ (say) $P_{A \text{ and/or } B} = P(2 - P)$ which can be generalized as:

$$P_{\text{system}} = 1 - Q^n$$

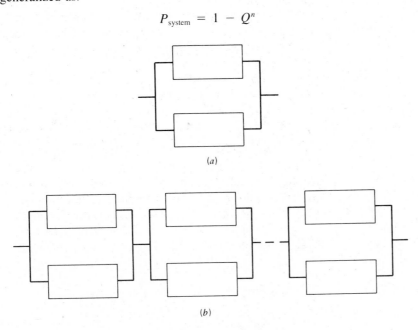

(a)

(b)

FIGURE 8.15   Arrangement of components in parallel redundancy. (a) Parallel arrangement. (b) Combination arrangement.

where:

P_{system} = Reliability of the system.
Q = Unreliability of each parallel subsystem = $1 - P_{\text{system}}$
n = Number of parallel subsystems.

Note that during the discussion on the binomial probability distribution (Chapter 6) the terms p and q were used to denote successfulness (equivalent to survival of the system) and unsuccessfulness of an event (equivalent to failure). If we have n number of parallel and identical subsystems, $(p + q)^n$ would include all possible events, and its sum would equal 1.0. Therefore, $P_{\text{system}} = 1 - Q^n$ represents all the binomial terms representing success (probability of survival or reliability) of the parallel subsystem.

Series-Parallel Systems

When a combination of series and parallel configuration is used as shown in Figure 8.15(b) which includes n identical pairs, the overall reliability of the system is:

$$P_{\text{system}} = [1 - (1 - P)^2]^n$$

where:

P = Reliability of the individual elements.
n = Number of identical pairs.

Figure 8.14(b) shows the variation in overall reliability (R) of this system.

Example: Suppose that an electronic system has a relay that opens a circuit under certain conditions with a reliability of 99.5% for a specified mission time. We may place another design of a relay from a different supplier with the same purpose in parallel redundancy, but having a reliability of 0.996. The overall reliability of the system is $R = 1 - (1 - 0.995)(1 - 0.996) = 0.99998$ or, 99.998%.

Now we place the second relay in series with the first one, such that under critical undesirable conditions, if either one or both relays open the circuit, mission is accomplished.

The overall reliability of the system in this case is given by the condition of both failing to open the circuit. Assuming that the failure of the two are independent events, probability of failure is $(1 - 0.995) \times (1 - 0.996) = 0.00020$; so that reliability of the two in series is 0.99998.

We have seen in the above discussion that the prediction of reliability is done on the basis of probability theory and prior information on failure rate data for various circuits, configurations and so forth. Thus, when the failure rate is experimentally or otherwise found constant, the exponential distribution applies, and the reliability prediction of a system can be made, simply by adding the parts' failure rates. This can be deduced by writing:

$$P_s = e^{-(t_1\lambda_1 + t_2\lambda_2 + \ldots + t_n\lambda_n)}$$

$$= e^{-\Sigma t\lambda}$$

If t is the same, and failure rates are given in terms of MTBF, then:

$$P_s = e^{-t\Sigma 1/\text{MTBF}}$$

Standby Systems: R = Constant

If one of the subsystems in parallel configuration is designed for standby purposes (so that one or more subsystems are standing by to take over operation when the unit fails) and its R is the same as the basic unit, Poisson distribution may be used to determine the reliability. Recalling the general formula for the probability equation with n standby units:

$$P_n = e^{-\lambda t}\left(1 + \lambda t + \frac{(\lambda t)^2}{2} + \frac{(\lambda t)^3}{6} + \frac{(\lambda t)^4}{24} + \ldots\right)$$

where:

$$e^{-\lambda t} = \text{Probability of no failures (zero failure).}$$

$$e^{-\lambda t} + \lambda t \times e^{-\lambda t} = \text{Probability of 0 and 1 failures.}$$

$$e^{-\lambda t} + \lambda t \times e^{-\lambda t} + \frac{(\lambda t)^2}{2} \times e^{-\lambda t} = \text{Probability of 0, 1, and 2 failures.}$$

$$e^{-\lambda t} + \lambda t \times e^{-\lambda t} + \frac{(\lambda t)^2}{2} \times e^{-\lambda t} + \frac{(\lambda t)^3}{6} \times e^{-\lambda t} = \text{Probability of 0, 1, 2, and 3 failures.}$$

Using the λt characteristic of Poisson distribution, the probability of more failures can be computed. In terms of values given in MTBF, (MTBF $= 1/\lambda$) when $n - 1$ units are in standby we can calculate the MTBF of the system from MTBF $= n/\lambda$ where $n = 2$ for one standby unit and $n = 3$ for two standby units.

If the MTBF between successive events are independent and identically distributed according to an exponential distribution, then the number of events that occur during the interval $(0,t)$ has a Poisson distribution with a mean value of λt, where λ is the failure rate.

Example: Suppose that the failure of an electronic module in a computer system is such that the intervals between successive events are independent and identically distributed, with a failure rate, $\lambda = 4$ per hour. Since the exponential distribution applies, the number of events in $(0,2)$ time interval has a Poisson distribution with expected number of events, $\lambda t = 4/h \times 2h = 8$. Writing the general Poisson expression as:

$$P(n) = \frac{(\lambda t)^n \times e^{-\lambda t}}{n!}$$

a probability distribution can be plotted for this case by computing $P(n)$ for $n = 0, 1, 2, 3, \ldots$, and taking $\lambda = 4$ and $T = 2$. This plot is shown in Figure 8.16.

Example: Consider a nickel-cadmium rechargeable battery used in an equipment setup to monitor leakage of ammonia gas in a refinery. The low charge in the battery is indicated by a flash of red light built into the system. Let us assume that the battery is put into operation at time $t = 0$. Each time a failure occurs the failed battery is replaced by a new one and the old one recharged. If the life lengths of the batteries are independent and exponentially distributed with failure rate $\lambda = 4/\text{year}$, then the probability that the number of failures, n, in one year will be less than or equal to 4 is $P(n \leq 4) = P_{(n=0)} + P_{(n=1)} + P_{(n=2)} + P_{(n=3)} + P_{(n=4)} = 0.639$.

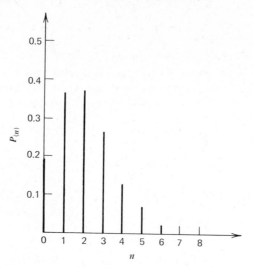

FIGURE 8.16 A plot of the probability distribution for failures of an electronic module.

Standby Systems: Varying R

In the above formulations we assumed that the basic and the standby unit(s) have the same λ, MTBF or reliability values. This may not be true in actual practice since parts are supplied by many different manufacturers using their own production practices and also since human factors are involved at every stage. For these reasons, it is essential to determine confidence limits for the known individual subsystem's reliability. When it cannot be assumed that the subsystems have 100% stated reliability, the equations for constant R given in the previous section can be modified to incorporate the known reliability value. For example, the probability of 0 and 1 failures with one single standby unit with a known reliability of R_1 will be arrived at by:

$$P = e^{-\lambda t} + (R_1)(\lambda t) \times e^{-\lambda t}$$

The Weibull Distribution for Reliability

When the number of failures that occur in equal time intervals does not remain constant, predictions based upon simple additions of components' failure rates are not truly valid. Instead, an alternative approach based on the Weibull distribution is used. Some common examples where the Weibull distribution has been used are for ball bearing failures, electronic equipment failures, fatigue failures, and vacuum tube failures. In analyzing items such as these where one is definite about the possibility of the bathtub curve behavior (Figure 8.12), predictions of reliability based on Weibull distribution serve the purpose quite adequately. Considering the bathtub curves one would, for example, find a period of low but improving reliability during the infant mortality period. Reliability would be expected to improve and indeed be high during middle age and to reduce again during the old age period. Analysis of reliability during these periods necessitates the use of a common formula to describe these three different curves. The Weibull distribution is often used for this purpose.

The Weibull density function is given by (Figure 8.17a):

$$f(t) = \alpha\beta t^{\beta-1} \times e^{-\alpha\beta t}; \quad f(t) = f(x - \gamma)$$

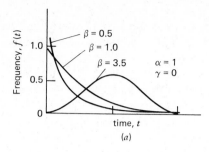

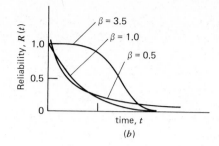

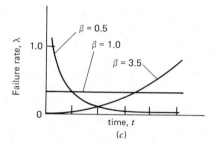

FIGURE 8.17   Typical Weibull plots for (a) Weibull frequency, (b) reliability, and (c) failure rates as a function of time. The curve varies greatly, depending upon the numerical value of the parameters α, β, and γ. This makes it popular in practice because it helps decide the best fit of data with rather easy manipulation.

where:

α = A scale parameter (related to the peakedness of the curve such that as α increases, the curves become more peaked).

β = A shape parameter (called Weibull slope, which reflects the shape of the curve).

γ = A location parameter (a constant on the time axis, assumed often as zero to simplify the equation).

Its generalization for reliability is equivalent to:

$$R(t) = e^{-\alpha\beta t} \qquad \text{(Figure 8.17b)}$$

and to:

$$\lambda(t) = \alpha\beta t^{\beta-1} \qquad \text{(Figure 8.17c)}$$

for failure rates.

Suitable choices of the three parameters (α, β, and γ) enable the Weibull distribution to be used to fit or simulate each of the three phases of failure curves. One such choice is that shown in Figure 8.17a. Note that when β is 1.0, the Weibull function reduces to the exponential and that when β is 3.5 (and $\alpha = 1$, $\gamma = 0$), the Weibull closely approximates the normal distribution.

Making predictions based on the Weibull distribution first requires the determination of the three constants α, β, and γ. Although procedures for graphical determinations of these constants are available,[6] it is most easily done by the use of Weibull probability graph paper, which has a logarithmic scale on both axes. Plotting the cumulative percent failure numbers on the y-axis (vertical) and the time taken in such a number of test parameters on the x-axis (horizontal), one expects a straight-line relationship.

For an acceptable level of confidence in the validity of the Weibull plot, at least seven points are needed. When $\gamma = 0$, the cumulative percent failures plotted on the graph will fall on a straight line, the slope of which gives the value of β. A knowledge of β gives the shape of the distribution curve which indicates whether failures are occurring in the infant mortality phase, or middle or old age periods. A value of $\beta < 1$ indicates the infant mortality period, whereas if $\beta = 1$, it represents constant failure stage or middle age and $\beta > 1$ is an indication of the wear out period or old age. If, in plotting points, a straight line does not result ($\gamma \neq 0$), the points must then be replotted against a new time axis. A quality engineer's hope that a straight line will always be obtained merely reflects a single stable population that may have come out from one known lot of production or one source. However, many times a nonstraight line gives clues to the hidden quality problems. Still, on other occasions, an approach called extrapolation (of a straight line) with only a few points available, is used to forecast early failures in a warranty of only vital parts in a complex system.

Example: To illustrate the use of the Weibull plot, consider the data of the following table from a system test for a specific type of electronic package.

Failure Number	Cumulative Percent Frequency	Number of Cycles to Failures ($\times 10^3$)
1	20	2.3
2	22	3.0
3	23	4.0
4	25	5.0
5	30	9.8
6	32	12.5
7	36	19.0

The cycles to failure are plotted on a piece of Weibull graph paper against the corresponding values of cumulative percent failures in Figure 8.18 from which a Weibull slope of $\beta = 1.0$ is obtained suggesting that the Weibull distribution applies and that the failure pattern for the product corresponds to middle age. In practice, β may vary from 1/3 to 5 depending upon the period or stage of the failure pattern.

8.9 STATISTICAL PROCEDURES FOR RELIABILITY TESTING

From basic reliability and safety theory, if failures occur randomly then, as we have seen, the time between failures follow an exponential distribution. As long as random failures occur and have no assignable cause as to their occurrence, all extractable information pertaining to the exponential distribution is applicable. As a result, most of the statistical procedures for reliability testing are based on the exponential form of distribution. If the

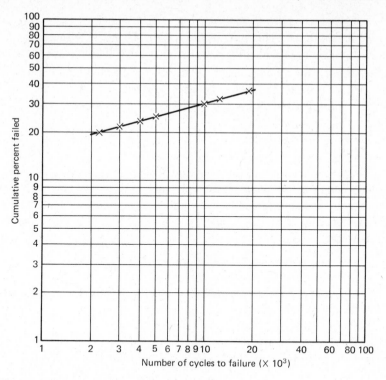

FIGURE 8.18   The Weibull plot of failure data.

failures are due to some abnormal environmental condition (such as heat build-up) then the curve between frequency and time between failures (TBF) will show some form of peak characteristics in its distribution (Figure 8.19). The statistical tests about to be described are essentially tests for an accept/reject criterion that screens the expected perturbations from the unexpected ones.

Reliability tests are basically of three types.

1.   *Failure-terminated tests:* In these testing schemes, reliability life tests are terminated upon occurrence of a preassigned number of failures.

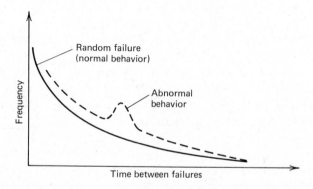

FIGURE 8.19   Schematic illustration of "random" and "caused" failures.

2. *Time-terminated tests:* This type of life testing scheme is terminated upon reaching a predetermined test time, provided the specified number of failures did not occur before.

3. *Sequential tests:* In sequential testing plans neither the number of failures nor the time required to reach a decision are fixed in advance. Instead, the accumulated test results for the life test are used to reach a decision whether to accept the product, reject it, or continue testing it for more confidence.

Failure-Terminated Tests

The failure-terminated tests in which the reliability life testing are terminated upon occurrence of a preassigned number of failures fall into two categories: testing conducted with replacement of failed unit, and testing conducted without replacement of failed units. For tests conducted with replacement of failed items, the acceptability of a lot is ascertained when $\Theta' \geq k \times \Theta_0$ where:

$\Theta' =$ The estimated mean life of the lot $= nt_F/r$.
$\Theta_0 =$ The mean life specified for the lot.
$k =$ A factor obtainable from published tables, for example, Department of Defense Handbook, H108.
$n =$ Sample size.
$t_F =$ Time when the preassigned number of failures occurred.
$r =$ Number of failures which terminated the life test.

The larger the termination number r, the greater the assurance of having accepted more reliable units.

When tests are conducted until failure but without replacement of failed units, the acceptability criteria remains the same except that the Θ' is calculated by using a different formula.

$$\Theta' = \frac{1}{r} [t_1 + t_2 + t_3 + , \ldots , + (n - r)t_F]$$

where $t_1, t_2, t_3, \ldots$ are the times respectively of first, second, and third failures.[7]

Time-Terminated Tests

The acceptability of a lot in this type of test is determined by the time required for a predetermined number of failures, r_F. A lot is acceptable if the predetermined number of failures has not yet occurred before termination time, T is reached.

The termination time, T is determined by $T = k \cdot \Theta_0$ where k is a constant to be found from tables in the military handbook H-108 (see note seven for this chapter) for known values of a producer's risk factor, α. When the probability of acceptance of lots, having different ratios of actual mean life to acceptable mean life, is to be found, one can use the operating characteristic curves given in the handbook 108A.

The *alpha* risk, as shown on the OC curve applied in acceptance sampling (see Chapter 12) measures probability of rejecting a lot on the basis of the sampling plan that should have been accepted; it is called the *producer's risk*. The *consumer's* or *beta* risk measures acceptance of a lot that should have been rejected.

Sequential Tests

The sequential reliability tests are planned with producers' and consumers' risks (α and β, respectively) in mind. The method of arriving for a decision using this plan can be best illustrated with the help of Figure 8.20.

The test begins at the origin of the graph. Starting with the first failure, as the test continues consecutive failures are plotted on the graph versus time in a continuous stairlike fashion. The testing is continued until the line crosses either the accept or reject line and a decision is made accordingly whether to accept the lot or reject it.

To determine the acceptability of the MTBF of equipment, four numerical quantities are selected.

1. Θ_0, The specified MTBF (note the different meaning of the symbol Θ_0 which was called "mean life" in the previous section).

2. Θ_1, The minimum acceptable MTBF.

3. α, The producer's risk corresponding to the probability that the plan will reject the unit with an acceptable MTBF value (Θ_0).

4. β, The consumer's risk, corresponding to the probability of accepting equipment with the minimum acceptable MTBF value (Θ_1).

The accept line and reject line are defined by the following equations.

$$
t_1 = \frac{\ln\left(\dfrac{\Theta_0}{\Theta_1}\right)}{\dfrac{1}{\Theta_1} - \dfrac{1}{\Theta_0}}(r) - \frac{\ln\left(\dfrac{\beta}{1-\alpha}\right)}{\dfrac{1}{\Theta_1} - \dfrac{1}{\Theta_0}} \qquad \text{(accept line)}
$$

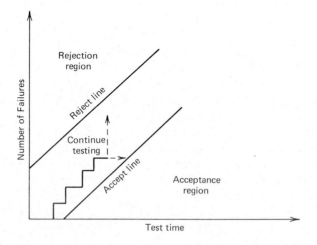

FIGURE 8.20   Illustration of a sequential testing plan.

and

$$t_2 = \frac{\ln\left(\frac{\Theta_0}{\Theta_1}\right)}{\frac{1}{\Theta_1} - \frac{1}{\Theta_0}} (r) - \frac{\ln\left(\frac{1-\beta}{\alpha}\right)}{\frac{1}{\Theta_1} - \frac{1}{\Theta_0}} \qquad \text{(reject line)}$$

where:

r = Cumulative number of failures.

$\dfrac{\ln\left(\dfrac{\Theta_0}{\Theta_1}\right)}{\dfrac{1}{\Theta_1} - \dfrac{1}{\Theta_0}}$ = Slope of the accept and reject lines.

$\dfrac{\ln\left(\dfrac{\beta}{1-\alpha}\right)}{\dfrac{1}{\Theta_1} - \dfrac{1}{\Theta_0}}$ = Time-axis intercept of the accept line.

$\dfrac{\ln\left(\dfrac{1-\beta}{\alpha}\right)}{\dfrac{1}{\Theta_1} - \dfrac{1}{\Theta_0}}$ = Time-axis intercept of the reject line.

Some common values of $\ln(\Theta_0/\Theta_1)$, $\ln[\beta/(1-\alpha)]$ and $\ln[(1-\beta)/\alpha]$ are given in Tables 8.1 and 8.2.

8.10 COMPUTER ASSISTED DESIGN, MANUFACTURING, AND QUALITY ASSURANCE

Computers have been employed in product and production design for many years. Numerical control (NC) of machinery, the early application of electronics on the factory floor, has now evolved into comprehensive computerized product design and production planning. Quality assurance becomes a natural component in these modern systems.

Computer-assisted design and computer-assisted manufacturing (CAD/CAM) integrates all functions from the product idea to the delivery and servicing of the product, that is, it accompanies the entire life cycle of a product. All planning and control activities focus, with this computer information support, on the maintenance and improvement of product quality and adequate performance in all major life stages.

This information and support process leads to greater creativity, adaptation to customer needs, and new competitive strength. Substituting manual drawing of prototypes and myriads of parts through computer graphics is not only more cost- and time-effective, it also relieves designers and planners. Transfer from graphics to the physical production of the prototype

TABLE 8.1

Table of Values of $\ln\left(\dfrac{\Theta_0}{\Theta_1}\right)$

$\dfrac{\Theta_0}{\Theta_1}$	$\ln\left(\dfrac{\Theta_0}{\Theta_1}\right)$
1.0	0.000
1.2	0.182
1.4	0.365
1.5	0.405
1.6	0.470
1.8	0.588
2.0	0.693
2.2	0.789
2.4	0.876
2.6	0.956
2.8	1.030
3.0	1.099
3.5	1.253
4.0	1.387

Θ_0 = specified MTBF.
Θ_1 = minimum acceptable MTBF.

can be done with the push of a button. Testing can then determine the actual adequacy of the new or modified product. The cycle can be repeated until the product appears to meet all requirements and anticipated hazards and risks have been dealt with.

Human beings create this powerful software and hardware, are served by it, and apply and modify it. Nevertheless these systems gradually expand and become widely self-sup-

TABLE 8.2

Table of Common Values for $\ln\left(\dfrac{\beta}{1-\alpha}\right)$ and $\ln\left(\dfrac{1-\beta}{\alpha}\right)$

	$\ln\left(\dfrac{\beta}{1-\alpha}\right)$	$\ln\left(\dfrac{1-\beta}{\alpha}\right)$
$\alpha = 0.05$ $\beta = 0.05$	-2.960	2.944
$\alpha = 0.05$ $\beta = 0.10$	-2.255	2.890
$\alpha = 0.10$ $\beta = 0.05$	-2.905	2.251
$\alpha = 0.10$ $\beta = 0.10$	-2.197	2.197

α = producer's risk.
β = consumer's risk.

porting. It can be expected that the assistance they render to designers and planners will nibble away traditional manual activities and individual decision making. However, greater variety in products, faster modifications, and rising customer quality demands, will always require human design and planning participation. This is particularly true with regard to assuring quality of the greater product and service spectrum.

Quality of products and services will potentially and actually be enhanced through computer assisted design and production, because proper functioning of the software and hardware is a prerequisite for quality assurance. Once the quality and adequacy of the computer system is assured, scope for human errors in the computer assisted design and planning is also obviously reduced. The computer as assistant, takes care of error avoidance to a large degree.

When human error and deviation from set design and production standards should still occur, the computer will readily monitor these, and possibly correct the errors automatically. The wider the computer assistance penetrates from design to actual production and product servicing activities, the more quickly and effectively are nonconformances monitored, corrected, and recurrences prevented. The quality of the software and hardware, and the participation of quality assurance experts at this stage, will largely determine the degree to which quality assurance is built into the design and production.

Data on specification, for instance, generated on the CAD system's video screens are automatically and immediately available, so that they can be verified by all concerned in design, production, and servicing. Once modifications become necessary in view of defects and failure, or even before they do, they can readily move through the various decision-making and action channels. Specifications are always up-to-date and accessible.

The communication at all times is multidirectional and the information and decision systems need to incorporate the quality assurance system. Thus, the development and existence of a well-designed quality assurance system with integrated procedures for assuring quality of design, production, and performance is a prerequisite, before computer assistant can function properly. As a result of the natural evolution of integrated CAD/CAM, explicit quality assurance using computers, or better termed computer assisted quality assurance (CAQ) has emerged as a strong design and planning partner.

Computer assisted quality assurance will not only focus on monitoring nonconformances on the shop floor and initiate corrective action; starting from the assurance of the software and hardware; it will assure the quality and integrity of the design, as has already been mentioned. Yet, the CAQ component should not be thought of as implicit only to design and production, as it has often been understood in the past; CAQ allows quality assurance to develop as an independent, albeit fully integrated, computerized subsystem and special information and control vehicle.

As a subsystem, CAQ will incorporate all traditional quality assurance techniques, and in particular, statistical process capability evaluation during design and process control, and during production and servicing. Moreover, for the first time, CAQ induces the application of other special management methods, such as forecasting, simulation, material requirement planning, "just-in-time" modes of operation and so on. The close interface of CAQ with the parallel CAD and CAM systems will have a considerable impact on management of quality assurance. Total quality assurance in all phases of the operation and at all functions and managerial levels will evolve, with quality assurance experts playing the role of quality assurance program designer and maintainer.

8.11 KEY TERMS AND CONCEPTS

Design of product and service are all specifications designed to satisfy customer needs within the limitations of technology and economy. The design is developed through systematic design procedures and activities.

Products or **services** are defined through specified attributes and measurements, which individually and jointly have utility and value for the customer and therefore have a price in the market. Product or service quality is defined through specifications.

Quality of design is the degree to which the product or service meets the expectations of customers and creates satisfaction.

Design assurance is quality assurance of the design during all development stages. Design is the early development stage in the product life cycle.

Specifying quality means to establish desired, measurable, or assessable attributes of a product or service. A specification (sometimes called quality characteristic) describes as clearly and completely as possible all the related hazards and defects, and tests for verification of the specification in the product or service. Design review and production capability studies help to arrive at technically and economically sound specification.

Standards related to a product and its specifications are those that are approved and generally accepted specifications, partly determined by mandatory impositions.

Tolerances are acceptable variations around a nominal value specified for a product or service.

A **defect** is an unsatisfactory attribute of a product or service. It can lead to failure in performance. A defect being a deviation from a specification is also called a nonconformance. Design assurance aims at anticipating defects and hazards and preventing or counteracting them.

A **failure** is a defect that relates to reliability. Reliability is failure-free performance of an item during a specified period under required conditions.

Failure mode and effect analysis and **fault free analysis** are two important techniques in a reliability engineering and failure prevention scheme.

A **design review** confirms the adequate quality and integrity of a design and design process, or initiates modifications. The review is a formal and systematic evaluation conducted by independent, knowledgeable staff, reporting to senior management. The review often includes a contract review in order to confirm ability to comply with contractual obligations.

8.12 SUPPLEMENTARY READING

Japanese and Western Quality—A Contrast[8]

Color TV set quality depends heavily on component quality. All manufacturers know this, and all go to great lengths to structure their vendor relations in ways which will assure the adequacy of component quality. The methods used for attaining this assurance have been widely published, but there remain important differences in emphasis between Japan and the West.

The most fundamental of these differences is in the policy followed as to the basic relationship between vendor and buyer. This relationship can range over a wide spectrum. At one extreme is the adversary concept, i.e., mutual doubt if not mistrust; reliance on contract provisions, documents, penalties and other elements of an arm's length relationship. At the other extreme is the teamwork concept, i.e., mutual trust and confidence, cooperation, joint commitment to the consumer, etc. This policy on basic relationship is of the utmost importance. It is decisive on such questions as extent of joint planning, mutual visits, technical assistance, exchange of data, etc.

The large Japanese manufacturers are heavily committed to the teamwork concept. So, to a considerable degree, are the Europeans. In the U.S. the manufacturers are emerging from an era in which the adversary concept was the rule. The directions for the future all seem to point to a policy of treating the vendor as a member of the team. In my judgment, the adversary policy is no longer a viable basis for vendor relations on products such as the color TV set.

Beyond this basic policy question, there remain numerous important elements of practice and technique.

1. *Design qualification.* Before component designs are "listed" they are commonly required to undergo environmental tests and life tests. These tests are similar in concept to those discussed above under "Qualification Testing." (Such testing is done whether the designs originate with the buyer or the vendor.) The resulting design qualification is quite separate from the problems of vendor or process qualification.

2. *Vendor selection.* To a considerable extent, color TV set manufacturers "buy" components from sister divisions in the same company. However, the residue which must be bought in the competitive market place is also considerable, and many independent component makers compete with each other for the available business. To reduce these many to a manageable few, the color TV set manufacturers make use of a vendor selection process and apply criteria in accordance with their policies and objectives. Wide use is made of questionnaires and preliminary screening to eliminate would-be suppliers who are clearly deficient on grounds of finance, facilities, etc. Following this screening, it is common to conduct "surveys", i.e., visits to vendors to determine in greater detail whether the vendors meet the buyer's criteria for quality, delivery schedule, price, etc.

3. *Criteria for predicting quality performance.* All surveys involve a look at production and laboratory facilities to judge their adequacy for production and test. Beyond these and other technological essentials, there is a divergence. In the U.S. great emphasis is placed on organization form, written procedures, manuals, audits, documentary proof, etc. In contrast, the Japanese emphasis is on the following:

(a) *Process "validation,"* i.e., quantifying process capabilities so that is is known in advance whether the vendor's processes can hold the tolerance.

(b) *Process controls,* i.e., the plans of control which will be used to keep the processes doing good work.

(c) *Management policies,* i.e., the vendor's willingness to operate in a teamwork atmosphere requiring mutual confidence, exchange visits, technical assistance, improvement programs, etc.

(d) *Training in quality control,* i.e., the extent to which the vendor's managers, supervisors, and the workforce have been trained in quality control methodology. (Deficiencies are remedied by conducting training courses for the vendor's people, in all levels.)

(e) *Quality of prior deliveries* for similar products.

These differences in emphasis are of course traceable back to the basic policy on vendor relations, i.e., adversary vs. teamwork.

4. *Joint quality planning.* The ingredients of joint planning are universal, but again there are important differences in emphasis. Generally there is little difference in practice with respect to such matters as measurement correlation, lot traceability, or exchange of test data. The differences are mainly in the following:

(a) *Use of AQLs.* In the U.S. the contracts embody agreed levels of AQL ("acceptable quality level"). These AQLs, once agreed on, serve to unify the parties on the sampling to be used as well as providing a basis for adjudicating of disputes, e.g., if the AQLs are exceeded, the buyer has a clear basis for a claim. Under this approach, many U.S. manufacturers tend also to view the AQLs as tolerable limits. This view is reinforced by some of the efforts which have been made to standardize AQLs in the industry.

The Japanese approach is quite different. They take the view that the AQLs tend to reduce the incentive for quality improvement. Hence while for contract purposes they may quantify AQLs, their real purpose is to find the best vendor, not merely one whose AQL is competitive.

The reasoning behind the Japanese view is evident from two major imperatives of color TV set reliability. In the first place, component failures have been the major cause of color TV set field failures (and of in-house "fall off" rates). Secondly, the number of components in the color TV set runs to over 1000. Such numbers can convert component defect rates of the order of 0.1% into set failure rates of the order of 100%. This multiplying effect has been a factor in the Japanese use of "ppm" (parts per million) as a unit of measure for component quality.

Clearly, for defect rates in the range of ppm, the AQL concept is useless. In the design qualification stage, resort must be had to overstressing, environmental testing, etc., to bring the component weaknesses out into the open. (In addition, there must be a strict discipline of discovering the precise failure modes so that remedial action can be taken.) Subsequently, during production, defects at the ppm level can be dealt with only on an automated sorting basis.

(b) *Inspection data feedback to vendors.* In the U.S. wide use is made of computerized data systems to provide manufacturers with summaries of vendor performance. These data are then used for tracking corrective action as well as for stimulating vendor improvement. In addition, there is some use of vendor rating systems for judging the comparative quality performance of vendors. The resulting ratings are usually made known to vendors and become the basis of awards for superior performance as well as penalties for inferior performance. Vendors are expected to share in the initiative to make the quality improvements needed to improve their ratings.

(c) *Product performance feedback.* The Japanese emphasis is on failure analysis, during assembly and in the field. These analyses go down to the precise failure mode. The resulting findings, along with samples of failed components, are presented to the vendor as part of the program of technical assistance. There is little doubt that the Japanese have secured much improvement in quality of components through their close working with vendors in failure analysis.

Manufacture

There are hard data available to suggest that as to internal failures (discovered in the factory) the Japanese have an overwhelming superiority over the West.

Prior to coming under Japanese management, the Motorola factory ran at a "fall off" rate of 150 to 180 per 100 sets packed. This means the 150 to 180 defects were found for every 100 sets packed, or 1.5 to 1.8 per set. Three years later the fall off rate at Quasar (the new name of the factory) had gone down to a level of about 3 or 4 per 100 sets, or only about one-fortieth of the previous level. (In Japan the fall off rates are about 0.5 per 100 sets, which is more than two orders of magnitude lower than the Motorola performance.)

I have seen other, unpublished data which tend to support the above example as being typical rather than extreme.

There are also hard data to suggest that Japanese productivity in color TV set manufacture likewise exceeds that of the West, though by margins much less spectacular.

On closer examination, these two performances are seen to be complementary. They are traceable to the practices set out below:

1. *The prior design for producibility.* The benefit conferred by such design is obvious.
2. *Component quality.* This quality is derived not merely from the program of vendor relations; there is an extensive added program of *sorting purchased components* to keep the defectives out of the assembly lines (and out of the market place). I observed such a program at Quasar. It involves a considerable investment in automated testing machinery and occupies the time of numerous people to set up and tend the machines. Similar practices have been observed in other plants by visiting journalists.

   While it takes substantial investment and personnel to sort the components, the benefits are spectacular. The Motorola assembly lines required about 15 employes per assembly line to perform line inspection and repair work. At the reduced fall off rates, this number is in the range of 1 or 2. This reduction, when applied to about eight assembly lines, is far greater than the number of people associated with component sorting.
3. *Process design.* The processes used for color TV

set manufacture are quite similar from one company to another. There are differences in emphasis, however, and it is clear that the Japanese have put their prime emphasis on quality. Examples of this emphasis include:

(a) More extensive use of the "sequencer" process for automatic insertion of discrete electronic components into printed circuit boards.
(b) Design of assembly conveyors to permit each assembler to disengage the set from the conveyor. This tends to avoid the quality problems created when the worker is unable to keep up with the conveyor.
(c) Design to automate and to minimize human handling, adjustment, repair, etc. This design is evident throughout, even as to packing the final product.

4. *Process control.* Here again there are many commonalities among all manufacturers. There is a worldwide trend toward returning process inspection back to the workforce. There is some, but not conclusive, evidence that Japanese study and control of critical process variables, e.g., the soldering bath, may be more complete than in the West. In the case of feedback to the workforce, the evidence is stronger. Of course, this aspect of manufacture is strongly influenced by the general state of employe relations (see below).

5. *Final test.* All color TV sets go through a final adjustment operation and thereby each demonstrates its ability to perform its basic function. These final adjustments are typically made after operating temperatures have been stabilized. In addition, there are programs of audit on a sampling basis, consisting of extensive testing. It is usual to keep the entire day's product on hold to await the results of this audit. I do not have enough comparative data on final test practices to be able to make a useful judgment on contrasts.

8.13 DISCUSSION AND REVIEW QUESTIONS

(8–1) *A product or service must be designed with the consumer in mind.* Explain and discuss.

(8–2) *Quality assurance specialists should always participate in the design.* Explain.

(8–3) *Quality defined as "fitness for use" means that the consumer must also be involved in the design phase.* Explain and discuss.

(8–4) *Product design is a project with several distinct phases.* Explain.

(8–5) *Quality of design and quality of conformance are related and distinct.* Explain.

(8–6) *Design assurance with customer's satisfaction in mind is really only possible for custom made items.* Discuss.

(8–7) *Quality specifications are an important prerequisite for quality control and quality assurance.* Explain.

(8–8) Define "major quality specification."

(8–9) *With any quality specification, measurement, or attribute, at least one defect is associated.* Explain.

(8–10) *Design defects are the most expensive and should therefore be avoided.* Discuss and include techniques for defect analysis.

(8–11) *There are certain principles managers must adopt for arriving at a valid quality specification.* Explain.

(8–12) Define "reliability" and "failure." What does reliability mean in the cases of: a smoke alarm, kitchen stove, an automobile tire, aspirin, bus service in the city, telephone service. Differentiate between the terms "liability" and "reliability."

(8–13) *Failure mode and effect analysis* is a special technique of design assurance. Explain; be specific by using an example.

(8–14) *Specifications should be standardized whenever feasible and possible.* Explain.

(8–15) *Any design of products or services should be properly documented; these documents should also reflect design changes.* Explain and discuss.

(8–16)   *The most simple design document should include information on defects and assurance measures.* Explain.

(8–17)   *Good quality is built-in, rather than inspected in.* Discuss.

(8–18)   a.   What are the criteria of a sound design?
         b.   What are the differences in designing a product and a service?
         c.   *Product complexity and impacts of defects on quality and performance dictate the degree of quality assurance during the design phase.* Explain.

(8–19)   a.   Define "specification of quality" and contrast with "standard" and "tolerance."
         b.   *Faulty designs create the most costly defects.* Discuss.
         c.   Define "design assurance."
         d.   *Scope and objectives of design reviews vary and depend largely on preceding design assurance.* Explain.

(8–20)   a.   List major principles to be observed by design assurance.
         b.   Define and contrast: "defect," "defective," and "failure."
         c.   *Nonconformances are significant deviations from specifications.* Explain with reference to the word "significant."
         d.   *Quality control should actually be called defect control.* Explain.
         e.   *Customers should participate in a failure mode and effect analysis.* Discuss.

(8–21)   a.   *Defect and failure costs increase when not detected during the design phase.* Discuss and illustrate by means of a graph.
         b.   *Design is a continuous task until the product or service is discontinued.* Discuss.
         c.   Built-in obsolescence is no longer a viable policy; competitive benchmarking is. Discuss.
         d.   By means of a break-even model illustrate limits to design costs due to expected output.

(8–22)   a.   By means of a life-cycle curve, illustrate the time when either a redesign should be considered or other counteracting measures must be taken. What would such measures be?
         b.   List important design documents. State reasons for their importance.
         c.   Describe major steps of design change and change control.

8.14   PROBLEMS

(8–1)   a.   Describe the types of specifications commonly used in manufacturing processes.
        b.   What relationships exist between a process and its specification? How are the specification conflicts best resolved?
        c.   A machine shop produces shafts about 2 inches in diameter with a standard deviation $\sigma_a = 0.0002$ in., and another process for bearings with $\sigma_b = 0.0004$ in. Supposing that the machinists exercise good control over the quality of their output, determine the maximum and minimum diameters for shafts and bearings for a normal distribution with $\pm 3\sigma$ specification limits.

(8–2)   A steel washer and a polyethylene washer are to be assembled to make a bearing vacuum tight, one on top of the other. The average thickness and the standard deviation of the former were determined as $\mu_s = 0.1235$, $\sigma_s = 0.00035$, where as for the latter; $\mu_p = 0.0918$, $\sigma_p = 0.00153$ (all in inches). Calculate μ and σ for the combined thickness. Assuming normal distribution, find $\pm 3\sigma$ limits for the combined thickness.

(8–3)   Make a fault tree analysis of the following products.
        a.   A pen or pencil you write with
        b.   Table lamp
        c.   Freezer
        d.   Restaurant service
        e.   Any product or service of your choice

(8–4) Make a failure mode effect analysis for one of the products mentioned in Problem 8–3.

(8–5) For any one of the products mentioned in Problem 8–3, describe the design function for guidance to assuring quality of design. Make a list of all critical components (parts) for the chosen product that can fail most frequently and describe methods of design assurance. Assume you as a design reviewer, are asked to report to top management.

(8–6) a. Define "failure rate." What factors influence part failure rates?

   b. Five tungsten filaments especially made for electron microscopes were tested at the maximum voltage for $1000h$ each. Three filaments failed after the following test time periods: $250h$, $600h$, and $750h$. Determine the failure rate for this type of filament in terms of (i) number of failures per hour, (ii) percent failure per $1000h$, and (iii) failure per 10^6h.

(8–7) Determine the MTBF of the filament mentioned in Problem 8–6. Explain why MTBF is not considered a guaranteed failure-free period.

(8–8) a. Draw the bathtub curve and discuss the three periods of equipment life.

   b. The failure rate on account of corrosion of a certain type of secondary boiler valve has been established as 25 failures for every month in a continuous operation. There are 600 valves in service for the hydro heating plant. After these valves get jammed and fail to open up, they are not replaced. How long will it be until only 50% of the original 600 valves will be left in service?

(8–9) An electrical motor is guaranteed to operate for one year in continuous operation. Its calculated failure rate is 0.00001 failure per hour. What is the probability of its survival?

(8–10) Determine the MTBF for the electrical motor if the probability of its survival is at least 90% for one year of continuous operation.

(8–11) Calculate the reliability of a standby system that consists of three identical operating subsystems in parallel configuration for a period of 1000 hours. The failure rate for each subsystem is known as 0.001 failure/hour.

(8–12) The following data applies to failure analysis of a certain type of ball bearing assembly.

Failure Number	Cumulative Percent Frequency	Number of Operation Cycles to Failure ($\times 10^3$)
1	18	2.1
2	21	2.8
3	23	3.9
4	26	4.8
5	31	10.0
6	33	12.5
7	36	18.8
8	42	21.0
9	46	24.2

Assuming a Weibull distribution, plot the data graphically, and calculate the Weibull slope. Comment on the value of the slope obtained.

(8–13) Hairdryers that can be used safely in the bath and even under water have been developed. They have a new switch that would break the circuit rather than electrocuting the user. Many accidents with electrical appliances occur when used under wet conditions, for instance with electric mowers. The patent office, having accepted the application by the designer of such a switch, required an exorbitantly long time for approval.

Outline a design assurance approach that can be taken by the inventor and those that must approve the design. Assume that technical subject matters will not be dealt with because administrators do not usually have sufficient expertise.

(8–14) Car buyers expect that "extras" become standard features of the newer models. What are the reasons for this trend? Why were these accessories not included in the car for the regular price in the first place?

(8–15) Customers demand greater variety in order to exercise their individuality. Describe the general economic conditions and the impact on design assurance of such customer demand.

(8–16) Design defects vary with the life-cycle phases of a product or service. Select one product you are familiar with and describe such changes in defects.

(8–17) Explain customers basic attitude towards taking new medications and the protection they require. What role do design and quality assurance play? How can major negative side effects be minimized? What about packaging and dispensing?

(8–18) Consumers buy bulk and ''no label'' items not only for the savings in price. What other reasons are there, and what role does quality assurance play?

(8–19) *Modern computer assisted design (CAD) can create new kinds of defects not known under the manual design approach.* Explain. *Automation and use of robots in production and assembly places new demands on design assurance and quality assurance roles in design.* Explain.

(8–20) Describe the use of simulation, forecasting, waiting line models, and cost models in conjunction with design assurance.

8.15 NOTES

1. For more discussion on different methodologies of statistical tolerancing, the reader should consult *Quality Control Handbook*, J. M. Juran (editor-in-chief), McGraw-Hill Book Co., 3rd ed., New York, 1974.

2. ANSI/ASQC, Z 1.15 Standard. ''Generic Guidelines for Quality Systems,'' is a joint effort by the American National Standards Institute (ANSI) and American Society for Quality Control (ASQC).

3. AGREE Report, named from the initials of the committees that prepared it, report by *Advisory Group on Reliability of Electronic Equipment*, Office of the Assistant Secretary of Defense (Research and Engineering), Superintendent of Documents, Government Printing Office, Washington, D.C.

4. A failure rate data bank is maintained by the government–industry data exchange program (GIDEP) at Operations Center, Corona, California, 91720. For representative mechanical parts, data are available through the Bureau of Naval Weapons, *Failure Rate Data Handbook*, SP-63-470.

5. MIL-HDBK-271C, Military Standardization Handbook, *Reliability Prediction of Electronic Equipment*.

6. King, J. R., *Probability Charts for Decision Making*, The Industrial Press, New York, 1971.

7. Readers with more interest in details of the life-cycle testing plans are advised to consult the appropriate military standard and the specifications and standards referenced therein. Handbook H-108, from which these test plans are mentioned, is one of the earliest documents published by the Department of Defense, the other two being the Military Standards MIL-STD-781B (*Military Standard Reliability Tests: Exponential Distribution*) and MIL-STD-690 (*Failure Rate Sampling Plans and Procedures*). MIL-STD-781C contains both sequential and fixed-time reliability test plans, and also introduces environmental stresses in the testing.

8. Juran, J. M., ''Japanese and Western Quality—A Contrast,'' *Quality Progress*, December 1978, pp. 13–18. Reprinted by permission of the author and the American Society for Quality Control, Inc.

8.16 SELECTED BIBLIOGRAPHY

Bogaty, H., ''Development of New Consumer Products—Ways to Improve Your Chances of Success,'' *Research Management*, Vol. XVII, July 1974, pp. 26–30.

Carpenter, James K., ''Configuration Control and the Quality Department,'' *Quality Progress*, May 1978, p. 30.

Gitlow, Howard S. and Hertz, Paul T., ''Product defects and productivity,'' *Harvard Business Review*, September–October 1983, p. 131.

Holt, Knut, ''Generating Creativity, Ideas, and Inventions—Information and Needs Analysis in Idea Generation,'' *Research Management*, Vol. 18, May 1975, pp. 24–7.

Hopkins, David, ''New Product Planning,'' *Quality Progress*, May 1980, p. 24.

Juran, Joseph M. and Gryna Jr., Frank M., *Quality Planning and Analysis*, 2nd ed., McGraw-Hill Book Company, New York, 1980.

Mills, Charles A., "Contract Review—Quality Input," *Quality Progress,* May 1976, p. 22.

Rosamilia, Vincent J., "Quantitative Specification Evaluation," *Quality Progress,* November 1979, p. 28.

Shocker, A. D. and Shrinivasan, V., "A Consumer-Based Methodology for the Identification of New Product Ideas," *Management Science,* Vol. 20b, February 1974, pp. 921–37.

Von Hipple, E., "Has a Customer Already Developed Your Next Product," *Sloan Management Review,* Vol. 18, No. 2, Winter 1977, pp. 63–74.

CHAPTER 9

QUALITY OF DESIGN: PRODUCTION AND OPERATIONS

Modern quality assurance management implies that the product is designed for attainment of the required quality. Production processes and operations that are predetermined by the design, influence, to a large extent, the actual quality of products and services. Therefore, designing quality must extend as a managerial activity into the production sphere. Quality of design permeates through the entire productive system and further into the external supplies of material, production equipment, manpower, technological knowledge, and so on.

In this chapter we shall discuss the planning of production with quality assurance in mind. The task is now to translate the product design into technologically and economically sound production plans. Again, as in the product design, we shall not separate production planning from quality assurance planning. We have to remember that quality assurance, under the view of a productive system, ranks as a subsystem of the overriding productive system. Following this logic, we shall first review briefly the general aspects of production planning. Then we shall proceed to develop the associated plan for quality control and inspection activities.

9.1 MAJOR CONCEPTS

Two major concepts should be made clear to the reader: First, the concept and view of a productive system, as described earlier, is to be applied in a rigorous manner. This view leads us to move from the product design to production design. Here, the question of "what" turns into the question of "how." Moreover, we must remember that any productive system has three basic dimensions expressed in the questions "what," "when," and "where"; that is, the material, chronological, and spatial dimensions.

Second, according to the planning cycle as described in Figure 7.2, we deal in this chapter with general, overall production and quality control planning. The resulting plans are master plans which allow the preparation of detailed operational procedures. For instance, an inspection plan outlines at what point in the production process an inspection is to take place. The respective inspection directives are usually not as yet included in the plan, because such technically detailed work instructions are solely directed to the inspector and are operational in nature. Preparation of such procedures will be covered in subsequent chapters.

It might seem that this separation of general planning and detailed procedure writing relates only to a large scale production system with a complex product and technology. However, this stepwise planning approach, from the general basics to the specifics, is also a sound management practice in small business. The work of every individual worker and production unit must be aligned with that of others. This coordination and cooperation towards full attainment of production goals, including required quality, needs understanding and acceptance of an overriding production plan. Only in exceptional and rapidly disappearing circumstances can the product design be directly and without further planning assigned to workers and production units.

Available production capacity should not dictate the product design. Given a particular production capacity with its well-established processes, facilities, machines, and manpower, management would be foolish to turn out products or offer services without giving due consideration to actual customer demands and needs. Logically and realistically in a free-enterprise economy, production design cannot precede product design, as this can mean that the customer is not given a choice and actual quality expectations are not met. Recent history in the automobile industry shows examples of producing car models that did not stand up to competition, mainly because production capacities are not aligned with customer demands and products were not reviewed and redesigned with developing quality changes in the car market.

9.2 PRODUCTION PLANNING: AN OVERVIEW

Production means the creation of value that has a demand and price in the market. It is the internal sphere of a productive system, in which inputs are transformed to the desired outputs. As was pointed out earlier, outputs in the form of products and services are based upon the production facilities and required resources. A prerequisite for a productive and sound enterprise, for example, a factory, restaurant, or insurance agency, is that those in charge of planning, production, and operations know the specifications of the output in sufficient detail. It is one decision to stipulate the product and its components, applications, and so forth, and another to determine the actual making of it.

The actual production capacity must match with the requirements of the product design. Production capacity is uniquely determined by product design. Each major quality characteristic needs to be conformed with, meaning that processes, material, and manpower must permit the production accordingly. This being the logic, only the degree of product complexity dictates the intensity of production planning and its difficulty. The more the product design matches already available production capacity, and the clearer the design documents describe the major characteristics of the product, the easier is the task of production planning. We therefore must assume that during the design activities production aspects were given due consideration. This principle is also important for quality assurance. Participation of quality assurance specialists in product design ensures that the final design meets customer's requirements and allows control of conformance during the production. In brief, the effectiveness of production planning depends on the quality of the product design.

Product designing results in drawings and other documentation, listing of major characteristics, hazards, possible defects and their causes, and many other detailed stipulations. In situations where the customer is directly involved in product design, a contract is signed that binds production. On the other hand, the producer who supplies a market, rather than an individual customer, can act more independently. A contract review and possible change is, of course, more difficult, as the production capabilities may not allow conformance in all situations. It might happen that production can meet major quality specifications but not conflicting delivery times and points. Quality assurance extends beyond the mere material specification to proper timing and delivery as well as servicing after delivery.

Such comprehensive production planning, with the customer's ultimate full satisfaction in mind, translates the product design into components and operational elements. For a pipe valve, (Figure 9.1(*a*)) for instance, the assembly chart in Figure 9.1(*b*) shows such elements. This chart does not, of course, provide all necessary information for the production department, but demonstrates general product features. One can easily imagine the information a production engineer would require before actual production could commence.

Steps in production planning are shown in Figure 9.2. These cover the problem areas of deciding on the technical processes and operations. the requirements for machines or other productive facilities for manpower, the layout of plants and workplaces, plant location, schedules for purchasing, subcontracting, and internal production.

Each decision in production planning as a relation to quality of the production output. For instance, technology can be chosen for the same production process from different alternatives. One might require more human labor while the other alternatives may allow the employment of machines and even robots. Costs differ with the availability and quality of such production agents. A conflict can easily arise when robots produce better quality, possibly at a lower cost than conventional manpower employment. Can stringent demands for quality as supplied through robots really justify disregarding a more socially desirable alternative? Short- and long-run aspects might lead in such a case to different decisions. However, if the quality-wise considerations are adhered to, technology will become the choice in the long run. Quality of the end product will still largely depend on the workmanship. However, a tight time schedule, poor working conditions, inefficient machines (that remain mostly out of statistical control and unnoticed), poor facilities, and so on, do not allow an otherwise well-motivated worker to perform with proper quality assurance. The pleasure of meeting the production deadline and required quantity is shorter than the negative effects of inadequate quality supplied to the customers.

9.3 QUALITY CONTROL AND INSPECTION PLANNING

Once the general production plan has been assessed under quality assurance considerations, specific planning of quality control and inspection can commence. The inspection plan is an essential element of the wider-ranging quality control plan; it is not identical with it. As shown in Figure 9.2, quality impact alignment of all elements in the production is the first

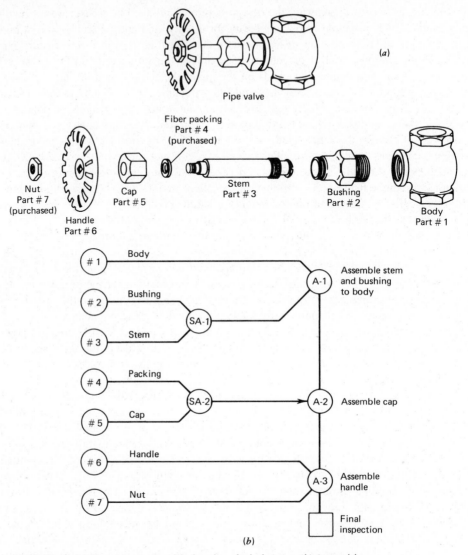

FIGURE 9.1 (a) Pipe valve, assembled and exploded views. (b) Assembly chart for pipe valve. Source: Buffa, Elwood S.; *Modern Production Management,* 2nd ed., John Wiley & Sons, Inc., New York, 1965, p. 190, Reprinted by permission.

logical step in quality control planning. Quality control as opposed to quality assurance, embraces conceptually only those activities that are to assure conformance to quality specifications and to all principles of good workmanship during production.

Quality control plans describe and determine the evaluation and development of technologies and production processes for a given product design. Moreover, quality conducive production facilities, layouts, locations, and workplace designs, as has been pointed out earlier, will be outlined with regard to selection, design, and maintenance. Special attention

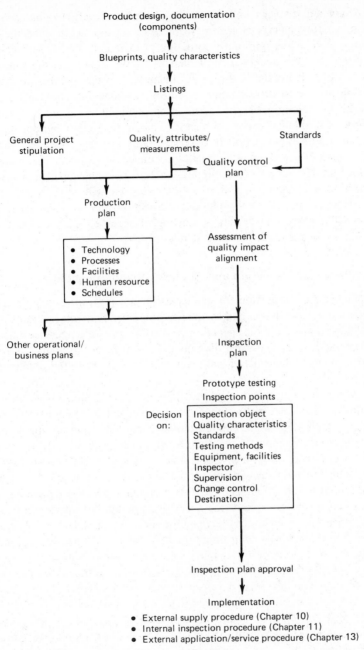

FIGURE 9.2   Production planning overview.

can also be given to quality attainment through respective staff training and support services, including those for supervisory management. As tight production schedules and work standards, along with sequencing and prioritizing of different orders can invalidate quality assurance principles, the quality control plan will have to address the timing in production activities as well. Lastly, the quality control plan will ensure that the production information and decision making system includes reporting schemes and communications designed to maximize quality, along with other goals of production.

The purpose of the quality control plan is to assist production and operational supervisory management and operational staff to observe quality assurance aspects. The quality control plan itself is part of a wide-ranging quality assurance program, that embraces the entire life cycle of the product. In this context quality control planning, as it involves all members of an organization, the suppliers, and also the customers, will be discussed in later chapters. Here the focus is on inspection planning. Inspection is a specially designed independent verification of required facts, such as meeting of specified quality characteristics or compliance with process and performance standards.

Interdependency of Inspection and Actual Production Functions

The independence of inspection for quality assurance from the actual production function, as it is required by convention and by quality program standards, warrants the separating of general quality control from the special inspection planning. Figure 9.2 indicates the position of inspection plans under the overriding quality control plan. Similarly, at the same level as inspection planning, other operational plans help to implement the production plan and the product design.

Before we discuss inspection in more details we should realize that small businesses and organizations with simple, well-established products and production systems might not formally conduct such planning in explicit terms. Naturally, there are many different ways of preparing for production and for the controlling and inspection of quality attainment during production. Given such differences, one should recognize that the ultimate goal of quality assurance cannot materialize without special attention given to quality in connection with any preparation for production. The fact that production needs to be planned and controlled, under any circumstances, explains the need for some kind of inspection planning.

In recent years, many customers have required of their suppliers and contractors, that they submit *production and inspection plans* for their approval. Quality program *standards,* to which suppliers have to conform as part of their contractual obligations, also include preparation of inspection plans. The main reason for this is that the inspection plan provides the customer with some product related overview of the intended verification of quality attainment during production. Customers might want to verify themselves certain quality characteristics and process requirements at an intermediate production stage and inspection point. However, even when the product is designed for a market, prudent production management will prepare special inspection plans, if for no other reason than to optimally coordinate and integrate all inspections during production. Subsequently, the inspection plan can be implemented through detailed inspection procedures, as will be shown in the following sections.

To begin with, we shall survey concepts, principles, and methods of inspection planning, before we trace the quality control during production in more technical and operational terms.

The fact that we relate this discussion to manufacturing industries, or might appear to do so, should not be construed to mean that inspection planning is not relevant in service establishments, resource industries, and so on. In any productive system, verification of quality attainment during production or operation is principally required if quality of the ultimate product or service is to be assured.

A jet engine model is bought by the Air Force based on performance of the prototype (as most military products are), with the assumption that tolerances are held to print in the model. When made with manufacturing tooling, many parts are rejected, requiring Material Review Board action, many delays, and many changes. Little consideration is given to whether the Production model does not match the print tolerances. . . . The problem in U.S. companies grows during the design stage. Many American companies do commit to a pilot or engineering model—but few wait to test it or incorporate the results into the manufacturing design. We don't have the time to do it right, but plenty of time to do it over and over.[1]

9.4 INSPECTION PLAN

Steps of inspection planning and content of resulting inspection plans are shown in Figure 9.3. Some inspection planning has usually already been conducted in conjunction with product designing and prototype testing. However, inspection planning and actual inspection prior to production planning serve mainly to validate the design and production capabilities. Now inspection planning is associated with production planning and the purpose is to establish means and procedures for quality assurance during actual future production. Prototype inspection plans and inspection results serve as a useful basis for further planning that proceeds along with the life cycle of the product. The main purpose of inspection plans for actual production is to gain approval from senior management and customers, and to allow further determination of special inspection and test procedures.

Inspection planning is understood as the determination of inspection object and inspection points relative to the production sequence and processes. For each inspection point, quality characteristics, applicable standards, kind of inspection method, inspector, and supervision are to be indicated.

Figure 9.4 shows a general production plan for the pipe valve example presented earlier (Figure 9.1a). The square symbol stands for an inspection point. Such a flowchart can rarely include all the inspection information required by those who are to receive the inspection plan. Supplemental listings will have to specify major quality characteristics and the object for inspection at this point; or inspection will be done to the drawing. An object can be each individual item, or a batch or sample. Figure 9.5 is another process flowchart, showing inspection points for a critical part made out of ductile iron casting which must be machined and inspected for defects. Standards are internal or external mandatory specifications and requirements which the inspection will have to verify. The inspection method describes, in general terms, how acceptance or rejection is decided upon; for instance on the basis of a sample, process control, external laboratory testing, or in participation with customers. Inspection planning, giving an overview of total approach from purchasing of materials to shipping of finished goods, is shown in a schematic drawing applicable to fashion dress-shirt manufacturing company in Figure 9.6.

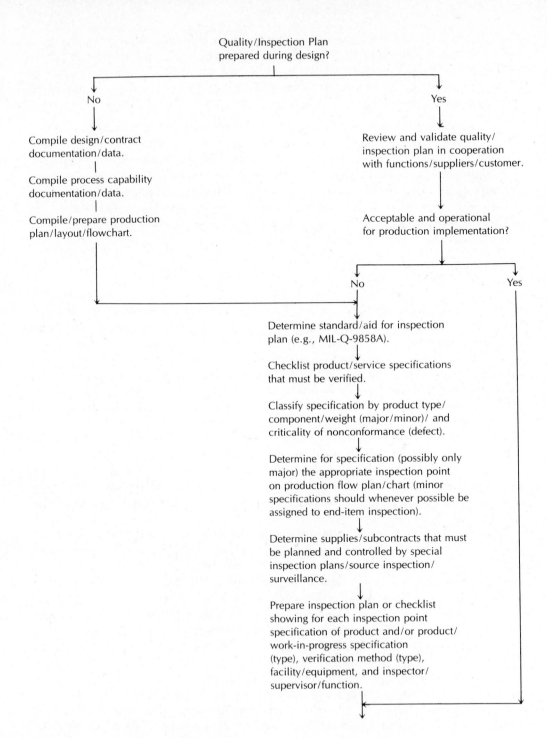

Quality/Inspection Plan
prepared during design?

No

Yes

Compile design/contract
documentation/data.

Compile process capability
documentation/data.

Compile/prepare production
plan/layout/flowchart.

Review and validate quality/
inspection plan in cooperation
with functions/suppliers/customer.

Acceptable and operational
for production implementation?

No

Yes

Determine standard/aid for inspection
plan (e.g., MIL-Q-9858A).

Checklist product/service specifications
that must be verified.

Classify specification by product type/
component/weight (major/minor)/ and
criticality of nonconformance (defect).

Determine for specification (possibly only
major) the appropriate inspection point
on production flow plan/chart (minor
specifications should whenever possible be
assigned to end-item inspection).

Determine supplies/subcontracts that must
be planned and controlled by special
inspection plans/source inspection/
surveillance.

Prepare inspection plan or checklist
showing for each inspection point
specification of product and/or product/
work-in-progress specification
(type), verification method (type),
facility/equipment, and inspector/
supervisor/function.

FIGURE 9.3 Inspection planning overview.

226

Prepare combined production/inspection
plan/flowchart; coordinate with other
internal and external functions and
possibly prepare an inspection schedule.

$\downarrow$

Attain approval and release the inspection
plan for possible complementation
by special inspection procedures, test
instructions, and modification of plan.

$\downarrow$

Conduct start-up training, briefing, and
plan/procedure adjustments; verify
reliability and validity of inspection/
test results.

$\downarrow$

Release inspection plan for regular
production run(s).

$\downarrow$

Supervise/audit inspection/testing;
prepare reports; if necessary take
corrective action and adjust plan
and procedures.

$\downarrow$

Standardize inspection plan and procedures
for repeated production runs and ensure
traceability of plan/procedure.

$\downarrow$

Control and facilitate change of plan and
procedures in conjunction with change in
design and production plans.

Fig. 9.3   (*Continued*)

Self-inspection

A brief description of the inspection method explains the inspector as a decision maker for each inspection point. For instance, inspection might be performed entirely or in part by the operator or production supervisor, rather than the official independent inspector. Such *self-inspection,* once properly controlled through quality assurance management, helps to involve qualified and trustworthy staff in quality assurance and to promote in a meaningful manner total quality commitment and sound workmanship in the entire organization. Adequate supervision of such self-inspection, and also of the regular inspection force, is attained through a quality audit initiated and controlled by senior management.

In addition to the information pertaining to inspection points, the inspection plan should describe procedures for changing inspection plans, the distribution of such plans, requirements for approval, and other stipulations that govern inspection procedure writing.

It should be noted that inspectors usually only decide on acceptance or rejection of an object or process on the basis of clear decision criteria. Cause-and-effect analysis of complicated matters and its subsequent remedial actions must normally be left to technical experts and qualified review boards. The inspection plan can only indicate such additional measures for preventive quality assurance.

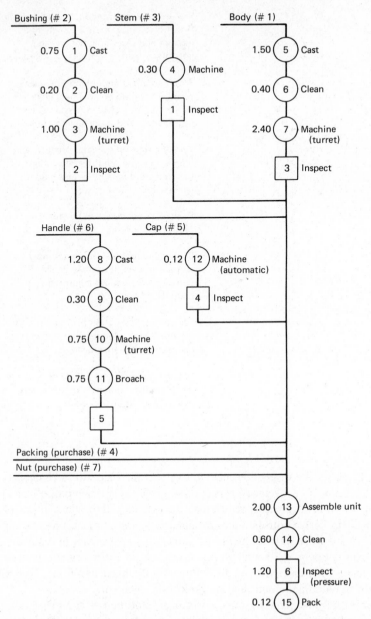

FIGURE 9.4 Operation process chart for the pipe valve example shown in Figure 9.1. Source: Buffa, Elwood S., *Instructor's Manual for Modern Production Management*, 2nd ed., John Wiley & Sons, Inc., New York, 1965. Reprinted by permission.

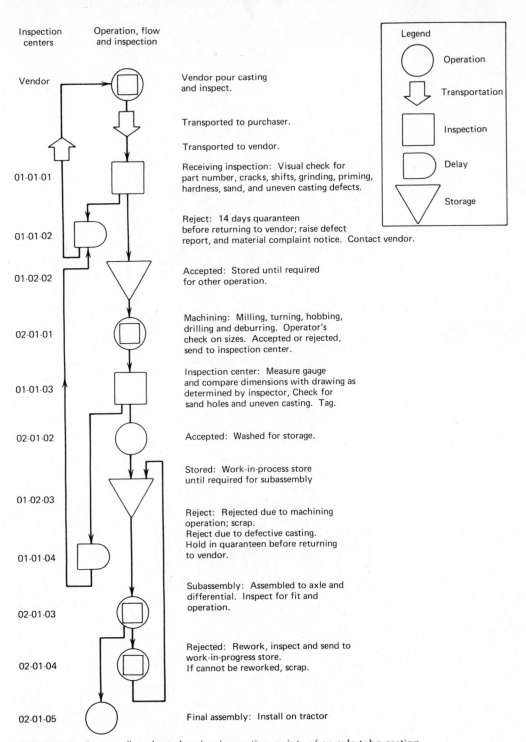

Inspection centers	Operation, flow and inspection	

Legend

○ Operation

⇩ Transportation

□ Inspection

D Delay

▽ Storage

Vendor — Vendor pour casting and inspect.

Transported to purchaser.

Transported to vendor.

01-01-01 — Receiving inspection: Visual check for part number, cracks, shifts, grinding, priming, hardness, sand, and uneven casting defects.

01-01-02 — Reject: 14 days quaranteen before returning to vendor; raise defect report, and material complaint notice. Contact vendor.

01-02-02 — Accepted: Stored until required for other operation.

02-01-01 — Machining: Milling, turning, hobbing, drilling and deburring. Operator's check on sizes. Accepted or rejected, send to inspection center.

01-01-03 — Inspection center: Measure gauge and compare dimensions with drawing as determined by inspector, Check for sand holes and uneven casting. Tag.

02-01-02 — Accepted: Washed for storage.

01-02-03 — Stored: Work-in-process store until required for subassembly

01-01-04 — Reject: Rejected due to machining operation; scrap. Reject due to defective casting. Hold in quaranteen before returning to vendor.

02-01-03 — Subassembly: Assembled to axle and differential. Inspect for fit and operation.

02-01-04 — Rejected: Rework, inspect and send to work-in-progress store. If cannot be reworked, scrap.

02-01-05 — Final assembly: Install on tractor

FIGURE 9.5   Process flowchart showing inspection points of an axle tube casting.

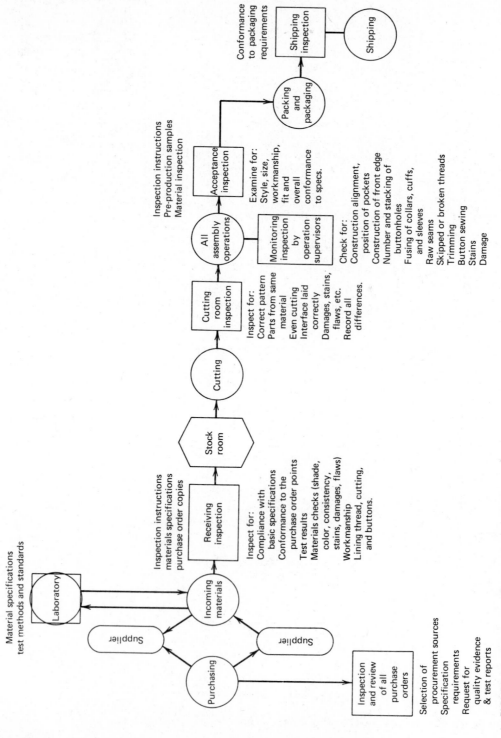

FIGURE 9.6 Inspection plan schematic for fashion dress shirt, from purchasing to shipping of the end-item.

9.5 INSPECTION PLANNING PROCEDURE

The general sequence of an inspection planning procedure is described in Figure 9.7. This scheme includes a cost analysis, because inspection itself should only be approved and conducted with proven cost effectiveness.

For sound inspection planning that results in valid and workable plans, principles such as the following should be observed.

1. A primary purpose of inspection is prevention of defects.

2. Costs for eliminating defects (inspection costs) must be less than costs due to defective items passed at a particular point of production. (Inspection must pay at least for itself and should be a profit center.)

3. The inspection principles spelled out or implied in the standard must be adhered to for adequate product liability protection and meaningful quality assurance to the customer.

4. Critical specifications have priority in any inspection.

5. Inspection and test plans must be clear and complete.

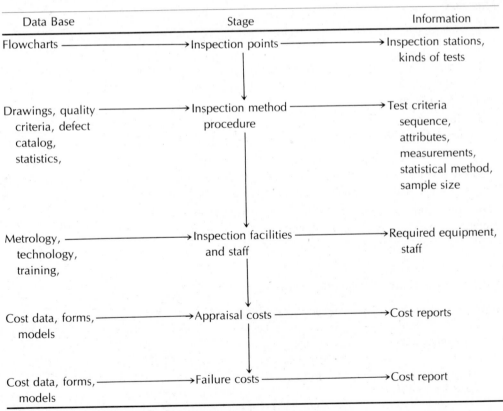

FIGURE 9.7 Inspection planning including quality costs.

6. *Inspection points should be considered:*
   a. When raw materials are received.
   b. When raw materials enter the production process.
   c. Before costly processes.
   d. Before irreversible processes.
   e. Before processes which cover defects.
   f. When finished products emerge from the production process.

7. *One hundred percent inspection is used where:*
   a. The probability of quality variation is quite high.
   b. Human error is prevalent.
   c. The liability risks of unacceptable quality are high.
   d. The product must be checked before delivery to the customer to see if it functions properly.

8. *Inspection of samples are to be used:*
   a. Where mechanical production methods are used.
   b. Where destructive testing is involved.
   c. Where large quantities of bulk items must be inspected.

More details on sampling inspections are described in Chapter 12. Some other important principles of inspection planning are as follows.

1. Inspection planning should not be done in isolation by only the quality assurance function.

2. Inspection plans should be reviewed at regular intervals, and more frequently for new product designs and in cases of other relevant major changes. Such reviews can take the form of quality audits.

3. Inspection of external suppliers and subcontractors should be explained in detail either in the inspection plan or in special supplements to the plan.

4. Published quality program standards should be used as a guide in inspection planning.

5. References to special inspection procedures, standards, and so forth should be made. Only in exceptional cases should such procedures be included in the plan.

6. Inspection planning should not be restricted to physical items and respective manufacturing processes. Verification of conformance and quality assurance is also necessary and useful in many service and administrative operations.

Form and Contents of Inspection Plans

Inspection plans in real use vary considerably by form and content. A decisive aspect of effective inspection planning is that it serve its designed and stated purpose as outlined. Prudent management will not cast any doubt on careful planning of inspection, because this is still the most widely applied and known mean of general quality assurance. Moreover, any careful worker or decision maker will objectively verify that required performance standards have been met.

An example of a simple inspection plan that does not require much expertise or effort is a simple trouble report form (see Figure 8.11) or defect alert scheme. Another quite effective plan is represented by the type of error cause removal (ECR) form shown in Figure 9.8. Table 9.1 demonstrates salient features of an inspection plan for a nonmanufacturing case.

Inspection planning should never degenerate into unnecessary decision making and paperwork. This can be avoided through careful explaining and reviewing of the intended purpose. Before implementing an inspection plan, those who are to receive and use the information from the plan should evaluate and approve it. This screening of inspection planning leads those in charge to seek reasonable participation during the planning stage and to apply generally accepted standards.

The implementation of an approved inspection plan will be covered in the following three chapters, much as they are outlined in Figure 9.2. In Chapter 10 discussion of the quality of design will apply product design and general production and quality control or inspection plans to quality assurance of external supplies and subcontracts. In Chapter 10, discussion of the design phase of the product or service is considered to be concluded, and discussion of actual production and operations in accordance with quality of conformance and quality of performance product and production design can commence. Actual inspection procedure writing and implementation will be considered as an activity in the realm of current operational planning. Inspection and test procedures must be reviewed and changed more often than the more long-term master planning, in response to frequently changing production and market conditions. Finally, the inspection plan, along with the overriding quality control and production plan, will be followed up in Chapter 13 where we will discuss what is relevant once the item has been sold, or the service rendered, to the customer.

TABLE 9.1
Features of Inspection Plan for a Restaurant

Points of Inspection	Examples of What to Look For	Consequences of Deviations	Possible Method of Inspection
Reception desk	Correct handling of reservations	Overbooking, lost customers	Observing at random
Bar service	Proper measuring of drinks	Dissatisfied customers never come back	Check with complaint logbook and random test
Meal service	Waiter following procedure for receiving orders	Customer not properly informed of what has been ordered	Questioning of waiters and random observation
Facilities	Cleanliness of room, tables, and so on	Failure to pass health inspection, lost customers, accidents	Visual check at random before and during business hours
Kitchen	General cleanliness and sanitary standards	Closure through health inspector, poor food preparation	Daily checks immediately before opening, defects to be entered in special logbook

Error Cause Removal Form (ECR)

From: _____ Submittal Dept.: _____

Date: _____

1. I have discovered a problem which I feel is causing me or my fellow workers to have less than perfect performance. The problem is:

_____

_____

_____

_____

Signed: _____

2. Supervisor's answer or comments on above problem identification.

_____

_____

_____

_____

_____

_____

Signed: _____

Date: _____

3. Action taken on problem and date action is to be completed (review committee).

Date Received: _____

_____

_____

_____

_____

_____

_____

4. Answer discussed with employee.

Date: _____ Signed (employee): _____

Answer to be discussed with employee no later than ten days after submittal

FIGURE 9.8 Error cause removal (ECR) form. In both situations the solution must be coordinated with the employee no later than 10 days from the submittal date.

Procedure For Use of Error Cause Removal

Action One is used by the employee to identify any problem which he or she feels is causing poor performance and is to be signed and dated and turned over directly to the supervisor.

Action Two is used by the Supervisor to answer the problem if possible, or to elaborate on the problem if required. If the supervisor is not able to answer the problem, copy number 3 is removed and the form (1 + 2) is deposited in the storage box for pick up by the review committee.

Action Three is used by the review committee for explanation of the corrective action taken and date the action will be completed. This section is the employee's answer. The committee will be responsible for insuring that the employee receives a clearly defined answer. The review committee will route the form to the department responsible for solution and will review the answer after a solution is determined. The responsible department may be asked by the review committee to discuss the answer directly with the employee.

Action Four is signed by the employee after an adequate answer has been received from either the supervisor, the review committee or the responsible department. The employee's answer should be received no later than ten days after submittal.

Form Routing

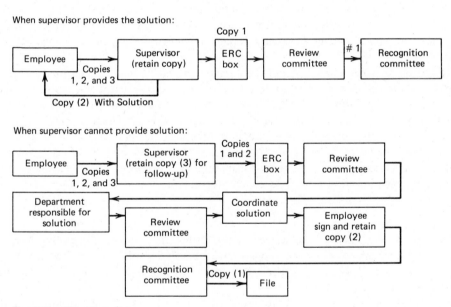

Fig. 9.8   (*Continued*)

9.6   QUANTITATIVE METHODS APPLICABLE TO QUALITY OF PRODUCTION

Waiting Line Models

The waiting line model is one of the models most applicable to quality assurance. One measure of quality of service is in terms of waiting for service and length of service. For instance, management can reduce waiting time of customers in a restaurant, by raising the service capacity in terms of numbers, tables, waiters, and cooks. The tradeoff, however, is higher idle capacity and respective costs. While waiting and idle time cannot entirely be avoided, the waiting line model helps to determine the capacity where the total of waiting and idle costs are at a minimum. The model is summarized in the following sections.

Main Features

Waiting line models provide answer to questions such as: How many loading docks should we have for our trucks? How many toll booths? How many inspection stations? How many inspectors per station? The last question formulates the problem for a waiting line application. The model, or "theory," is applied to analyze the feasibility of adding facilities and for assessing the amount and cost of waiting time for arriving items to be tested. The theory is stochastic (statistical) by nature in that it assumes a specific distribution of arrivals or services rendered per period; or the related mean interarrival or service time.

Definition of Queuing Terms

The *queuing process* is centered around an inspection station which has one or more *service channels*. Item *arrivals* are drawn from an *input source,* the production line. In waiting line models, the arrivals from the input source are generally characterized by a probability distribution. The symbol used to represent the *mean arrival rate* of customers is the Greek letter *lambda* (λ).

The customer entering the system joins a *queue* or *waiting line*. The customer is selected from the queue for inspection according to a *queue discipline* or a *priority rule*. Usually service times follow some probability distribution and the *average inspection rate* is represented by the Greek letter *mu* (μ). In order to have a stable queuing process, the average service rate (μ) must be greater than or equal to the mean arrival rate (λ). After the service is completed, the customer exits the system. Generally, a queuing system is characterized by its arrival pattern, its service time distribution, its queue discipline, and its layout, or flow-pattern. See Figure 9.9 for a schematic illustration.

Arrival Patterns

Arrivals are of the items to the service facility, since they are the things that need to be processed. Time intervals between consecutive arrivals are often distributed according to negative exponential distribution (i.e., longer time intervals have a lower probability of occurance).

In this case, the number of arrivals expected forms a Poisson distribution. The Poisson distribution corresponds to completely random arrivals since it is assumed that an arrival is

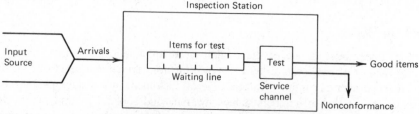

FIGURE 9.9 Single queue–single-channel, single-phase queuing process.

completely independent of other arrivals, as well as of any condition of the waiting line. Mean arrival rates per time unit $= \lambda$. Mean arrival time $= 1/\lambda$.

Service Time Distribution

The simplest situation is one in which each arrival requires the same time for service as every other arrival (i.e., a constant service time). Service rates follow a Poisson process with mean service rate $= \mu$. The service time distribution follows the negative exponential distribution. Mean service time $= 1/\mu$.

Queue Disciplines

The *queue discipline* is the priority rule by which waiting jobs are selected from the queue for service. The most common priority rule is the *first-come-first-served rule* (i.e., the first job to arrive in the queue will be the first job to be serviced).

Another priority rule is the *random rule* which selects the job which has the smallest value of a random priority assigned at the time of its arrival. The *shortest operation time rule* selects from the queue the job which requires the least processing time at that service center.

Example 1: Donald McRonald, a university student, is tired of standing in waiting lines when he goes for lunch at the campus food service. To help eliminate his problem, and the problem of many other students who experience this daily noontime dilemma, Donald has plans to open a fast food service offering four varieties of hamburgers. With his new Panasonic microwave oven, Donald speculates that he can offer hungry students a quick, hot, and tasty hamburger for lunch. Donald is confident that his hamburgers will be hot and tasty, however, he is concerned that his noontime service eliminate waiting lines. Donald knows he can estimate the probability of a lengthy waiting line by using the waiting line formula.

Donald's marketing research efforts have turned up the following information.

1. The distribution of customer arrivals will most likely (assumption one) follow a Poisson distribution with a mean arrival rate of 60 per hour, $\lambda =$ one person per minute.

2. His microwave oven takes 1.5 hamburgers per minute. Therefore his mean service time is 90 hamburgers per hour following a $1/\mu = 1/1.5$ (assumption two, negative exponential distribution).

3. Donald has planned that each customer, when arriving, will take a sequential number and await his turn for service (assumption three). This facilitates servicing customers on a first-come-first-served basis. Donald's marketing research shows him that with his plan, the service rate is larger than the arrival rate (assumption four).

Being satisfied that his case follows the Poisson distribution, Donald will use the waiting line model and work with arrival rates and service rates or Poisson and negative exponential distribution to determine the probability of a waiting line.

Mean number in waiting line is arrived at by:

$$L_q = \frac{\lambda^2}{\mu(\mu - \lambda)} = \frac{1}{1.5(1.5 - 1)} = 1.33$$

The formula for mean number in system—including the one being served is:

$$L = \frac{\lambda}{\mu - \lambda} = \frac{1}{1.5 - 1} = 2$$

The formula for mean waiting time is:

$$W_q = \frac{\lambda}{\mu(\mu - \lambda)} = \frac{1}{1.5(1.5 - 1)} = 1.33 \text{ minutes}$$

Mean time in system—including service is arrived at by:

$$W = \frac{1}{\mu - \lambda} = \frac{1}{1.5 - 1} = 2 \text{ minutes}$$

Probability of units in system is expressed as:

$$P_n = \left(1 - \frac{\lambda}{\mu}\right)\left(\frac{\lambda}{\mu}\right)^n$$

$$P_0 = \left(1 - \frac{1}{1.5}\right)\left(\frac{1}{1.5}\right)^0 = 0.33; \ 33\% \text{ chance of } 0$$

$$P_1 = \left(1 - \frac{1}{1.5}\right)\left(\frac{1}{1.5}\right)^1 = 0.22; \ 22\% \text{ chance of } 1$$

Utilization rate, $\rho = \dfrac{\lambda}{\mu} = \dfrac{1}{1.5} = 0.66$; Idle service $= 1 - \rho = 0.34$.

The most simple waiting line model assumes one waiting line, and one service facility (channel) with one server (phase). Customer arrival and service rates (persons served per time period) are assumed to follow the Poisson distribution. The time between two arrivals and the service time per customer are described by a negative exponential distribution. Other assumptions of the simple model are that service is given on a first-come-first-served basis and that an infinitely long buildup of a waiting line is theoretically possible. As shown in the following example, the mean arrival rate and the mean service time have been estimated. Using the respective equations, the following can now be calculated: system capacity utilization, percent of time the server will be idle, the expected number of customers waiting, and the average time a customer will spend in the system.

Example 2: The Bank of Manitoba has installed an automatic teller at its main branch. Since its installation, it has been measured that customers arrive at the teller at the rate of 20 per hour. The average time required to service a customer is 90 seconds. Service is on a *first-come-first-served* basis. The source is infinite (very patient customers). The bank management does not wish to impose inordinately on the patience of its customers, and has decided that a time of 4.0 minutes in the system represents adequate quality service with respect to time. Is their present system adequate, or should they install a second automatic teller?

We want to know the total expected time spent for a customer in the system. This is given by $W = 1/(\text{mean service rate—mean arrival rate})$. Service rate is 90 seconds per customer which is 40 customers per hour. Therefore: $W = 1/(40 - 20) = 1/20 = 0.05$ hour $= 3$ minutes, which is an acceptable amount of time.

An inspection station also represents a waiting line system, as illustrated in the following example. The arrivals are items to be inspected and actual inspection is the service to be rendered. The literature on waiting lines describes more complicated systems with multiple lines, multiple servers, fixed service times, and so forth. In the stringent assumptions of these models lies a certain limitation of usefulness to the practitioner. Wherever the validity of the waiting line models must be questioned in view of the real or assumed situation, the simulation technique, often called Monte Carlo simulation, might help in the decision making process. In a discrete event simulation, the clock moves from the occurrence of one event to the next, that is; from an arrival to the start of a service, completion of service, closure of shop, and so forth. The relatively easy and convenient access to computer simulation suggests the use of simulation as a decision making tool.

Example 3: Laminated automobile windshields come off a production line and await (in a queue) for the attention of an inspector who checks them for quality (defects—surface scratches, bubbles, stains). The arrival rate is defined by the speed of the line, and is X units per hour. Service time is the inspection time. If the arrival rate is too high, the inspector tends to speed up the inspection process to control the length of the queue, thereby decreasing the effectiveness of the inspection.

1. If service time averages 45 seconds per windshield, what should the arrival rate be to ensure a steady supply of windshields for the inspector, yet to ensure that the queue length does not build up too much? Assume that this desired situation has a 0.05 probability of the system being empty.

2. What is the expected length of the queue?

3. If the arrival rate doubles, what must the service rate adjust to maintain a 0.05 probability of an empty system?

Solutions

1. Probability of an empty system is:

$$P(0) = 1 - \frac{\text{arrival rate}}{\text{service rate}}$$

Service rate = 45 seconds per windshield = 80 windshields per hour.

$$0.05 = 1 - \frac{\text{Arrival rate}}{80}$$

Arrival rate = $(1 - 0.05)80 = (0.95)(80) = 76$ windshields per hour.

2. Expected queue length is arrived at by:

$$L_q = \frac{(\text{arrival rate})^2}{[\text{Service rate (service rate} - \text{arrival rate})]}$$

$$= \frac{76^2}{80(80 - 76)}$$

$$= \frac{5776}{320}$$

$$= 18.05 \text{ or } 18$$

3. $P(0) = 1 - \dfrac{\text{Arrival rate}}{\text{Service rate}}$

$$0.05 = 1 - \frac{(76.2)}{\text{Service rate}}$$

or,

$$\frac{152}{\text{Service rate}} = 0.95$$

$$\text{Service rate} = \frac{152}{0.95} = 160/\text{hr}$$

Hence the service rate also must double.

Example 4: The new fun resort, Splash City, will feature a water slide. The slide winds and snakes its way down and around the hillside. Patrons enter hot, sweltering, and lethargic at the top of the slide and emerge cool, refreshed, and exhilarated at the end of the run. This is quality service. The patron then climbs the hill and joins the queue to get back on the slide.

In a steady state operation this is a single queue, single server, queuing system with an infinite calling population and an infinite queue length. Service time is the time spent on the slide. Note that service time will overlap, as at any one time there will be many people on the slide. People are to enter the slide at eight second intervals on average.

1. Market research indicates expected arrivals of 445 per hour. What is the expected waiting time in the queue? On a sunny, hot, day, does this represent quality service?

2. Splash City's design engineer, the renowned Splash Gordon, has calculated that by increasing the angle of the slope of the slide and increasing its length, customers can spend the same time on the slide, while entering the slide at five second intervals, being much more exhilarated in the process. What is the expected waiting time in the queue for this situation?

Solutions

1. Service rate is eight seconds per person, or 450 people per hour. Waiting time in queue is given by:

$$W_q = \frac{\text{arrival rate}}{[\text{service rate (service rate} - \text{arrival rate)}]}$$

$$= \frac{445}{450(450 - 445)}$$

$$= \frac{445}{2250}$$

$$= 0.19777 \text{ hr}$$

$$= 11 \text{ minutes } 52 \text{ seconds.}$$

This time seems excessive.

2. Service rate if five seconds per person or 720 people per hour. Now:

$$W_q = \frac{\text{arrival rate}}{[(\text{service rate (service rate} - \text{arrival rate)}]}$$

$$= \frac{445}{(720)(720 - 445)}$$

$$= \frac{445}{198,000}$$

$$= 0.0022474 \text{ hours}$$

$$\text{or } 8.1 \text{ seconds}$$

This is much more satisfactory.

Simulation Technique

Simulation technique has been called the laboratory for management. It permits the building of a model of a real process in logical and numerical terms, and use of the model for experimentation and decision making. This approach is similar to physical simulation in prototype testing and pilot training, in that it captures the real situation to a fairly high degree. The advantages of digital simulation are that it is relatively easy to design, perform, and use, particularly when computer facilities are to be used. As it is very versatile in application to productive systems and thus in quality assurance, it can aid managers in studying the behaviour of real systems and in predicting outcomes of crucial decisions.

Once the problem under study has been sufficiently defined and delineated and a goal or criterion for the desired and expected outcome has been established, a simulation model can be developed. The model is still an abstraction from reality and should entail only those features of the real system that bear on the problem and objective of study. Once the model appears to be valid, the experiment using the model can be designed and performed. Modifications on the model and experiment are made until the results from simulation runs appear realistic and valid.

These simulation experiments normally use statistical, probabilistic data, such as an empirical probability distribution of arrivals. This distribution is then used to interpret a random number from zero to one hundred as a real (simulated) interarrival time, service time, or whatever is applicable. The random numbers must have certain known and well-assured properties, such as a uniform distribution. Random number tables should be used.[2] In one of the following examples (Example 2) simulation is carried out for an inspection station where the probability distribution for arrivals and inspection times has been compiled from the real system. The sequence of random numbers is taken from a table. In discrete event simulation the clock in a run is advanced from one current event to the next, such as arrival, test completion, and so on.

Validity of simulation results depends largely on the size of the sample and length of the simulation run. However, the example obviously suffers by the small sample size. For a computerized model, large sample sizes and relatively long runs are not a problem anymore. Therefore, a computer should be used for this and other reasons; the complexity of the model, the analysis of results, the interpretation of these, and many other "housekeeping" functions can conveniently be performed by the computer. Individual computer models, especially prepared for quality assurance management, must often be integrated with other simulation models, using the same software.

General Purpose Simulation System

The General Purpose Simulation System (GPSS) from IBM is one of many pieces of user oriented software for this kind of digital simulation. The clock during a simulation run will move from one event, such as an arrival, to the next event, such as start of the inspection. The GPSS defines four kinds of entities: transactions, facilities, statistical processing, and operational routines. Each of them has specific functions and automatically calls forth corollary functions commonly associated with simulation of the system internal flow. Blocks, which are programed subroutine packages, determine the logic of the system as well as the logic of transaction flow through the simulation model. There are about 50 blocks in GPSS that provide for flexibility in modeling and capturing the actual flow of transactions through processes, such as defects (transaction) through a rework station (facility).

Transactions (i.e., a customer order) are the dynamic entities that are generated, moved through the system as in the real world, and then terminated at the end. The nature of each transaction, such as its parameters, can be marked and changed.

The facilities used by the system are specified as entities and they service the transactions, just like a work station in a plant. The GPSS processor keeps track of the utilization and other statistics on the facility and its interaction with other facilities. Statistical entities may be called on for analysis of results, such as the waiting line statistics.

In principle, real world production systems consist of sequences of arrivals, waiting, service, departure to the next station, and so on, until the process is completed. This is simulated and recreated in the simulation model, including the statistical properties of cause-and-effect relationship as they occur during a timespan. Changing the conditions and generating respectively different output statistics reflect the system behaviour and lead to experimentation, learning, and rational decision making.

Example 1: In the following example, it is required to simulate the process of adoption used by a child-care agency in placing a child into a permanent home. We concentrate on the quality control aspect of adoption. Quality assurance of adoption of a child might appear an unusual concept. However, a mismatch is obviously a failure that should, if at all possible, be defined and prevented. The purpose of this example is to demonstrate application of quality assurance in government services and also to demonstrate the close relationship of quality of service to quality of life and welfare. After six months, a postplacement evaluation is made, in order to determine if the match between the child and the family was successful. This evaluation is performed by the child-care worker who approved the family. If the match was deemed successful, as shown by the adjustment of both the family and the child, steps will begin to grant an order of adoption. If the match was unsuccessful in the opinion of the child-care worker, the child may be removed from the home and they will attempt to find another home for the child.

Thus, quality would refer to a good adjustment between a child and a home in a reasonable length of time. If, generally, the postplacement review resulted in a large number of cases where children were removed from homes, a more thorough investigation would be performed.

Approach taken in building the model: A facility was used to simulate the inspector or child-care worker and a one-line, one-server system was used. The queue *wait* was used to gather statistics on the queue of matched families and children waiting for their postplacement evaluation.

The unit of time used was days and the simulation was run for the equivalent of 1000 days. Information on the control is given in Table 9.2.

Model implementation: This was done in the following sequence, where the technical terms are explained in GPSS/IBM user's manual:

TABLE 9.2
Table of Definitions

GPSS Entity	Interpretation
Transactions	
Model segment 1	Children
Model segment 2	A timer.
Facilities	
INSP	The child-care worker.
Queues	
WAIT	The Queue used to gather statistics on the waiting experience of matched children and homes.
Save values	
TIMER	Duration of the simulation.
CHILD	Average number of children in the system.
NUM	Total number of children in the system.
Variables	
AVG	Average number of children in the system.

Defects = mismatches.

1. A number 1 was put as the D operand on the START card in order to have a chain printout to refer to. An example of the chain printout is shown in Table 9.3.

2. A SAVE VALUE (a technical term) named NUM was used to store the computed average number of children being serviced by the child-care agency. The SAVE VALUE NUM was increased by 1 as a child came into the care of the agency and was decreased by 1 as the match was completed. It was felt that this statistic would be useful to illustrate the caseload of the agency.

3. It was decided that 4% of the matches would be found to be unsuccessful from the postplacement interview and the child would then be removed from the home. After counseling the child to emotionally prepare him or her, another match would be attempted. If the postplacement evaluation found the match to be successful, the child would be adopted by the family and leave the jurisdiction of the agency.

Program output: From the output, it was seen that 22 children came into the care of the agency during the 1000 days simulated. The utilization of the inspector was extremely low (0.21) but the inspector is in fact a child-care worker with a great many other responsibilities. Not all of the 22 children had to wait for the service of the inspector or child-care worker. The average number of children in the care of the child-care agency at one time was nine.

A second simulation was run to see what the effect would be if the percentage of unsuccessful matches was increased to 25%. The number of entries or children to come into the care of the agency was 24 in this case and once again there was not much waiting time for the inspector or child-care worker. The increased percentage of unsuccessful matches resulted in an increase in the average number of children in the agency's care from 9 to 13. It can be seen that the increase in unsuccessful matches may have further detrimental effects on

TABLE 9.3
Table of Definitions

GPSS Entity[a]	Interpretation
Transactions	
Model segment 1	Quality control inspector.
Model segment 2	Malfunction of machine.
Model segment 3	The timer.
Functions	
ERR1	Number of rejects when machine is running properly (per 15 min).
ERR2	Number of rejects when machine is malfunctioning (per 15 min).
Logic switches	Simulating the malfunctioning of the
MALF	machine.
SAVE VALUES	
HALT	Number of machine stoppages.
TGWIL	Malfunction duration.
LGWIL	Loss of good will cost.
SAMPL	Sampling cost.
LTIME	Stoppage and adjustment cost.
Variables	
JOE	Determine if time to sample.
GLOSS	Loss of good will cost (calculation).
SCOST	Calculation of sampling cost.
DOWN	Calculation of downtime cost.

[a]Implicit time unit = one minute.

the quality of the matches. An increase in the caseload would lead to less time to adequately prepare the child and family for adoption or less time to find the family most suited to adopt the child.

The above simulation could certainly be made more complex and realistic. The relationship discussed between the agency caseload and the quality of the match, for example, could be shown through a function. Also, the steps in the matching process could be broken down and a MATCH block used to simulate the adoption process.

Special simulation languages, such as GPSS, greatly add to the simplicity of simulation in the hands of practitioners. The General Purpose Simulation System is especially adaptable for waiting line systems, but can also be used for other situations. Once a flowchart for the real system has been prepared, showing the various entities and determinants, and once the transaction, moving through the model from event to event, has been defined, the programming is straightforward. This makes the language user oriented.

The best and most valid simulation model and experiment does not substitute for management's critical assessment of the results in view of the reality as they see and expect it. In other words, simulation does not provide an optimum solution within a set of conditions such as linear programming would do. It serves mainly as a trial-and-error approach, either where more powerful mathematical and analytical models do not exist or where the situation is too complex for such modeling. However, through simulation managers learn and gain considerable experience regarding the problem and the real productive system under their care. As a result, their decisions tend to be more sound and realistic. An example of a more complex simulation exercise in the quality assurance area, using GPSS, is given in the following.

Example 2: A certain machine in a mattress factory produces coils (springs) at a rate of 30 per minute. Periodically, the machine malfunctions, producing a large percentage of faulty coils, and as a result needs to be adjusted. The machine malfunctions on an average of 480 ± 240 minutes. When the machine malfunctions, the number of rejects produced per 15 minute period follows this distribution.

Number of Rejects	Cumulative Probability	Probability
10	0.05	0.05
20	0.10	0.05
30	0.20	0.10
40	0.55	0.35
45	0.85	0.30
50	1.00	0.15

When the machine is functioning properly, some coils are still defective, following this distribution.

Number of Rejects	Cumulative Probability	Probability
5	0.30	0.30
10	0.50	0.20
20	0.65	0.15
30	0.85	0.20
40	0.95	0.10
45	1.00	0.05

The company would like to determine when the process is out of control and when it should be adjusted at the lowest possible cost. Here, the cost of sampling, the cost of shutting down the machine for adjustment, and the cost of the loss of customer good will must be considered and balanced. The company has determined that the cost associated with the loss of customer good will is about $100 per defective mattress produced (450 coils; 15 min). The cost of taking a sample is $50 and the cost of shutting down the machine is also $50.

Using a combination of different sampling distributions (times) and various tolerances of defective coils in a sample, the company would like to determine what gives the lowest total cost. The company would also like to experiment with sampling every hour, every two hours, or every four hours, to determine if the process is out of control. In addition, the company would like to experiment with using 30 and 40 as rejection numbers (number defective in sample) to verify if the actual rejection $\geq$ rejection number, in order to adjust the process. Simulation is to be done for a 5 day work week.

Solution

Figure 9.10 shows a block diagram for the various model segments and Table 9.3 gives the definitions.

In the first model segment, a quality control inspection station is simulated. The number of rejects per 15 minute period is computed and a test is made to determine if a sample is to be taken. In the first configuration, a sample is to be taken every hour; in the second configuration, every two hours; and in the third configuration, every four hours. If the number of rejects found is greater than or equal to 30, then the machine is stopped for adjustment (even if the process happens to be in control). This same sequence is repeated with a rejection number of 40. A save value is used to determine the number of times the machine is stopped for adjustment.

The second model segment simulates the malfunctioning of the machine. A logic switch is turned on when the machine is malfunctioning and is not turned off until the machine is reset (after the quality control inspector determines that the process is out of control). A save value is also used to calculate the duration from the time the machine malfunctions to the time when the situation is corrected.

The third model segment is the timer transaction. Here save values are used to determine sampling cost, adjustment cost, and loss of customer good will.

Program output: The output from the model can be summarized in the following cost table. Figure 9.11 shows the program.

Configuration	Number of Rejects	Good Will Cost	Sampling Cost	Downtime Cost	Total
1-hr sampling	30	$1886	$2000	$550	$4436
2-hr sampling	30	2286	1000	350	3636
4-hr sampling	30	5279	500	200	5979
1-hr sampling	40	1886	2000	250	4136
2-hr sampling	40	4479	1000	200	5679
4-hr sampling	40	5279	500	150	5929

As one might expect, as the time between samples increases, the cost of lost good will increases, since the machine is left malfunctioning for a longer period of time. Conversely, the sampling and downtime costs decrease with an increase in the time between samples (as less samples are taken).

When the number of rejects is allowed to increase, the good will cost increases in the two hour sampling distribution. If the simulation had run longer, all would have likely had higher good will costs. On the other hand, with a higher rejection number, the downtime cost is decreased because the machine is stopped less frequently for adjustment.

Based on the results of the simulation, the configuration of sampling every two hours and concluding that the machine is out of control when 30 or more rejects appear in the sample provides the lowest total cost. A longer simulation might provide a more conclusive result.

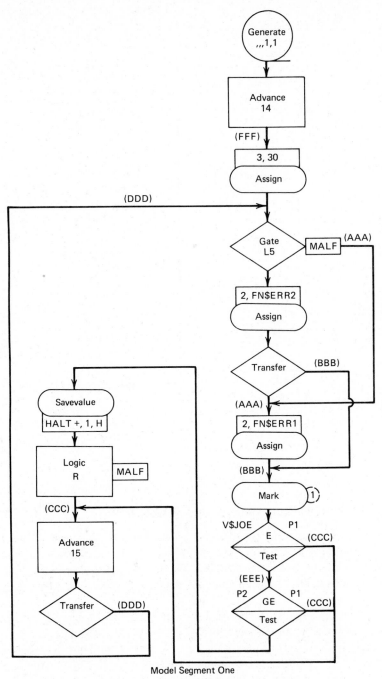

Model Segment One

FIGURE 9.10 GPSS block diagram showing inspection station and system flow.

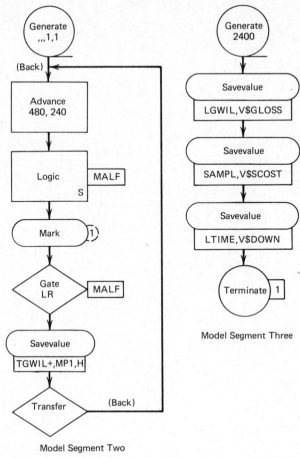

Model Segment Two

Model Segment Three

Fig. 9.10 *(Continued)*

Location and Facility Layout: Methods and Models

Quality assurance effectiveness also depends on the proper plant location and the internal workplace environment and layout. All major locational factors, such as available work force, suppliers, material resources, management aids, government services, and distribution channels, influence quality of design and conformance, and thus the resulting products and services.

Most analytical techniques can be used for evaluating alternative locations and layout patterns, such as linear programming, simulation, and cost models. There are only a few methods especially designed for this problem area. These use heuristic approaches, where predetermined rules generate near-optimum solutions.

Location factors are mostly not quantifiable. As, in the long run, management must minimize costs and maximize profits, quality assurance remains as one of the major factors considered in terms of location. Special additional checklists stressing the quality aspects of a location should be prepared when other lists do not give sufficient attention to quality. As

```
BLOCK
NUMBER  *LOC     OPERATION   A,B,C,D,E,F,G              COMMENTS
                 SIMULATE
        *
                 RMULT       611,511                   SET RN SEQUENCE
        *
        *        FUNCTIONS
        *
        ERR1     FUNCTION    RN1,C6                    REJECTS WHEN NOT MALFUNCTIONING
        .30,5/.5,10/.65,20/.85,30/.95,40/1,45
        ERR2     FUNCTION    RN2,C6                    REJECTS WHEN MALFUNCTIONING
        .05,10/.10,20/.20,30/.55,40/.85,45/1,50
        *
        *        VARIABLES
        *
        JOE      VARIABLE    P1/60*60                  TIME DETERMINANT
        GLOSS    FVARIABLE   XH$TGWIL/15*100           LOSS OF GOODWILL COST
        SCOST    VARIABLE    N$EEE*50                  SAMPLING COST
        DOWN     VARIABLE    XH$HALT*50                RESETTING COST
        *
        *        MODEL SEGMENT 1
        *
 1               GENERATE    ,,,1,1                    QUALITY INSPECTOR ARRIVES
 2               ADVANCE     14                        ADVANCE CLOCK TO TIME 15
 3      FFF      ASSIGN      3,30                      SET 'REJECTION' NO. INTO PARAMETER 3
 4      DDD      GATE LS     MALF,AAA                  SYSTEM CHECKS IF MACHINE IS MALFUNCT
 5               ASSIGN      2,FN$ERR2                 YES, DETERMINE # OF REJECTS
 6               TRANSFER    ,BBB                      GO TO DETERMINE TIME
 7      AAA      ASSIGN      2,FN$ERR1                 NO, DETERMINE # OF REJECTS
 8      BBB      MARK        1                         RECORD TIME IN PARAMETER 1
 9               TEST E      V$JOE,P1,CCC              IS IT TIME TO TAKE A SAMPLE?
10      EEE      TEST GE     P2,P3,CCC                 YES,IS # OF REJECTS >= REJECTION NO.
11               SAVEVALUE   HALT+,1,H                 YES, UPDATE # OF MACHINE REPAIRS
12               LOGIC R     MALF                      RESET THE MACHINE
13      CCC      ADVANCE     15                        NO, ADVANCE CLOCK 15 MINUTES
14               TRANSFER    ,DDD                      RETURN TO DETERMINE MALFUNCTION
        *
        *        MODEL SEGMENT 2
        *
15               GENERATE    ,,,1,1                    'MALFUNCTION' ARRIVES
16      BACK     ADVANCE     280,240                   DETERMINE MACHINE MALFUNCTION TIME
17               LOGIC S     MALF                      SET MALFUNCTION SIGNAL
18               MARK        1                         RECORD TIME IN PARAMETER 1
19               GATE LR     MALF                      WAIT UNTIL MACHINE IS RESET
20               SAVEVALUE   TGWIL+,MP1,H              DETERMINE MALFUNCTION TIME DURATION
21               TRANSFER    ,BACK                     START RUNNING UNTIL NEXT MALFUNCTION
        *
        *        MODEL SEGMENT 3
        *
22               GENERATE    2440                      TIME ARRIVES
23               SAVEVALUE   LGWIL,V$GLOSS             DETERMINE COST OF LOST GOODWILL
24               SAVEVALUE   SAMPL,V$SCOST             DETERMINE SAMPLING COST
25               SAVEVALUE   LTIME,V$DOWN              DETERMINE RESETTING COST
26               TERMINATE   1                         END OF RUN
        *
        *        CONTROL CARDS
```

```
RELATIVE CLOCK        2440  ABSOLUTE CLOCK        2440
BLOCK COUNTS
BLOCK CURRENT   TOTAL    BLOCK CURRENT   TOTAL    BLOCK CURRENT   TOTAL
  1       0        1       11       0       13      21       0        7
  2       0        1       12       0       13      22       0        1
  3       0        1       13       1      162      23       0        1
  4       0      162       14       0      161      24       0        1
  5       0       22       15       0        1      25       0
  6       0       22       16       1        8      26       0        1
  7       0      140       17       0        7
  8       0      162       18       0        7
  9       0      162       19       0        7
 10       0       40       20       0        7
```

```
CONTENTS OF FULLWORD SAVEVALUES (NON-ZERO)
SAVEVALUE  NR,        VALUE        NR,        VALUE        NR,        VALUE
       LGWIL          1839      SAMPL         2000      LTIME          650

SAVEVALUE  NR, VALUE            NR, VALUE
       TGWIL     276       HALT      13
```

FIGURE 9.11   GPSS extended program listing; inspection station simultation model.

all factors do not have the same weight in the final decision, weights can be attached and each alternative compared on the basis of the sum of such weights or ratings.

Cost models sometimes help when fixed and variable costs can be calculated for each alternative location. The various break-even points help to set decision rules depending on expected output capacity.

When quality costs and performance indicators for branches in different locations are compiled and compared, the transportation model of linear programming can, for example, be used to assign available capacity or to decide on the addition or deletion of branches. Some locations are simply more conducive to quality assurance than others and these facts should enter into locational decision-making models.

The location of workplaces within the plant can be analyzed by means of special layout and line balancing models. Changes in location of the entire plant or part of it, changes in products, processes, quality assurance programs, and so on, always require an assessment of the layout with respect to quality of work performance. The environment for worker and machine influences quality of work, and in turn quality of products and services.

Two basic types or models of facility and workplace layout are the product layout for continuous production of standardized items, and the process layout for the job shop situation with custom-made items. The heuristic approaches for determining the most desirable layout are designed to maximize flow and output in a given time. Quality assurance aspects are usually overshadowed by this quantitative criteria.

If an inspection has to be performed in an unfavorable environment, results might well be affected and distorted.

Figure 9.12 is an example of a product, or line, layout, in which stations are the assigned elements, or the tasks. Existing constraints of time demand and so forth are considered before the output is maximized and this idle time minimized. The special inspection points have been kept separate in the model as external inspection augments self-inspection, as assigned directly to the operator.

In a balanced line, the remaining and unavoidable idle time in the work stations allows the additional assignment of sampling inspection. The special inspection stations are assumed in this example to conduct 100% inspection. If sampling is done, then the model must be modified accordingly.

Downtime will also occur when operators detect actual defects and conditions creating such defects. The cycle time should be raised for promoting such assigned or voluntary inspection, if quality of performance has not already been considered in the time standards for individual tasks. The trade-off for a too tightly balanced production line is more waste and less output of saleable items. Time allowances to workers included in the cycle time during the early learning and start-up phase and a review when the steady state has been attained makes sense from a quality viewpoint. For such a dynamic line layout and balancing approach, the simple heuristic method should be augmented with a more dynamic simulation approach.

The Heuristic Method

The heuristic method for the layout of processes is slightly different from the line layout. The objective is the same; that is, maximizing of output and minimizing of idle capacity. The quality of products suffers when processes such as drilling, painting, assembling, testing, and so forth are located in a way that results in unnecessary handling and transportation of the items. The measure of efficiency for these process layout models is the minimum load time distance between nonadjacent departments and processes.

The stepwise analytical approach to the process layout is demonstrated in the following example. The test laboratory and other production departments are categorized in locational relationship to each other. Some must be located adjacent to each other, others definitely should not be, and still others have other layout conditions. The load matrix is then used

Product Layout
Situation

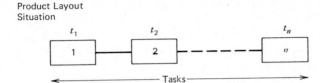

Σt = Total task time
OT = Operating time per day
D = Desired output (in number of items per day)
$\frac{OT}{D}$ = Cycle time (CT)
$\frac{OT}{D}$ = Interval (time) between units coming off
the line, or the time allowed per
station per cycle

Production Line

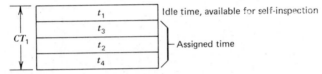

N = Minimum number of stations necessary

$$= \frac{D\,(\Sigma t)}{OT} = \frac{\text{Total time required (lots/day)}}{\text{Total work time available per day}}$$

Idle time = ΣCT_i − Assigned task time to station i.
Objective: Minimize idle time or assign it for slef-inspection.

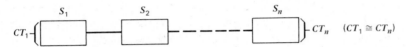

FIGURE 9.12 Line (product) layout; line balancing.

for determining the layout pattern that maximizes the criterion for an optimal and quality-conducive layout.

Example: Layout of processes, including a testing laboratory and the Department C is shown typically in the following table.

To Dept.	A	B	C	D	E	F	G	H	
A		20					40		
B			30		30				
C						70	50	40	
D			40						
E		40							
F									
G	20			40					
H				60					Σ loads = 480

From Department

Department C is the test laboratory which receives a total of 160 loads from departments like F, G, and H for final testing or inspection.

Figure 9.13 (diagram a) shows the location of the departments in relation to each other. The actual dimensions of the areas are not indicated, because the distance factor is measured by intermediate departments or processes.

The broken lines indicate nonadjacent loads (NAL). These loads are:

$$\begin{array}{rcl}
\text{AG } 60 \ (1) & = & 60 \\
\text{BE } 70 \ (1) & = & 70 \\
\text{HD } 60 \ (1) & = & 60 \\
\text{GD } 40 \ (1) & = & \underline{40} \\
\text{Total NAL} & = & 230; \text{ adjacent loads are } 480 - 230 = 250
\end{array}$$

Objective: Minimize load time distance factor. Distance is measured by number of processes between the sender and receiver of loads. One area, indicated with an "*" cannot be used for any processes, as it is an area reserved for nonproductive activities. Quality assurance considerations must be included in the decision on final layout.

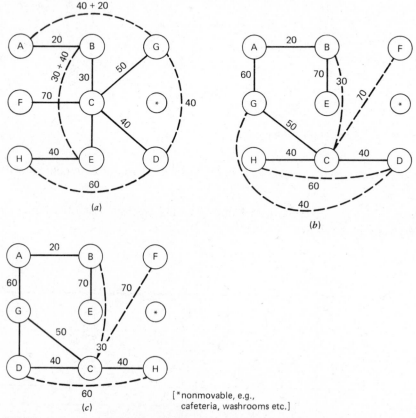

[*nonmovable, e.g., cafeteria, washrooms etc.]

FIGURE 9.13 Process (function) layout.

Approach: A heuristic program with different rules for exploring feasible and near-optimal solutions is to be applied. This method allows adaptations to different situations in which locations for processes and departments can be arranged and changed in relation to each other.

Data: To be compiled from observations during one week; loads must be considered average and representative for a sufficiently long period. Relayout might involve considerable cost. The following matrix are loads moved between processes. Note the nonadjacent loads, which in the solution will have to be minimized.

An attempt will be made to reduce nonadjacent load by interchanging the departments and/or processes. Visual inspection seems to improve the solution by interchanging GF and CE; normally only one pair at a time should be changed.

The drawing in Figure 9.13 (diagram b) shows the new solution.

Resulting NAL are now:

$$
\begin{array}{llll}
\text{BC} & 30\ (1) & = & 30 \\
\text{CF} & 70\ (1) & = & 70 \\
\text{HD} & 60\ (1) & = & 60 \\
\text{GD} & 40\ (1) & = & \underline{40} \\
\multicolumn{3}{l}{\text{Total NAL}} & = 200.
\end{array}
$$

This solution has saved 30 NAL and will be accepted.

Further improvement is sought by interchanging DH. The new layout pattern is shown in Figure 9.13 (diagram c).

Nonadjacent loads are now further reduced to 160. At this point the heuristic process of searching for an optimal solution with no NAL left is stopped. The solution appears satisfactory and will now have to be evaluated for hidden impacts on production effectiveness and quality assurance. The chart can be refined by showing the area dimensions for each process or department, so that distances can be measured in terms of actual distance measurement (feet or meters).

Computerized programs such as Automatic Layout Design Program (ALDEP) and Computerized Relative Allocation of Facilities (CRAFT) are available for handling a large scale layout pattern. A hospital or school has many more processes to arrange than those shown in the example just given. The adequacy of a layout rests in its conduciveness to a proper work environment, to quality of the workplace for people and machines, and thus also in its contribution to quality assurance.

Computerized heuristic programs, such as CRAFT and ALDEP, interchange departments systematically, evaluate the criteria of efficiency, and then accept improvements and reject negative results. In this way a near-optimum solution is gradually produced.

These layout models facilitate quality assurance in that proper physical environments for inspection and/or testing stations and critical processes can be created. Too often existing layouts can only be explained by historical developments, instead of by a rational analysis. The same models should also be used for the internal layout of large-scale testing facilities.

In summary, layout models assist management in assessing existing layouts critically and systematically, possibly with computer usage. Alternative layout patterns can also be evaluated on the basis of the criteria of efficiency. One of these criteria is quality assurance, measured in quality costs. The layout pattern that minimizes quality costs in the long run is normally identical with the one that satisfies all other criteria, such as maximization of saleable output. Moreover, a quality conducive layout of the workplace and environment is safe, pleasant, and amenable to good work performance. The quality of the workplace influences the quality of work.

Example: An inspection laboratory has to perform a set of given tests. Tasks cannot be split and must follow the given sequence conditions outlined in the procedure. Line layout with one tester per station is to be arranged. Work time per day is seven hours, and 42 tests must be performed per day, as outlined in the following table.

Task	Performance Time in Minutes	Task that Must Precede
A	3	—
B	6	A
C	7	A
D	5	A
E	2	A
F	4	A,C,B
G	5	A,C
H	5	A,B,C,D,E,F,G

1. Prepare the precedence chart.
2. Compute the cycle time.
3. Determine the minimum number of stations.
4. Assign tasks to stations.
5. Compute the percentage idle time.

Solutions

1. Precedence Chart:

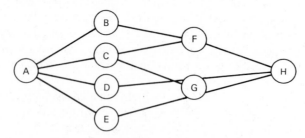

2. $CT = \dfrac{OT}{D} = \dfrac{7(60)}{42} = 10$ minutes

3. $N = \dfrac{D(t)}{T} = \dfrac{42(37)}{420} = 3.7 \cong 4$ stations

4. Use the criteria (rule) of ''largest number of following tasks'' and tie-breaker rule: ''longest task time.''

Station	Task	Task Time	Remaining Unassigned Time	Feasible Remaining Tasks	Task with Largest Number of Followers	Task with Longest Operation Time	Tasks Assigned to Station
1	A	3	7	B,C,D,E	C		A,C
	C	7	0				
2	B	6	4	F,E	F,E	F	B,F
	F	4	0				
3	D	5	5	E,G	E,G	G	D,H
	G	5	0				
4	E	2	8	H			E,H
	H	5	3				

5. Percent idle time $= \dfrac{\text{idle time per cycle}}{(\text{no. of station}) \, (\text{cycle time})}$

$= \dfrac{3 \text{ min}}{4(10)} = 7.5\%$

Example: This is an example that demonstrates line balancing method with 100% inspection assigned to work stations. Using the information contained in the table below, do each of the following:

1. Assuming an eight-hour work day, compute the cycle time per station.

2. Determine the minimum number of stations required.

3. Draw a precedence chart.

4. Assign tasks to inspection stations and compute the percentage idle time.

Tasks	Time (t) in min	Immediate Predecessor
a	3	—
b	5	—
c	1	a,b
d	4	a
e	7	I_1
f	6	d,e
g	2	f
h	1	f
i	4	h
j	3	i,g
Inspection,		
I_1	2	c
I_2	3	f,I_1
I_3	5	j,I_2

Total task time including inspection is $T = 46$ min.

Solutions

1. Plant works eight hours per day, production is assumed continuous during breaks;

$OT = 8(60) = 480$

Cycle time per station is $OT/T = 10.4$ min.

2. Minimum number of stations (N_{min}) required is $T/CT = 46/10.4 = 4.4$ or 5 stations.

3. Precedence Chart:

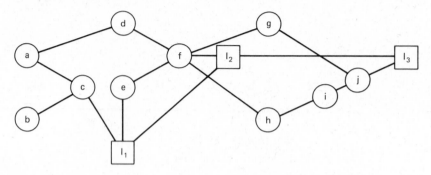

4.

Station	Task	Task Time	Remaining Unassigned Time	Feasible Remaining Tasks	Task with Longest Operation Time	Task with Largest Number of Followers	Tasks Assigned to Station
1	b	5	5.4	a,c	a		a
	a	3	2.4	c,			c
	c	1	1.4	—			
2	I_1	2	8.4	d,e	e		e
	e	7	1.4	—			
3	d	4	6.4	f,			f
	f	6	0.4	—			
4	I_2	3	7.4	g			g
	g	2	5.4	h			h
	h	1	4.4	i			i
	i	4	0.4				
5	j	3	7.4	I_3			I_3
	I_3	5	2.4	—			

$\Sigma t = 46$

$$\text{Percent idle time} = \frac{1.4 + 1.4 + 0.4 + 0.4 + 2.4}{46} = 13\%$$

Note: other heuristic rules might generate a better solution.

Aggregate Production Planning

In *aggregate production planning* various alternative plans for a given demand forecast are evaluated, and the least costly plan selected. Quality cost will vary relatively in the plans as, for instance, more defects can be expected during overtime production, inventory holding, labor force fluctuation, and subcontracting.

In the following example a demand pattern for one year has been forecasted. Beginning inventory is 100 units and holding costs are $1 per unit per period. Regular production capacity is 400 units per period. Overtime is feasible up to 100 units per period. Shortage costs are calculated as $5 per item. No back-ordering is possible. Subcontracting will involve a $6 cost per unit. Work force variation costs $3 per unit.

Quality costs will have to be added as follows: inventory $1.50 per unit per period, overtime $2 per unit, work force fluctuation $1 per unit; subcontracting will not involve extra quality costs because defective units are returned to the supplier. Under the above conditions, a plan with the lowest total cost is to be selected given the various cost figures assumed in the illustration.

Demand is as follows.

Period	Demand (units)
1	400
2	450
3	500
4	480
5	420
6	430
7	380
8	320
9	280
10	360
11	380
12	400

Consider the two alternative plans, A and B, as given below.

Plan A: Level production; no workforce fluctuation.

Period	Demand	Production			Inventory		
		Regular	Over/Under	Short	Beginning	End	Averaged
1	400	400			100	100	100
2	450	400			500	50	275
3	500	400		50	50	0	25
4	480	400		80	0	0	0
5	420	400		20	0	0	0
6	430	400		30	0	0	0
7	380	400			0	20	10
8	320	400			20	100	60
9	280	400			100	220	160
10	360	400			220	260	240
11	380	400			260	280	270
12	400	400			280	280	280

Incremental cost summary for the above plan A becomes:

Inventory holding	1220 ($1)	=	1220
Shortage	180 ($5)	=	900
Quality cost inventory	1220 ($1.50)	=	1830
Total cost			$3950

Plan B: Total labor force fluctuation, overtime, subcontracting.

Period	Demand	Production Regular	Production Over/Under	Inventory Beginning	Inventory End	Inventory Averaged
1	400	400		100	100	100
2	450	350	50	100	0	50
3	500	500	100	0	0	0
4	480	480	80	0	0	0
5	420	420	20	0	0	0
6	430	430	30	0	0	0
7	380	380	20	0	0	0
8	320	320	80	0	0	0
9	280	300	100	0	20	10
10	360	340	20	20	0	10
11	380	380	20	0	0	0
12	400	400		0	0	0

Incremental cost summary for the Plan B is:

Inventory holding	170 ($1)	=	170
Work force variation	520 ($3)	=	1560
Quality cost overtime	230 ($2)	=	460
Work force variation (quality)	520 ($1)	=	520
Inventory	170 ($1.50)	=	255
Total costs			$2965

Therefore, plan B is selected.

9.7 MAINTAINABILITY AND AVAILABILITY

It should be recalled that two of the major parameters that were mentioned concerning the definition of quality in Chapter 1 were maintainability and availability. The fact is that no matter how reliable a product is designed, failures will occur sooner or later. On such occasions failed parts have to be repaired or replaced. The need to have an effective maintenance program for the purpose of maximizing performance of production equipment and to prevent breakdowns or failures is quite obvious.

By definition, maintainability is a characteristic of design and installation which is expressed as the probability that an item will conform to specified conditions within a given period of time, when maintenance action is performed in accordance with prescribed procedures and resources.

Whether the failure is simple or complex, when the equipment is down there is always some time spent in active repair and/or waiting for spare parts. Often production planning and scheduling have to be developed using this concept of availability of spare parts to reduce the downtime (when repairs must be made to reduce production losses). Thus, an important aspect of the quality of design, when considering reliability of a product (or service), must include maintainability and availability. Programs under the titles of *maintenance engineering* and *spare part provisioning* have received great attention from military personnel organizing for quality from the very beginning.[3] Judging from the high incidence of failure of many pieces of equipment and machinery and of service promises, programs of maintainability assurance become of vital importance to all manufacturing and service operations.

The designing and planning of a maintainability program follows the same pattern as that of a reliability program. The *Maintainability Design Criteria Handbook* (see note 3 for this chapter) describes the essential tasks and major outputs of an equipment maintainability design program for military equipment based upon the following factors.

1. Reducing complexity of maintenance.
2. Reducing the need for and frequency of design-dictated maintenance activities.
3. Reducing maintenance downtime by design.
4. Reducing design-dictated maintenance support cost.
5. Limiting maintenance personnel requirements.
6. Reducing the potential for maintenance error by design.
7. Accessibility of parts.
8. Identification.
9. Standardization.
10. Safety.
11. Cables and wiring.
12. Connectors.
13. Controls.
14. Displays.
15. Handles.
16. Mounting.

Approaches for designing a maintainability program or for improving the maintainability of a design can be apportioned in the above manner to any product in question. Many designers find it convenient to work with the use of checklists under each of the major headings.

Quantification of Maintainability and Availability Requirements

To achieve many of the maintainability design objectives, it is helpful if the important requirements are quantified. For example, a typical maintainability specification may include the following figures: availability (A) requirements; dependability (D) requirements; and, mean-time-to-repair ($MTTR$) requirements.

Equipment availability (A) can be further measured in terms of operational availability (A_0) and inherent availability (A_i); which can be quantified (approximately) to:[4]

$$A_0 = \frac{MTBF}{MTBF + MDT}, A_i = \frac{MTBF}{MTBF + MTTR}$$

where:

$MTBF$ = Mean time between failures.
MDT = Mean downtime.
$MTTR$ = Mean time to repair.

Operational availability (A_0) is defined as the probability that an equipment, when used under stated conditions, will operate satisfactorily at any time. Its importance to designers lies in the planning stage when changes to design are still possible. The term *inherent availability,* on the other hand, is defined as the probability that an equipment, when used under stated conditions (in an ideal support environment), will operate satisfactorily at any given time. Its value is more important to designers since it can provide an early measurement of equipment effectiveness.

9.8   KEY TERMS AND CONCEPTS

Inspection planning is associated with the planning of production and operations; both are closely related to the respective product or service design.

Quality control as opposed to **quality assurance** is a narrow concept, because the focus is on conformance to specification and on good workmanship.

Inspection is an independent verification of required specification. Normally it is performed by a specially trained and assigned staff member, but it can also be organized as **self-inspection.** Self-inspection is performed by the operator who is responsible for conformance to specification and the related workmanship.

An **inspection plan** describes for any product or service and respective production design the various inspection points, as related to the production or process sequence. For each inspection point the kind of inspection method and object for inspection is indicated. Special inspection and/or testing procedures are prepared for more complex inspection tasks.

A **quality control plan** is more comprehensive than an inspection plan, because it includes other aspects of the production phase, such as distribution and servicing. It might also describe follow-up of inspection, such as control of nonconformances and remedial action.

Waiting line models quantify operational features of systems with arrivals that require service items to be inspected. This model helps to determine optimal size of inspection station.

Simulation technique quantifies variables of a system, such as a productive system with inspection points, and observes the behavior during a period, expressed in statistical terms.

Heuristic models interrelate a set of variables and outcomes in a predetermined systematic manner, thus generating feasible alternatives to a problem, such as the layout of an inspection station.

Aggregate production planning or scheduling determines the cost-effective production plan for a given demand pattern in general aggregated terms. Quality control and respective costs can be shown explicitly in the aggregated production plan.

9.9 SUPPLEMENTARY READING

Operator Inspection at a Japanese Steelmaker[5]

A generally held notion is that inspection work must be kept independent from operation work. At Kawasaki Steel, however, we have found that inspection by operators contributes to achieving greater quality improvement and quality assurance.

Kawasaki Steel annually manufactures approximately 12 million tons of steel products (plates, sheets, pipe, shapes, etc.) One of these products is tinplate for food cans and other cans. The entire series of production processes for manufacturing tinplate consists of various processes, from steelmaking to the tin mill process. The time required for each process ranges from a few minutes to 15 days, and it takes a long time—65 days on the average—from order entry to shipping. From the standpoint of mass production capacity, a massive flow of products goes through each process every hour. A production rate of 40 t/h is achieved in the final process of tinplating alone.

It is impossible to avoid some degree of nonconformance in these processes. Furthermore, since each process is part of continuous production, a nonconformance must be discovered immediately after it occurs. The cause must be traced and investigated, and corrective action must be taken. Otherwise, a mass of nonconforming products will follow and it will become difficult to trace the cause of the problem. It is, therefore, absolutely necessary for these mass production processes to be covered by a system to ensure that nonconforming products are stopped from entering the next process.

Our reasoning was that this problem could be solved by having the operator inspect the performance of his operation at each process. The operator is given quality targets to accomplish at each process, and has responsibility for performing inspection work. In this system, an operator's motivation for improving quality can be strengthened and less reliance on inspection can be expected.

QC Targets

We designed the Operator Inspection System to give a quality control target for each process. The control target for each process is set at a process capability (C_p) of

$$C_p \geq 1.33, \quad C_p = \frac{T}{6s}$$

where T = allowance of process criteria and s = standard deviation. This means that the operator must keep the standard deviation (s) smaller than the previous deviation.

The operator checks to find out if the item conforms with the control target. If it does, the product goes on to the next process. In case of nonconformance, immediate corrective action is taken and the product is treated as "Special Attention Material." This product is to be specially checked by another person responsible for inspection according to the specification—the guaranteed standard.

For example, if we set the allowance of temper rolling reduction within ±0.2%, the control target will be set within ±0.15%. The elongation of the sheet is measured by the extensometer and recorded. The operator is permitted to pass the product to the next process if production is controlled within ±0.15%. If elongation is off the control target but within the limits (±0.2 ∼ +0.15% or −0.15 ∼ −0.2%), the product is subjected to a mechanical test.

Through such constant checking of the control target, operators obtain firsthand information on work results, which motivates them to ensure product quality, and they will often try to improve their techniques. Of course, control targets are shown in terms of quality characteristics or operational conditions, and information regarding some control targets of quality characteristics checked by succeeding processes is fed back to the operation site affecting the quality characteristics.

Automation

In designing the plan, we also attempted to mechanize and automate inspection procedures, so that operator inspection can be easily incorporated as a part of the manufacturing system. The Operator Inspection System results in quicker corrective action; however, inspection accuracy tends to be poor, when compared with inspector inspection. Developing inspection equipment or an automatic operation method cannot only solve this problem, but also extend the application of inspection, which has previously been limited by human capabilities. (See Figure 1.)

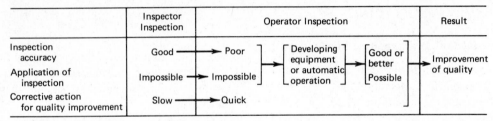

FIGURE 1   Schematic flow to establish the operator inspection system.

Inspection work for nearly all control items is being mechanized and automated at Kawasaki Steel. In particular, surface inspection equipment for cold-rolled sheets and tinplate was developed by the company in the initial period. Mechanization of visual inspection, such as surface flaw inspection, contributes not only to improving inspection accuracy but also to reducing labor costs, speeding up data collection, and raising productive capacity.

Other Improvements

The company also sought to improve on related conditions affecting the system. For example, a program of education, training and qualification of operators has been established in order to maintain and improve the levels of operator inspection. The Operator Inspection System depends upon the abilities and skills of the operators. The company has a comprehensive scheme of educating and training its operators, from newcomers to foremen.

The company has also clearly defined each operator's duty in the Operator Inspection System. It is essential to state that the operator has direct responsibility for product quality. Furthermore, job assignments must be made properly so that the operator can spare time for inspection. Even under these circumstances, the Operator Inspection System will bring some savings in total manpower.

Inspection procedures for operators have also been set up. Mechanized and automated inspection has been fully standardized and is constantly reviewed in order to keep abreast of the latest technological progress.

Furthermore, to assure that the Operator Inspection System is effective, the QA Department performs mandatory audits.

Results

The Operator Inspection System has achieved these results.

- Operators tend to strive for improvement. As a result of giving the operator responsibility for inspecting the products he makes, the amount of corrective action taken in his operation has gradually increased. The number of defects under operator inspection is running 40% less than under inspector inspection. This indicates to us that the operator's tendency to improve his work and prevent defects has been strengthened by the Operator Inspection System. (See Figure 2.)

- Inspection equipment has been developed for operator inspection. Cross-sectional profile is a critical item in quality control of thin-gage sheets, such as tinplate, and the profile of tinplate is strongly affected by the profile of the mother hot-rolled sheet. Therefore, even if information on tinplate profile is provided by operator inspection, it is not useful for control purposes unless information on the hot-rolled sheet profile is also provided. So, when the Operator Inspection System was adopted, it was desirable to enable the operator to measure hot-rolled sheet profile during hot rolling. Subsequently, an instrument which could make continuous measurement of the sheet's profile during hot rolling was developed. The hot rolling operator currently operates according to the control target, recording the profile by using a profile meter and, at the same time, making inspections. As a result, the profile of the tinplate is now greatly improved.

- Surface flaws are now detected more accurately. In the conventional process of inspecting cold-rolled products (including tinplate), the operator inspects the travelling strip visually at the final process of

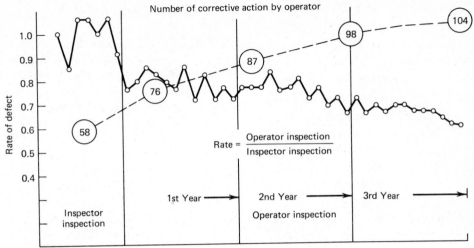

FIGURE 2   Decrease of defects due to preventive/corrective action by operator.

shearing and rejects the product if any defect is found. There is, of course, a limit to an inspector's ability to spot microdefects in the travelling strip. Consequently, Kawasaki Steel developed surface flaw inspection equipment. On-line installation of this inspection equipment resulted in a 20% reduction in misinspection of defects in travelling strip. (See Figure 3.) After the new system was adopted, misinspection was further decreased by 30%. An operator's suggestion helped achieve this sharp drop. After it was discovered that the equipment's accuracy was affected by the way oil was applied to the unit, oiling techniques and detecting sensitivity were improved, resulting in more accurate evaluations.

The Operator Inspection System has contributed to improving quality; operators' motivation to strive for quality improvement has been also enhanced. Yield improvements in the past three years were 1.4% for hot-rolled products, 1.0% for cold-rolled products, and 0.6% for tin mill products. Furthermore, surface defect complaints about cold-rolled products and tin mill products have declined sharply.

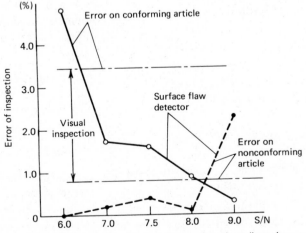

FIGURE 3   Error of visual inspection and surface flow detector.

9.10 DISCUSSION AND REVIEW QUESTIONS

(9–1) *An inspection plan is different from an inspection procedure.* Explain.

(9–2) *There are certain advantages to a combined production and inspection plan.* Explain.

(9–3) *Production planning is performed in certain logical steps and quality assurance is involved in most of them.* Explain.

(9–4) *While an inspection plan format is not standardized, symbols for flowcharts do apply.* Explain.

(9–5) *Self-inspection should be allowed under carefully prepared conditions.* Discuss.

(9–6) *Inspection points should be considered under certain conditions.* Explain.

(9–7) *One hundred percent inspection is seldom the most effective.* Discuss.

(9–8) *Waiting line models and simulation are particularly applicable for inspection planning.* Discuss.

(9–9) *When heuristic models are used for determining the layout of inspection facilities with relation to other productive units, a certain measure of effectiveness must be determined.* Explain.

(9–10) *Self-inspection in work stations along an assembly line is more difficult to arrange than in a process layout.* Explain and discuss.

(9–11) *Inspection should always be designed to prevent, rather than to weed out, defectives.* Discuss.

(9–12) *Inspection plans should be reviewed in regular intervals and changed only under controlled conditions.* Discuss.

(9–13) *Production managers should approve inspection plans.* Discuss.

(9–14) Prepare an inspection plan of the type shown in Table 9.1 for a gas service station.

(9–15) Select an item or service of your choice and prepare an inspection plan using the flowchart format.

9.11 PROBLEMS

(9–1) Select a product of your choice which has major measurements and attributes. Assuming that if any one of these critical quality characteristics is nonconforming to specification in the end item the item is defective:

a. Prepare a listing that identifies in the heading of the record the item and other related data (in the body of the record list) the major quality characteristics (about five to ten), the respective possible defect(s) for each characteristic, and the general inspection methods, such as visual inspection. Give further related information in the footnote of the record.

b. Write a procedure for preparing the design document as outlined above. This procedure must identify the steps taken for determining and compiling the data, and also the purpose and application of the document.

(9–2) For a common household lamp, conduct a simple failure mode and effects analysis. The record should show the following: name of component, failure mode, cause of failure, effect of failure, correction of the problem. It is assumed that you are not an expert in household lamps.

(9–3) Select an item of your choice and conduct a failure mode and effects analysis. Explain the purpose and use of the information from this analysis.

(9–4) Prepare a fault tree for a simple item of your choice and describe the use of the document with regard to quality assurance.

(9–5) Prepare a checklist for inspection purposes from problem (9–3) or (9–4).

(9–6) A small woodworking shop wants to control frequent changes in drawings supplied by customers. Prepare a procedure of about five steps that is designed to assure working with the latest drawing.

(9–7)   Simulation: Set up a flowchart for simulating the following situation. Acme Widget Works produces high quality widgets for the aerospace industry. Because of tight industry standards, widgets must be 1000 ± 10 microns in size. To maintain its excellent reputation Acme wishes all widgets to fall within the limits.

Acme's process produces units with a mean size of 1000 microns normally distributed with a standard deviation of three. The quality control team has set control limits at $X \pm 30$, or 1000 ± 9 microns. All units must be tested. They are produced at a rate of one every 10 minutes and go to the tester, who takes 10 ± 3 minutes to perform the measurement. Defective units must be analyzed and this results in their destruction. As soon as there have been two defects, production must be shut down. It takes 60 minutes to restore the system. Management wants to simulate the system in order to determine an adequate number of testers and adequate inspection capacity.

(9–8)   In a brewery, two inspector/repairmen are used at the carton inspection point. Each inspector performs two functions; watching the cartons go by on the belt and removing those cartons that are defective to repair them, if possible. The repair function is a two-step process; examination of the carton to determine the type of repair that is necessary and doing the actual repair.

The cartons come off the wraparound machine two at a time, and then pass by the inspection station. The cartons are either taken off and put into the repair area or they continue on to packing and loading. After repair, cartons are again put on the line; some cartons, however, are beyond repair and are returned.

The quality assurance manager wonders if only one inspector would be sufficient. Possibly a second inspector would help to reduce the number of defective cartons passing by

undetected since there have been some complaints recently.

Prepare a flowchart for a simulation model.

(9–9)   A large industrial company has just received a shipment of 15 lots of 1000 units each that it requires in the assembly of its final product. In recent months the supplier has been suffering from the economic recession and thus has had to trim their budget. The units they manufacture are constructed according to strict specifications. However, since they have been cutting corners, rumors have been circulating that some of their machines have become worn out, resulting in a deterioration of quality. As a result, the receiver of these 15,000 units has decided to carefully inspect the units and, based upon this inspection, will make a decision whether to accept or reject the shipment. Since the company needs these units now, it will inspect the units by lot rather than to inspect the entire shipment. Therefore, it will only return the lots which are found to be unacceptable. Management has decided that a random sampling of about 5% will be taken from each lot. Since they have agreed that about 4% defectives is an acceptable level, they will reject any lot that has more than two defectives in the lot. One further constraint that is faced by management is the machine that they use to inspect the units. The machine has a fairly short lifespan, and once the machine has been used for a certain amount of time the results of the inspection are not very accurate. Thus, management wants to ensure that once fifty hours of machine use has been recorded they will be notified at which point the machine should be replaced. The machine takes 6 ± 4 minutes to inspect each unit.

Prepare a flowchart for a simulation experiment, in order to determine the reliability of the receiving inspection and the acceptance sampling procedure.

(9–10) Simulation: A television repair shop wants to estimate the normal service time per order so that they can quote more realistic times to their customers. The reputation this shop wants to build in the community is that the unit is properly repaired and inspected within the time quoted, so that customers can rely on having the unit back when their favorite program of the week is on. In the shop the first technician inspects the unit to see that parts are properly replaced or the unit serviced. Next he performs the first general inspection to see if the unit is functioning properly. If minor adjustments are required, the technician will do them immediately. Then he puts the unit on a table where it is left running for 30 minutes. Finally, once the unit has functioned for 30 minutes the technician inspects it once again, checking for flaws. The unit then leaves the store. The simulation is to determine the time the technician must dedicate to this quality inspection and the number of units which are rejected at each of the three points in the inspection sequence.

Prepare a flowchart and the simulation exercise. Outline specific data which would have to be compiled for this experiment.

(9–11)   Waiting line problem: At a test station in an electronic assembly plant, units for testing arrive at the rate of 15 per hour. The tester can test one of the units every three minutes.

Assuming Poisson arrivals and exponential service, find:
a.   The utilization rate of the tester.
b.   The average number of units waiting for testing.
c.   The average waiting time.

(9–12)   A print shop has four copying machines, one of which occasionally breaks down, producing defective copies, or no copies at all. The working machines are automatically serviced about twice each hour. Service time, on the average, is five minutes. Equipment downtime costs $25.00 per hour and the attendant servicing the machines is paid $3.00 per hour.
a.   What is the average number of machines waiting for service?

b.   Should the firm hire another attendant at the same rate?

(9–13)   Line balancing: Students' applications for university scholarships are processed through several interdependent work stations, where the application forms are transported by a belt from one station to the next. Transportation time is constant between all stations. The line is to be balanced and the idle time is then to be used for intermediate inspection. Cycle time is the minimum length possible, as shown in this table.

Work Element	Time (minutes)	Predecessor
a	1	—
b	2	a
c	1	—
d	2	b,c
e	1	—
f	6	e
g	8	d,f
h	12	g

a.   Draw the precedence chart.
b.   Assign tasks to stations, by taking longest times first.
c.   Determine the idle time in stations and make comments concerning feasible inspection.

(9–14)   Location/layout analysis: A hospital wishes to arrange six departments into the following floor plan.

| 1 | 2 | 3 | 60 feet |
| 4 | 5 | 6 | |

90 feet

Each of the locations, one through six, is thirty feet square. The criterion for allocating departments A–F to locations one through six is minimizing the total sum of patient trip distance traveled between the departments.

This is the sum of the products of the number of patient trips between departments and the distance between departments. Thus, the quality of the solution is a function of the total patient distance traveled.

Assume that records show the following to be the average number of patient trips between departments per week:

To Department

	A	B	C	D	E	F
A	—	15	20	16	24	
B	30	—		55		15
C	100	22	—			
D				—	80	60
E	60	14	100		—	
F			5		12	—

(From Department at left)

Assume also that the following table gives the distances between locations.

	1	2	3	4	5	6
1	—	30	60	30	43	68
2	30	—	30	43	30	43
3	60	30	—	68	43	30
4	30	43	68	—	30	60
5	43	30	43	30	—	30
6	68	43	30	60	30	—

To provide adequate patient care quality, departments A and C must be located adjacent to each other. To minimize risk of infection, departments C and F must be located as far as possible from each other. Arrange the six departments.

(9–15) A bicycle manufacturer produces two kinds of bicycles. Inspection can handle 1000 units of type A and 700 units of type B. The stamping department can handle 1200 type A units and 1400 type B units. The assembly department handles 1000 units of A and 900 units of B. The paint department can handle 800 and 1200 units respectively of types A and B. The profit per bike is $20 per unit for A and $35 for B. How many of each should be produced in order to maximize the profit?

(9–16) A company making golf carts has arranged three inspection stations for its three product lines. Inspection costs vary with each individual inspector at the stations and the respective product line, as shown in the matrix below. The inspection station can handle any of the three product lines. Assign inspectors to the product lines for inspection so that total inspection costs are at a minimum.

Inspection station/ Inspector	Product Line			Inspection Time Available
	One	Two	Three	
A	8	15	3	15
B	5	10	9	7
C	6	12	10	6
Required inspection time	12	8	8	

(9–17) A quality control laboratory manager must decide on the purchase of expensive testing apparatus. The alternatives are to buy only one apparatus which would probably be insufficient or buy two and allow for excess capacity. Naturally, one machine can be bought first and then the second later, but by buying two now there is a considerable discount.

The estimates for low demand is 0.3 and for high demand is 0.7. The present net value for purchasing two machines immediately is $90,000 if demand is low and 150,000 if demand is high.

The present net value for one machine at low demand is $100,000 and the options in this case are to do nothing (net present value [NPV] $60,000), to subcontract (NPV 120,000) or to purchase a second apparatus ($110,000).

How many of these testing machines should the manager purchase initially? (Hint: a decision tree must be used for this problem.)

9.12 NOTES

1. Putnam, A. O., "Three Quality Issues American Management Still Avoids," *Quality Progress,* December 1983, p. 12.

2. Many textbooks on statistics give tables for random numbers.

3. The reader would immensely benefit from a review of the selected materials contained in the various handbooks and literature of the U.S. Military and Navy publications. Department of the Army Pamphlet 705-1, Research and Development of Materials, *Maintainability Engineering,* Headquarters, Department of the Army, Issued June 1966; *Maintainability Design Criteria Handbook for Designers of Shipboard Electronic Equipment* (Change 4), Document No. NAVSHIPS 0967-312-8010, Department of the Navy, July 1972; MIL-STD-471A Military Standard, "Maintainability Verification/Demonstration/Evaluation," March 27, 1973; MIL-HDBK-472 Military Standardization Handbook, "Maintainability Prediction," May 24, 1966; MIL-STD-756A, "Reliability Prediction," May 15, 1963, Defense Supply Agency, Washington, D.C.; DoD. Directive 4100.35, *Development of Integrated Logistic Support for Systems and Equipments.*

4. The Department of the Navy document, NAVSHIPS 0967-312-8010, contains availability nomographs to be used without a need to solve the equations.

5. Kondo, Isao et al., "Operator Inspection at a Japanese Steelmaker," *Quality Progress,* December 1982, pp. 14–16. Copyright © American Society for Quality Control Inc. Reprinted by permission.

9.13 SELECTED BIBLIOGRAPHY

Barker, Eugene M., "Why Plan Your Inspection," *Quality Progress,* July 1975, p. 22.

Dodge, H. F., "Inspection for Quality Assurance," *Journal of Quality Technology,* July 1977, p. 99.

Jensen, Ronald J., "Evinrude's Computerized Quality Control Productivity," *Quality Progress,* September 1977, p. 12.

Lester, R. H., Enrick, N. L. and Mottley, H. E., *Quality Control for Profit,* Industrial Press Inc., New York, 1977.

Perry, Sandra L., "Quality Control and the Design of Jobs," *Quality Progress,* February 1979, p. 26.

Schrock, Edward M., "How to Manufacture a Quality Product," *Quality Progress,* August 1977, p. 25.

Troxell, Joseph R., "Standards for Quality Control in Service Industries," *Quality Progress,* January 1979, pp. 32–4; "Service Time Quality Standards," *Quality Progress,* September 1981, pp. 35–7.

CHAPTER 10

QUALITY OF DESIGN: RESOURCES AND SUPPLIES

After having specified the quality of a particular product or service and then prepared the production and inspection plans, we complete the quality of design stage by planning the provision of adequate resources and supplies. Any company, and for that matter, any productive system, needs input from outside resources, such as material, insurance services, manpower, and finances. This is true for a retail store, a manufacturing plant, a government department, or a gold mine. While certain materials and services are directly used for ongoing production and have a significant quality impact, others, such as available staff and equipment, are already presumed available as internal resources on a more long-term basis. Nevertheless, all resources used for production for a particular item or for a product line determine ultimate product quality, because they are production inputs.

The production and inspection plans outline quality specifications along with required quantities, delivery dates and conditions, and delivery destinations.

After having decided on the quality specification of the product and of the respective production processes, the company is now ready to purchase supplies. The quality objectives that are set by the customer and the producer must now be further translated into objectives for the supplier, or the subcontractor. Only when the supplier meets quality specifications can we expect that defects will be minimized and avoided as early as possible.

The ideal of *zero defects* in supplies and perfect conformance with specifications is an objective which can seldom be achieved. However, there is a great deal which the producer can do to assure the quality of resources and supplies before he actually allocates orders, and receives them. The principle of prevention of defects requires that the producer keep a list of qualified vendors and verify resource and supplier capabilities before the actual ordering takes place.

269

10.1 ASSESSING RESOURCE CAPABILITIES

In the preceding chapters we have stressed the short-term aspects of production and productive systems, with regard to product and production design, as they relate to quality assurance. The focus remains on a particular product or service and its customer. Longer-term aspects of production design, and establishment of basically sound resource bases and supplier relationships, will be discussed later under quality assurance programming. In the short run, most companies do have an existing production capacity and also entertain business relationships with many suppliers as an ongoing concern.

Given a particular product design, adequacy of the available process capabilities must be confirmed during product design and production planning. Under normal circumstances, internal facilities, process capacity, and manpower know-how should suffice, or supplements be made available within a relatively short time. Once the production and inspection plans have been approved by senior management, and possibly by the customer, quality assurance for external suppliers must still be dealt with. More and more however, customers will not be willing to approve plans without first having attained inspection plans from the suppliers. Producers will also be held liable for quality of external supplies and subcontracts. For many obvious reasons, the producer must therefore:

1. Exert special care regarding suppliers' quality and relationships. These otherwise independent enterprises with separate management must learn to communicate and cooperate on equal levels.

2. Rely more on indirect quality controls as opposed to internal operations.

3. Delegate the responsibility for quality to the suppliers' quality assurance systems.

The objective and task for the producer, as assumed in this chapter, is to establish quality assurance of major supplies and subcontracts in conjunction with the overall product and production design. Only when this goal is achieved can the actual internal production with its operational planning commence. Proper quality assurance of supplies is designed in principle to prevent defects and failures as early as possible in the life of a product or service. This general policy has evolved forcefully in recent years with growing quality awareness, and is expressed in the following.

While in the past there was a tendency to "make parts fit" today every occurrence of a poor fit is dutifully recorded and made known to the supplier if the problem persists. In the past, all of the problems were solved in the plant; now we are making the vendors responsible for getting it done right the first time.

A typical problem producers have with their suppliers is that the actual capabilities of vendors do not suffice, quality-wise. Quality obligations are not really understood by the suppliers and are often overshadowed by other obligations and contract stipulations. This is often the case when the lowest bid is accepted. Meeting delivery dates, rather than quality specifications, is often considered more important. Subsequent changes in the contract create instability and often have a negative quality impact.

In order to avoid contracting with unreliable and unqualified suppliers and subcontractors, producers now impose compliance with official quality program standards. Standards, such as the MIL-Q-9858A or the Canadian Standard Association's Z299 series outline quality assurance requirements for major suppliers in conjunction with purchasing. The ANSI Standard *Generic Guidelines for Quality Systems*, Z-1.15, (see note 2 for Chapter 8) under

"Control of Purchased Materials" outlines that a quality system should include procedures to assure effective supplier quality management for all purchased materials. Supplier quality management as described includes:

1. Supplier selection methods.
2. Material quality requirements.
3. Supplier quality system requirements.
4. Material quality assessment.
5. Inspection facilities and procedures.
6. Supplier performance evaluation.
7. Nonconforming material control.
8. Supplier quality evidence, use of statistical techniques.
9. Quality control, or surveillance at source.
10. Receiving inspection.

These topics describe steps in establishing a suitable supplier's quality management system.

10.2 QUALITY ASSURANCE IN PURCHASING OR PROCUREMENT

The importance of quality assurance in purchasing lies in designing relationships with quality assurance disciplines properly and truly considered by the producer as well as by the supplier. Figure 10.1 illustrates a general purchasing or procurement scheme and associated special

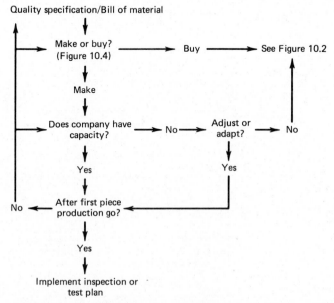

FIGURE 10.1   Establishing a general purchasing or procurement scheme for associated quality assurance measures.

quality assurance measures. The production and inspection plans, together with quality specifications derived from the product design documents, initiate the purchasing or procurement activities (Figure 10.2). A typical purchasing process flowchart is shown in Figure 10.3. What follows are:

1. Requisitions preparation (possibly after a "make or buy" analysis).
2. Contacting.
3. Appraising.
4. Selection of the supplier (based on quality, price, and delivery).
5. Negotiating and concluding of the legal contract.
6. Communicating and initiating of quality control.
7. Inspection plans.
8. Surveillance and other performance controls of the supplier.
9. Receiving of the product or service.
10. The completion of the contract.

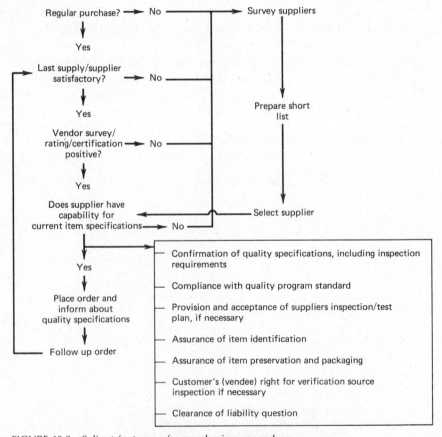

FIGURE 10.2   Salient features of a purchasing procedure.

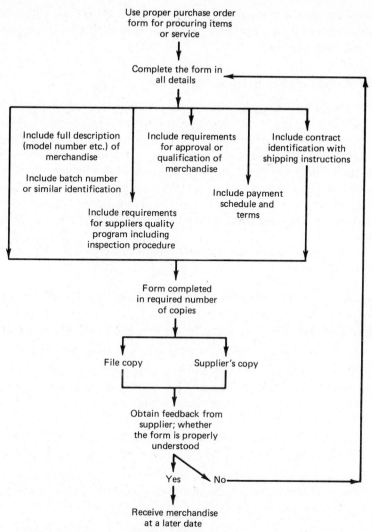

FIGURE 10.3 Purchasing process flowchart.

This basic scheme applies not only for producer–supplier relationships in the manufacturing industries, but in other nonmanufacturing industries as well. For instance, quality assurance of supplies in retailing and wholesaling is as important as in manufacturing. Hospitals also require careful planning and control of their medical and other supplies and of contract services, in addition to the hospital's own facility capabilities and staff. Even in a theater, the best actor can not perform without being supported by supplies and services which the theater company contracts for from external sources.

In the following we shall concentrate on the quality assurance of supplies and services from outside the company for major items which have an important quality impact on the product in question. Using production and inspection plans as the basis for the discussion, we shall first list some principles instrumental for sound quality assurance of supplies. Then the individual steps of the general purchasing cycle will be followed through. This will complete the quality of design cycle for all phases of the productive system: the output, throughput, and input phases.

Principles of Sound Supplier Relationships

Principles that should be observed in maintaining relationships with suppliers are well formulated in standards for quality assurance programs. Such guides and standards can help both producer and supplier to establish and maintain positive business relationships. Therefore, these documents should be agreed upon wherever quality assurance is an important feature of the contract and the ultimate product or service. Modes for selecting and surveillance of suppliers and subcontractors are then regulated in a generally acceptable and consistent manner. This fact strengthens defense for possible future liability claims, as the producer can prove having invoked and applied proper measures concerning quality assurance of critical supplies.

With the use of quality assurance standards, important principles will be automatically observed. These are that:

1. Quality specification of the contract will be sufficiently and clearly defined and communicated.

2. Requirements for approval and acceptance of the item by the producer will be clarified.

3. Information which allows the producer to survey supplier performance will be forthcoming from the supplier, including test data or control charts forwarded with the shipment.

4. The supplier's quality assurance program will be defined in the program standard imposed by the buyer and user, and a verification from the producer, prior to assigning the contract, will be made possible.

5. The records of all agreements and assurance decisions will be kept, including changes in the contract and required corrective actions during the contract performance.

6. For future references, supplier performance will be assessed and rated fairly and objectively. This will allow the maintenance of a qualified vendors list and selection of suppliers only from the list.

7. In case of major supply contracts, further appraisal of supplier quality assurance will be conducted, such as: plant visitation, approval of inspection plans and source inspections, verification of important inspection and test equipment of the supplier, and so forth.

With discriminating and careful imposition and appraisal of quality assurance on suppliers, the producers, in the role of customers, shift responsibility for quality and quality assurance to where it actually belongs: with the supplier. Not only will the probability of defects be minimized but the producer can reduce their own receiving inspections. In order to reap the benefits of sound quality assurance from suppliers, many major producers have to invest much time and energy in such relationships (educating suppliers, creating an awareness of their needs), particularly when dealing with a great number of small suppliers. Producers with their own internal quality assurance procedures can demonstrate much leadership. The producer's purchasing agents and quality assurance specialists need to coordinate all measures systematically and effectively.

Specification of Quality of Supply

After the major quality characteristics are known by the producer and the producer's customer, and once quality control and inspection plans are completed, the task remains to translate these also for suppliers. When broken down into specifics, one finds the essential elements to be as follows.

1. Contracts and purchase order documents must communicate explicitly all major and critical specifications and pertinent quality assurance measures and obligations to the supplier in an unambiguous manner. This may sometimes require new formulation and documentation of quality specifications with additional information and explanations.

2. Simple order forms with standardized entries and fine prints do not necessarily ensure that suppliers and subcontractors know and understand precisely the quality they are expected to deliver. The more complex the item or service in question and the more unknown the supplier, the more care in specifying and communicating quality characteristics becomes a necessary part of management planning.

3. Often the producer must rely on the supplier's technical knowledge and experience, so that the imposition of a standardized quality assurance program becomes relatively important. However, the producer, in the role of a prudent consumer, should verify not only that the supplier does in fact understand quality specifications, but that adequate supply capacity and quality assurance measures also exist.

4. Before the producer decides on buying and subcontracting, much information is needed. It might be that the producer would be better off making their own components and relying on internal resources, rather than becoming dependent on other independent enterprises and supply bases. A thorough make-or-buy study or value analysis should be undertaken in cases where such an alternative does reasonably exist.

Figure 10.4 exemplifies questions to be addressed in a make or buy study. This study should not be undertaken without all quality specifications being known for the item or service in question. The decision might easily be made by emphasizing considerations other than quality, such as price, delivery dates, reciprocity, reliability, maintainability, and possibly a host of other items.

10.3 SUPPLIER SELECTION

Once the final decision to buy has been made suppliers have to be contracted, their suitability and capability have to be appraised, and the best one has to be selected. Again, quality assurance has to be given adequate weight in these decisions.[1]

The first contact with suppliers might have been earlier, when questions were raised during the design and production planning phases. Producers, particularly in small business environments, may have long time personal business relationships with most of their major suppliers. However, contracting for supplies and services, particularly when critical quality impact decisions need to be made, demands an objectivity analysis. This principle demands some formality, if for no other reason than to prove in case of future liability claims that proper care had been observed. Quality assurance standards, whether or not they are imposed by the producer, also require formal supplier selection procedures.

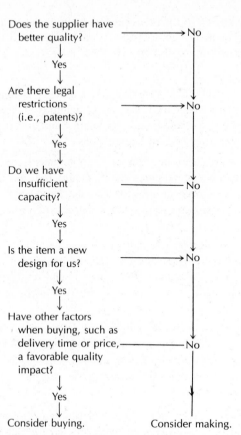

Key Factors

Does the supplier have better quality? ————→ No

Consistency in the overall quality throughout the life cycle; profit considerations.

↓ Yes ↓

Are there legal restrictions (i.e., patents)? ————→ No

Patent or legal restrictions, safety or environmental laws.

↓ Yes ↓

Do we have insufficient capacity? ————→ No

Long-term considerations, supplier relationships, skills

↓ Yes ↓

Is the item a new design for us? ————→ No

Design capability of supplier, optimum utilization of internal resources, quantity.

↓ Yes ↓

Have other factors when buying, such as delivery time or price, a favorable quality impact? ————→ No

Other intangible factors such as seasonal demands, reliability of outside suppliers.

↓ Yes ↓

Consider buying.   Consider making.

FIGURE 10.4   Make or buy analysis.

Many factors play a role in supplier selection, including the actual ready availability of suitable and acceptable suppliers from which the producer can in fact select. However, even in a so-called ''seller's'' rather than ''buyer's market,'' the buyer must beware, and assure quality of supplies. Without this possibility the producer does not have the capacity to assure quality to the ultimate customer. In the long run, the producer should try to substitute suitable suppliers for inferior ones or to extend in house production capacities accordingly.

Selection Criteria: Purchasers' Point of View

Suppliers to be contacted are usually already preselected. They must be known to have adequate supply capacities and possibly be registered on the producer's qualified vendors list. Many companies apply for status of qualified supplier to the major public and private organizations and procurement agencies. Even simply being accepted (without receiving any supply contract) has many advantages for the applicant. For instance, the fact of being a ''qualified'' supplier for a major company proves to others that one's quality assurance does meet certain standards. Major customers, in their own interest of maintaining business relationships with certified suppliers, provide valuable management aids to applicants, which are often small businesses.

Selection Criteria: Suppliers' Point of View

Suppliers who contact producers will have to meet certain quality assurance standards that vary with the degree of quality sensitivity of the item in question. After having received the application the producer will then inform suppliers about further procedures and required quality program standards. A quality manual, describing the procedures and other quality assurance measures of the supplier will normally be appraised before an in-plant visit or audit by the producer's representative will be made to verify proper implementation of the program.

This manual development and the compliance with such quality program standards will be discussed in a later chapter in conjunction with quality assurance programming. Here, it is of relevance whether or not a particular supplier is in fact known to the producer to be acceptable. Once the production and inspection plans for a particular item have been finalized suppliers will have to be contracted. A list of qualified suppliers is, of course, the major information source for decisions regarding suppliers.

The Conventional Approach

The conventional approach is to make the tender documents available to the general public, or to a restricted group of suppliers with certain predetermined and described qualifications. This tendering approach provides the opportunity to attain necessary quality assurance that can facilitate a fair and objective selection of a supplier.

Usually, after the initial contact has been made, the actual selection might be, as in the tendering scheme, more or less automatic. However, selection of a supplier does not as yet mean that a contract has been signed. The producer will have to appraise the actual supply capacity. The performance of a vendor is conveniently measured in terms of *quality, cost,* and *delivery* factors. A fourth factor can also normally be introduced to include *service rating*. A typical form for use to keep a record of supplier's quality control performance is shown in Figure 10.5.

A typical, frequently used set of weights for rating is given below:

Quality . . . 30 percent

Cost . . . 40 percent

Delivery . . . 30 percent

As shown in Figure 10.5, quality is measured in terms of the percentage of shipments accepted. Cost is given in the form of unit price, and delivery is evaluated as the percentage of time when promises for deadlines were actually met. The service rating may include percentage of times when interruptions or unnecessary troubles (solely because of the vendor's internal problems) lead to unpleasant surprises. Based upon this scheme, many companies give accreditations or certificates to their vendors. For example, a marginal rating is given to a supplier who is barely meeting the requirements. An acceptable rating goes to a supplier who meets requirements, but still needs much improvement to reach a qualified rating.

Other types of rating schemes use what is known as a point system.[2] A scale of zero to five is applied to each item to allow for a quantitative comparison among vendors similar to

Supplier Quality Control Record

Supplier:					Product/Part/Service:					
Date										
Quality rating										
Price rating										
Delivery rating										
Service rating										
Total rating										

Defects: (by code number)

Quality History

Date	Purchase Order No.	Lot size	Sample size	Accepted	Rejected	Percent of defectives

Comments:

FIGURE 10.5   Supplier quality control record form.

278

grading of students in a class. A commonly used rating scale is: A. four to five points—very good, B. three to four points—good, C. two to three points—average, and D. below two points—poor.

Supplier Quality System Survey

As was mentioned, the producer will require further information from the supplier specifically on quality assurance capabilities. The approach can vary from a simple mailing of a questionnaire, called a desk audit, to an elaborate supplier survey. In Table 10.1 some of the questions asked by the producer are listed. Preparing for such surveys and conducting them is a topic that will have to be discussed later. Usually, when a particular production and inspection plan is to be implemented with regard to a selected supplier, the product-related survey, based upon agreed principles, is a fairly straightforward matter. Once the quality characteristics and related obligations have been communicated the contract has been negotiated, and the supplier's ability to supply has been assessed to the extent possible, then the producer will most likely request only an inspection plan before approving actual commencement of the work by the supplier.

Supply contracts might include additional quality assurance stipulations to regular purchase orders. When quality characteristics are of relatively high importance, contracts require some kind of surveillance of supplier performance during production rather than only at delivery through receiving inspection.

TABLE 10.1
Potential Supplier Questionnaire

1. Have you a production controller?
   Name (_____)   Telephone (_____)
2. Have you a quality manager?
   Name (_____)   Telephone (_____)
3. Have you a chief inspector?
   Name (_____)   Telephone (_____)
4. Please describe your quality control organization with a family tree.
5. Have you a quality control manual?
6. Have you a bonded and quarantine store?
7. Are instruments and test equipment regularly calibrated?
8. Are all processes controlled by specification?
9. Are processes verified to stay in the state of statistical control?
10. Is documentary evidence of process control, inspection, and test maintained?
11. Can all materials and components be traced to the original batch at any point in the manufacturing cycle?
12. Are cost effectiveness studies carried out?
13. What are the company's normal products?
14. Is a brief description of production facilities available?
15. Are modern planning techniques used? (Critical path method, programme evaluation and review techniques, line of balance.)
16. Do you use vendor rating schemes?
17. Have you a R&D laboratory with environmental test facilities?

Contracted agreements that involve surveillance are, for instance:

1. Confirmation of attainment of major quality specifications at inspection ''hold points'' during production.

2. Examination of compliance with quality program standards.

3. Assessment of supplier's inspection and/or test plans.

4. Assessment and reliability of testing capability.

5. Assurance of item identification.

6. Question of traceability.

7. Clearance of nonconformances and corrective actions.

The *surveillance,* in whatever form, has mutual benefits for both producer and supplier in that it fosters closer cooperation, particularly in quality matters. It allows for ready clarification of questions, and intermediate verification for confirming that performance standards have either been met or corrective actions have been taken.

One important principle to be observed in such surveillance is that the producer does not interfere unduly with the independent management and decision making of the supplier. Normally such intermediate checks by the producers are well defined in the contract and the quality assurance and performance standards agreed upon by the parties. The agreements may also specify the kind of surveillance (authorized inspection agency, in-plant personnel, etc.).

A *surveillance* is defined as ''monitoring or observation to verify whether an item or activity conforms to specified requirements.''[3] One can further differentiate between purchase order audit, surveys, and source inspections. As a matter of wise policy, a purchase order should be checked regarding quality assurance.

Such an *order review* confirms that the supplier has been approved, that instructions are adequate, and other quality assurance stipulations are properly documented and communicated. Surveys and source inspections verify suppliers capabilities and performance in more detail and more frequently.

Surveys can be initiated by the supplier or by the producer. The supplier might want to attain qualified supplier status or bid for a contract. These precontract appraisal surveys were already mentioned. The object of such surveys, or audits, is the entire quality assurance program and the organization, rather than only the work in progress. In cases where the supply contract deals with one particular item, surveillance is usually restricted to this object.

The example of the pipe valve described in a previous chapter included an inspection plan. Under this plan some items were required to be purchased. Although the item in this example would probably involve only a regular purchase order audit, one can imagine a more complex part or subcontract. Here, the subcontractor would also have to submit an inspection plan for approval. This plan would designate inspection hold points where the producer could conduct source inspections.

This *source inspection* would be initiated by the supplier on the basis of the approved inspection plan and would be done at the time when production has progressed to this stage.

Naturally, the producer often undertakes inspections or more general surveys and audits independent from the inspection plan. In cases of major changes in the order or contract, a representative of the producer will inspect quality assurance in general or for certain elements such as personnel, nonconformance report, declining supplier rating, and so forth. Normally, surveys are announced to the supplier, so that staff will be available. Major producers provide supplier handbooks in which details of surveillance are outlined; this is in addition to the quality program standards mentioned.[4]

Source inspections are greatly influenced by the outcome of more general surveys and audits of the suppliers. The examination of a supplier's quality assurance program and the cooperation at the level of management will be discussed later. Moreover, the documentations and ratings of past supplier performances guide the frequency and intensity of special item and work-in-progress inspections. Wherever possible and reasonable, producers will tend to minimize costs through reliance on the quality assurance program of suppliers and the confirmation of good quality performance through product inspection.

10.4 RECEIVING INSPECTION AND SUPPLIER PERFORMANCE EVALUATION

The purchasing cycle ends once the item has been accepted by the producer. Source inspection cannot substitute entirely for the receiving inspection when packaging, transportation, and assembly, had been involved. Yet, reliable and positive verification of quality at the suppliers plant and the respective prevention of defects and nonconformances give sufficient confidence for reducing receiving inspection. Statistical acceptance sampling is usually the proper acceptance procedure in which risks for the supplier and the buyer can be quantified (see Chapter 12). Such sampling plans should be agreed upon in the original supply contract. Published standards for acceptance sampling, such as the widely accepted MIL-STD-105D,[5] is an easily applied and valid decision-making aid.

Receiving inspection is an important element in most quality assurance programs. Quality program standards, once invoked by the producer in the contract, also describe quality assurance safe-guards at the receiving point. The U.S. Small Business Administration has published a management aid under the title, "Setting Up a Quality Control System."[6] The following list[7] presents the features and a procedure for receiving inspection, which can easily be adopted.

1. All parts and materials will be received and logged in by the Receiving Department.

2. All parts and materials will be sent to Receiving Inspection after logging in.

3. Receiving Inspection will assure that proper certification, physical and chemical test data, special process certifications, or source inspection certifications are with the items to be inspected.

4. The receiving inspector must document the complete results of all inspections and tests.

5. Inspection will identify accepted lots and send them to stock.

6. Rejected lots will be identified and set aside in Receiving Inspection until the buyer and Production Control decide on disposition.

7. The Receiving Department will send a copy of each rejection report to the Purchasing Department and the supplier.

8. The Purchasing Department has the responsibility of assuring that a pattern of continually receiving faulty items from any supplier doesn't develop and assuring supplier corrective action.

9. The Quality Department will follow-up to see that a supplier who has sent us items we reject has effectively corrected what it has been doing wrong.

10. Receiving Inspection instructions will be written with consideration given to the complexity of the parts, material received, and customer requirements. Follow customer instructions (if any) for inspection.

11. Sample according to customer requirements (if any) or MIL-STD-105D (available from Superintendent of Documents, Washington, DC 20402).

12. The Quality Department will review Receiving Inspection records periodically to see if any suppliers are consistently failing to meet standards.

13. All inspection records will show the number inspected, the number rejected, and the name of the inspector.

14. Inspection records will also show the disposition of supplier-provided records and data.

Producers need to maintain harmonious business relationships with all of their suppliers, especially with those that have performed well. Compiling performance data, even for suppliers with less critical quality impact, should be a matter of routine in any organization. (A simple form for this purpose is shown in Figure 10.5.) Other simpler recordings might be entered with the company's purchase documents, such as the purchase order or supply contract.

In order to keep the supplier evaluation fair and objective, a formal scheme, consistently applied, will be required. In addition to survey and source inspection reports, ratings for monitoring of receiving inspection results and customer complaints traced to suppliers help to quantify objective evidence. These numerical values can then be conveniently compared with those for other suppliers, with past performance periods in the form of *indexes* and with criteria other than just the quality criteria.

Ratings might not suffice as information and as the sole decision aid in major supply contracts with sensitive quality assurance requirements. The supplier appraisal will require more current and possibly specialized evidence, rather than data on past performance alone. In an ever-changing and dynamic technological and business environment, new issues and problems emerge that might invalidate the supplier quality assurance capability.

Changes during and after the negotiation and execution of a supply contract have to be handled with utmost care by both parties. It will be shown later that new developments might require adaptation and flexibility. Strict compliance to contract stipulation, quality design specifications, and agreed quality assurance procedures might become counterproductive. Ready monitoring of needs for changes aids in further decision making and control of corrective actions. A well-designed reporting and information system which is very crucial in maintaining good relationships between otherwise independently managed and operating entities will also be discussed later in a separate chapter.

Modern computer facilities have great usefulness in providing producers and suppliers with timely and satisfactory data in all phases of the purchasing cycle and the quality assurance aspects of this cycle. The producer's customer, to whom both producer and supplier are liable, and the general public as well, are dependent on effective and harmonious cooperation between production and external supply resources.

10.5 BASE AUDIT OF POTENTIAL SUPPLIERS

The sequence of steps in the supplier selection process will now be traced by an example. A major federal government department was experiencing problems with the suppliers of complex electronic equipment, requiring the control of work operations and manufacturing processes to achieve the specified level of quality, together with the inspection and tests necessary to ensure conformance.

The specification booklet established the requirements for the contractor's quality program. It also identified the elements of the program to be established and maintained by the contractor for the purpose of ensuring that materials and services fulfilled specified quality requirements.

Contractors who have demonstrated that their facility meets the rigid requirements of this specification are listed in the appropriate section of a separate publication and, as such, are considered to be qualified suppliers for these complex products. Being a qualified supplier has numerous advantages for a contractor. It makes him eligible to be awarded a contract for a complex product (it may even give him preference in the award of a contract, particularly if the available time is short) and it shows to the rest of the business community that he is a high-quality contractor, which increases his opportunity for securing additional business from the commercial industries.

Within the government department, the Director General of Quality Assurance (DGQA) is the final authority on quality assurance (QA), and has the responsibility of determining whether or not a contractor meets the department's quality requirements. To make these determinations the DGQA conducts QA audits on the contractor. The entire activity procedure for this task is shown in Figure 10.6. The details of steps involved are as follows.

Steps in Performing a Quality Assurance Audit

STEP 1: *Initial request from contractor.* The first step in the quality assurance *audit* process is the request (usually written) from a contractor who wishes to participate in the acquisition program. This request may emanate through various channels. It may be submitted directly to DGQA, through another government department, or more often, through the Technical Services Agency (TSA) in the locale of the contractor.

STEP 2: *Visit contractor.* The TSA, upon receipt of the request, will visit the contractor and invite him to complete the application form.

STEP 3: *Complete application form.* A contractor who wishes to participate in the program is required to submit a written request (via the application form) which should include the particular type of program that he wishes to participate in and the specific capabilities that he is willing to demonstrate.

STEP 4: *Forward application form.* The completed application form is forwarded to DGQA with the local TSA recommendations and comments.

STEP 5: *Determine whether to proceed.* The DGQA will review the request, consult with other government departments as necessary, and will determine whether to proceed with evaluation and recognition action. Requests for recognition will be considered only if the contractor is willing to demonstrate that he is capable of meeting, and will subsequently continue to meet the requirements, and that participation is in the interest of the Federal

Activity	Description	Predecessor
A	Initial request from contractor	—
B	Visit contractor	A
C	Complete application form	B
D	Forward application form	C
E	Determine whether to proceed	D
F	Explain requirements to contractor	E
G	Draft quality manual	F
H	Develop procedures	F
I	Arrange for preliminary evaluation	F
J	Conduct preliminary evaluation	I
K	Establish target dates for submission of draft quality manual	F
L	Submit draft quality manual and supporting procedures	G,H,J,K
M	Review draft quality manual and supporting procedures	L
N	Advise DGQA of draft manual completion	M
O	Decision of formal plant evaluation	N
P	Inform company	O
Q	Arrange for evaluation	P
R	Arrange for manpower	P
S	Review files	P
T	Conduct formal QA audit	Q,R,S
U	Informally debrief contractor	T
V	Submit and write up checklists	U
W	Correct deficiencies	U
X	Complete final quality manual	V
Y	Complete final report	V
Z	Approval	W,X,Y
AA	Inform company	Z
BB	List in DNDP 19	Z
CC	Annual reevaluation	AA,BB

Network (partial)

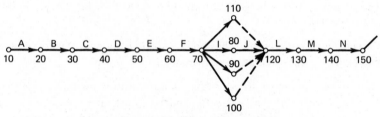

FIGURE 10.6 Activity procedure for base-audit of potential supplier.

Government. If it is considered that the contractor is only aiming to increase his commercial business, it is doubtful that the evaluation will be undertaken.

(For the remainder of this case example, it will be assumed that a positive decision was made to evaluate the contractor.)

STEP **6:** *Explain requirements to contractor.* When directed by the DGQA to proceed with recognition action, the TSA will explain to the contractor the requirements of the government specifications.

STEP **7:** *Draft quality manual.* One of the prerequisites of recognition is that a manual be produced by the contractor which describes the company's quality system and illustrates the conformity of the system to the requirements. The quality system manual provides the government department with a contractor's commitment to and acceptance of the requirements. This manual is considered by the agency to be very important and a suggested format for this manual and the items described are given to all contractors (Annex A).

STEP **8:** *Develop procedures.* The contractor will develop procedures to provide for the control of the quality-related functions of his facilities. These procedures will include, as applicable to his product, the operating functions listed in Annex B.

STEP **9:** *Arrange for preliminary evaluation.* This preliminary evaluation will be of the contractor's facilities, quality system, quality manual, and supporting control of procedures.

STEP **10:** *Conduct preliminary evaluation.* The main intent of this evaluation is to determine the level of effort required by a contractor to meet the requirements, and to provide assistance to the contractor where required.

STEP **11:** *Establish target dates for the submission of the contractor's quality manual.*

STEP **12:** *Submit draft quality manual and supporting procedures.*

STEP **13:** *Review draft manual and supporting procedures.* The local TSA undertakes this review in concert with the contractor and provides assistance and guidance where required.

STEP **14:** *Advise the DGQA of draft manual completion.* When the manual is in satisfactory form, the local TSA will advise the DGQA.

STEP **15:** *Decision on formal plant evaluation.* Upon receipt of the notification of the acceptability of the draft manual, the DGQA will decide whether to conduct the formal plant evaluation or direct the TSA to do so. The latter is the usual method; however, if they do not have experience in this product, the DGQA will do the evaluation.

STEP **16:** *Inform the company.* The company is informed who will conduct the formal evaluation.

STEP **17:** *Arrange for evaluation.* The TSA will make the arrangements with the contractor with respect to time, agenda, personnel attending, and other requirements for the formal evaluation.

STEP **18:** *Arrange for manpower.* The TSA or the DGQA will determine how many men are required for the evaluation, and whether any specialists are needed. They will also make the necessary arrangements to obtain the required manpower and/or specialists.

STEP **19:** *Review the file.* The TSA will review the file on the contractor. This will be particularly applicable if the company has had a previous qualification or if the company has had problems in their past performance.

STEP **20:** *Conduct the formal QA audit.* The Allied Quality Assurance Publication, AQAP-1 Standard of North Atlantic Treaty Organization (NATO) is used as a guideline and the 24 items contained in Annex C form the basis of this audit.

STEP **21:** *Informally debrief the contractor.* Upon completion of the audit, the TSA will informally debrief the contractor on the results. Any deficiencies will be identified to the contractor so he can commence his corrective action.

STEP **22:** *Submit and write up checklists.* After the contractor has been informally debriefed, the members of the audit team will write up their portions of the audit, paying particular attention to the deficiencies identified.

STEP **23:** *Correct deficiencies.* The contractor will correct all deficiencies identified in the audit and demonstrate to the TSA that the deficiencies have been corrected.

STEP **24:** *Complete final quality manual.*

STEP **25:** *Complete final report.* A final evaluation report with TSA recommendations will be prepared and submitted with a finalized copy of the contractor's quality manual and appropriate certification pages for signature by the DGQA. The report will include all aspects of the evaluation and will show that each facet of the contractor's system has been evaluated.

STEP **26:** *Approval.* Upon approval of the final report and correction of all deficiencies, the contractor will be approved as a AQAP-1 (NATO) qualified supplier.

A formal letter from the DGQA to the contractor will be forwarded to the TSA with signed certification pages for presentation to the contractor. The letter will include a statement directing the contractor's attention to the requirement for periodic surveillance and reevaluation.

Listing in Government Document

Once the contractor has been approved, he is entered into the Department of National Defense (DND) Publication 19, the list of qualified AQAP-1 suppliers. A contractor will be audited at least once a year to ensure that he continues to meet the requirements of AQAP-1.

10.6 QUANTITATIVE METHODS OF RESOURCE PLANNING FOR QUALITY

Inventory Models

Inventory of completed items, work in progress, or materials and supplies are subject to quality change. Many factors reduce quality, such as obsolescence, damage, spoilage, and wear and tear. Only in exceptional cases, as for instance with certain wines, does quality for the consumer improve with age. Mainly, quality assurance considerations are simply implied in the various inventory models. The purpose of inventory is to ensure ready and timely service to the customer, to facilitate smooth production and capacity utilization, and to coordinate required supplies with independent suppliers. Minimization of costs results from optimizing order quantities and timing of ordering and delivery.

Quality costs are usually subsumed under ordering or set-up costs, holding costs, shortage costs, excess costs, and other related cost elements. When ordering a lot or an individual item, vendors must be approached, selected, and contracted. This involves specification of quality, communication, and assessment of quality assurance programs and capacity. The more important the supply is for the quality of processes and end items, the more cost, time, and effort can be afforded by both the purchaser and the supplier to prevent rejection of the delivery.

The general trend of relying more on vendor's data supplied with shipment or on source inspections and audits of supplier quality assurance programs, rather than on customer's internal receiving inspection alone, shifts many ordering costs from the customer to the supplier. On the other hand, the trend toward greater quality awareness and significance leads to more inspection, auditing, and other quality assurance activities, with respective cost increases.

The effect of quality cost variation within ordering costs affects the *economic order quantity* as computed by the EOQ model. In the following example inspection and defect prevention costs have been separated from ordering costs. If the quality cost component, (i.e., inspection and defect prevention costs) rise, the order quantity will rise also, and vice versa. A trade-off for higher inspection and defect prevention costs is a long term reduction in defect or failure costs with the respective nonlinear trade-off cost model showing the minimum in the total cost; that is, the quality costs.

In the EOQ model, defect or failure costs can be related to holding and shortage costs. Unacceptable quality might be rejected at the receiving point and thus create a shortage, or be detected later during holding. It must be assumed that the demand for the item, for which the optimal order quantity is calculated, is constant over, say, a year. The impact on the holding cost varies, with rejection of items reducing them and increasing shortage costs, and with detection during holding increasing both holding and shortage costs. In order to show the impact of defects and failures of items more clearly, they should be kept separate from holding and shortage costs. As the following example demonstrates, successful quality assurance activities in purchasing and stockkeeping that reduces defects and failures increases the optimal order quantity. With the given demand for the item, the number of orders declines with the concentration of the suppliers on better quality. Greater lot sizes allow relatively smaller samples in acceptance sampling.

The impact of better quality on shortage and excess costs can similarly be demonstrated. Less rejection because of inadequate quality will reduce shortage costs, safety stocks, relative order quantities, and inventory holdings in the long run. Only in the short run will less rejection increase holdings, because the improved quality assurance of supplies will allow rationalization.

Example: Assuming that the *economic order quantity* (EOQ) were calculated according to the following formula with numerical values of the symbols as given below for a machine shop that uses drill bits:

$$EOQ = \sqrt{\frac{2\,DS}{H}} \times \sqrt{\frac{A}{F}}$$

where:

D = Demand for 1000 drills per year.
S = Ordering cost $35 per order.
H = Holding cost $2 per unit per year.
A = Supply quality assurance costs $20 per order.
F = Defect cost = 0.3% of total cost (PD) $20,000, where P is the price of $20.00.

Determine and interpret EOQ.

Solution

$$EOQ = \sqrt{\frac{2\,DS}{H}} \times \sqrt{\frac{A}{F}}$$

$$= \sqrt{\frac{(2)(1000)(35)}{2}} \times \sqrt{\frac{1}{3}}$$

$$= \sqrt{3500} \times \sqrt{\frac{1}{3}}$$

$$= (59.16)(0.577)$$

$$= 34.14 \text{ drills}$$

Because of the relative high defect cost, a smaller lot is ordered; more inspection and/or selection of a better supplier is considered.

Economic order quantity models applied to internal production lot sizes include set-up rather than ordering costs. Quality of production depends to a large extent on the quality of set-ups in the broadest sense of that term. "Getting ready for production" under the quality assurance aspect should be understood as "getting it right the first time." A sound and effective design of product and process, coupled with well trained personnel and modern automated equipment, reduces the cost of set-up in the long run. This not only decreases the economic lot size, with the other components in the model unchanged, but allows for more frequent production changeovers. Such flexibility enhances the ability of a company to adapt to customer demands more readily, to utilize available production capacity more effectively, to partially reduce inventory holdings and excess stock, and to enrich jobs and motivate workers positively.

The difference of assuming *deterministic* or *probabilistic* conditions only refines these inventory models. Improved quality assurance creates greater stability in production and operation in the long run, reduces cost levels, and develops the demand. Greater stability will tend to reduce variances and increase the average in demand. Statistically determined safety stocks, shortages, and excesses will therefore tend to decline with less variance and only increase in the long run due to higher average demand.

Timing of ordering, or new set-ups in these basic inventory models is determined by demand during lead time plus allowance for safety stock. *Lead time* again entails quality-assurance-related determinants that vary in impact over the short and long run. If the supply base is new to the purchasing company, source inspections, and negotiations for proper quality specifications and quality assurance, will delay the delivery time and thus extend the lead time. This is also true when requisitions for supplies are done within the company between one work unit and another. This delay can reflect positive efforts to do the work right the first time and to eliminate later failures, down times, and other costly delays.

However, once the relationships with suppliers and work units have attained the steady state of adequate control, lead time will be shortened. This not only results in lower inventory levels and safety stocks; production becomes more closely synchronized between independently operating internal and external units. Shorter lead times also create conditions for more rapid and smoother changeovers and greater flexibility in the use of available production capacity. This then can lead to ''just in time'' inventory systems.

Simulation can demonstrate in detail the behaviour of a productive system in response to different quality assurance strategies as outlined above. The measure of effectiveness that evaluates different inventory strategies is the sum of costs as defined in these models. If the implied quality costs are defined separately and integrated into these models, impacts of various quality assurance strategies emerge. In the most general terms, all measures of the ''right the first time'' principle minimize inventory and maximize profit contribution.

This improvement of inventory planning and control is still restricted to the ''A'' items in the ABC or Pareto type of classification. It is this classification which extracts roughly those 20 percent of all items that together constitute about 80 percent of the investment in inventory. Quality assurance aspects might well add ''B,'' and even ''C'' items into the ''A'' category. A low cost item can create considerable quality costs when tested, and the consequence of defects and failures of such items will be accordingly high. The objective in the use of inventory models for quality assurance management is not only to reduce costs in the various ways, but to work in conjunction with conventional inventory planning and control. A major contribution must be made to develop this methodology into a comprehensive and integrated productive system with minimum inventory and maximum ability to adapt to demands of the market and competition. The systems and models that offers the greatest challenge to quality assurance are modern production planning under material requirement planning and related ''just in time'' systems.

Material Requirement Planning (MRP)

Modern computerized management information systems have made it possible to order parts and components of end items at a calculated order release and delivery time. Inputs into the computer program of the material requirement planning (MRP) are the master production schedule, showing the demand for end items, parts, and components, in a structured bill of material file with stock-on-hand data, and lead times for ordering. An example given at the end of this section illustrates the material requirement plan for a simple product.

The various reports produced by MRP can entail erroneous information if insufficient attention is given to quality assurance aspects. The forecast and master schedule of the independent demand for end items should cover only saleable items that meet expectations of customers. In order to avoid costly shortages and thus poor customer service, an allowance for defectives in end items should be added, reflecting expectations and experience. Repercussions of defective end items are severe when the demand of complementary products is also directly affected. The end item of one company might be an essential assembled component in another product produced by the customer. Moreover, if the design of the end item is not properly tested and quality assured, the inputs to MRP are invalidated accordingly. This creates considerable disturbances in the production, including production planning.

The *bill of material* needs to be aligned and possibly adjusted to actual available supply capacities. Make or buy decisions have to be made with quality assurance being properly accounted for. Orders will have to be placed with suppliers that meet specification of quality and delivery time. Vendor rating becomes an important source of information for the selection of reliable suppliers. Good vendor/buyer relationships grow from careful coordination of bilateral quality assurance efforts. If, for instance, compliance with quality specification appears to be fairly sure because of documented quality assurance programs or past experience, but meeting delivery time is doubtful, remedial action must be taken. In the extreme case, delay of delivery under the MRP system will cause costly downtime and other ripple effects, including delay of the purchaser's own deliveries to customers. Safety stocks and additional lead time adds costs which proper quality assurance of supplies can and must avoid as much as is feasible.

In the final instance, all components listed in the bill of material are of equal importance and substitutes are seldom available because of the high degree of synchronization. Even if an acceptance sampling plan was agreed upon with suppliers, safety stock is needed to cushion unfilled MRP supply requirements.

Once all deliveries meet quality and time specifications, the individual internal production units must be able to perform without unexpectedly high defectives and failures. Any such event will, of course, multiply into delays of subsequent processes. As in a fairly tight and originally accepted and approved production plan, not much time, if any, is left for corrective action. An essential prerequisite of MRP is a preventive quality assurance program. There is no stock, other than the minimal safety stock, that can replace major defects and failures in time. Success of MRP depends upon an adequate quality assurance program.

Of course, quality cost levels decide the optimum of defect prevention and defect correction costs. Low defect prevention costs, partly due to a well-established and steady state kind of quality assurance program, allow for relatively greater efforts in further eliminating defects and failures. On the other hand, if defects remain high, and thus production disturbances and delays remain high, short run remedial actions of increasing inspection, safety stocks, and lead times, are among the necessary countermeasures.

In the long run, the interrelationships between MRP and the quality assurance program remain close and are mutually inclusive. The benefit of a quality assurance program that is integrated with MRP is that the time perspective, as an important quality determinant, moves into focus. An MRP plan that is derived from a possible independent demand and exceeds a company's capacity to deliver quality as required, at the time it is required, presents senior management with a decision between at least two alternatives; either abandon MRP or establish an adequate quality assurance program. With the availability of computer facilities and the need to rationalize operations, the price of a poor quality assurance program rises continuously.

"Just In Time" and Quality Assurance

Very similar to material requirement planning is the supply of relatively small quantities to individual work stations from internal suppliers "just in time," (JIT) when it is actually required; normally by the hour. This JIT system can be linked with the MRP computerized ordering system. The KANBAN[8] system uses a similar, more mechanical "pull" of supplies by downstream work centers from upstream fabrication via a KANBAN card.

The primary advantages of a JIT system are lower cost, greater stability in production level, better work performance, and thus higher quality. Communication, updating of plans and schedules, order releases, and so forth are computerized. This, once firmly established the JIT system reduces paperwork and scope for human error. Specific advantages from a quality perspective are:

1. Less stock and continuous flow of work in progress reduces risk of defects, breakage, spoilage, and other special holding costs.

2. Closer cooperation during production, and higher degree of automation, preventive maintenance, vigilance, and training reduces scope for errors.

3. Work stoppage due to defective supplies must be ruled out because of severe repercussions. Quality assurance is a prerequisite for the functioning of stockless, and possibly lotless, continuous production flow.

4. Schedule, with frequent changeover during the day, is frozen for about 10 days, providing for repetitive work pattern and creating routine and steady state conditions.

5. Interdependence of work stations motivates supervisors and operators towards defect-free performance. Poor performance becomes immediately and widely visible. All workers are interested in attaining proper performance after changeover as quickly as possible and in seeking improvements.

6. More intensive and comprehensive planning integrates production and quality control. Higher planning costs are more than compensated for by lower holding and set-up costs, which in turn allow smaller lot sizes.

7. Frequent changeover provide flexibility for modifications and improvements, leading to greater adaptations to customer and market needs, greater variety, less break in routine, and less error.

The disadvantages of the JIT system are less than those of MRP, because the JIT system is an internal combined production and inventory quality assurance control system which is supervised by in-house management. The potential negative impact of defects and failures under this tight production flow and changeover is more an advantage than a disadvantage, as it demands absolute preventive quality assurance. Wherever management lacks skill and experience in quality assurance and modern production management methods, often coupled with poor cooperation and work attitudes, the JIT system is an unfeasible mode of operation. A quality assurance program offers such a management and organization an opportunity to start to change.

A *combination of MRP and a quality assurance program* can take many forms. Making allowances for defectives in supplies, production, and final delivery, and planning for higher safety stocks and longer lead times comprises a relatively passive strategy. An example later in this chapter demonstrates a simple MRP system with such special allowances made. One should base these allowances on well-founded quality control data. If the establishment of an MRP precedes the introduction of a more preventive quality assurance program, defect allowances might be justifiable for a certain start-up period. A special output report evaluating quality costs of the MRP system, (i.e., compiling inspection, prevention, and defect costs) will soon show whether this passive strategy of defect correction and substitution is cost-effective. It will also pinpoint the weak links causing the major costs. This information can then be used to direct defect preventive measures.

Defects in end items multiply as allowance for defective components. It can be shown that component supplies with relatively high quality costs and unnecessarily long lead times due to unreliable service generally are purchased in larger lot sizes. The alternative of ordering from a better supplier is often not feasible. In cases of unnecessarily high lot sizes and other quality costs, demands for special discounts can reasonably be made.

In summary, MRP offers to quality assurance the impetus for preventing, rather than correcting, defects. It also clarifies actually existing capacity to meet requirements of quality and time jointly, and to select supply and production units with quality assurance considerations. Both MRP and a quality assurance program, once well-coordinated and mutually integrated can be further developed into a comprehensive "just in time" production planning and control system.

Example: A product is assembled from three main subassemblies; A, B, and C. Each unit of A requires three units of D, and two units of E. Each unit of B requires one unit of F and one unit of G. Also, each unit of G requires five units of part H. Assume that current inventory includes: 200 units of D, 100 units of G, and 800 units of H; and that in ordering parts D, E, F, and H an allowance of 3% must be included for defective parts. What are the net requirements of A, B, C, D, E, F, G, and H needed to produce 500 units of the end-product?

Solution

The product structure tree for the end-product is shown below.

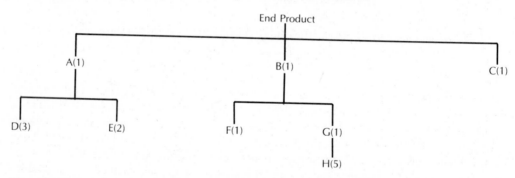

The quantities which are generated by diagraming the bill of materials are tabulated in the following table.

Item	Inventory	Gross Requirement	Net Requirement	Allowance for Defectives	Total Requirements
A	0	500	500	—	500
B	0	500	500	—	500
C	—	500	500	—	500
D	200	3 × 500 = 1500	1300	39	1339
E	0	2 × 500 = 1000	1000	30	1030
F	0	500	500	15	515
G	100	500	400	—	400
H	800	(5)(500 − 100) = 2000	1200	36	1236

10.7   COOPERATION AND COMMUNICATION WITH VENDORS FOR NONCONFORMANCE CONTROL

While cooperation is an essential element in all vendor/buyer transactions, the need for an open communication line with each other cannot be overemphasized. Both parties always wonder:

1. What actually happens when a nonconforming item is identified?
2. Are items promptly identified in a positive manner to prevent unauthorized use?
3. Are items tagged or marked to make them easily recognized?
4. Are defective items segregated away from conforming items?
5. Are special precautions taken to prevent inadvertent acceptance or further processing?
6. Are proper forms used for the disposition of nonconforming items?
7. Are records of the number of such nonconformances kept for corrective action?
8. Is corrective action documented and reported to the appropriate levels of management?
9. Is someone responsible for the evaluation and disposition of nonconforming items?

All of the above questions can be integrated into the firms' contractual agreement. They are all relevant to nonconformance control and corrective action. All of these questions and others as well are necessities for the proper implementation of nonconformance control and corrective action.

There is an increasing awareness of this problem, when it comes to communicating to the buyers. Vendors should not be given bad news all the time. Rather, an atmosphere of constructive improvement and mutual understanding should be the general approach.

Example: The experience of a small pocket calculator manufacturer is pertinent here. This manufacturer has had five suppliers from day one who are all considered quite reliable for supplying tiny chip transistors. Although some problems occurred once in a while, these were not considered serious enough to threaten the contract termination of any one of the five suppliers. Whenever quality got worse, the president normally had talks with that particular vendor's representative over lunch; that was all. The quality manager, however, took exception to this attitude and wrote to the vendors' executives. In part the manager's letter read:

> Think of the damage a faulty 1-cent transistor can do. If you catch the transistor before it is used and throw it away, you loose one cent. If you don't find it until it has been soldered into the circuit board, it may cost one dollar to find it from the completed and assembled calculator. If you don't catch the faulty calculator until it is in the hands of a consumer, the shipping and repair will cost more than ten dollars. Indeed, if this is the case, then the expense exceeds the manufacturing cost.

10.8   WORK SYSTEM PLANNING AND CONTROL

Work systems (as opposed to plants, offices, shops, etc.) are microproductive units, such as jobs, group assignments, and small projects. As a system, each unit has a particular objective, method or process, and/or productive agent, (like worker and machine, resource and material base) that interfaces with other work systems in a hierarchical and lateral respect, and for a given physical, social, and technological environment. These and other determinants of system performance also influence quality of work and resulting products and services. Quality assurance is an inherent task in work system planning and control.

Analogous to the distinction between continuous and intermittent, (or job shop, productive systems, and their related product and process layouts) work systems can also be divided into those with a repetitive work cycle and those with changing work content and methods. Quality assurance in an assembly line type of job differs from that in the job shop setting.

Standardization of work all along the line permits specialization and automation, including in inspection and testing. However, in order to counteract inherent monotony of work, the operator might need a job enrichment, rotation or enlargement. Self-inspection and quality control feedback, along with participation in planning and control of work inputs and outputs can have a motivating effect.

In the job shop situation, the variety of work assignments requires greater knowledge and skill on part of the human operator. Frequent set-ups and other changes in the work routine demand particular efforts for assuring quality of work. Operators in these jobs participate more directly in the planning and control of the work and have quality assurance responsibilities.

Methods and models, for work system design and performance measurement are relatively simple and in, fact permit, supervisors and operators to participate in the decision making.

Work systems with repetitive work content and methods are schematically described in operation as: flow process; worker–machine; and activity charts. These analytical devices often include symbols and information on special inspection and quality assurance work elements. Integrated inspection plans show schematically the interface of production and quality assurance.

The following two examples involve *repetitive work cycle* and *nonrepetitive work cycle* for inspection and test stations work planning. The examples show the calculation of standardized time to perform the job and to serve as the main information link between the designer and the job holder.

Example 1: The testing of an assembly was observed and the following cycle times noted in minutes: 4.4, 4.8, 3.8, 3.6, 4.2, 4.4, 3.6, 4.3, 4.7, and 3.3. Performance rate was 1.2; allowance was given as 10% and 5% respectively for verifying test results (self-inspection) and proper calibration of measuring equipment 10 times per work day at random intervals. Allowance is based on normal time characteristics. Determine time standard.

Solution

$$\overline{X} = \frac{\Sigma t_i}{n} = 4.11 \text{ min}; \quad s = 0.48; \quad \text{Normal time} = (\text{avg})(\text{perf.rate}) = 4.11 \times 1.2 = 4.93$$

min.

Standard time = normal time $\times$ (1 + allowance rate) = 4.93 min $\times$ 1.15 = 4.67 min.

Example 2: Self-inspection is to be assigned to a worker and the current idle time is therefore to be determined, for 90% confidence. The idle time is estimated at 20% of total maximum time. Error allowed is $\pm 5\%$; total time was observed and a normal distribution confirmed with $\overline{X} = 44$ min; s = 8 min.

Solution

Total maximum time $= Z_{0.90}(s) + \bar{X} = (1.29 \times 8) + 44 = 54.32$ min

Idle time $\qquad = .2$ (Total maximum time) $= 0.2 \times 54.32$ min

$\qquad\qquad\quad = 10.86$ min

Self-inspection should not require more than about 10 min.

There is practically no work system, or rather, there *should* be none in a company, without a proper definition and delineation. Any system description must include, for example, the system's purpose and objectives, approaches, resources, and environmental factors.

Schematic models of work systems become very complex in automated and electronically-controlled productive systems. Drawings and other schematic descriptions of such systems require special expertise, such as that of engineers, system analysts, and machine designers. Data provided from such designs might permit the use line balancing and similar models to establish and integrate the technologically complex work systems. Simulation can also provide the information that allows management to assess efficiency and, in particular, quality assurance impacts.

Automated systems normally yield greater uniformity of quality with faster speed over longer timespans than systems run by human operators. Emphasis on design, rather than on correction during performance, reduces the chance of defects once the testing period has been successfully completed. A design defect, however, can have serious repercussions in automated systems. Job descriptions for operators in highly automated systems should include well-defined authority for stopping production in the event of apparent or suspected failures. Job descriptions should either include cross-references to special operational quality assurance procedures, or directly include such procedures.

Example: The inspection of a computer disk-drive is conducted by means of a checklist. Time required per inspection of one disc-drive depends on the number of defects detected, because nonconformances must be analyzed and recorded in some detail. Arrival of items for inspection does not follow a time pattern. Management wants to assure that inspectors are not affected by undue time pressure and, therefore, want to measure the idle time as an indicator for time pressure. A 98% confidence assessment is stipulated that the resulting estimate for the idle time at the inspection station is within 5% of the true value ($e = 0.05$). What is the required sample size assuming a Poisson distribution?

Solution

Given that Z = number of standard deviations (normalized) defined by the confidence level (i.e., 98%); $\bar{p}$ = sample proportion, in this example idle time is about 10% of total work time; and n = sample size; we use the formula:

$$e = Z \sqrt{\frac{\bar{p}(1 - \bar{p})}{n}}; \quad n = \left(\frac{Z}{e}\right)^2 \times \bar{p} \times (1 - \bar{p})$$

$$n = \left(\frac{2.33}{0.05}\right)^2 \times 0.1(1 - 0.1) = 195.48$$

In case the 10% idle time was a preliminary estimate, the sample size should be recalculated after about 100 observations with a revised proportion ($\bar{p}$).

Procedures, like any other descriptive model, are of obvious importance to quality assurance. Standardized content and format of procedures insure that these guidelines and directives are actually applied and adhered to by those concerned. Schematic drawings, graphs, listings, and augmenting verbal descriptions enhance the communication and thus the effectiveness of procedures.

As no procedure directed to a human operator can cover all factors and eventual events in the work system, well-defined allowance should be made for the necessity of adapting to unforeseen situations. Of course, in practice this is a very difficult principle. Yet, user ingenuity, decisiveness, courage, and competence often exceed that of those who prepared the procedure. Quality assurance effectiveness requires both adherence to procedure and adaptation to situations.

Work system designs, reviews, and modifications need the input of the operator concerned as well as that of the specialists. Methods and models describing such systems and their changes need to be relatively simple in order to permit wide participation in planning and control. This principle extends also to the related performance measurement.

Time standards for performance measurement in repetitive work cycle systems have a significant impact on quality. Standards that are too tight obviously cause a relatively high number of defects. The same is also often true for standards that are too generous and allow for interruptions of work routines. Quality assurance aspects demand an optimal work standard that stresses adequate quantitative and qualitative output. This can be achieved in various ways. For instance, the performance rating can also be related to measurements for failure-free output.

If inspection is an element of the work assignment, the actual inspection reliability should be assessed before the work standard is finalized. When sampling inspection is more effective than 100% inspection, a special time allowance for self-inspection might be more useful. Special allowance for quality assurance activities, such as checking machines, materials, inspections and so on, help to reduce defective work and reminds operators of their duty to achieve quality output.

Work sampling applied to nonrepetitive jobs uses statistical techniques similar, if not identical, to other quality control methods. Listings of job related activities for work sampling will include special quality assurance tasks, whenever these are part of the respective job description. The distinction in a work study between busy and idle states often lacks definition of quality assurance related activities that appear to belong to the idle category. This leads to a work standard that does not support observation of quality assurance in work performance.

10.9   LEARNING CURVE MODEL

The *learning curve* is the relationship between the time required per unit and the number of repetitions of the identical production task by the same productive unit. The relationship is usually negative exponential with a constant rate of time reduction (learning) as repetitions double, (Figure 10.7).

Learning curves postulate that quality remains constant. Learning also means that the same work cycle can be performed in less time with repetition. Therefore, it is reasonable to assume a quality impact as well. The better, more intensive, and more effectively the

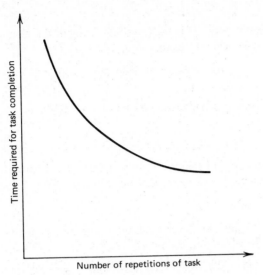

FIGURE 10.7   Schematic of a learning curve.

production of the first unit is prepared, the more adequate will be the quality of design and production in actuality. As learning curves are usually applied for major complex items consisting of many components and operations, performance can be measured in terms of defects or failures, as well as in time required.

The rate of decline in defects with subsequent work units, or cycles, can be compiled, and thus the model expresses quality improvements. The examples given are such applications of the learning curves in quality assurance. The improvement in terms of defects and/or defectives depends on the relative quality of the first item. When the number of defects is relatively small because of quality assurance and other measures, the rate of decrease in defectives will be respectively smaller.

The learning curve can be used to compare performances, set performance standards, relate quality costs to work cycle frequency, demonstrate the effect of quality assurance programs, and share results of better quality performances.

As a motivational device, learning curves supplement quality control charts in that they measure *improvements*, rather than performance within established control limits. Learning curves need to be applied with caution, as the data base must remain valid. Hidden factors might have an influence, and thus causes for improvements or failures cannot be readily recognized. Perhaps cause/effect analysis, and similar simple analysis techniques in conjunction with quality circles and so on, may prove helpful in assessing paths for further quality improvements.

Example 1: Routine inspections are to be performed on production equipment every 1000 hours of use. Study of time logs of technicians performing the inspection indicates that a 90% learning curve applies to this situation. If a technician requires six hours to complete his first inspection, what is his expected time to complete the inspection for the sixteenth time?

Solution

The sixteenth inspection represents the fourth doubling of the technician's first inspection: $(1 \rightarrow 2 \rightarrow 4 \rightarrow 8 \rightarrow 16)$. Therefore, time required equals:

$$(0.9)^4(6) = (0.6561)(6) = 3.9366$$

or *3.94 hours*

Example 2: Final inspection and testing of the first bus of the new production line requires 42 hours.

1.   What is the time requirement for the eighth bus, given a learning percentage of 80%?

2.   The first bus showed 40 major and minor defects. Given an improvement standard of 10 each time production doubles, at what unit should defects be reduced to about 20?

Solutions

1.   The eighth bus represents the third doubling of the original production of one bus. Therefore, total time for the eighth bus will be $(42)(0.8)^3 = (42)(0.512) = 21.5$ hours.

2.   The first unit had 40 defects. First doubling = 2nd unit will have $40 - 10 = 30$ defects; second doubling = 4th unit will have $30 - 10 = 20$ defects; therefore by the fourth unit, defects should be reduced to about 20.

10.10   CASE: SAFELIFT HYDRAULICS

Tom Bach, Vice-President of Safelift Hydraulics, faced rapidly climbing production and distribution costs, undermining the company's otherwise strong position in the national market. Major investments for reequipping the plant and rationalizing processes utilizing modern technology were temporarily ruled out because of high interest costs. It also appeared that potential for reducing costs through more streamlined operation and staff cooperation in reducing scrap, waste, and so forth was also already exhausted.

Tom discussed the urgent matter with the senior staff once the annual figures for financial and operational developments were available. The marketing manager pointed out that Safelift could and should bid for a lucrative contract, yet the quality assurance program had to be modified in order to meet the Standard MIL-9858A. This was not to be a major problem. However, the contract stipulations for assuring adequate supplies from outside suppliers was not as easy to comply with.

Safelift was under pressure from major local customers to give preference to local sub-contractors and small suppliers. The purchasing agent for Safelift was quick in pointing this out to Tom. The plant manager felt that supervisors and operators in the plant could not further improve their performance and workmanship. The receiving inspector's record indicated that most of the defects originated with suppliers; end-item inspection, as well as customer complaints, confirmed that fact.

Tom was not quite convinced that the production manager was right in putting the blame for high defect costs entirely on suppliers. This seemed to be too easy a way out. Nevertheless, negotiations with the union were to commence soon and, therefore, it was felt that tensions in the plant should be avoided. Once supplies were improved through decisive and prudent action, downstream work in the plant would also reduce defects simultaneously. Tom remembered having listened to a plant manager of Westinghouse explaining their Rejection Improvement Program (RIP) that had successfully changed poor supplier performance for the better.[9]

In the past, Tom had tried to contract only with those suppliers that had proven reliable and satisfactory performance. However, whenever a supplier was dropped, influential members of the local business community exerted extreme pressure on Tom. All too often he felt he had to give in. Moreover, some of these suppliers were also customers of Safelift. So Tom decided that the RIP plan of Westinghouse was something he should try. The following is a brief listing of the major features of the RIP:

1. The underlying principle of "management by exception" places the focus of the improvement program on rejection of incoming lots.

2. Receiving inspection continues to gather data on suppliers' actual performance; initially simple reject counts, which then are analyzed and compared. Suppliers that would come under the RIP are then selected on this basis. The vendor rating system is gradually refined.

3. The receiving inspection system is reviewed and made more integrated with purchasing. Another important action is to recognize and correct errors, ambiguities, or contradictions in drawings and specifications that have been transmitted to suppliers.

4. All suppliers are informed about the new RIP by letter and at a one-day supplier seminar.

5. About 10 suppliers with the most rejected lots and the poorest records are selected and informed about special rejection improvement measures. These suppliers already know of the program. They are requested to analyze their problems and to report on planned corrective action by a certain deadline.

6. The responses from suppliers under the program are also reviewed by the members of senior management. In all cases of such response a representative of the buyer visits the supplier in order to verify the actions and to provide additional aid.

7. Those suppliers that have not responded, are given a second notification about inadequate supplies, and, if failing to respond within seven days, they are automatically taken off the supplier list.

8. Supplier performances are to be reassessed once a year and if improvement warrants they will be recognized and taken out of the RIP. Others might then enter the program; but gradually, this should not anymore be the case.

Tom decided to develop and institute an improvement program similar to that of Westinghouse under the name "Supplier Aid, Improvement, and Development" (SAID).

In consideration of this case, the reader should consider the following questions.

1. What are the advantages and disadvantages of this improvement program?

2. What other action should Tom consider for improving quality costs?

3. Should the receiving inspection continue under a MRP mode of operation? Should it continue once no supplier is left under the SAID program?

10.11 KEY TERMS AND CONCEPTS

A **supplier** is an external source of supply. A contract governs quality assurance requirements and obligations between the supplier and purchaser. Under modern quality assurance standards, the purchaser and producer of the specific design will be held responsible for the quality of purchased supplies.

Supplier quality management consists of all planning and control activities on the part of a purchaser, for assuring that suppliers comply with their quality and quality assurance obligations. Purchasers normally support suppliers in their efforts to deliver adequate quality.

Supplier selection is to be based on careful assessment of a supplier's capability to meet requirements for quality and quality assurance as stipulated by the purchaser and the purchaser's own customer and/or regulatory bodies. Suppliers deemed to be qualified are listed within the purchasing company for reference by purchasing agents.

A **supplier's quality system survey** is an formal and systematic evaluation of a supplier's capability to meet a purchaser's quality and quality assurance requirements. The survey, sometimes called an audit, is normally based on quality system standards and specific product or service specifications.

Supplier surveillance consists of frequent verification by the purchaser that the supplier meets contractual requirements. It normally leads to close cooperation between purchaser and supplier, with mutual benefits.

Source inspections are performed by the purchaser at the supplier's plant in accordance with contractual agreements and possibly under approved inspection plans. Supplier surveillance, as opposed to source inspection, is more concentrated on the verification of product and process specification, rather than on evaluation of the general quality assurance system.

A **supplier performance record** monitors results of surveys, surveillance, source inspection, receiving inspection, possible corrective action, and other performance criteria. The resulting performance status is used as data in supplier qualification and selection procedures.

Inventory models can assist in planning and controlling cost-effective supply assurance in quantitative and qualitative terms. Economic order quantities, safety stocks, service levels, reorder points, and so on, permit explicit inclusion of quality assurance factors.

Material requirement planning is a time-phased ordering system based on a schedule for end-items, a *bill of material* (*product tree*) showing components (parts) of the end-item, and stock-on-hand records. This computerized information system permits inclusion of quality assurance factors, and thus can contribute to improve the quality of products and services.

A **work system** is a systematic design of work units or tasks that are assigned to workers and/or automatic processes. The quality of a work system design influences workmanship and quality of performance.

A **learning curve** measures resource requirements for subsequent identical units of work (cycles, products, services). Resource requirements, such as manpower and/or time requirements, will normally and proportionally decrease with frequency of items or cycles. This "learning" is related to quality of performance.

10.12 DISCUSSION AND REVIEW QUESTIONS

(10–1) *All purchase orders should be scrutinized for quality assurance.* Discuss.

(10–2) *You get what you pay for.* Discuss.

(10–3) Is receiving inspection normally a fairly ineffective measure of quality assurance of supplies? Explain and discuss.

(10–4) *A Pareto analysis should be applied for determining supplies that must be closely controlled.* Explain.

(10–5) *For complex items it is best to impose quality assurance program standards and verify compliance by suppliers.* Explain.

(10–6) According to some quality assurance program standards, suppliers must submit an inspection plan before work on the contract can commence. What are the reasons for this?

(10–7) *Once an order is placed with a supplier it is the supplier's responsibility to assure compliance with specifications.* Discuss.

(10–8) Quality assurance of supplies consists of several steps. What are these?

(10–9) Col. John Glenn, first American astronaut said, in a television interview, "The thought that governments always accept the lowest bid price disturbed me deeply." Discuss.

(10–10) *The quality manual of a supplier is an important document in the supplier selection process.* Explain.

(10–11) *Major suppliers should be required to conduct a contract review and process capability study.* Explain.

(10–12) *Source inspection is often the only way to verify compliance with specifications.* Explain.

(10–13) *Quality of supplies does not only refer to compliance with quality specifications.* Explain and discuss.

(10–14) *Rejection or acceptance of a delivery requires thorough follow-up measures.* Explain.

(10–15) "If I were a banker, I would not lend money for new equipment unless the company that asked for the loan could demonstrate by statistical evidence that they are using present equipment to reasonably full capacity." So says Dr. W. Edwards Deming, a world renowned quality consultant. Do you agree? Give your reasons.

(10–16) *Material requirement planning models and systems imply quality assurance programs.* Explain.

(10–17) *The risk of rejecting a lot in conjunction with statistical acceptance sampling is more serious than the risk of shortage of inventory during lead time.* Explain.

10.13 PROBLEMS

(10–1) You are to conduct a source inspection for an electric power generating corporation at the plant of one of their transformer suppliers.

a. List the information you will require on the supplier's quality assurance program, with reference to the inspection.

b. List other information you will require with reference to the inspection to be performed by you.

c. Describe steps to be taken in the source inspection.

d. Prepare a checklist for conducting the source inspection; omit any technical details related to transformers.

(10–2) You are the quality assurance manager in a robot manufacturing company. Quality of supplies is crucial and therefore, a list of reliable suppliers is to be prepared.

a. Outline steps to be taken in preparing such a list.

b. Prepare a standardized record form for such a list.

c. Set up a procedure that ensures selection of only suppliers that are on the list.

(10–3) In an application of the Pareto analysis for determining those suppliers that must be relatively tightly controlled by quality assurance;

a. describe factors used to categorize suppliers in these three categories.

b. establish a guideline for classifying suppliers into any of these three categories.

c. establish a procedure for quality assurance of suppliers in the category of prime importance.

(10–4) In a make-or-buy-analysis a company is assumed to have several alternatives.

a. List advantages of buying, rather than making, from the viewpoint of quality assurance.

b. Prepare a list of factors you want to have considered in a make-or-buy decision, as representative of quality assurance.

c. Describe the approach for including considerations of this list in any future make-or-buy decision.

(10–5) You are to decide whether a supply order is

to be inspected by receiving inspection or by source inspection at the suppliers plant.

a. Prepare a checklist of factors you want to have considered. Include the heading of this checklist.

b. List the steps to be taken by a quality assurance department in preparing such a checklist.

c. Describe measures to ensure that decisions for source inspection are made with reference to your checklist.

(10–6) The manager of an electronic security manufacturing company wants to control parts on stock using Pareto analysis. Given the monthly usage and unit costs, classify the items in the first three major categories according to the dollar usage.

Item Number	Usage	Unit Costs
356	100	2800
287	600	24
576	80	1400
630	300	40
560	20	2040
428	160	280
410	4000	30
518	800	80
534	9000	10

(10–7) MRP problem: A product structure tree is shown below.

quality, these are very important items. The quality assurance manager wants to use the information to schedule proper source inspections. One hundred and twenty units of end-item A are required at the start of the week. There is no stock on hand.

(10–8) "While pursuits of the learning curve can reap great benefits, the manufacturer will find it has a 'bottom' unless he maintains flexibility"[10]. The impact of a cost reduction strategy with repetitive and increased production are, for example, higher fixed costs and less scope for innovation.

a. List the positive and negative aspects of an interrelationship between the above strategy and quality assurance.

b. Formulate and solve a problem, using the learning curve model, to demonstrate a major point made under a.

10.14 NOTES

1. A booklet that discusses many aspects of vendor selection is *How to Conduct a Supplier Survey*, American Society for Quality Control, Milwaukee, Wisconsin, 53203, 1977.

2. Pettit, R. E., "Vendor Evaluation Made Simple," *Quality Progress*, March 1984, pp. 19–22.

3. ANSI/ASQC Standard A3-1978, "Quality System Terminology."

4. The U.S. Small Business Administration has published

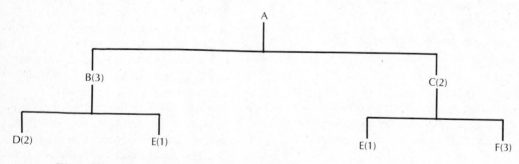

The quality assurance manager wants to know when parts E and F will be ordered and scheduled to be reviewed, because, in terms of

respective management aids. See, for instance, MA No. 215, *How to Prepare for a Pre-Award Survey,* 1978.

5. U.S. Government Military Standard MIL-STD-105D, "Sampling Procedures and Tables for Inspection by Attributes," 1963.

6. Management Aid No. 243, "Setting Up a Quality Control System," U.S. Small Business Administration, Washington, D.C., 1979.

7. Management Aids No. 243, U.S. Small Business Administration, Washington, D.C.

8. KANBAN is a Japanese word for a card that accompanies each bin of parts. When a worker starts drawing parts from a new bin, the KANBAN is removed and returned to the supplier.

9. *Burgess,* J. A., "R.I.P.—A Program for Supplier Quality Improvement," *ASQC Annual Quality Congress Transaction,* Boston, 1983.

10. Abernathy, W. J. and Wayne, K., "Limits of the learning curve," *Harvard Business Review,* Sept.–Oct. 1974.

10.15   SELECTED BIBLIOGRAPHY

ASQC, Vendor—Vendee Technical Committee, Procurement Quality Control, 2nd ed., 1976.

ASQC, *How to Conduct A Supplier Survey,* 1977.

ASQC, *How to Evaluate a Supplier's Product,* 1981.

Burgess, J. A., "Measuring and Upgrading Supplier Performance," *Quality Progress,* October 1980, p. 30.

DePriest, Douglas J. and Lauber, Robert L., *Reliability in the Acquisitions Process,* ASQC, 1983.

Falvo, Vincent A., "A Computerized Vendor Rating System," Quality Progress, June 1977 p. 20.

Hall, R. W. and Vollman, T. E., "Planning Your Material Requirements," *Harvard Business Review,* September–October 1978, p. 105.

Heinritz, S. F. and Farrell, P. V., *Purchasing: Principles and Practices,* 5th ed., Prentice-Hall, Englewood Cliffs, N.J., 1971.

Kraljic, Peter, "Purchasing must become supply management," *Harvard Business Review,* September–October 1983, p. 109.

Mihalasky, J., "The Vendor: A Neglected Quality Improvement Tool," *ASQC Quality Congress Transactions,* Detroit, 1982.

Peterson, R. and Silver, E. A., *Decision Systems for Inventory Management and Production Planning,* John Wiley & Sons, Toronto, 1979.

Schonberger, R. J., "The Transfer of Japanese Manufacturing Management Approaches to U.S. Industry," *Academy of Management Review,* Vol. 7, No. 3, pp. 479–487, 1982.

Shiliff, John W. and Bodis, Milan A., "How to Pick the Right Vendor," *Quality Progress,* January 1975, p. 12.

Stockbower, E. A., "Customer–Supplier: An Advantageous Relationship," *Quality Progress,* January 1978, p. 34.

11 CHAPTER

QUALITY OF CONFORMANCE: INSPECTION AND PROCESS CONTROL

Quality of conformance means to meet all specifications, including those of quantity, at the predetermined delivery time and place. However, internal quality assurance planning for separate dimensions and performance standards of the output of a certain expected quantity remains hidden without a true knowledge of the process behavior. In other instances, if time standards are too tight, if workplace conditions are inadequate and no measures exist to see that machines are operating as expected, quality of conformance becomes a difficult task even for the most motivated and qualified worker. Although we focus here on the quality of conformance, we must continue recognizing the context in which this is to be attained. In practical terms this means, for example, that a production manager's responsibility, for assuring high productivity and for meeting delivery dates, must also be translated to mean responsibility for maintenance of good process control for quality attainment.

In preceding chapters quality had been described in terms of specific attributes and measurements. Subsequently, preparation of the production plan and verification of the necessary production and supply capacities were discussed. Once the actual production starts, special process control techniques and inspection plans must be utilized to assure that all quality objectives will, in fact, be attained. At this point we also assume that quality of supplies from outside sources is assured. Now the task for the producer is to assure the quality of its own production.

The production plan and the associated inspection component is not as yet operational, as the actual creation of value is largely identifiable with good process control. We must not forget that any production design is to guide workers in preparing and conducting work; their own, that of others, or, that of machines and *systems*. To different degrees this is true in simple operations as well as in large-scale and complex operations. The higher the relative automation and computerization of production and the more indirectly people use automated production processes, the more important becomes the preparatory process control planning.

Therefore, production plans must be developed into *operational details*. Any production system—in total and in its parts—is ultimately designed by management and directed and controlled by management. Therefore, the quality of outputs depends greatly on the workmanship of planners. This entails the setting of clear and workable process control standards so that the quality of conformance can be monitored at each production stage day by day, week after week, and from one year to the next.

Control operations are to be continued until all significant deviations and nonconformances are detected and improvements begin to show. The more costly nonconformances are, the more affordable an intensive and preventive control system is. The interplay between process control and correction, with possible revisions of production plans and procedures, must continue until satisfactory output, measured in improved articles, stabilizes.

The implementation of the inspection plan, through detailed inspection and test procedures at the designated inspection points, can then be clarified against the general background of ongoing production activity. We shall deal with inspection procedures and facilities, designed to verify quality of conformance at crucial production stages. However, complete prevention or elimination will not normally be feasible; thus, statistical methods seem to be more realistic, effective, and economical. Therefore, the use of *process control techniques* is an essential component in inspection plan implementation.

11.1 THE PRINCIPLE OF PERFECT PRODUCTION

Like the concept and practice of zero-defect, an ideal must be firmly established in the mind of every person involved as to how production will run, smoothly and flawlessly, once operations have started. In a well-rehearsed concert, individual musicians know and have practiced their role. Because of this, the conductor can masterly combine and coordinate individual performances into an overall harmony and rhythm. Perfect artistry in performance can be reached through basic training and intensive practice and preparation. Even with the best support services, working conditions, and leadership, which are instrumental to satisfactory performances, improvement in quality will be realized only through respective human talents and ingenuity.

The ideal situation on the production floor would have everybody and everything well-prepared for the role to be played. This ideal is not difficult to realize when certain principles of good workmanship are observed. After all, defects occurring after production give an indication of a fault in the production preparation. Hayes, in an article titled "Why Japanese Factories Work,"[1] traces their success to paying attention to manufacturing basics. Keeping the workplace clean and tidy, minimizing work-in-progress inventory and handling, rejection of the attitude inherent in *Murphy's Law* (according to which "anything that can go wrong, will go wrong"), preventing machine overloads and a crisis atmosphere, and analyzing every question and problem in sufficient detail, are among the measures and conditions of a well-planned production system.

As far as the assurance of operations is concerned, several factors are of importance. The production plan is the important first step for well-initiated and directed individual operations.

Each individual job and operation has to be designed sufficiently, depending upon the operators' ability for the task on hand, the process capability, and the complexity of the task itself.

Work system planning that includes job design, work standardization, and performance measurement, has an essential impact on workmanship and quality. Emphasis must still be given to planning of quality assurance through independent inspection and testing. However, the quality control plan should closely follow the production operations through all the preparatory stages.

Job and work system design is similar to planning of large-scale productive systems. Once the tasks have been defined and delineated from other related production capabilities in terms of operations, work methods, tools, equipments, and environmental conditions, operator qualifications must also be determined. Quality assurance specialists contribute through assessing the work design explicitly from a quality conformance perspective. Once they are satisfied that the tasks and capabilities (both for the present and future) are well-formulated, documented, and communicated to the operator, verification procedures for inspection at each of the already designated inspection points can proceed.

Principles in job analysis that encompass the work system, design, and worker performance are the following.

1. The design must specify objective, method, tools, and operator qualification and performance criteria.

2. The operator must be informed, oriented, trained, and supervised in a helpful and constructive manner.

3. The operator must be able to measure his or her performance in qualitative terms, and then become innovative and creative in improving the performance further.

4. The worker must participate in the planning and implementing of the design as much as is feasible.

5. The workers must know their responsibilities and authorities. They should be encouraged to strive for more responsibilities and respective authorities on the job and in terms of their future career, so they have "pride" in what they do.

The First Piece Inspection

The preparation with *first piece* inspection, the release and initiation of the operation, the actual execution with implied work controls, and the completion of the operation are all interrelated. Through all these operational phases the operator has to comply with certain procedures and, in addition, is expected to observe due care. The inspection possibly undertaken by the operator as part of the work procedure (self-inspection) is to independently verify that everything is in order, that is, the product or service conforms to the requirements or specifications.

Perfect production can only be realized when quality assurance has been considered in elementary work system design and is further founded in well-conceived process control systems.

11.2 INSPECTION AND TEST PROCEDURES

A *procedure* must be understood as a mandatory directive for performance. A procedure stipulates and usually outlines in step form what the person to whom it is directed is expected to do, in order to achieve certain outcomes and results. A *practice* as opposed to a procedure, is what people actually *do*. This might not necessarily conform with the respective valid procedure. Operational and inspection or test procedures serve as important guides in attaining performance standards. When operators or inspectors can not properly comply with their procedures and therefore if their practices deviate, the procedure itself might be defective. In the same meaning, if a machine or system is faulty, the performance can be no better. Performances and the resulting degree of quality of conformance depend to a large extent on the quality of the procedure and the process control system.

Quality program *standards* and *guides,* such as ANSI Z-1.15,[2] stipulate and describe inspection procedures for work in progress. Sufficient leeway remains for adaptation to prevailing production conditions in each instance and existing inspection plan.

The practice of in-process inspection of one's own company's work varies with the kind of production system utilized. The inspector has to confirm that the operator is capable of meeting standard requirements. Such inspection situations usually exist in job shops with small lots, or in the manufacture of custom-made products. In mass production systems, inspection procedures are much more complex, involving the sampling and continuous adjustment of the production processes. Three types of inspection systems are schematically shown in Figure 11.1. The superiority of having a closed-loop end-product inspection can be easily comprehended by a comparison of these three types.

The set-up or first piece inspection is crucial (see example in Figure 11.2).

In smaller plants, *patrol inspection* can take care of various inspection points at random intervals. Often the patrol inspector handles the setting-up and first piece inspection points (Figure 11.3).

Final inspection of the finished product is basically an in-process inspection. Often it is the only one. Comparison of costs and benefits derived from inspection decides whether or not a system is efficient and optimal. Finished-item inspection can be the most thorough and intensive method, depending on the levels of preceding in-process and receiving inspection. It can include inspection of packaging and installation, and can take place at the customer's location (*source-inspection*).

The American National Standards Institute (ANSI) Guide Z-1.15 (see note two for this chapter) suggests that special controls be used for processes having parameters that affect product acceptance difficult to determine after the fact (e.g., painting, plating, heat treatment, etc.). These are implemented in addition to those controls at the set-up and first piece stage, those used by machine operators, those at fixed inspection stations, and the other controls of production quality.

The inspection status of any work in process must be clearly identified on the item or associated documents, tags, and so forth.

A general scheme of the *in-process* inspection procedure is suggested in Figure 11.4. These, together with the inspection plan prepared for a particular item or project, provide a master plan for the person in charge of writing more specific inspection procedures for specific inspection points. Valid and reliable decisions by the inspector depend on a careful

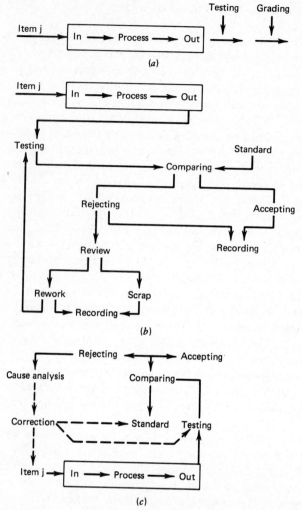

FIGURE 11.1 Schematic of the three types of end-item based inspection system. (a) Open loop sorting. (b) Open loop end-item inspection. (c) Closed-loop end-item inspection.

and systematic preparation of the procedure and its implementation and surveillance. Figure 11.5 describes the essential features needed in determining inspection or test procedures.

Deciding Inspection Points

The inspection plan, unlike a master plan, usually includes all technical details that must have the identification criterion filled in for each inspection point and ''fail point.'' This also concerns the proposed inspection method, which should be appraised mainly in order to confirm whether it is optimal with respect to available inspection and testing capacity. Once criteria for acceptance have been specified, the verification test itself can now be selected and described in technical terms.

Set Up Inspection

- Set-up inspection to be carried out in conjunction with first-off inspection.

- Set-up inspection will be carried out at machine station.

- First-off inspection to be carried out in inspection office and only to be carried out with numerical readout instruments.

- Per following inspection instructions.

(Example: a machined hub casting)

Characteristics	Specifications	Tools	Records to Be Kept
"Go-No-Go" plug gauges	All plug gauges must be on size	Micrometer	Record and tag defective gauges
Drill bushings and fixtures	Drill bushing to be on size and correct and drill fixture being used	Hole gauges and micrometer	Record and tag defect drill bushings and fixtures
First off of machine hub casting	One hundred percent inspection of all machined dimensions	Bore mics. vernier caliper and thread gauge	All previous recordings to be included in inspection records

FIGURE 11.2   Salient features and associated details of a set-up inspection.

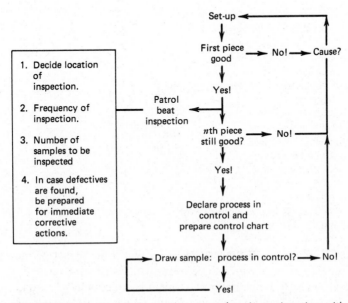

FIGURE 11.3   In-process inspection procedure in conjunction with patrol beat inspection.

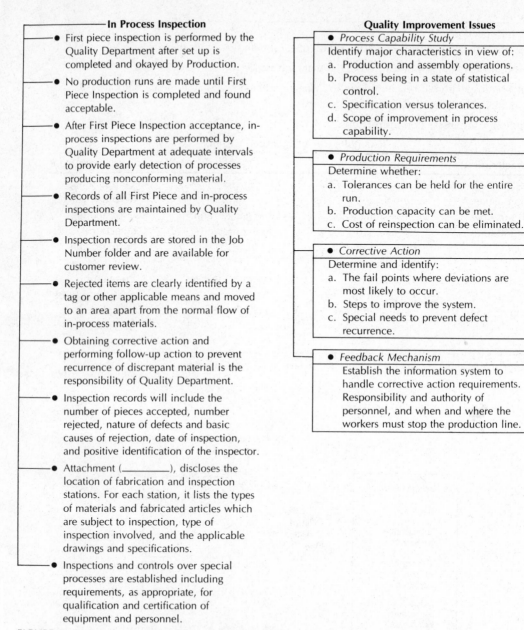

In Process Inspection

- First piece inspection is performed by the Quality Department after set up is completed and okayed by Production.

- No production runs are made until First Piece Inspection is completed and found acceptable.

- After First Piece Inspection acceptance, in-process inspections are performed by Quality Department at adequate intervals to provide early detection of processes producing nonconforming material.

- Records of all First Piece and in-process inspections are maintained by Quality Department.

- Inspection records are stored in the Job Number folder and are available for customer review.

- Rejected items are clearly identified by a tag or other applicable means and moved to an area apart from the normal flow of in-process materials.

- Obtaining corrective action and performing follow-up action to prevent recurrence of discrepant material is the responsibility of Quality Department.

- Inspection records will include the number of pieces accepted, number rejected, nature of defects and basic causes of rejection, date of inspection, and positive identification of the inspector.

- Attachment (_____), discloses the location of fabrication and inspection stations. For each station, it lists the types of materials and fabricated articles which are subject to inspection, type of inspection involved, and the applicable drawings and specifications.

- Inspections and controls over special processes are established including requirements, as appropriate, for qualification and certification of equipment and personnel.

Quality Improvement Issues

- *Process Capability Study*

  Identify major characteristics in view of:
  a. Production and assembly operations.
  b. Process being in a state of statistical control.
  c. Specification versus tolerances.
  d. Scope of improvement in process capability.

- *Production Requirements*

  Determine whether:
  a. Tolerances can be held for the entire run.
  b. Production capacity can be met.
  c. Cost of reinspection can be eliminated.

- *Corrective Action*

  Determine and identify:
  a. The fail points where deviations are most likely to occur.
  b. Steps to improve the system.
  c. Special needs to prevent defect recurrence.

- *Feedback Mechanism*

  Establish the information system to handle corrective action requirements. Responsibility and authority of personnel, and when and where the workers must stop the production line.

FIGURE 11.4   Details of in-process inspection procedure. Adapted from, *Management Aids* No. 243, U.S. Small Business Administration, Washington, D.C.

The schematic in Figure 11.6 suggests steps by which the person writing the procedure determines the quality characteristics to be inspected, tested, and verified. Once the test is technically sound, and can be carried out as required, the entire procedure must be documented. Again, the actual start-up of the inspection, which coincides with the commencement of production, will have to be carefully controlled.

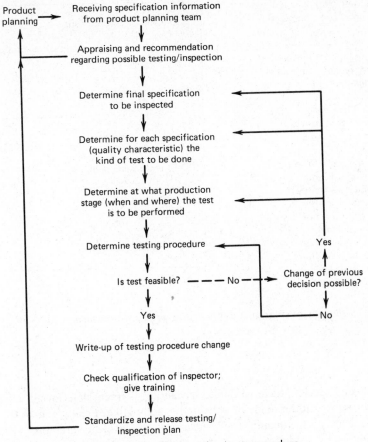

FIGURE 11.5 Determination of inspection/test procedure.

In-process Control

Subsequent to the decisions made on inspection plans and procedures, it becomes necessary to emphasize quality, not only during production, but also prior to and after the work has been completed. *In-process* implies that the variables affecting quality characteristics are to be controlled *during* the manufacturing process. Thus, the real usefulness of *control techniques* lies in being able to *predict* nonconformances of the product *yet to be produced*. Inspection must be a *profit center* in practice and in theory!

Sampling versus 100 Percent Inspection

The so-called 100% inspection, under which each passing item is subjected to testing, does not, in practice, weed out all defects. Variation and error are a fact of life. The practical usefulness of statistical quality control methods as an inspection and decision-making tool is based on the recognition of this very fact of variability as a major factor in all production and operations.

1. Attain inspection plan and determine inspection points.
2. Determine quality characteristic to be verified.
3. Appraise proposed inspection method, process capabilities.
4. Determine possible nonconformance.
5. Determine decision criteria for acceptance.
6. Determine method for inspection.
7. Establish inspection/test verification procedure.
8. Provide testing facility.
9. Select and train inspector.
10. Document inspection/test procedure; plan for recording system.
11. Finalize the procedure.
12. Appraise and supervise performance (inspector and system).

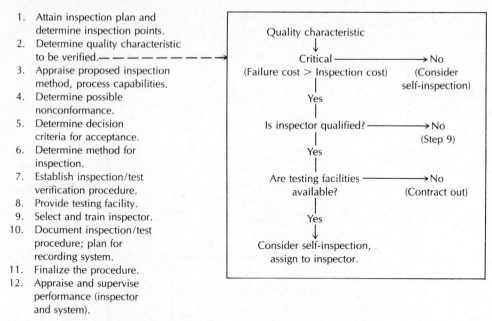

FIGURE 11.6 Schematic of steps vital to procedure writing for inspection.

All products and services have variability because it is impossible for all units (or items) of a product or service to be made or delivered exactly alike. Quality cannot be controlled unless there is a means for quantitatively measuring the level of variation (whether caused by management or by workers) and for distinguishing *controllable* variables from *uncontrollable* variables.

Figure 11.7 summarizes the typical approaches used in quality control or inspection techniques. In the application of one or the other, or even of a combination of the two approaches, the following general principles are worthy of mention.

1. Apply these methods only when they are properly understood by those compiling the statistics and by those using the information.

2. Contrast these methods with nonstatistical methods such as 100% inspection or a constant percentage of a lot as a sample.

3. Use these methods not only for reducing inspection load and costs but also for improving efficiency of workers and systems.

4. Simplify the application of methods by using sampling tables, prepared control charts, and other aids in conjunction with the fact-finding tools earlier described (see Chapters 5 and 6).

5. The fact that statistical methods are used is not sufficient evidence of an up-to-date quality control system. The *actual use* of the data must lead to *improvement* by finding the real cause of the problem.

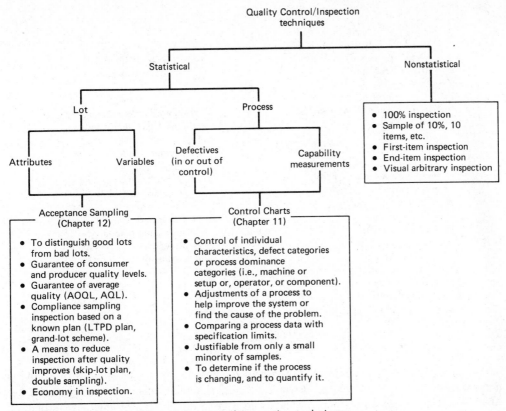

FIGURE 11.7   Summary of the quality control/inspection techniques.

6.   These methods, as any others, should be used only when the benefits exceed the costs involved. However, the benefits may often not be fully recognized and realized if the focus remains on short term goals due to failure to recognize the causes of variability and to general management apathy towards the data's indications.

7.   Statistical sampling also refers to the probability of finding a certain number of defectives in a lot. Moreover, as inferences drawn from a sample are subject to error, these techniques allow the producer to specify a certain level of confidence that the inference in the quality of the lot is correct within defined acceptance or rejection limits.

11.3   CONTROLLING THE PROCESS: BASIC CONCEPTS

Because of the fact that no production processes can produce any two items that are exactly identical, there is a need not only to locate and segregate defective products, but also to control the variations. The use of statistical methods has proven to be so effective in controlling variability, that its proper uses in quality assurance management deserve a detailed description. It must be understood however, that any such statistical technique is not a magic wand. The important idea here is *preventive action*. Preventive action is dedicated to *defect prevention*. Therefore, quality of conformance strategies must be planned as a part of the process control system (Figure 11.8).

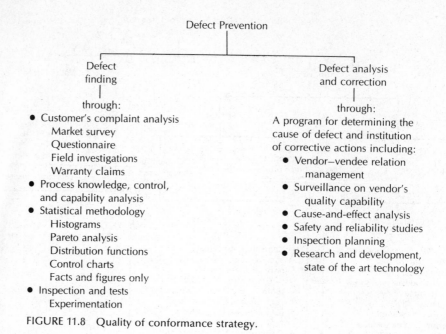

FIGURE 11.8   Quality of conformance strategy.

Although we shall be examining more details of process control techniques by studying a set of tools called *control charts,* it would be quite ridiculous to suggest that the mere use of a control chart can turn a poor product into a good quality end-product. Nevertheless, good control of manufactured products using statistical techniques always aids the confidence of management. A record of the use of such control devices also becomes very helpful later on in dealing with customer complaints or, at the extreme, in liability cases.

Causes of Variability

Causes of variability in manufacturing processes are always almost unlimited. There is variability in machines, materials, and production processes, so much so as amongst the operators and between the actions of one management group and another. The capabilities of a process, for example, may vary with the types of tools and materials used, machine vibrations, work holding devices, positioning, electrical fluctuations, and so on and so forth. There can also be piece-to-piece variation, within-piece variation, time-to-time variation and variations between one defect and another produced under one fixed set of conditions.

What is unusual? This is the basic question that must be answered in any planning for process control. Beyond those reasons that direct straightforward questions regarding product specifications and machine behavior, one may also very well ask; "Which factor dominates over all?" Depending upon the complexity of products and processes, identifying the dominant factor or component in the entire manufacturing system has great value. It enables production planners to concentrate on the most important variables by using the right process control techniques available.

Concept of Dominance

According to the concept of dominance, the variations in a given quality characteristic of a manufactured product are grouped into several forms.

1. *Machine-dominant* (or machine dependent) processes are those which cause quality characteristic variations if suitable preventive measures are not instituted in time; (e.g., printing ink decay, liquid bottle filling, tool wear, wear of refractory lining in melting furnaces, etc.).

2. *Setup-dominant* operations are those in which the quality of products depend on the initial setup (adjustments etc.) of machine and related accessories. Examples are molding, labeling, automatic metal punching operations, data processing, and drilling.

3. *Operator-dominant* processes are dependent on the skill, competence, motivation, and workmanship of the operators, such as manual arc welding, soldering or brazing, brush painting, file sorting, sales, and any other job done with a large accountability to individual performances.

4. *Vendor-dependent;* where the quality of manufactured products depends, to a great extent, on the quality of purchased parts from vendors. Automotive assembly, food and beverages processing plants and a host of other assembly plants or subcontracting assignments can be cited as examples of this type of dominance.

5. *Management-dependent* factors are those where the management team, by way of their policies and actions, contribute a great deal to a company's quality problems. Lack of management commitment to quality, bad quality coordination, absence of written quality control manuals and procedures, and incorrect application of quality standards are some of the most common examples, constituting management apathy in general.

Broadly speaking, variability can be attributed to one of two types of causes. These are either *chance* causes, or *assignable* causes.

Chance variations do not occur in any repetitive fashion and are seldom produced by one cause only. In other words, their occurrence is not predictable and we can do nothing about these random variations, except assume that the process behaviour is normal.

Assignable variations are those that result from discrepancies and show larger variations than normal. The common assignable causes in a production process result from differences among materials, machines, or workers, as well as being due to interaction between any two or all three of these major factors.

When a process is in a state of statistical control, variations that occur in quality characteristics are due only to chance causes. Assignable causes for product variability usually signal an *out of control* condition.

Meaning of the State of Statistical Control

A process that is operating without evidence of assignable causes of variations is said to be in a *state of statistical control*. When a process is in state of control with all or any assignable causes having been eliminated, the data points show a natural (random) pattern of variation as shown in Figure 11.9. The control limits used on the chart Figure 11.9 were calculated from actual data of inspection. The rules of calculating the limits are discussed later in this

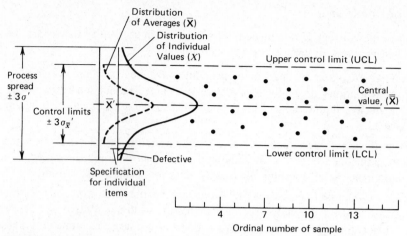

FIGURE 11.9   Random variation of data points within the control limits.
The distribution curves at left describe the relationship of limits with
individual measurements and averages.

chapter. However, the important point to remember at this stage is that control limits are not specification limits. Control limits, as based upon actual data, are obtained from a process (presumed) to be in a state of statistical control; it is wrong to show specification limits on the control chart. When the data points (being averages of a certain quality characteristic) are distributed randomly around the central value, they will tend to follow the shape of the normal distribution.

If something goes wrong (temporarily) with the process of manufacture or the measuring instrument, the operator should be able to discover (or at least be trained to discover) the assignable cause, and to remove it. Getting involved in a fact-finding mission regarding larger variations, if they have a trend or if their occurrence seems to be more than a chance, is obviously a management responsibility. The point is always to make sure what is really meant by assignable cause.

Example 1: In a chemical plant that manufactured Liquid Paper (correction fluid), the control chart for the average concentration of one of the solvents showed that the process was out of control. To the best of anyone's knowledge, all that was done by the operators was to add (or not add) an extra amount of solvent, depending upon what the control points (measured sample averages) showed in a previous test.

Clearly, the operators were confused. What they were told to do was to adjust and readjust. It seemed an easy way out for all. No one recognized the *true* assignable cause. In this case it was found that a faulty temperature control system caused all the trouble. It indicated lower temperatures than the actual ones, and, in fact, the temperature was high enough, causing evaporation losses. The control chart in this case obviously failed to serve its real purpose.

Examples such as the one cited above can be numerous. The students are encouraged to think of other examples appropriate to their own conditions or workplaces; (compare the minisheet of problems cited in the beginning of Chapter 5 to lead to a discussion on this topic).

Example 2: On many occasions the data points obtained are such that they tend to accumulate in a fashion of a definite trend. The example of such variations shown in Figure 11.10 was found in a food processing plant where the final test of the percentage composition of one critical preservative chemical did in fact stay within the upper control limit, UCL (3.7%) and the lower control limit, LCL (3.0%). However, when the entire pattern for 24 hour checks was plotted together, the situation became serious enough to warrant the attention of the chief inspector. The matter obviously gained attention for the three distinct patterns (marked as (*a*), (*b*) and (*c*) in Figure 11.10) found in data that varied during three work shifts in one day. Note that assignable causes can be present even though all points are within control limits. Therefore the real assignable cause must be found and corrected before a normal, stable, process can continue.

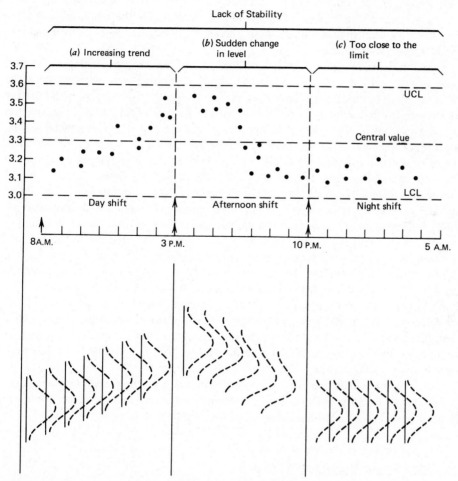

FIGURE 11.10 Erratic variation of data points indicating lack of stability in the controlling process. The bottom portion shows the shifts in the distribution profiles that the process is creating in each case.

Control Limits

Control limits are usually established at $\pm 3\sigma$ (three times standard deviation), best applied to sample averages or sample ranges. A range, as previously defined is the arithmetic difference between the maximum and minimum values in the data. Using averages, rather than individual values, forms a basis for its distribution being normal, which in turn means that it complies with the statistical laws of probability, whether or not an actual industrial process strictly follows this behavior. By establishing $\pm 3\sigma$ limits, we restrict the population up to 99.73% as being dependent on chance variation and the remaining 0.27% are left to the assignable cause.

Defining process capability as $\pm 3\sigma$ (a total of six standard deviation) is quite common, primarily from a cost point of view. However, it need not be so liberal. For example, if the tolerance limit for inspection is exceeded beyond the $\pm 3\sigma$ control limit (say, to $\pm 4\sigma$, to include 99.94% of the population), it would require having a superior manufacturing process; to be able to control the 99.94% of outcoming population. On the other hand, if a process having the $\pm 4\sigma$ limit remains inside these tolerances, it will still be in control and unlikely to produce nonconformances.

The final decision in selection of control limits must involve the calculated cost based on the size of risk and the total cost of making errors[3] depending upon whether the process is capable or not. The first error, *Type I* referred to by statisticians occurs when the process involved remains in control but a point falls outside the control limits due to chance (chance is 0.27%, or 3 out of 1000) and a cost is incurred because of incorrect conclusion that the process is out of control. The second error, called *Type II*, occurs when the process involved is out of control but the data point falls within the control limits giving the impression that the process is under control. Type II cost is associated with the nonconforming output.

The readers should recall the two previous examples where operators merely did adjustments and readjustments whenever an assignable cause was found. With reference to Type I and Type II errors, such actions can be termed respectively as *over-adjustment* and *under-adjustment* of the process or the machine. At times it is advisable not to jump to a quick conclusion, because, if a real cause exists, it will arise again to provide another chance for detection. Rather than finding the real cause of the trouble, simple over-adjustment and under-adjustment simply guarantees that the method will never be improved.

Furthermore, a process, even if in a state of statistical control, may not guarantee freedom from defectives. In its true sense of meaning, when a process is in the state of statistical control, the random (up and down) variations in the data within the set limits are predictable. It is quite possible that the fault then may lie with the setting up of limits which may have been unrealistic to begin with.

The main point is that, whether the control limits are to be reduced or revised, no one firm rule exists. Every problem is different from every other, and the respective aspects of each need to be considered, depending upon the situation at hand.

11.4   PROCESS CAPABILITY STUDIES

Before a mass-scale production is started, it is necessary and important in planning for quality to make an analysis of the capability of the intended process. This concept of *process capability*, which provides a quantified measure adequacy, is learned indirectly by quality control checks done on the actual products.

There are numerous worthwhile reasons in knowing the capability of intended process, including those of the operators, the systems, and so forth.

1. As a basis of specifying quality characteristics or specification of purchased machines, materials, or other supplies for the purpose at hand.

2. To make predictions about the extent to which the processes will be able to hold to tolerances in the course of time and thereupon to improve them.

3. To make reliable decisions about product specifications as supplied by the customers; whether these are feasibly achieved by the plant's production capabilities.

4. To decide among many competitive processes those that will give the most desired tolerances. Interchangeability criteria of components demands tight tolerances which the process must be capable of producing adequately.

5. To plan future sequences of production in view of customer requirements. If the capabilities of several processes are quantified, it becomes easier to plan with these, both in terms of reproducibility and of time-to-time variability, and to know the output precisely.

6. To find various causes of defects during production, based on the capability history of different processes, in order to improve the process or the system.

Evaluation of Process Capability

Conducting capability analysis for a given process or machine needs a good deal of planning. The analyst's aim is to determine a trial run, which, in all likelihood, will be representative of the normal production conditions as much as is possible.

Work planning for process capability follows a logical sequence: define the purpose of chosen process capability; choose the machine or process; then, delineate and write down process conditions that strongly influences the quantification of process capability. Variability may depend on operator's skill, machine loading conditions, workplace safety and health factors, temperature–pressure interrelations, and on other similar factors. Technical know-how may be necessary.

Next select the operator, raw materials, and other consumables to help make uninterrupted runs or trials under the known conditions.

Start the process or machine work. Keep track of all pertinent data related to each and every batch or lot of sample produced, with special notes referring to any unusual matter noticed, and the measuring techniques used by the operator or inspector. Ask yourself, "Is everything under a state of statistical control?"

Lastly, analyze the data and then set the limits and watch for the assignable causes.

The word *process* here has a wide connotation. It applies not only to the functioning of machines, but also to its operators, the testing and measuring equipment, inspection plans and procedures, process of receiving inspection and purchase control, scheduling and inventory control, process capability of vendors, and so on. By knowing where the capability of each part of the process stands, efforts should then be directed toward improving them; if not all at once, then one by one, and ultimately, the entire situation. Simply meeting specification limits has nothing to do with quality!

Averages and Individual Measurements

The control limits, as discussed before, are established as a function of averages from several individual readings. If there are a large number of individual measurements (X's), one may very well find their subgroup averages ($\overline{X}$'s) to be the same. The distribution curve for individual measurements, however, may not always look close to a normal distribution shape as would the distribution for their averages. When population values are available for the standard deviation from individual measurements (σ'), the relationship that describes its standard deviation for averages ($\sigma'_{\overline{x}}$) is given by:

$$\sigma'_{\overline{x}} = \frac{\sigma'}{\sqrt{n}}$$

where:

$\sigma'_{\overline{x}}$ = Population standard deviation of subgroup averages ($\overline{X}$'s).
σ' = Population standard deviation of individual measurements (X's).
n = Subgroup size.

The distinction between the distributions of individual measurements and averages using the same data can be seen from Figure 11.9.

The control limits and specification limits should not be misunderstood to be the same. The specification limits are the permissible variations established by design engineering. Thus, the location of specification limits (or the tolerances), if as shown, can only be optional.

Process Spread and Specification Limits

In order to study the specific process characteristics of particular interest, (for example, the diameter of machined rods) measurement are necessary on large number of samples. Assuming process capability as $\pm 3\sigma$ has been a common practice. However, the trend is towards assuming $\pm 4\sigma$. Since $\pm 3\sigma$ theoretically represents 99.73% of the total population, the process in this case can be quite safe, if 6σ does not exceed the total specification spread. However, it will still not be 100% defect-free.

There are three situations: $6\sigma' = U - L$; $6\sigma' < U - L$; or, $6\sigma' > U - L$; where U and L are abbreviations for upper specification limit and lower specification limit respectively. The three situations are depicted in Figure 11.11 (diagrams a, b, and c). In case a where $6\sigma' = U - L$ (Figure 11.11 diagram a), as long as the process remains in control, indicated by the distribution curve in column I, process capability meets the specification requirements. When the process is *out of control* (columns II and III) because of either shift or spread, a defective product is being produced. In case b, since the specifications are appreciably wider than either the shift or the spread, the assumption will be that no apparent difficulties exist.

The obvious question here is to find out why the specification limits are so wide apart. Unless the limits are narrowed down to be made feasible, defectives are, in all likelihood, being produced and simply not being detected under the circumstances.

Two further exemplifications of similar situations can be seen in Figure 11.11 (diagram d), where a plot of percent defectives on simple run charts in both cases show definite trends. In the top plot, the points mostly stay above the center line, and in the bottom plot a continuous decreasing trend is observed. Plots with only a few readings would have completely failed to signal these trends. However, remember that such charts do not by themselves disclose what the *cause* is; for this one has to look a little deeper.

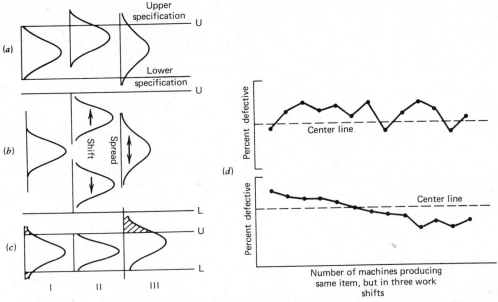

FIGURE 11.11 Illustrating the changes in process average and dispersion for the conditions. (a) $6\sigma' = U - L$. (b) $6\sigma' < U - L$. (c) $6\sigma' > U - L$. (d) A special case where the data points show good control, still a trend is obvious waiting for action.

In case c, where $6\sigma' > U - L$, even though a natural variation pattern is occurring to start with, some products are being made without defects. This situation should remind one that process is not producing what is required by the specification. The term process capability index, C_{pk} is generally used (C_{pk} = specification width divided by process width) to measure the process variation around a target or central value of the specification. A paper by Sullivan (see end-of-chapter bibliography) gives details of C_{pk} analysis. There are several alternatives to use in looking for the solutions to the given cases.

1. Improve the process, so that a more peaked distribution occurs.

2. Narrow the specification limits if it still meets the design criteria and does not interfere with reliability of the system.

3. Change over to another facility or machine more capable of producing within the specification limits.

4. Provide the production workers with some basic knowledge in statistical terms to help them see where and how they should improve.

5. Find out whether the specifications are met on all operations. If not, find out why.

6. Find out whether any abnormality in the workplace environment itself is causing variations, such as humidity, noise, vibration, poor ventilation, confusion between operators and inspectors, parts and materials being mishandled, or other factors.

7. If none of the above solutions seem practical, consider staying with what is feasible; the added cost in doing 100% inspection to eliminate the defective parts (but not on a permanent basis!).

Example: As an illustration, let us assume that the specification submitted to a machine shop for drilling holes in gray iron hubcaps calls for a size of 0.500 ± 0.004 in. Therefore, the total tolerance is $0.004 \times 2 = 0.008$ in. Based upon 75% criterion,[4] the total variability to 6σ of the drilling process will be $0.75 \times 0.008 = 0.006$ in. The machine operator must relate this capability to tolerance. The capability ratio is given by:

$$\frac{6\sigma \text{ variation}}{\text{total tolerance}}$$

In order that the total variability for the process does not exceed 6σ; we have $6\sigma = 0.006$ in. or, $\sigma = 0.006/6 = 0.001$. Capability ratio will be $0.006/0.006 = 1$. Thus, the 75% rule of thumb criteria seems applicable in the case of bilateral tolerance. In unilateral tolerance, the approximation of this rule is 88%. In any case, students should realize that the process capability is a property of the process and is independent of the tolerance.

11.5 STATISTICAL CONTROL CHART TECHNIQUE

Just as there are various types of processes, so also are there various types of control charts. These can be categorized in two general classes, depending upon the characteristic being tested, as control charts for *attributes,* or control charts for *variables.*

Control charts for attributes are based on the inspection of an item using *go* and *no-go* criteria. For example, an item is either good or no good, the lot received is either accepted or rejected. In these situations the quality characteristics are expressed only as conforming or not conforming with the specifications. Commonly used control charts for attributes are: p charts, np charts, c charts, and u charts.

Control charts for variables, on the other hand, focus on one or more variable parameters or individual readings on samples which truly determine the quality of the product. Examples of such critical parameter are, for instance, tolerance measurement in physical dimension, tensile strength properties, chemical composition of steels, and other factors. Commonly used control charts for variables include $\overline{X}$ charts, R charts, and moving-range charts.

The control chart techniques are to be used on a continuous basis to check a process.[5] Thus they remain a powerful device used to make a large number of tests of significance in a systematic manner. However, in all cases it is important to discover promptly when the process has gone out of control. This constitutes a warning that a hypothesis should not be accepted.

The purpose or usefulness of control charts are as follows.

1. As a source of information in regard to product specifications and tolerances in production processes or inspection procedures, to help decide if the items produced will meet customer expectations.

2. As a continuous tool for production and supervisory personnel to help them take corrective actions when and if necessary to eliminate assignable causes and reduce variations.

3. To provide information to senior management for current decisions on acceptance or rejection of a lot. If assignable variations are present, the process must be modified or changed, otherwise it is not operating at its best.

The true achievement in using statistical charting techniques comes in the elimination (or reduction) of assignable causes leaving only chance variations to occur.

11.6 CONTROL CHARTS FOR ATTRIBUTES

Control Chart for Fraction Defective—*p* Chart

Control charts for fraction defective may be based on samples or on 100% inspection. In the former case the effective sample size may well vary, while in the latter case it should be constant (provided that lot sizes are constant). If sample data are used, it is important that a large sample be used.

The fraction defective, *p* is given by:

$$p = \frac{\text{Number of defectives}}{\text{Total number inspected}}$$

Percent defective is simply 100 times fraction defective. If the mean fraction defective ($\bar{p}$) remains constant, then the fraction defective (*p*) in each sample will vary from sample to sample and will vary according to a binomial distribution, such that standard deviation of fraction defective:

$$\sigma = \sqrt{\frac{\bar{p}\bar{q}}{n}}$$

where:

$\bar{p}$ = Mean of the fraction (or proportion) defective = $\dfrac{\Sigma np}{\Sigma n}$.

$\bar{q} = 1 - \bar{p}$.

n = Sample size.

Since the pieces or items inspected are either defective or not defective, using a binomial distribution assures that the probability of occurrence of a defective can be assumed to be constant. If sample size *n* is large, the upper and lower control limits can be spaced from the average fraction defective $\bar{p}$ by $\pm 3\sigma$, so that UCL = $\bar{p} + 3\sigma$ and LCL = $\bar{p} - 3\sigma$

where:

$$\sigma = \sqrt{\frac{\bar{p}\bar{q}}{n}} = \sqrt{\frac{\bar{p}(1 - \bar{p})}{n}}$$

Example: The data of Table 11.1 will be used to illustrate the use of the *p* chart. This tabulation shows the record of inspection checked in 100 valve lots with the number of defectives listed.

The average fraction defective, $\bar{p}$ is:

$$\bar{p} = \frac{214}{20 \times 100} = 0.107$$

The standard deviation, σ is:

$$\sqrt{\frac{(0.107)(0.893)}{100}} = \sqrt{0.000955} = 0.0309$$

$$\text{UCL} = \bar{p} + 3\sigma = 0.107 + 3\,(0.0309) = 0.199$$
$$\text{LCL} = \bar{p} - 3\sigma = 0.107 - 3\,(0.0309) = 0.014$$

TABLE 11.1
Record of Defective Valves in 20 Lots of 100

Lot Number	Number of Defectives, np	Fraction Defective, p	Number Inspected
1	6	0.06	100
2	14	0.14	100
3	7	0.07	100
4	11	0.11	100
5	18	0.18	100
6	10	0.10	100
7	11	0.11	100
8	19	0.19	100
9	2	0.02	100
10	9	0.09	100
11	19	0.19	100
12	20	0.20	100
13	4	0.04	100
14	3	0.03	100
15	16	0.16	100
16	8	0.08	100
17	9	0.09	100
18	15	0.15	100
19	7	0.07	100
20	6	0.06	100
Total	214		2000

Using the above values, a control chart, as shown in Figure 11.12, was constructed by making point plots for each of the 20 lots; center line at $\bar{p} = 0.107$, UCL at 0.199 ($\cong 0.20$) and LCL at 0.014.

Inspection of this control chart reveals that sample lot 12 is right at the upper control limit. All other lots have a fraction defective within the UCL and LCL with the exception of the readings of lots 9 and 14, which are a bit closer to LCL. In order to safeguard against any chances of producing defectives, a new center line and new revised control limits need to be established, and an effort made to keep the manufacturing process in control at this improved quality level. For this new level, we may recalculate the revised parameters for the p chart by eliminating lot 12 from the computation. The points close to or outside of the lower control limit need not be eliminated.

$$\bar{p}_{\text{revised}} = \frac{189}{17 \times 100} = 0.111$$

$$\sigma = \sqrt{\frac{(0.111)(0.889)}{100}} = \sqrt{0.000987} = 0.0314$$

$$\text{UCL}_{\text{revised}} = 0.111 + 3(0.0314) = 0.205$$
$$\text{UCL}_{\text{revised}} = 0.111 - 3(0.0314) = 0.017$$

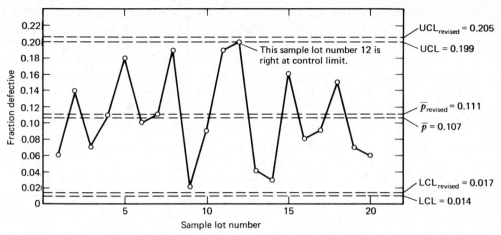

FIGURE 11.12   Control chart for fraction defective.

The new values of $\bar{p}$ and the new control limits are now drawn in the control chart and labeled as $\bar{p}_{revised}$, $UCL_{revised}$ and $LCL_{revised}$ for attaining better quality level.

Adaptation to the p Chart

Constant sample size: The formulas for UCL and LCL in the previous example have been based upon $\pm 3\sigma$ variations as a common practice for quality control inspections. Still, some practitioners choose $\pm 2\sigma$ or $\pm 2.5\sigma$ as indicative of *warning* limits or *action* limits to make such charting efforts more meaningful and quality control more preventive in nature. This is particularly helpful to inspectors and control personnel who must know in advance what action to take in order to prevent defect recurrences. For example, in the preceding case of defective valve production:

$$\text{Standard deviation} = \sqrt{\frac{\bar{p}\bar{q}}{n}}$$

With the original value of $\bar{p} = 0.107$

$$3\sqrt{\frac{\bar{p}\bar{q}}{n}} = 0.0927$$

$$2\sqrt{\frac{\bar{p}\bar{q}}{n}} = 0.0618$$

$$\text{and } 2.5\sqrt{\frac{\bar{p}\bar{q}}{n}} = 0.0773$$

Hence, a warning limit can be posted in the control chart of Figure 11.12 at $\pm 2\sigma$ (UCL = 0.169 and LCL = 0.045) where the operators must start thinking about the assignable causes likely to come, and similarly an action limit at ± 2.50 (UCL = 0.184, LCL = 0.029) where corrective actions must be taken if the fraction defective has kept on increasing. Lloyd Nelson[6] describes the test for early warning for use with the Shewhart p control chart.

The main reason behind using the $\pm 3\sigma$ (or for that matter the $\pm 4\sigma$) limit lies in the statistical reasoning that follows from the probability theory discussed earlier. If the system of chance cause produces a variation in samples that follow the normal curve, then the 0.001 (or one in one thousand) probability limits are practically equivalent to 3σ limits. It should be remembered that $\pm 3\sigma$ limits correspond to 99.73% of the population within which area the normal curve falls. The probability that a deviation from the mean will exceed 3σ in one direction is 0.00135 (1/2 of 100% $-$ 99.73% $=$ 1/2 of 0.27% $=$ 1/2 $\times$ 0.0027 $=$ 0.00135), that is, one in one thousand.

Varying Sample Size. Up until now, we have assumed a constant sample size (n) in each lot that was inspected. If the sample size varies considerably, then the value of the standard deviation will vary as well—increasing as n decreases and vice-versa. As a result, the UCL and LCL values will depend on the sample size,[7] where ($\bar{p} = \sqrt{\bar{p}\bar{q}/n}$, that is, $\bar{p} \propto 1/\sqrt{n}$), and the p chart will no longer consist of a set of parallel lines; rather we will have curved lines. If the sample sizes vary only slightly, using average sample size, $\bar{n}$ is still advantageous instead of individually varying n.

Control Chart for the Number of Defectives—np Chart In the case of an np chart, the number of defectives are plotted, instead of the fraction defective. Number of defectives $=$ (sample size) $\times$ (fraction defective) $= np$. The standard deviation of the binomial distribution for this case becomes:

$$\sigma = \sqrt{n\bar{p}(1 - \bar{p})}$$

where $n\bar{p}$ = average number of defectives.

Control Chart for the Number of Defects—c Chart

The c chart is applicable in instances where one is concerned about the number of defects per item or in each product: for example, judging a typist's ability on the basis of typographical errors per page typed; condemning jobs done by a pipe welding contractor by counting the number of poor welds; imperfections in a roll of paper; number of pin holes in castings; number of surface scratches on a finished painting; and other similar defects where the product is a large unit.

The computation of a c chart is based on Poisson distribution and, therefore, it is assumed that the frequency of defects will follow the Poisson law. Usually the conditions of a small $\bar{p}$ and a large n are satisfied when one is dealing with number of defectives per unit. No matter how the sample size is defined, it must always represent the same area of opportunity for the likelihood of occurrence of defects. As an example, assume that we are asked to inspect a truckload of microwave ovens for some kind of given defect. Let us further assume that we adopt an inspection unit of four ovens. Theoretically, this would mean the possibility of finding as many as 20 defects in that particular oven. However, one does not know offhand of such probability, and the minimum number of ovens to be inspected may be more than just one.

The average of number of defects is:

$$\bar{c} = \frac{\text{Total number of defects}}{\text{Total number of units inspected}}$$

The standard deviation, $\sigma = \sqrt{\bar{c}}$

TABLE 11.2
Record of Number of Defects in Paper Rolls

Roll number	1	2	3	4	5	6	7	8	9	10	11	12	13	14	15
Number of defects	12	13	10	15	10	9	11	15	13	12	23	11	9	10	7

(Total number of defects = 180)

As before, it is quite common to place the limit values at $\pm 3\sigma$ away from $\bar{c}$. Hence:

$$UCL = \bar{c} + 3\sqrt{\bar{c}}$$
$$LCL = \bar{c} - 3\sqrt{\bar{c}}$$

Example: Fifteen rolls of book quality printing paper produced by a paper mill company were inspected for brownish spot defects. The rolls with identical lengths were picked up at random, based on a day's production. The total number of defects were recorded as shown in Table 11.2. In order to plot a c chart, we first calculate the average and the standard deviation of the Poisson distribution. The average number of defects per roll is $\bar{c} = 180/15 = 12$.

$$\text{The standard deviation, } \sigma = \sqrt{\bar{c}} = \sqrt{12} = 3.46$$

$$UCL = \bar{c} + 3\sqrt{\bar{c}} = 12.0 + 3(3.46) = 22.38$$
$$LCL = \bar{c} - 3\sqrt{\bar{c}} = 12.0 - 3(3.46) = 1.62$$

Figure 11.13 shows the c chart. It is seen that lot 11 is outside the control limits and needs further investigation to determine the cause for this condition. Note that a decrease in c means improvement in quality; consequently, effort should be made in decreasing the c values.

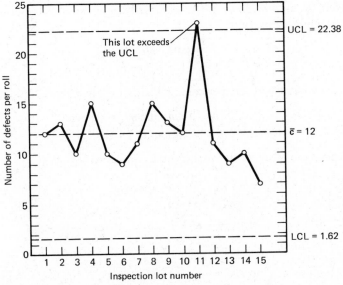

FIGURE 11.13 · c chart for the number of defects.

As with other control charts, an improvement in quality can be monitored by watching the trend and revising the 3σ control limits with new values for the number of defects $\bar{c}'$ and the standard deviation. For points falling below LCL out of control values are not discarded, since low values represent exceptionally good quality. Because the number of defects can not be less than zero, the lower control limit is set at zero. In that case finding zero percent defective or zero number of defects in a lot is to be expected from chance variations alone.

Selection of Samples

As pointed out before, the selection of sample sizes is a very important factor in avoiding wrong conclusions. The sample may consist of one or more items, but this must remain constant. If the sample size varies, this chart is not applicable. In the case of large and expensive subassemblies or continuously manufactured bulk products (such as paper rolls, coils of wire, etc.), location of samples must be identified with the source so that causes of defects may be traced. For example, the frequency of sampling at the start of every run, at the beginning or end of each work shift, from batch to batch operations, or every hour or so, must also be decided upon prior to sampling.

Control Chart with a Varying Number of Defect—*u* Chart

When the sample lot sizes vary, it is necessary to use a *u* chart in order to have a constant central line. For instance, if *c* is the total number of defects found in any sample and *k* is the number of inspection units in a sample, in setting up a *u* chart we would plot the quantity $u = c/k$. The purpose of bringing up the *k* factor is to be able to compare the number of defects observed in the ''unequal'' sizes of sample lots.

As an example, if a 100 unit length (feet, meters etc.) of wire is taken as the common basis lot size for inspecting coils of wire which actually vary from 90 to 200 units in length, we must convert the total number of defects actually observed in each lot to the number of defects per 100 unit length. Thus:

$$k = \frac{\text{Size of inspection lot}}{\text{Size of common basis lot}}$$

The average number of defects per common basis lot size then becomes:

$$\bar{u} = \frac{\Sigma c}{\Sigma k} = \frac{\text{Total number of defects}}{\text{Sum of } k \text{ values}}$$

Hence, for a sample of *k* units, the control limits are given by:

$$UCL = \bar{u} + 3\sqrt{\frac{\bar{u}}{k}}$$

$$LCL = \bar{u} - 3\sqrt{\frac{\bar{u}}{k}}$$

Example: To illustrate, suppose we wish to plot a *u* chart for controlling the production quality of coils of wire made continuously in a steel mill. The following tabulation (Table

TABLE 11.3
Inspection Data for Coils of Wire

Inspection Lot Number	Length of Wire (in meters) Per Inspection Lot	Number of Defect, c Per Lot
1	105	5
2	85	2
3	100	4
4	80	3
5	90	6
		Total = 20

11.3) lists inspection data for five sample inspection lots. It was decided that the common basis lot size would be 100 meters.

The first step is to calculate separate k and u values for each inspection lot. From this the average number of defects per 100 meter basis for the five inspection lots can be determined. The center line on our u chart will be at $\bar{u}$ value. The control limits are then calculated by using the appropriate formula. Summarizing the calculations in tabular form makes it easier to plot the chart as shown in Table 11.4.

$$\bar{u} = \frac{\Sigma c}{\Sigma k} = \frac{20}{4.60} = 4.35$$

When the data are given in the form of a series of subgroups of values of u with their respective sample sizes then:

$$\bar{u} = \frac{n_1 u_1 + n_2 u_2 + \ldots + n_k u_k}{n_1 + n_2 + \ldots + n_k}$$

Figure 11.14 shows the u chart based on calculations of the above table.

Table 11.5 summarizes the formulas of central lines and control limits used thus far for p, np, u, and c control charts.

TABLE 11.4
Tabulation of Calculated Data for Plotting the u Chart

Inspection Lot Number	k	u	$3\sqrt{\dfrac{\bar{u}}{k}}$	UCL $= \bar{u} + 3\sqrt{\bar{u}/k}$	LCL $= \bar{u} - 3\sqrt{\bar{u}/k}$
1	1.05	4.76	6.38	10.73	Negative[a](0)
2	0.85	2.35	3.24	7.59	1.11
3	1.00	4.00	6.00	10.35	Negative (0)
4	0.80	3.75	6.49	10.84	Negative (0)
5	0.90	6.67	8.17	12.52	Negative (0)
	$k = 4.60$				

[a]Negative data can be assumed zero.

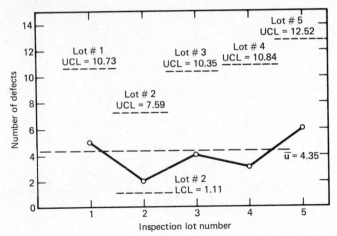

FIGURE 11.14   Plot of u chart.

11.7   CONTROL CHARTS FOR VARIABLES

Basic Principles

As discussed before, when a quality characteristic is measurable on a continuous scale, its variation in the form of a distribution of specific type can be predicted. If the output of a process forms say, a normal frequency distribution, it can be completely described by knowing its mean and standard deviation. When control charts are undertaken using variables rather than attributes, it is most common to use the average or $\overline{X}$ charts, and range or R charts. The primary advantage is that control charts based on variables are often more economical means of controlling quality than attribute control charts, since a smaller number of samples are required.[8] The basic premise of both $\overline{X}$ and R charts is that, when a process is under control, both the sample statistics, its mean, and the range, should fall within the control limits. If not, then undesirable changes may be assumed to have taken place in the process.

TABLE 11.5
Summary of Control Charts for Attributes Data

Type of Control Chart	Central Line	Standard Deviation	Control Limits
p chart (fraction defective)	$\overline{p}$	$\sqrt{\dfrac{\overline{p}(1-\overline{p})}{n}}$	$\overline{p} \pm 3\sqrt{\dfrac{\overline{p}(1-\overline{p})}{n}}$
np chart (number of defectives)	$n\overline{p}$	$\sqrt{n\overline{p}(1-\overline{p})}$	$n\overline{p} \pm 3\sqrt{n\overline{p}(1-\overline{p})}$
c chart (number of defects)	$\overline{c}$	$\sqrt{\overline{c}}$	$\overline{c} \pm 3\sqrt{\overline{c}}$
u chart (defects per unit)	$\overline{u}$	$\sqrt{\overline{u}}$	$\overline{u} \pm 3\sqrt{\dfrac{\overline{u}}{k}}$

While most of the quality control activity based upon control charts for variables is concerned with $\overline{X}$ and R charts, some production personnel use other charts which are equally simple without mathematical complicacy in understanding the relationships between standard deviation, control limits, specifications, and other such related terms. These include charts for *median,* and *mode* or *moving averages.* For a description of these charts, the reader is referred to other suitable texts on statistical quality control.

Establishing an $\overline{X}$ and R Control Chart

For many technical comparisons made on certain test results, it is often convenient to examine the mean and the range separately, since the factors that influence the mean of values are different than those that affect the range. For example, in melting operations encountered in steel foundries, a change in the mean values of base metal chemical composition (such as % of carbon or % of silicon) frequently requires a correction of the charge put into the furnace; whereas any increase in the range is corrected by technical adjustments to the melting practice. An $\overline{X}$ and R chart, used in conjunction with good practices, supplements quality control decisions and helps operational staff (melters) to correctly judge the situation.

From the test results on each and every sample, it is possible to calculate the mean value and range value for every batch (value for each day or each work shift taken out from say, five or six random sample tests). The mean of the sample means, $\overline{\overline{X}}$ is the central line on the plot. The UCL and LCL are $3\sigma_{\bar{x}}$ limits on either side of $\overline{\overline{X}}$, where $\overline{\sigma}$ is the grand average of σ'_s, the standard deviation of the sample lots. The $\overline{X}$ and R values are plotted on separate charts against their $\pm 3\sigma_{\bar{x}}$ limits.

The calculation of average mean is easy; but for standard deviation the task becomes quite time-consuming if done manually. However, this problem has been solved using modern calculators and shortcut formulas are available, such as UCL $= \overline{\overline{X}} + A_2\overline{R}$ and LCL $= \overline{\overline{X}} - A_2\overline{R}$ where A_2 is a constant factor available through published tables and the center line of the control chart is at $\overline{\overline{X}}$. Table 11.6 provides the values of factor A_2 along with D_3 and D_4 factors for the R chart discussed next.

The R chart is used to show fluctuations of the range values of sample lots drawn from production. Establishing its control limits follows the same principle as the $\overline{X}$ chart, using the mean of the ranges, $\overline{R}$ as the center line and $\pm 3\sigma_R$ as the control limits: UCL $= D_4\overline{R}$, and LCL $= D_3\overline{R}$, where D_3 and D_4 are constant factors determined from Table 11.6.

For a production process to be in a state of control, both $\overline{X}$ and R charts must have all points inside the control limits. However, in practice the distribution of the ranges of all possible small samples from a normal population may not always be normal. In some cases the lower control limits for the ranges may even be negative. It is likewise possible for the process to go out of control only by exceeding the UCL. In such a case, an R chart would give warning of an assignable cause which could be missed from an $\overline{X}$ chart. For these reasons, it is a common practice to construct both charts, side by side, and use them as complementary.

Use of $\pm 3\sigma$ as Control Limits

The recent trend in many automotive industries is to use 4σ limits, although the use of $\pm 3\sigma$ control limits is almost universal. Many statisticians however, consider this to be playing the game safe. The primary disadvantage of this is that the $\pm 3\sigma$ limits are slow in warning,

TABLE 11.6
Tabulation of Factors A_2 for $\overline{X}$ Charts and D_3 and D_4 for R Charts.

Size of Inspection Sample Lot n	Factor for $\overline{X}$ Chart: A_2	Factors for R Chart D_3	D_4
2	1.88	0	3.27
3	1.02	0	2.57
4	0.73	0	2.28
5	0.58	0	2.11
6	0.48	0	2.00
7	0.42	0.08	1.92
8	0.37	0.14	1.86
9	0.34	0.18	1.82
10	0.31	0.22	1.78
11	0.29	0.26	1.73
12	0.27	0.28	1.72
13	0.25	0.31	1.69
14	0.24	0.33	1.67
15	0.23	0.35	1.65

Source: Adapted from, Grant, E. L. and Leavenworth, R. S., *Statistical Quality Control,* 4th ed. McGraw-Hill Book Company, New York, Copyright ©1972. Reprinted by permission of publisher.

TABLE 11.7
Measurement of Percentage Carbon in the Molten Base Iron Samples Withdrawn from an Electric Furnace

Subgroup	Subgroup Number	Measurement of Individual Items	$\overline{X}$	R
A (first week)	1	3.21,3.25,3.28,3.20	3.24	0.08
	2	3.05,3.07,3.14,3.18	3.11	0.13
	3	3.05,3.10,3.10,3.22	3.12	0.17
	4	3.14,3.17,3.20,3.23	3.19	0.06
	5	3.05,3.08,3.18,3.20	3.13	0.15
B (second week)	1	3.20,3.18,3.22,3.23	3.21	0.05
	2	3.21,3.05,3.14,3.05	3.11	0.16
	3	3.25,3.07,3.10,3.17	3.15	0.18
	4	3.28,3.14,3.21,3.05	3.17	0.23
	5	3.20,3.05,3.17,3.10	3.13	0.15
C (third week)	1	3.10,3.20,3.21,3.17	3.17	0.11
	2	3.16,3.15,3.19,3.22	3.18	0.07
	3	3.05,3.20,3.17,3.19	3.15	0.15
	4	3.14,3.21,3.00,3.12	3.12	0.21
	5	3.17,3.21,3.15,3.21	3.19	0.06
		Total	47.37	1.96

especially when the shift in the population mean is small. When the process is under control the probability of exceeding $\pm 3\sigma$ limit is only 0.27% and choosing these safe working limits saves one from unnecessary fire fighting. However the probability of making a Type II error becomes high.

The reader will recall the two types of error, Type I and Type II mentioned earlier. Rejecting a hypothesis as false that is in fact true is called a Type I error (which is 0.27% in the present case). On the other hand, when the stated hypothesis is not true but is accepted as true, a Type II error occurs. This means that with $\pm 3\sigma$ limit, there is a small probability of looking for trouble when none exists but a high probability of failing to detect a shift (out of control situation) that remains.

Example: To illustrate the use and construction of an $\overline{X}$ and R chart as an example, the data of Table 11.7 will be used out of the daily control chart of a foundry's melting operation (Figure 11.15).

The measurements in Table 11.7 are given for 15 subgroups, each containing four individual measured readings representing the percentages of carbon content for the base metal chemical composition in a ductile iron foundry melting operation.

Table 11.7 also gives the mean of sample means calculated individually where $\overline{X}$ will be used as an estimate of the population mean. From individual means the values of $\overline{\overline{X}}$ and $\overline{R}$ are calculated, using the following relations.

$$\overline{\overline{X}} = \frac{\text{Sum of all } \overline{X}\text{s}}{\text{Number of inspection sample lots}} = \frac{47.37}{15} = 3.16; \text{ and}$$

$$\overline{R} = \frac{\text{Sum of all } R\text{s}}{\text{Number of inspection sample lots}} = \frac{1.96}{15} = 0.13$$

Next, the UCL and LCL for the R chart are determined using D_4 and D_3 factors from Table 11.6 for $n = 4$; when $D_4 = 2.28$ and $D_3 = 0$; UCL $= D_4\overline{R} = (2.28)(0.13) = 0.30$; and, LCL $= D_3\overline{R} = (0)(0.13) = 0$.

Similarly, for $n = 4$, the value of A_2 from Table 11.6 is 0.73, the UCL and LCL for the $\overline{X}$ chart can be determined by: UCL $= \overline{\overline{X}} + A_2\overline{R} = 3.16 + (0.73)(0.13) = 3.26$, and LCL $= \overline{\overline{X}} - A_2\overline{R} = 3.16 - (0.73)(0.13) = 3.07$.

Figure 11.16 shows the plots of the R chart and the $\overline{X}$ chart. An examination of this figure suggests that no points fall outside the limits. Compared with the charts showing actual values (Figure 11.15), the $\overline{X}$-R charts compacted in this way, possess the advantage that long term trends can be recognized early. In addition, they are clear and precise in respect of both the collation and extraction of information. As with other types of control charts, warning limits can be established on the basis of operational conditions and the capability of the process.

11.8   GENERAL RULES FOR SETTING UP CONTROL CHARTS

One of the fundamental requirements of process quality control by means of control charts is that the data plotted must be reliable and accurate. The entire interpretation of the observed results depends on whether or not there is a justification for people believing in those data. The following recommendations apply when data are collected, summarized, and presented in the form of control charts.

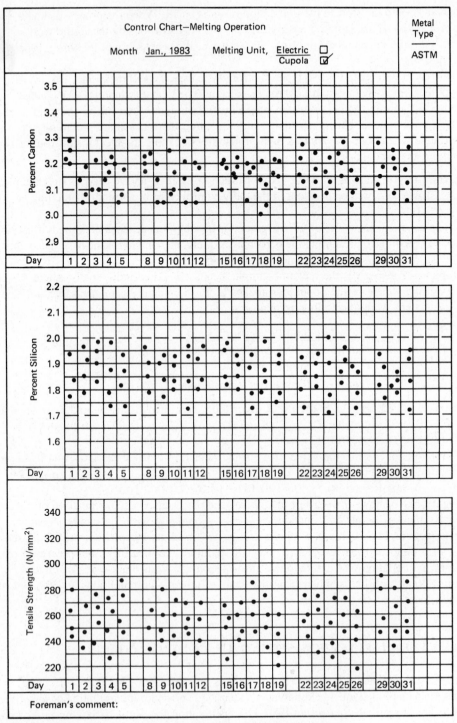

FIGURE 11.15   Control chart for the melting operation used in the case example.

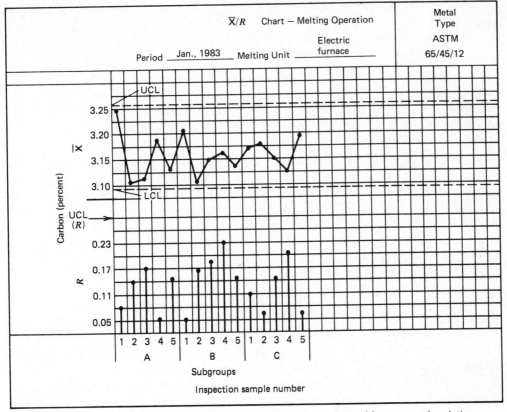

FIGURE 11.16 $\overline{X}$ and R charts confirm the process being free of assignable causes of variation.

1.   Observe precautions in the precision of data gathering, sample collection, analysis methods, mode and conditions of production, and anything that has a bearing on the correctness of data.

2.   Choose the quality characteristic to be charted and the charting technique itself that will best display whatever is desired.

3.   Decide on the control limits. The usual practice is $\pm 3\sigma$ for the control limits, but other multiples may be chosen as warning or action limits.

4.   The inspection sample units or lots must be selected with care. The decision as to the number of samples should be understood with respect to the particular control chart being used.

5.   The selection of a control plan must depend on a balance between the cost of investigation and cost of letting a process that is out of control continue. Such decisions need support of the study of risk analysis keeping such factors in mind as the sample size (or the rational subgroup size), the interval between the samples, and the limits of control.

6.   The entire scheme or provision of data collection and charting techniques should be treated as part of the overall quality improvement effort. If control charts are to serve their purpose, their use must be treated as a permanent activity.

Again, it is the responsibility of the supervisory and management personnel to teach the operators the proper use of control charts as an ongoing activity, if they are to be expected to play a part in quality control activities.

For each of the control charts described above, there are two distinct situations.

1. *Control charts where no standards are given:* In this case, control charts based entirely on the data from samples are used for detecting lack of constancy of the cause system. The purpose is to discover whether the observed values (of variables or attributes, $\overline{X}$, s, p, etc.) for several samples vary among themselves, by an amount greater than what should be attributed to chance.

2. *Control charts with respect to a given standard:* When standard values (such as $\overline{X}_0$, σ_0, p_0, etc.) are given or known (based upon experience or prior knowledge), the purpose becomes to identify whether the observed values or $\overline{X}$, s, p, and so forth, for several samples of n observations each, differ from their respective standard values by amounts greater than that expected to be due only to chance causes.

11.9 PRECONTROL CHART TECHNIQUE

Precontrol chart technique is particularly applicable in high volume production where one change in production conditions can give rise to major defects.

The basic principle of the precontrol chart technique is illustrated in Figure 11.17. It starts out with a process located at the central value of the normal distribution with precontrol limits midway between the specification limits and the central value. Assuming that the design tolerance and process capability are closely matched, the width of the tolerance band is equal to 6σ (σ = standard deviation). The bands (space) between the specification limit

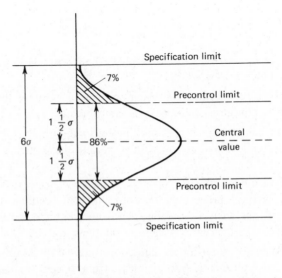

FIGURE 11.17   Basic precontrol chart.

and adjacent precontrol limits is known as *warning* signal bands and contains 7%, (or a chance of one piece in 14), to lie on either band. This, when it happens, is taken as an indication that a shift in a machine setting is the most likely explanation.

For any two parts falling in the signal band simultaneously, the chance is $1/14 \times 1/14 = 1/196$, (i.e., only one chance in 200 pieces). On the other hand, in this situation, the chance that a process has shifted (from original settings) is much larger, $= 1 - 1/96 = 195/196 = 0.9949 = 99.5\%$, suggesting that the process must be stopped and reset. In the case where one piece gets beyond one precontrol limit and the next piece outside the other precontrol limit, the process must again be stopped. This situation suggests excessive variability, meaning that the pattern has been widened and defectives are being produced.

Selection of Samples

The selection of samples is very important in order that preventive action may be taken in time. The frequency of sampling must be decided from the knowledge of process capability. At the start of the run however, it should be very frequent, and, depending upon the production quality, may then be gradually reduced. Various rules covering different quality levels and process dispersion are described by D. Shainin.[9]

Modification of Precontrol Charts

We have seen that in its simplest form, the precontrol charting technique is based on the assumption that the process capability and drawing tolerances are closely matched. However, if the process is more capable of meeting drawing tolerances, then the system of precontrols in its basic form remains undoubtedly safe, but would result in premature stopping of the machine for resetting. To overcome this, a modified chart can be prepared to make the most use of available conditions.

The procedure involves establishing the standard deviation first, and then drawing up *modified* precontrol limit lines at 1.5σ inside the upper and lower drawing limit instead of just at one quarter the width of the band (see Figure 11.18). For instance, if the shift is due

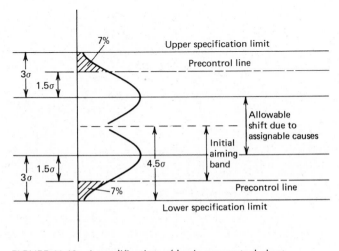

FIGURE 11.18   A modification of basic precontrol chart.

to tool wear (an assignable cause) giving defective pieces from undersize to oversize, then another line would be drawn at 4.5σ above the lower specification limit which will give a band (from 1.5σ to 4.5σ from the lower limit) as the aiming band for sample inspection. In cases where the shift is due to some other type of assignable cause that produces undersize pieces, then the aiming band would be identical but above the center of the chart in Figure 11.18.

11.10 CUMULATIVE SUM CONTROL CHARTS

In discussing the use of control charts in previous sections, the use of warning limits or the action limits were both suggested. From these developments a proposal came from a British statistician, E. S. Page[10], to adopt a rule for action that was based on *all* the data and not only the last few samples. The chart that facilitate this has been called a *Cumulative Sum* or *Cusum* control chart. The advantage of the cusum chart lies in the possibility of picking up a sudden and persistent change in the process average more rapidly than with a comparable Shewhart chart, particularly if the change is not large.

Cusum control is based on the cumulative sum of the differences between individual points and the aim (or action) points. A statistical test then determines whether the process is still *on aim* or *off aim*, in a similar manner to that of the Shewhart $\pm 3\sigma$ limits.

As an example, consider the observations (measurements) given in Table 11.8. To illustrate the cusum technique, the differences of the average sample readings from the desired central value are calculated and the cumulative sum of these differences is tabulated as shown.

TABLE 11.8
Tabulation of Data for the Cusum Plot

Readings	Sample Average, X_i	Difference From Central value, Δ	Cumulative Sum, $\Sigma\Delta$
1	20.1	+0.1	+0.1
2	20.2	+0.2	+0.3
3	19.7	−0.3	−0.0
4	20.1	+0.1	+0.1
5	19.8	−0.2	−0.1
6	21.0	+1.0	+0.9
7	20.2	+0.2	+1.1
8	20.7	+0.7	+1.8
9	20.3	+0.3	+2.1
10	20.6	+0.6	+2.7
11	20.9	+0.9	+3.6
12	21.2	+1.2	+4.8
13	21.9	+1.9	+6.7
14	21.8	+1.8	+8.5
15	20.5	+0.5	+9.0
16	19.5	−0.5	+8.5
17	19.4	−0.6	+7.9

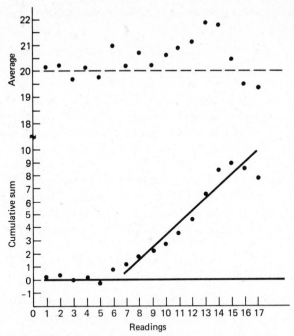

FIGURE 11.19   Variations in terms of simple averages as compared to cumulative sums.

A typical cusum plot of the data is shown in Figure 11.19. The top shows an ordinary plot of the averages, where a change becomes noticeable for readings 13 and 14 but is not sufficient to warrant any action. In the lower part of Figure 11.19 the same data are plotted as a cumulative sum which indicates in a general way the possible advantage of a cusum plot. The general applicability and advantages of cusum control charts has been discussed by Lucas[11] and Goodman.[12]

11.11   NONCONFORMANCE CONTROL AND SELF-INSPECTION

Nonconformance control procedures are the key activity of any quality control of production, as they are directly related to assuring quality of conformance. Procedures vary with degree of impact of a defect, complexity of the cause analysis, and remedial actions. Much can be said for the participation of the operator and inspector in this decision-making process. However, the following are some leading practical questions that must be answered in case of nonconformances.

1.   What happens when inspection has identified a nonconforming item, or lot?

2.   Are items or lots brought into quarantine in order to prevent unauthorized use?

3.   Are items tagged or marked?

4.   Are special precautions taken to prevent inadvertent acceptance or further processing?

5.   Are proper forms used for the disposition of nonconforming items?

6. Have immediate actions been taken to prevent any negative impact on quality, and has everyone who needs to know been alerted?

Questions such as these must be answered in the inspection procedure and their answers must be known to the inspector. The ANSI guideline Z-1.15 also suggests that after nonconformances are segregated and brought under control, a clearly defined authority, such as a material review board, should make further decisions and initiate preventive action.[13]

Figure 11.20 shows a typical action plan designed to take care of nonconformances.

All nonconformances should be documented in sufficient detail. Defects and causes are not usually new and unexpected, but can recur even after the cause has been rectified. Reaction to major nonconformances reveals the seriousness a company assigns to quality assurance and the competence of implementation of the inspection plans. A positive reaction also provides for participation and communication for all concerned in the production process. Its message is simply this: nonconformance control should be combined with learning experiences and constructive motivation for meeting workmanship standards and attaining product quality.

Workmanship and Self-inspection

One device for involving operators and supervisors directly in attaining quality of conformance at their workplace is to delegate the actual inspection, verification, and even possibly the corrective actions to them. In the small business setting with a relatively simple and nonhazardous item, a separate and independent inspector often can neither be afforded nor is necessary. The risk of unreliable self-inspection by the operator and the subsequent passing of defective work is relatively high compared with verification through an independent

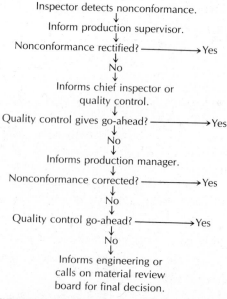

FIGURE 11.20   Flowchart of action plan for nonconformances.

inspector. However, the advantages and disadvantages must be weighed carefully. The *advantages* of self-inspection, where the operator checks his or her own work against predetermined performance standards are as follows.

1. Enrichment of job and task, more assigned responsibility and implied recognition.

2. More independence on the job and motivation for participating in quality assurance efforts.

3. Earlier detection of nonconformances and possible remedial actions, including operator learning for better job performance.

4. More variety in inspection methods and greater cost effectiveness.

Some *disadvantages* of self-inspection are that the operator might not have sufficient background and training for inspection tasks or that the simplicity of inspection would be a waste of the time of a technically higher-skilled operator. When quality program standards have been imposed, self-inspection is not permissible. Nevertheless, inspection tasks assigned to operators can still be controlled through less frequent separate independent inspections, and possibly through quality audits.

Limitations affecting self-inspection are technological difficulties exceeding worker inspection capability, safety consideration, jurisdictional requirements, and program standard restrictions for self-inspection. Inspection points might not necessarily coincide with the job in which self-inspection appears acceptable.

Workmanship and Resource Facilities

Given well-prepared inspection plans and procedures that are properly integrated with general production plans and workplace directives; given further, that jobs and tasks are designed with performance standards and work methods; that the workplace and the equipment is physically conducive to good workmanship; and finally, that the operator is well selected, trained, and supervised; management verification of quality of conformance will be relatively straightforward and can more likely and effectively be assigned to the operator.

Particularly worth mentioning is the maintenance of well calibrated and maintained inspection and test equipment. The best qualified inspector cannot make valid acceptances of work in process without reliable and functioning tools and facilities. The standards and guidelines for quality assurance already mentioned pay particular attention to inspection and test equipment. Again, the operator and inspector who has been assigned the inspection task in accordance with procedures is not usually in charge of the design, selection, and control of the test equipment. Those in charge of the design of a company-wide quality assurance program, (as will be studied later in Part III), will have to ensure proper production and inspection facilities, as well as other aspects of total quality of conformance, not as yet discussed in this chapter.

Determinants for Self-inspection

Inspection, as a special testing and verification of meeting quality specifications, can only be assigned to the operator when certain obvious prerequisites are fulfilled. Determinants for instituting operator conducted self-inspection are the qualification and proven reliability of the operator, the complexity of test and inspection procedures, the risk of false verification and subsequent passing of defects and defectives, the stipulation of mandatory quality assurance standards, the extent of subsequent special and external inspection, and the costs of self-inspection compared with other alternatives.

Once conditions appear to be favorable, a thorough feasibility study must be conducted, that will involve operators considered for such additional assignments.

The initial analysis should prove a high probability of attaining the many advantages from self-inspection. Such improvements are not only derived from an enriched job for the operator. Benefits for the organization are earlier detection and possible correction of defects and process out of control situations. Quality costs should be lowered in the long run, quality should be further improved, and general motivation for quality in the organization should be enhanced.

Once there is sufficient reason to expect positive results from introducing self-inspection, planning and introduction must be carried out with utmost care. It will be very difficult to continue delegated self-inspection, if it proves itself unreliable and harmful.

Planning for Self-inspection

Planning for self-inspection is best carried out in the form of an ad hoc project that is declared an experiment. In this pilot project senior management, together with quality control specialists, provide essential initiative and guidance. Supervisors and the current inspection force will naturally be more sceptical than operators interested in such job enrichment. Appendix D of this text describes major steps to be taken in developing and testing self-inspection. In addition, a checklist of questions that ought to be considered will also help to make correct decisions. Such a checklist will question adequate consideration and verification of major prerequisites, such as:

1. Is project management properly selected, authorized, and visibly supported by senior management and those responsible for the quality assurance program?

2. Are workers informed about opportunities, responsibilities, prerequisites and other major changes related to their current job?

3. Are the objectives for the project and approaches, budgets, and other factors properly prepared and approved?

4. Are inspection methods, points, equipment, and decision criteria, adequately formulated or adapted for self-inspection?

5. Have selections of workers and workplaces been made on the basis of systematic and careful approaches?

6. Have employment contracts, stipulations, and collective agreements been reviewed?

7. Have features of the project been discussed and explained to those concerned and also at general staff meetings?

8. Have goals, performance standards, procedures, deadlines, schedules, and revision points been established?

9. Has an advisory or steering committee been formed for supporting the project team?

10. Have members of the project team been properly selected and prepared for the task?

11. Have current quality assurance procedures and production plans and procedures been coordinated with the self-inspection project?

12. Have reports and communications been arranged for controlling implementation and outcome?

13. Is regular and independent inspection aligned with related preceding self-inspection?

14. Are operators selected for self-inspection adequately trained and supervised?

15. Have current inspection plans, approved by customers, been reviewed and possibly adjusted and approved?

After completion of the first experimental project stage, a report with sufficient documentation should be made either to the advisory and steering committee or directly to senior management. At this point probability of success when continuing and expanding self-inspection must be clearly established or otherwise correction must be made. If at this stage self-inspection appears impractical or unfeasible, the project can be discontinued without any repercussions.

11.12   KEY TERMS AND CONCEPTS

Quality of conformance refers to the attainment of specifications within set limits during the production and operation phase.

An **inspection** is the independent testing and verification of quality characteristics on the basis of standards and by predetermined inspection procedures.

The **inspection point** is the stage in the production process where the inspection is to be conducted.

Process control is the monitoring of process output by means of control charts or other forms of inspection and the analysis and correction of deviations from standards. A process is in a state of control when it produces good parts; (i.e., plots on the control chart indicate random variation within control limits).

A **control chart** is a graph in which the center line stands for a nominal value and the upper and lower control limits are set for acceptable confidence levels, using a statistical basis. Plots in the control chart are statistics from samples taken from a process. Plots outside the control limits, or showing nonrandomness, indicate assignable causes and require corrective measures, perhaps more than just overadjustment or underadjustment. The conclusion of out of control condition can be false if the real cause is not investigated. Control charts for variables are, for example, $\bar{X}/R$,

Cusum, and moving range charts. Charts for attributes are p, np, c, and u charts. For details see the text.

A **process capability study** is an application of control charts best performed during a preproduction stage where stipulated tolerances or statistical limits are compared with control limits calculated from actual production runs. Variation in the actual quality characteristics as measured must be equal, or less than, the stipulated specification tolerances. Improving quality, from an operational standpoint, means *reducing variability of the process*. Once the process capability index, C_{pk} is known, quality improvement would mean increasing C_{pk} values, from 1.00 (i.e., ± 3 sigma = specification tolerance) to C_{pk} values of 1.33 ($8\sigma/6\sigma$), 1.66 ($10\sigma/6\sigma$), 2.00, 5.00, and so on.

A **nonconformance** is an inspection result of significant deviation from the standard. For instance, a plot outside the control limit is assumed to indicate a nonconformance, and will require measures as outlined in the respective inspection procedure and/or special nonconformance procedure.

Self-inspection is performed by the operator and/or supervisor responsible for the work or process. This kind of inspection is specially planned and assigned in order to maintain validity and reliability of inspection results. Self-inspection promotes quality motivation and workmanship.

11.13 SUPPLEMENTARY READING

Supplier Process Control versus Product Control[14]

Historically, U.S. industry has depended almost exclusively upon product control to assure the quality of the products and materials we manufacture.

What do I mean by product control? Basically I use product control to describe the quality control approach whereby, using a sampling plan, a completed production lot of material or parts is evaluated for conformance to specification.

And what is wrong with that? After all, isn't that what quality control is all about? Isn't that what we do when we evaluate purchased materials via receiving inspection or source inspection?

Typically this has been our approach. However, if U.S. industry is to be competitive, this approach must change. The current approach of using a 1% AQL plan for evaluating incoming materials is a contract with your supplier to buy 1% defective material. The result of assembling a piece of equipment with (1,000) parts in it with each part being 1% defective results in excess of 99% of the finished assemblies having at least one defective part.

Quality control has fought and won the right to be independent from manufacturing operations. As a result, we have become defect reporters and have targeted industry to be defect producers by establishing AQL's, AOQL's, LTPD's, etc. We have become very proficient at utilizing electronic data reporting systems to report what the defect levels were that we or our suppliers were successful in producing last week!

We must recognize that our objectives must change to that of producing 100% acceptable materials, parts and finished products. If we are to accomplish this we must move away from our historical approach to product control. After a recent discussion with some of our suppliers on the need to supply 100% acceptable parts, I was asked what kind of sampling plan should be used to assure this level of quality. If you stop and think about it, there is no sampling plan including 100% inspection which will assure this level of quality.

If this is the case, then what is the answer? The answer is process control. Process control recognizes that regardless of how much time, money and effort we expend in inspecting the product once it has been manufactured, we will still have a defect escape rate. Process control assures that the product is produced to the specification at the point of manufacture and eliminates the need for the subsequent inspection, sorting and rework. Not only does it result in a higher level of quality, but it will result in lower costs due to the reduction in inspection, scrap and rework costs. When applied to purchased material, it will eliminate the need for receiving inspection which will allow purchased material to flow directly into the manufacturing cycle resulting in the elimination of costly delays for receiving inspection or source inspection.

Process control entails running a process capability study to assure that the manufacturing process is capable of producing a part to specification and then maintaining control charts to assure that the process is routinely running in control.

There are five key steps in establishing process control:

1. First piece inspection
2. Identification of process variables
3. Histogram analysis of the process variables
4. Conduct process capability studies of process variables
5. Develop and maintain control charts for the process variables

Figure 1 shows a typical histogram analysis whereas Figure 2 shows a typical control chart.

In applying this approach to supplier quality assurance activities, the objective is to have suppliers institute process controls in their manufacturing locations and have them provide you, their customer, with their process capability studies and their control charts on an ongoing basis as evidence that their process is being maintained in control. With this evidence, the supplier should be considered certified and there is no need for either source inspection or receiving inspection on an ongoing basis.

While process control techniques are understood by the quality professionals, these techniques are not understood or applied on a very widespread basis. When Xerox initiated this activity, less than 90% of our suppliers understood or applied basic process control techniques.

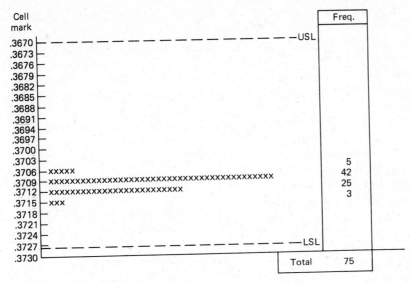

FIGURE 1   Typical histogram analysis.

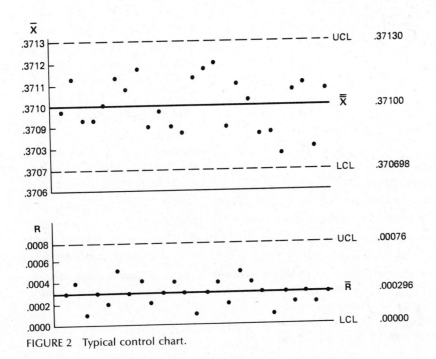

FIGURE 2   Typical control chart.

The biggest hurdle that has to be overcome is the training of suppliers and their personnel. This means that the management of the supplier base must understand the benefits that can be realized by them in terms of reduced scrap, rework and inspection costs as well as the benefit to their customers in terms of improved quality and reduced costs. It also means that the supplier's personnel (including production foremen and operators) must be trained in basic process control (statistical quality control) techniques.

With the limited levels of skill and understanding of basic statistical control techniques that exist in the average U.S. manufacturing operation, it is mandatory that the process control techniques be kept simple and unencumbered with all the statistical refinements. If the system is to be effective, the operator and foreman must be capable of interpreting and taking action based upon their results. They must not have to have a degree in statistics to do their jobs!

The steps that must be followed to effectively implement process control with your suppliers are:

1. Supplier management training on costs/benefits of process control.
2. Training of supplier operating personnel on process control techniques.
3. Assist suppliers with implementation of process control on a few parts:

   (a) first piece evaluation
   (b) identification of process variables
   (c) histogram analysis
   (d) process capability study
   (e) establishment of control charts

4. Once supplier has learned the approach, review results on all other parts.
5. Certify part/supplier combination based upon successful implementation of process control.
6. Routinely monitor supplier's process control charts.

While the concept and the techniques are very straightforward, the key is in the implementation. We are trying to change some very basic ways in which the suppliers assure the quality of their products. This requires a substantial investment in the training of not only the suppliers' operating personnel, but also the management personnel. Without the commitment on the part of the management of each and every supplier, this approach is doomed for failure.

However once fully implemented, this puts the control where it can have the largest impact on quality; at the point of manufacture. Remember, there is no sampling plan that can assure 100% acceptable quality after the material is manufactured. Only process control during manufacturing can assure that.

11.14 DISCUSSION AND REVIEW QUESTIONS

(11–1) *The higher the degree of automation and computerization the more important production and quality planning becomes.* Explain and discuss.

(11–2) *The principle of perfect production is similar but not identical to that of zero defect.* Explain.

(11–3) *Procedures are as important as practices, and have an important role to play in quality assurance.* Explain and discuss.

(11–4) *Inspection plans serve different, although related, purposes than inspection procedures.* Explain.

(11–5) *There are certain principles and rules that govern the position of an inspection point along a production line.* Explain.

(11–6) *Start-up inspection, or first piece inspection is of particular importance.* Give your reasons.

(11–7) *Normally, statistical quality control and inspection is more reliable than 100% inspection.* Explain.

(11–8) Define "Statistical Quality Control."

(11–9) *The best quality of conformance strategy is defect prevention.* Explain and discuss with reference to a particular product or process.

(11–10) *Statistical quality control techniques can be used separately or in combination with each other.* Discuss.

(11–11) *Quality characteristics of a manufactured product are grouped into several forms ac-*

cording to the concept of dominance. Explain.

(11–12) *Nonconformance procedures are the key activity of any quality control of production.* Explain.

(11–13) *Nonconformances are a brilliant opportunity for improvement.* Discuss.

(11–14) *Total quality of conformance depends on supporting inspection that is perceived as a service to operators.* Explain and discuss.

(11–15) *Self-inspection should be used wherever possible but it does not entirely substitute for external, independent inspection.* Explain and discuss.

(11–16) *Sound workmanship is the root of quality and it arises from more than the competence and the will of the worker.* Explain and discuss.

11.15 PROBLEMS

(11–1) a. What are the two possible causes of variability in manufactured products?
   b. What is meant by the term "process"?
   c. Check the following listing of words and distinguish which of these are attached with the meaning of chance causes and assignable causes.

Consistent	Normal	Mixed
Steady	Homogeneous	Inconsistent
Important	Nonhomogeneous	Different
Shifting	Erratic	Predictable
Stable	Unnatural	Unpredictable
Natural	Unstable	Disturbed

(11–2) a. What is the meaning of a process being in state of statistical control?
   b. It has been said that quality must be built into a product and that a control chart by itself does not cause a product to have high quality. In what ways does the control chart technique contribute toward improvement of quality?

   c. What level of variability (in terms of the standard deviation) is usually acceptable when a process is in a state of a statistical control? Give reasons.

(11–3) In a controlled process assuming normal distribution, what percentage of the population is:
   a. included in $\pm 2\sigma$, $\pm 2.5\sigma$, $\pm 3\sigma$ and $\pm 4\sigma$ limits?
   b. not included in the four limits specified in a.?

(11–4) A soft drink manufacturing company has an automatic filling machine for standard bottle sizes. Data collected for a given period show an average liquid volume after filling of 1.630 ml with a standard deviation of 0.043 ml. The specification for liquid volume is given as 1.670 ± 0.071 ml.
   a. What percentage will not meet the specification?
   b. If the nominal value and volume tolerance is 1.670 ml $\pm$ 10%; for a population of 10,000 filled bottles, how many will, on the average, be rejected?
   c. How many more bottles could be accepted if the tolerance in b is increased to $\pm 15\%$?
   d. What tolerance limits should be set which would be 95 percent certain of including 99% of production?
   e. If the label on the soft drink bottle states that it contains 1.680 ml; how many ml of beverage should the average bottle contain if the company is to be 95 percent certain that 99% of its production contains at least 1.680 ml?

(11–5) The following data pertain to an inspection result in which 10 sample lots were checked out of 100 units each.
Determine the upper and lower control limits and plot a *p* chart. Indicate if any data fall outside the control limit and in that case, compute the revised control limits.

Sample Lot Number	Number of Defects
1	7
2	5
3	12
4	16
5	2
6	19
7	8
8	7
9	2
10	5

(11–6) A paper mill inspector has checked 13 sample rolls for the presence of perforation spots. The number of perforation spots per roll ranged from 8 to 21. Complete data is given in the following table:

Roll number	1	2	3	4	5	6	7	8	9	10	11	12	13
Number of perforation spots	6	7	8	10	5	3	12	18	4	21	7	9	5

Determine the upper and lower control limits and plot a c chart.

(11–7) In an automatic spray painting process, 30 units are inspected for cosmetic defects (scratches, etc.) every eight hours of work shift. If the fraction defective is 0.05, assuming the process to be in a state of statistical control, determine the control limits and central value for an np chart.

(11–8) The following inspection data represent the number of defects found on each of a set of kitchen cabinets installed.

Sample Number	Number of Defects	Sample Number	Number of Defects
1	7	9	6
2	10	10	4
3	9	11	3
4	6	12	1
5	5	13	10
6	4	14	7
7	8	15	8
8	5	16	7

Plot a control chart with control limits.

(11–9) The percentage of moisture added regularly to a molding sand bin is an important characteristic to the quality of molds produced for iron casting production in foundries. Willsin Metal Works and Foundries Ltd. have always relied on the experience of its molder, who could tell the moisture content by grabbing a fistful of sand. New management took exception to this kind of measurement and installed a laboratory with moisture measuring equipment. It was decided to analyze the process with a control chart. Twenty samples were tested on one day with the following results.

Sample Number	$\overline{X}$	R	Sample Number	$\overline{X}$	R
D1	3.1	0.1	A11	4.0	0.1
D2	3.5	0.2	A12	3.5	0.3
D3	4.0	0.1	A13	3.7	0.3
D4	2.9	0.4	N14	3.1	0.6
D5	2.8	0.5	N15	3.5	0.4
D6	3.6	0.2	N16	2.8	0.8
A7	3.8	0.1	N17	2.9	0.6
A8	3.0	0.8	N18	3.7	0.2
A9	3.6	0.3	N19	3.8	0.3
A10	4.1	0.2	N20	4.0	0.1

a. Plot the data on an average and range control chart with control limits.

b. What points, if any, have gone out of control?

c. What can be surmised from the shapes of the curves on these charts?

11.16 NOTES

1. Hayes, Robert H., "Why Japanese Factories Work," *Harvard Business Review,* July–August, 1981, p. 57.

2. ANSI/ASQC Standard, Z-1.15-1980, *Generic Guidelines for Quality Systems.*

3. See for example, Mayer, R. R., "Selecting Control Chart Limits," *Quality Progress,* September 1983, p. 24;

Burr, Irving, "Specifying the desired distribution rather than maximum and minimum limits," *Industrial Quality Control*, Vol. 24, No. 2, 1967, p. 94.

4. Some companies use a rule of thumb for following the maximum for the capability ratio. A dimension with a bilateral tolerance is assigned to an existing process having process variation of less than 75% of the tolerance range.

5. For an in-depth discussion of various control chart techniques, see, for example, E. L. Grant and R. S. Leavenworth, *Statistical Quality Control*, 4th ed., McGraw-Hill Book Company, New York, 1972; *ASTM Manual on Presentation of Data and Control Chart Analysis*, STP 15D, published by American Society for Testing and Materials, Philadelphia, 1976. For an annotated bibliography of research papers in control chart techniques, see Vance, Lonnie C., "A Bibliography of Statistical Quality Control Chart Techniques, 1970–1980," *J. of Quality Technology*, Vol. 15, No. 2, April 1983, pp. 59–62.

6. Nelson, Lloyd S., "An Early Warning Test for Use with the Shewart p Control Chart," *J. of Quality Technology*, Vol. 15, No. 2, April 1983, p. 68.

7. The larger the value of n, the smaller the standard deviation, and the closer 3σ or 0.001 probability limits will be to the central line on the chart. The net effect of these tighter controls is to reduce the risk of not catching a defective in the inspection process.

8. For a good discussion on the advantages and disadvantages of attribute versus variable sampling practices, see, Duncan, Acheson J., *Quality Control and Industrial Statistics;* 4th ed., R. D. Irwin, Inc., Homewood, Illinois, 1974.

9. Shainin, D., "Techniques of Maintaining a Zero-Defect Program," *American Management Association Bulletin No. 71*, 1965; see also, Ott, Ellis R., *Process Quality Control*, McGraw-Hill Book Company, New York, 1975.

10. Page, E. S., "Continuous Inspection Schemes," *Biometrika*, Vol. XLI, 1954, pp. 100–15.

11. Lucas, J. M., "The Design and Use of V-mask Control System," *J. of Quality Technology*, Vol. 8, No. 1, 1976.

12. Goodman, Alan L., "Cumulative Sum Control and Continuous Process," *ASQC Quality Congress Transaction*, Detroit, 1982, p. 270.

13. The U.S. Small Business Administration has published a management aid, MA No. 242, entitled *Fixing Production Mistakes*. This contains a checklist that is helpful in tracing nonconformances to possible causes. It also explains the general characteristics of a corrective action plan and provides a form for reporting.

14. Tolman, Fay E., "Supplier Process Control versus Product Control," 1982 *ASQC Quality Congress Transaction*, Detroit, p. 209. Copyright © American Society for Quality Control Inc. Reprinted by permission.

11.17 SELECTED BIBLIOGRAPHY

Besterfield, D. H., *Quality Control*, Prentice-Hall, Inc., Englewood Cliffs, N.J., 1979.

Deming, W. E., *Quality, Productivity, and Competitive Position*, Massachusetts Institute of Technology, Center for Advanced Engineering Study, Cambridge, Mass., 1982.

Duncan, A. J., *Quality Control and Industrial Statistics*, Richard D. Irwin, Inc., Homewood, Ill., 1974.

Grant, E. L. and Leavenworth, R. S., *Statistical Quality Control*, McGraw-Hill Book Company, New York, 1972.

Halpern, S., *The Assurance Sciences, an Introduction to Quality Control and Reliability*, Prentice-Hall, Inc., Englewood Cliffs, N.J., 1978.

Hassan, M. Zia and Knowles, Thomas W., "An Optimal Quality Control Design for a Single-Product Serial Manufacturing System," *Journal of Quality Technology*, January 1979, p. 20.

Ishikawa, Kaoru, *Guide to Quality Control*, Asian Productivity Organization, Unipub., New York, 1981.

Juran, M. M. and Gryna Jr., F. M., *Quality Planning and Analysis, From Product Development Through Use*, 2nd ed., McGraw-Hill Book Company, New York, 1980.

Juran, J. M. (editor), *Quality Control Handbook*, 3rd ed., McGraw-Hill Book Company, New York, 1974.

Kondo, Isao W. K. and Hanawa, Namio, "Operator Inspection at a Japanese Steelmaker." *Quality Progress*, December 1982, p. 14.

Parker, H. V., "A Paper Mill Solves a QC Problem with Pre-Control Data," *Quality Progress*, March 1981, p. 18.

Sears, James A., "Changing Role of the Quality Inspection Function," *Quality Progress*, July 1983, p. 12.

Sullivan, L. P., "Reducing Variability: A New Approach to Quality," *Quality Progress*, July 1984, pp. 15–21.

Timestra, Peter J., "Measuring Performance of Inspection and Sorting Systems," *Journal of Quality Technology*, July 1981, p. 149.

12 CHAPTER

QUALITY OF CONFORMANCE: ACCEPTANCE SAMPLING TECHNIQUES

The acceptance sampling approach to quality planning and analysis makes good sense when management has to decide acceptance or rejection based on the quality of a few samples out of a large lot or population. The main advantage of sampling is in the economy gained by eliminating the cost of 100% inspection to achieve the level of quality prescribed. Clearly, the acceptance sampling approach is the only alternative between the two choices of either 100% inspection or no inspection at all.

It should be obvious, however, that any such type of acceptance sampling entails the risk of sometimes making a wrong decision. The proportion of defectives in a lot may be small enough for management to decide to accept the lot, but the reverse may also be true. It must be emphasized that an acceptance sampling merely determines acceptance or rejection of a lot. It is not truly a complete method of quality control, since there is always some risk involved in accepting inferior quality items. However, some of the indirect effects of acceptance sampling are quite important. For example, in a case where a supplier's product is being rejected quite often, it signals one of two possibilities; the buyer may ask the supplier to improve his quality assurance program or the buyer may decide to seek other, better sources of supply. Thus, acceptance sampling puts the responsibility (economically, psychologically, and politically) for quality in the hands of the supplier, where it actually belongs. For the benefit of both parties, a proper design of this sampling plan is required, to help in the overall quality planning. However, getting to the point of no inspection should always remain a real objective of acceptance control, when sufficient quality history and full confidence in the supplier's quality assurance program has been achieved.

12.1 TYPES OF ACCEPTANCE CONTROL

There are four major types of *acceptance control* in use: 100 percent inspection, spot-check inspection, certification, and acceptance sampling. Table 12.1 gives the various advantages and disadvantages of these types of acceptance control. Just as with statistical quality control methods, parallel acceptance sampling procedures are available both for when classifying items as good or bad (sampling by *attributes*), or where one makes an actual measurement of some kind that indicates how good or bad the item is (sampling by *variables*). Both situations are discussed in later sections.

12.2 THEORY OF ACCEPTANCE SAMPLING

Sampling for a basis of decision to accept or reject a lot is extremely essential for most quality control organizations, simply because of the high cost, as well as the impracticality of 100% inspection and test. Two steps are involved in the decision; the first step is to evaluate the material or items based on a few samples, and the second step is to decide on the acceptability of the lot. Lot acceptability decisions involve the use of a sampling plan and random selection of samples according to that plan. The American National Standards Institute and The American Society for Quality Control standard ANSI/ASQC A2-1978[1] defines a sampling plan as *a specific plan that states the sample size or sizes to be used, and the associated acceptance and non-acceptance criteria;* a sampling scheme is *a specific set of procedures which usually consists of acceptance sampling plans in which lot sizes, sample sizes and acceptance criteria, or the amount of 100 percent inspection and sampling are rated.* Before we discuss various schemes and different sampling plans, the terminology common to acceptance sampling must be made clear.

Terminology

In order to understand the terminology, let us consider the following example in which a supplier claims that his product contains no more than 1% defective items. Let us assume that the buyer decides to verify this claim, and wants to assure himself by randomly selecting 100 items from the lot received. If, say two, or less than two items were found defective, he would accept the lot; otherwise he would reject the lot. Here we have a situation dealing with samples where there is always a probability that a good lot will be rejected and a bad lot might get accepted. Both the supplier and the buyer would definitely like to know the probability of this happening (called *risk* analysis) as a means of protection and feedback for process control.

Sample Size, n and Acceptance Number, Ac
The sample size (represented by the symbol n) is the number of pieces selected to be inspected. The acceptance number (represented by the symbol Ac) is the largest number of defectives found in the sample that will still permit the lot to be accepted. The sample size and the acceptance number together make up the decision criteria for determining what proportion of product of a given quality will be accepted or rejected. As will be discussed later, sometimes

TABLE 12.1
Comparison of Types of Acceptance Control

Type of Acceptance Control	Approach	Advantages	Disadvantages
100% Inspection	Inspection of each and every product on a 100% basis.	Theoretically, one of the surest methods to achieve defect-free products.	In practice, depends on the inspector's accuracy. Monotony and repetition in work tend to create fatigue and boredom, giving rise to possibility of not eliminating 100% defects. Very costly if each and every aspect put through inspection (cost = $N \times I$, where N = number of item, I = cost of inspection per item) Automated inspection more reliable, but subject to machinery malfunction, mechanical or electric breakdown with additional overhead cost.
Spot-check inspection	Purely arbitrary. No scientific basis for selecting sample for inspection.	Used in circumstances where the result of inspection is not normally borne out to be of critical importance; e.g., surveillance inspection as a cursory glance (such as street cleanliness checks, violation of traffic rules, etc.) Least costly of all inspection methods.	The method of selection of sample is not defined, sometimes even not realistic. It does not relate to sampling risks. The results not always reliable for drawing definite conclusions about the population. Should a defective part be drawn, then the only remedy to assurance is to inspect all the parts between this inspection and the one previously drawn.

352

Certification method	Based on the receipt of an approved certification by a registered professional engineer or recognized laboratory notifying to the buyer that certain tests have been performed that meet the requirements.	It solely depends on the trust and confidence of the supplying company. When consumers have no other way of finding out the evidence, the label or certification mark of the nationally recognized laboratory come as a guarantee for being defect-free.	There is always certain risk involved. If the supplier has a bad reputation for quality, the buyer takes an enormous chance by continuing business with such a company.
Acceptance sampling	Based on sound mathematical theory widely used and accepted.	Relatively inexpensive. Less time consuming, less fatiguing. Based on well-established principles of probability theory. Less damage to the product; i.e., handling incidental to the inspection is itself a source of defects. Less problem with inspector's boredom and human error. The level of protection and the risk to producer and consumer as afforded by the various sampling plans are known, and possibly a matter of negotiation between the parties concerned. Particularly useful when: Inspection requires some form of destructive testing of product, or Possible losses by passing defective items, or data, are not too great, and the cost of audit is high.	It does not provide full details about quality. Always contains some risks.

353

a sampling plan may call for more than one sample size and acceptance number. Such cases are referred to as *double* or *multiple* sampling plans.

Probability of Acceptance, P_a

The probability of acceptance (represented by the symbol P_a) of a sampling plan is the percentage of samples out of a given series of samples that will cause the different lots of the product to be accepted. Probability of rejection, $P_r = 1 - P_a$.

Example: To illustrate the use of sample size, acceptance number, and probability of acceptance, let us consider the plot shown in Figure 12.1 representing the operation of a simple sampling plan. For lot Number 1 (supplier A), when the product is 4% defective, the inspector takes a sample size of say, 100 from each of the five boxes received and allows an acceptance number of 5. The third box, on the finding of having 6 defectives becomes nonacceptable. The probability of acceptance for supplier A is then determined to be 80% of the product. Using the same approach for supplier B when the product is 8% defective, the inspector will, in the long run, accept only 20% of the product.

Risks

The risk of accepting a bad lot in a given sampling scheme is of concern to the buyer and is called the *consumer's risk*. On the other hand, a supplier likewise wants to know the risk of having good lots rejected; this risk is identified as the *producer's risk*.

Producer's Risk, α.   The producer's risk, α of a sampling plan is defined as the probability of having a lot (or lots) rejected, even though it meets the specified quality requirements. In practice, this risk is usually established in the neighborhood of 5%; that is, $\alpha = 0.05$. To a producer, this would mean having $1 - 0.05 = 0.95$, or a 95% chance of acceptance of the production lots.

Consumer's Risk, β.   The sampling plan must also protect the consumer. While the producer is concerned with the possibility that a good lot will be rejected, the consumer is concerned with the possibility that he will accept a lot that is, in fact, defective.

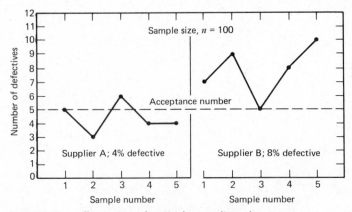

FIGURE 12.1   Illustration of a simple sampling plan.

The consumer's risk, β of a sampling plan is defined as the probability of having a lot accepted even though it does not meet the specified quality requirements. In practice, this risk is usually established in the neighborhood of 10%; that is, $\beta = 0.10$.

Acceptable Quality Level, AQL

The *acceptable quality level,* or AQL is defined as the maximum percent defective (or maximum number of defects per 100 units) that can be considered satisfactory as a process average for the purpose of acceptance sampling. Identified with a particular sampling plan, AQL corresponds approximately to the quality associated with the producer's risk. Accepting a particular AQL value as an agreement between producer and consumer means that the control of manufacturing quality must be such that the average percentage of defectives will not exceed the specified AQL value.

Lot Tolerance Percent Defective, LTPD

The *lot tolerance percent defective,* or LTPD is defined as the percent defective in the incoming lot that the consumer desires having protection against, while recognizing that consumer's risk (β) exists at about say, 10%. Such plans are used when it is desired to limit the acceptable quality level of individual lots within a certain percentage (e.g., 95 out of 100 times) and reject the nontolerable lots, for example, 90 out of 100 times. If the probability of accepting any one lot that deviates in respect to an important (quality) characteristic must be kept low, the LTPD of the plan should be specified. All of these terms are explained diagrammatically in Figure 12.2.

Average Outgoing Quality Limit, AOQL

Sampling procedures that determine the average quality of the outgoing product are referred to as AOQL schemes. If the incoming lot has a percent defective p, which is screened out

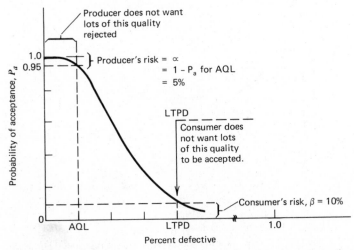

FIGURE 12.2 Description of an OC curve showing the conventional requirements of α (producer's risk) and β (consumer's risk). AQL = acceptable quality level, LTPD = lot tolerance percent defective.

by 100% inspection, the average outgoing quality is $P_a \times p$, where P_a = probability of acceptance. In cases where 100% inspection can not or will not be done on all rejected lots, sampling plans based on AOQL will not apply.

To illustrate, let us assume that an incoming lot which is 2% defective has an 80% probability of acceptance. Suppose that the buyer makes a rule that all lots rejected by this sampling plan must be 100% inspected, that all defective items found must be replaced with good items, and further, that the rejected lots that have had all defective items removed must then be considered together with the accepted lots in such a way as to make one total quantity. For this sampling plan then, about 80% of the lots will be accepted and 20% will be rejected and then subjected to 100% inspection. Since 20% of the product lot theoretically will not be defect-free, the balance of 80% will still be about 2% defective. To calculate the percentage of defectives that will be left in the lot, the following expression is used.

$$\text{AOQ} = P_a \times p \times \frac{N - n}{N}$$

where:

AOQ = The average outgoing quality.
P_a = Probability of acceptance.
p = Percent defective.
n = Size of the sample.
N = Size of the lot.

Alternatively, for one lot of known percent defective AOQ is simply $P_a \times p$, when sampling is done without disturbing the population.

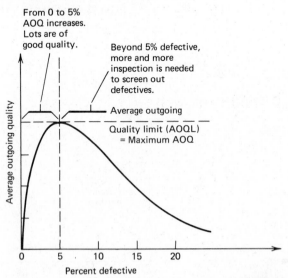

FIGURE 12.3   Illustrating the average outgoing quality curve for a given sampling plan. The AOQ curve rises until it reaches a certain maximum point, after which it falls off as a result of more and more product being 100% inspected.

From the above relationship, one can develop a curve for any given sampling plan to show the AOQ for any level of incoming quality. The probability of acceptance, P_a can be taken directly from the OC (operating characteristic) curve (see next section) or from tables, and then substituted in the formula to plot the AOQ against different p values. The plot will give a variation like the one shown in Figure 12.3. Typically, for a sampling plan where $n = 18$ and Ac = 0, AOQL is the maximum value of AOQ. A peak in the curve accounts for the fact that, initially, with less percent defective, the outgoing quality improves and then falls off again as a result of more and more products being 100% inspected and hence due to variation in P_a and p. If incoming quality is superb ($p < 5\%$), more lots will be accepted, seldom requiring 100% inspection.

Another method of calculating the AOQL of a sampling plan is by the use of the following equation:

$$AOQL = \frac{y}{n} - \frac{y}{N}$$

where y is a factor that depends on the acceptance number of the sampling plan (available in MIL-STD-105D tables and Dodge-Romig Tables).

12.3 QUANTIFICATION OF SAMPLING RISKS: THE OPERATING CHARACTERISTIC CURVE

When judging the suitability of a particular sampling plan, it is desirable to know the probability that a lot submitted with a certain percent defective will be accepted or rejected. This is important to help resolve the risks involved.

The *operating characteristic curve* (or OC curve, as it is frequently called) for a sampling plan quantifies these risks. The producer wants all good lots accepted and the consumer wants all bad lots rejected. Thus, if the probability of acceptance of a batch, P_a, is plotted against the proportion of defectives in that batch (p), the resulting curve is known as the operating characteristic curve. In the ideal situation mentioned above, only an ideal sampling plan which has an OC curve that is a vertical line (Figure 12.4) can satisfy both the producer and the consumer. This curve, however, is attainable only with 100% inspection (theoretically speaking) and, considering many of the disadvantages of 100% inspection (see Table 12.1), and the amount of risk involved (α and β), an actual OC curve deviates quite considerably from the ideal one.

Construction of A Typical OC Curve

It should be mentioned that an OC curve for a given (or specified) sample size and sampling plan states only the probability that a lot having p percent defectives will be accepted. It does not predict the confidence level in connection with the percent defective, nor does it give any indication of the final quality of the lots that have been accepted.

An OC curve is fully described by the sample size (n) and the acceptance number (Ac). The probability of acceptance is the probability that the number of defectives in the sample (sample size, n) is equal to or less than the acceptance number (Ac) for the sampling plan under consideration. The reject number (Re) is always equal to Ac + 1 and, when Ac = 0, no defectives will be permitted. Ac is sometimes also referred to simply as c.

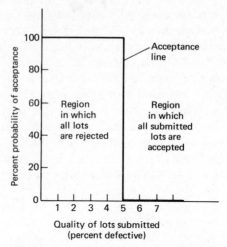

FIGURE 12.4   Ideal OC curve for an ideal sampling plan.

The construction of an OC curve requires the calculation of probabilities of acceptance. There are three probability distributions that we can use to find the probability of acceptance; namely the Poisson, the binomial, and the hypergeometric distributions. The details of each of these distribution functions have already been described in Chapter 6. To refresh the memory, it will be recalled that adhering to the hypergeometric method takes into account that the sample is taken without replacement from a finite lot. When the lot size is considered infinite, binomial probability distributions can be used. However, when sampling is without replacement and a sample size which approaches infinity is taken from an infinite population whose proportion of defectives approaches zero, these probabilities can be calculated by using the Poisson method.

For many practical purposes, the Poisson method yields satisfactory approximations to binomial. This is particularly so when:

1.   A fairly large lot size is used, in which case it is required that the population be infinite, and when sampling conditions without replacement are somewhat met.

2.   A fairly large sample size is taken, in which case the requirement that the sample size approaches infinity is satisfied.

3.   A lot fraction defective is small enough to be approximated as zero. Statistically, these conditions can be taken to be approximately valid when the lot size is at least 10 times bigger than the sample size, the latter being at least 16, and the fraction defective, p, being less than 0.1. The Poisson distribution is therefore commonly employed (except for small sizes, when the hypergeometric method is used) to simplify the task of designing an acceptance sampling plan.

It will be recalled that the formula for the Poisson probability distribution is given by:

$$P_{n,r,p} = \frac{(np)^r \times e^{-np}}{r!}$$

where:

  n = Sample size.
  p = Percent of defectives.
  r = Number of defectives.
  np = Expected number of defectives.

To construct an OC curve, we assume a range of values of percent defective for the incoming lot, and calculate the expected number of defectives by multiplying them individually with the sample size.

Example: Let us assume that a particular lot contains a fraction defective of 0.08, for which $n = 10$, and Ac = 1. The individual probabilities can be found by using the Poisson formula. The probability of accepting this lot will be equal to the probability of finding zero defectives plus the probability of finding one defective. Thus:

$$P(0) = \frac{(10 \times 0.08)^0}{0!} \times e^{-0.8} = 0.4493; \text{ and}$$

$$P(1) = \frac{(10 \times 0.08)^1}{1!} \times e^{-0.8} = 0.3596$$

Total probability, which is the sum of the above two probabilities becomes:

$$P_a = 0.4493 + 0.3596 = 0.8089.$$

A similar computation can be made for any lot with a different proportion of defectives. However, these individual calculations are not needed since, for convenience, all of the Poisson values are available through published tables. The Poisson equation can be solved using Table B in Appendix B, or by use of published curves such as those of Figure 12.5. The Table gives the probability of r defectives in a sample of n from a lot having a fraction defective of p.

Given the availability of the Poisson Table, it is easier to compute probabilities when designing an acceptance sampling plan. For instance, to develop a plan whose OC curve is required to pass through two preselected points, one can assume different combinations of n and Ac, and find the probabilities of accepting lots with various proportions of defectives from the Poisson Table. This exercise is continued until the correct combination of n and Ac is found.[2]

In cases where a firm becomes interested in a plan which has an OC curve passing through a single stipulated point (e.g., the plan may be required to provide a 0.80 probability of accepting a lot which is 3% defective), trial and error with the help of the Poisson Table or with the help of other special tables can give sample size, n, and acceptance number, Ac, directly. Such plans are discussed by Hamaker.[3]

In some situations, it is necessary to use binomial or hypergeometric functions rather than the Poisson. One useful source of binomial probability is "Applied Mathematics Series No. 6, Tables of the Binomial Probability Distribution," published by the U.S. Government Printing Office, Washington, D.C.

For still many other practical situations, one can use the OC curves which have already been plotted and published without going to the trouble of making calculations. Examples

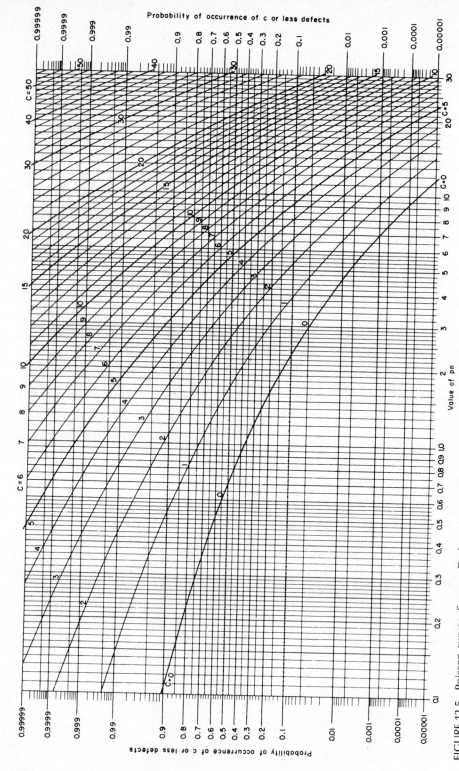

FIGURE 12.5 Poisson curves. Source: Dodge, H. F. and Romig, H. G., *Sampling Inspection Tables - Single and Double Sampling*, 2nd ed., John Wiley & Sons, Inc., New York, 1959. Reprinted by permission of the publisher and AT&T Bell Laboratories.

of some sources are the MIL-STD-105D Tables and the Dodge-Romig Tables. Both of these are discussed later.

OC Curve for Double and Multiple Sampling Plans

The construction of an OC curve for double or multiple sampling plans follows the same logical sequence as that for a single sampling plan, but needs somewhat more calculation. In the case of double sampling, two OC curves are to be plotted: one curve for the probability of acceptance on the first sample (n_1) and the second curve for the probability of acceptance on the combined samples. A multiple sampling plan involves even more calculations. Examples for a stated double and multiple sampling plan can be in the following form.

Double Sampling Plan	Multiple Sampling Plans
$N = 2000$	$N = 2000$
$n_1 = 150$, $Ac_1 = 0$	$n_1 = 50$, $Ac_1 = 0$, $Re_1 = 4$
$n_2 = 200$, $Ac_2 = 3$	$n_2 = 50$, $Ac_2 = 2$, $Re_2 = 5$
$n_1 + n_2$ number of pieces	$n_3 = 50$, $Ac_3 = 3$, $Re_3 = 5$
in the two samples	$n_4 = 50$, $Ac_4 = 4$, $Re_4 = 5$
combined.	

The above double sampling plan can be interpreted as follows.

1. First inspect the first sample of 150 from a lot of 2000.

2. Accept the lot if the first sample contains zero defectives, but reject it if the first sample contains four or more defectives.

3. Inspect a second sample of 200 if the first sample contains one, two, or three defectives.

4. Accept the lot if the combined sample of 350 contains three or less defectives, otherwise reject the lot for more than three defectives.

The probability of accepting the lot will be the sum of the probabilities of these different ways in which it may be accepted. The calculation of probabilities will follow using either the Poisson formula or the hypergeometric expression described before.[4]

The procedure in multiple sampling follows a similar approach to that of double sampling, except that the number of successive samples required to reach a decision to accept or reject the lot is more than two. The number of steps required to reach a decision depends on the cumulative number of defective items found in the samples taken progressively. Each step may give rise to its own accept or reject criteria—the acceptance decision is made where the cumulative number of defective items is equal to or less than the acceptance number. A reject decision is made where the number of defective items equals or exceeds the rejection number for that step.

Parameters Affecting the OC Curve

It was stated earlier that an OC curve is completely described by two parameters, namely, the sample size (n) and the acceptance number (Ac). One of the major advantages is that by defining the operating characteristic curve, the sampling risks become known. This is important, since for many safety-related products one wants to eliminate the consumer's risk as much as possible. The effect of varying n (sample size), N (lot size) and acceptance number (Ac) are such that one is able to manipulate the shape and location of OC curves for a desired sampling plan in terms of producer's risk (AQL) and consumer's risk (LTPD). In other words, the protection provided by a sampling plan can be altered by changing any of these parameters, either singly or in combination with one another.

Figure 12.6a–e, summarizes typical trends of variation in the OC curve for five different situations. Although these curves are only schematic, following the construction method outlined earlier, one should be able to easily verify these against a set of known quantitative parameters.

In Figure 12.6a, the lot size is changed but the acceptance number and the sample size are held constant. The magnitude of variation indicated in lot size (N) is such that it would incorporate N values from five hundred to infinity. Notice that the lot size has very little effect on the probability of acceptance. However, as the lot decreases, the curves fan out slightly in the bottom. Still, when $n < 10\%$ of N, it can be safely assumed that lot size variation has practically no influence on the OC curve.

When the sample size (n) is changed while keeping everything else constant, the trend of variation can be effected as shown in Figure 12.6b. As the sample size increases, the OC curve becomes steeper. Its slope increases, approaching becoming a vertical line (compare this behavior with that of the ideal OC curve of Figure 12.4). This means that sampling plans with larger sample sizes have better discriminating power to allow separation of good lots from bad lots. Therefore, with larger sample sizes the consumers would have fewer lots of bad quality accepted and producers would have a few lots of good quality rejected. Increasing the sample size (n) together with reducing the acceptance number (Ac) (Figure 12.6e) again gives the closest approach to the ideal OC curve. Notice that in this particular case, when two parameters are changed together, the curves have a tendency to intersect each other.

Figure 12.6c illustrates a situation where sample sizes are taken as a fixed percentage of lot size (N and n are both changed simultaneously). For example, if sample size $n = 10\%$ of N for three separate situations (i.e., $N = 1000$, $n = 100$, Ac $= 0$; $N = 800$, $n = 80$, Ac $= 0$; and $N = 500$, $n = 50$, Ac $= 0$), one will have three plans with three levels of protection—P_1, P_2, and P_3—assuming that the lots are from a production process which is known to be $P_2\%$ defective.

The effect of varying the acceptance number (Ac) on the shape of the OC curve is shown in Figure 12.6d. Assuming constant values of sample size and lot size, a decrease in Ac gives a concave shape that is typical of zero acceptance OC curves. As the acceptance number increases, the OC curves lose their original concave shape, bulge toward the top, and also lose some of their steepness. Increasing the acceptance number would therefore mean that inspection lots with a larger percent defective can be accepted by the plan. In other words, this means that the higher the acceptance number, the lower the protection.

In learning to think of *economic advantages* in terms of different shapes of the OC curves, one would be interested in three parts of these curves as we will describe in relation to Figure 12.6.

The Top Portion. Depending upon the particular sampling plan chosen, an OC curve may be a flat portion extending to a considerable distance horizontally (Figure 12.6b, d, and e) or it may have only a sharp peak (Figure 12.6a and c). The top portion is very important since it reflects the quality of a product that will be accepted by the given sampling

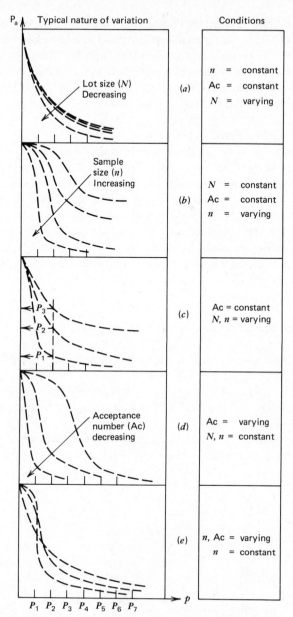

FIGURE 12.6 (a–e) Effect of changes in various parameters on OC curves.

plan without any question. For example, in Figure 12.6d, the third curve on the right will allow a greater percentage of defective items than the curves to its left.

The Middle Portion. In general, the middle portion of an OC curve constitutes about half of the total curve and, hence, reflects a probability of acceptance of about 50%. At the exact center of the curve, from a point where $P_a = 0.5$, a product of corresponding quality will have 50/50 chance of being accepted or rejected.

The Tail Portion. The tail or bottom portion of an OC curve may be short or sharp, or it may extend for a considerable distance horizontally. Depending upon the percent defectives (P_1, P_2, P_3, . . . etc., in Figure 12.6), a short and sharp tail of an OC will reject less items compared to one with an extended tail (compare the three curves of Figure 12.6d; $P_1 < P_2 < P_3$).

Maximum economy is likely to be realized when the running capability of the process matches the top portion of the OC curve to be used.

Rule-of-Thumb Sampling Plans

From a statistical point of view, a rule-of-thumb type of sampling plan is quite inadequate, although it is still used by some firms. For example, if the buyer and the supplier agree on a 10% sampling rule, then a sample equal to 10% of the lot is selected, and if no defectives are found, the lot is accepted; otherwise, if any defective is found, the lot is rejected.

An OC curve for rule-of-thumb type sampling plan is shown in Figure 12.7. The crudeness of estimation and high risks involved with such a plan can easily be seen. For example, if the lot submitted is known to contain as much as 80% defective, still there is a 59% probability of acceptance for this lot. In the case of statistical sampling utilizing OC curves, no such disparity in risk is present.

12.4 SAMPLING METHODS

By way of review, the reader should recall that the basis of any sampling theory is a study of the relationship between a population and a sample drawn from that population (compare with Chapter 6). The statistical footing of such relationships is used to ascertain the quality characteristics of an unknown population from a knowledge of sample parameters, including the variations from sample to sample if they are significant. A way of facilitating the chance for inclusion of any item in the sample to be predetermined was earlier described as *random sampling* (or probability sampling). However, many practical situations do not allow the

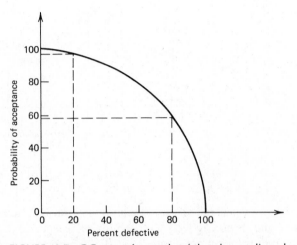

FIGURE 12.7 OC curve for a rule-of-thumb sampling plan.

possibility of random sampling. Nonrandom samples are influenced by several personal biases and are to be classified as *nonprobability* samples.

The two types of sampling techniques can also be referred to as *statistical* or *nonstatistical*. Some of the statistical techniques are *simple random sampling, stratified random sampling, systematic sampling,* and *cluster sampling.* Some of the nonstatistical sampling techniques and their features are:

1. *Convenience sampling:* This is based on the inspector's convenience and it is difficult to apply statistical thinking to its results.

2. *Restricted sampling:* This is similar to convenience sampling, but the restrictions can be natural, artificial, or due to accessibility problems of the situation at hand.

3. *Judgment sampling:* This is also called *experience sampling.* In this kind of sampling, the inspector's past experience is the guiding rule in the selection of samples.

4. *Quota sampling:* Here, quotas of the samples are specified in advance. Quota sampling can be based on statistical reasoning.

Another method of sampling is *bulk sampling,* used for solids, liquids, or gases, either in a static or dynamic state. Bulk sampling may or may not be based upon randomness of the statistical methods. Puri[5] describes various methods of sampling techniques in detail with insight and knowledge of broad scope.

Procedure for Choosing a Random Sample

Simple random samples can be drawn using tables of random numbers. A partial list of computer-generated random numbers is given in the following table.

77033	68325	10160
94372	06164	30700
48838	60935	70541
80235	21841	35545
88034	97765	35959

Suppose a random sample of 6 electronic watches is to be drawn from a lot of 70 watches where each has been labeled with a number from 1 to 70. Starting at the top of the first column of the table, the watches to be drawn for the sample are numbers 77, 03, 36, 83, 25, and 10.

The numbers 77 and 83 are ignored, since we know that these numbered watches are not in the lot. Therefore, proceeding along the row, we select two more numbers, 16 and 09. Thus, the final six random numbers selected to make up the sample of watches are 03, 36, 25, 10, 16, and 09.

Selection of Sample

The inspector assigned to the receiving area of the quality control process faces a variety of problems dealing with acceptance or rejection. The problem multiplies as the list of vendors grows. The following set of rules are presented for guidance in these situations.

1. Sample pieces selected for inspection or testing must be representative of the entire lot so that each piece in the lot receives an equal chance of being selected.

2. If the lot size is too big, dividing the lot into *sublots* or *strata* is a common practice. A typical example is an egg packing station where the boxes of eggs are kept on pallets according to the grade size, and the population is divided into strata. This is to ensure that chances of sample selection remain random, homogeneous, and unbiased; that the cost of conducting the actual sampling remains less; and that a separate estimate of population parameters can be obtained.

3. Inspector's personal biases must not occur in the sample selection process, no matter what the temptations may be (drawing should be done *blind folded*).

4. If the lot size is small, receiving inspection must make use of other supplementary information regarding process or manufacturing conditions used (certification of tests by the supplier, prior inspections, etc.).

5. Attempts should be made not to mix items from different lot shipments, even if they are small and products remain exactly identical. Unless rigorous sampling plans are followed, there will always be mix-up of one kind or the other. Keeping good records helps to avoid future troubles with traceability, call backs, recalls, and other corrective measures initiated for various reasons.

6. Inspector-to-inspector variations must be watched for. Think of the quality problems that arise when the decisions of one inspector vary from time to time, or when inspectors are inconsistent with each other, or when some of them don't know what is acceptable or what is defective.

12.5 INTERPRETATION AND ACTION BY MANAGEMENT

Just as we mentioned before that variation in a production process is natural, so is variation in the judgments and decisions of receiving inspection departmental personnel. The fact is, that by the time the inspection is made, the product is already acceptable or defective. No receiving inspection can give perfect results no matter how perfect the sampling plans are.

It is important to stress that the concept encompasses more than mere institution of a sampling plan. Top management's responsibility of creating a better system for inspection control cannot be overemphasized. The following aspects of this should be noted.

1. If management cannot precisely understand the real purpose of inspections, how can it reject them or not feel responsible after bad lots have been put into assembly lines?

2. If management does not have a good system of receiving inspection, one can be sure that even if the company has good statistical process methods, defects are creeping in from outside sources (suppliers).

3. To improve quality, top management must be able to distinguish between good suppliers and inferior ones. Without measuring and knowing the quality of supplies the delivery schedules, prices offered by the lowest bidders have no meaning. A vendor who can't be trusted to delivery quality from year to year or from lot to lot proves very costly to the buyer.

4. Putting too much emphasis on the effectiveness or performance of inspectors cannot alone uncover or correct faults in the management of purchasing. This is true even if the company has an excellent acceptance sampling plan, mutually satisfactory to both parties, that agrees with an established standard. Standards do not tell managers what actions to take or how to improve the purchasing policy or system. Mistakes are very seldom fixed and repaired without planning; they have to be identified, diagnosed, and then resolved through a corrective system.

5. Top management must view inspection less as an end in itself than a means to an end. A more appropriate solution would be, for example, creating a rigorous test for new suppliers, paying frequent plant visits to old suppliers who have known quality problems, and selecting suppliers based as much on their management philosophy, quality commitment, process capabilities, quality program (through a quality control manual) evaluation, and so forth, as on price and delivery aspects. Acceptance control and process control must go hand in hand in matters of improving quality. Acceptance inspections should be always terminated when they are no longer needed.

6. With regard to the selection of OC curves for the purpose of sampling, the management and engineering group should be aware that most costs are associated with the producer's risk. When the risk of rejecting a relatively good quality product increases, a variety of problems arise. The samples must not only be rechecked, but handled, stored, and accounted for on each and every occasion. This increases the cost of inspection, causes unnecessary delays in getting needed parts, and interrupts the normal routine.

7. With the chance that unsatisfactory products will be accepted and sent out (consumer's risks), typical costs incurred would include, for example, the cost of handling and answering complaints, the effect on customer goodwill, the cost of calling client meetings, additional expenses of reengineering, and so on. A remarkable insight on the question of accepting defective items while using sampling plans comes from W. Edwards Deming. ''A company that purchases items on AOQL of 3 percent is making it known to the vendor that the purchaser is in market for 97 good items and three bad ones out of 100. The vendor will be pleased to meet these requirements.''

12.6 TYPES OF SAMPLING PLANS

Sampling plans are of two types, *attributes plans* and *variables plans*. In attributes sampling plans, a decision to accept (or reject) is based on the sample size (n) and the acceptance number (Ac). For an allowable stated acceptance number in the plan, the number of defectives found in the random sample determines whether to accept the lot or reject it.

In variables sampling plans, a decision to accept (or reject) a lot is also based on the sample size (n), but a measurement of a stated quality characteristic is made on each item. The difference between the sample averages and the specified limiting values of the characteristics being measured, is compared with the allowable difference specified in the sampling plan. If a product has several critical quality characteristics, then each of these are evaluated independently and compared against given acceptance criteria.

12.7 ACCEPTANCE SAMPLING PLANS BY ATTRIBUTES

Acceptance sampling by attributes is generally applicable to lots of discrete items where the concept of percent defective is applicable and the items or some characteristics of the items can be classified as acceptable or not acceptable. By virtue of this simplification, sampling-by-attribute schemes have the following advantages.

1. They are simple and easier to apply, with no special skill or training required for inspectors.

2. Since the acceptance/rejection criteria is based on go, no-go decisions, time and money is saved. Decisions are not regarded as a matter of the "luck of the draw."

3. These plans can be applied to any number of quality characteristics.

The most commonly used sampling plan by attribute is that provided by the MIL-STD-105D standard of the U.S. Department of Defense.[6]

MIL-STD-105D

The military standard, MIL-STD-105D is a widely used document throughout the world, including even in civilian operations. The basic aim of this standard is the maintenance of the outgoing quality level, whether used for receiving inspection, in-house inspection, or final inspection. Following the last revision of this standard in 1963 by a team of American, Canadian, and British experts, it has become quite popular, to the extent that, in 1973, it was adopted by the International Organization for Standardization (designation number ISO/DIS-2859).

The plans provided by this standard can be classified into three categories: AQL plans, LTPD plans, and AOQL plans.

Acceptable Quality Level Sampling Plans

Acceptable quality level (AQL) plans are usually utilized when the product is submitted for inspection in discrete lots. As stated earlier, AQL is defined as the maximum percent defective which can be considered satisfactory as a process average. For example, to quality control engineers and receiving inspectors, a 1% AQL sampling plan would thus mean a plan which will regularly accept a maximum of 1% defective product. It does not tell anything about the rejected proportion. Considering this in terms of OC curves for a number of different sampling plans, it is possible to have acceptance curves fairly similar (to the shoulder portion) out of a number of given sampling plans. Figure 12.8 is an illustration of this point for a typical 2% AQL sampling plan, coinciding with the maximum probability of acceptance.

It should be noted that the AQL plans are basically producer oriented. With a normal rule, when the producer's risk is 5% the probability of acceptance of lots with a quality level equal to AQL will be 95%, whereas the risk of having lots rejected will be 5% (i.e., 1 lot in 20). When the MIL-STD-105D standard is used for percent defective plans, the AQLs range from 0.010% to 10.0%. For defect-per-unit plans, there are additional AQLs, from 0.010 to 1000.0 defects per 100 units[7].

The AQL is normally specified in some of the contracts of the government agency or by the responsible procuring authority. It is also sometimes common practice to designate AQLs

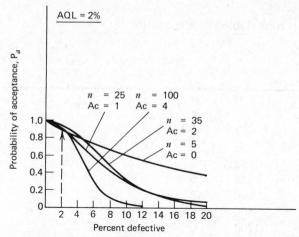

FIGURE 12.8   OC curves for some frequently used sampling plans showing similarity at the top portions for a typical AQL of 2%.

for groups of defects, either collectively or on an individual basis. The procuring authority may furthermore ask (or specify) different AQLs for critical, major, and minor defects, with lower AQLs for critical defects and higher values for the minor ones.

The most promising feature of the MIL-STD-105D standard is the provision of distinction in levels of inspection, as shown in Table 12.2. In this table, the sample size is determined by the lot size and the inspection level through code letters. Knowing the code letters, the AQL, and the type of sampling to be used (single, double, or multiple sampling), the sampling scheme can be read from other tables like the ones shown for single sampling (Tables 12.3A–C, 12.4A–C, and 12.5).

Example: Let us suppose that a government purchase contract specifies a 1.5% AQL and the vendor brings in the product in lot sizes of 2000. For a general inspection level of II, what will be the single sampling plans for normal, tightened, and reduced inspection?

Normal Inspection. Using the lot size $N = 2000$ and inspection level II, the sample size code letter K is obtained from Table 12.2. Now, using Table 12.3A for single sampling (normal inspection), we find the desired plan for code letter K and an AQL of 1.5%, as follows: sample size, $n = 125$; acceptance number, Ac = 5; and rejection number, Re = 6. Thus, from a lot of 2000 pieces, a random sample of 125 is inspected. If five or less defectives are found, the lot is accepted. If six or more defectives are found the lot is rejected.

Tightened Inspection. The sample size code letter is still the same; that is, K. However, we will use Table 12.3B (Single Sampling Plans for Tightened Inspection) to get the desired plan. It is: sample size, $n = 125$; acceptance number, Ac = 3; and rejection number, Re = 4. Thus, from a lot of 2000, a random sample of 125 is selected. If three or less defectives are found the lot is accepted. If four or more defectives are found the lot is rejected. (Note that a tightened procedure is used when quality deteriorates.)

Reduced Inspection. The sample size code letter is the same as previously determined, that is, K. However, the desired plan for code letter K and AQL = 1.5% is obtained from Table 12.3C (Single Sampling Plans for Reduced Inspection). The plan is: sample size, $n = 50$; acceptance number, Ac = 2; and rejection number, Re = 5. Thus, from a lot of

TABLE 12.2
Sample-Size Code Letters (Table I of MIL-STD-105D)

Lot or Batch Size			Special Inspection Levels				General Inspection Levels		
			S-1	S-2	S-3	S-4	I	II	III
2	to	8	A	A	A	A	A	A	B
9	to	15	A	A	A	A	A	B	C
16	to	25	A	A	B	B	B	C	D
26	to	50	A	B	B	C	C	D	E
51	to	90	B	B	C	C	C	E	F
91	to	150	B	B	C	D	D	F	G
151	to	280	B	C	D	E	E	G	H
281	to	500	B	C	D	E	F	H	J
501	to	1200	C	C	E	F	G	J	K
1201	to	3200	C	D	E	G	H	K	L
3201	to	10000	C	D	F	G	J	L	M
10001	to	35000	C	D	F	H	K	M	N
35001	to	150000	D	E	G	J	L	N	P
150001	to	500000	D	E	G	J	M	P	Q
500001	and	over	D	E	H	K	N	Q	R

Note.

Small sample inspection levels of MIL-STD-105C	Convert to these special inspection levels
L-1 and L-2	S-1
L-3 and L-4	S-2
L-5 and L-6	S-3
L-7 and L-8	S-4

2000, a random sample of 50 is selected. If two or less defectives are found the lot is accepted. If five or more defectives are found the lot is rejected. (Note that the reduced inspection is used where the supplier's quality has been good or when the process average is better than the AQL by 2σ or more.)

In comparing the three tables (Tables 12.3A, 12.3B, and 12.3C), notice the vertical arrows. If a vertical arrow is encountered, the first sampling plan above or below (as the case may be) the arrow is followed. The switching rules—*tightened*$\longleftrightarrow$*normal*$\longleftrightarrow$*reduced*—apply when the produced goods are submitted in a continuing series of lots, rather than in a few isolated lots.

Table 12.2 also identifies four additional special inspection levels (S-1, S-2, S-3, and S-4) that are used where relatively small sample sizes are necessary and larger sampling risks can or must be tolerated.

Provision for Double and Multiple Sampling

The provision for a double and multiple sampling plan is an extension of the single sampling plan. Although OC curves for each sample size code letter (code letters A through R) together

TABLE 12.3A
Single Sampling Plans for Normal Inspection (Table II-A of MIL-STD-105D)

Acceptable Quality Levels (normal inspection)

Sample size code letter	Sample size	0.010		0.015		0.025		0.040		0.065		0.10		0.15		0.25		0.40		0.65		1.0		1.5		2.5		4.0		6.5		10		15		25		40		65		100		150		250		400		650		1000	
		Ac	Re	Ac	Re	Ac	Re	Ac	Re	Ac	Re	Ac	Re	Ac	Re	Ac	Re	Ac	Re	Ac	Re	Ac	Re	Ac	Re	Ac	Re	Ac	Re	Ac	Re	Ac	Re	Ac	Re	Ac	Re	Ac	Re	Ac	Re	Ac	Re	Ac	Re	Ac	Re	Ac	Re	Ac	Re	Ac	Re
A	2	↓		↓		↓		↓		↓		↓		↓		↓		↓		↓		↓		↓		↓		↓		↓		↓		↓		↓		0	1	1	2	2	3	3	4	5	6	7	8	10	11
B	3	↓		↓		↓		↓		↓		↓		↓		↓		↓		↓		↓		↓		↓		↓		↓		↓		↓		↓		0	1	1	2	2	3	3	4	5	6	7	8	10	11	14	15
C	5	↓		↓		↓		↓		↓		↓		↓		↓		↓		↓		↓		↓		↓		↓		↓		↓		↓		0	1	1	2	2	3	3	4	5	6	7	8	10	11	14	15	21	22
D	8	↓		↓		↓		↓		↓		↓		↓		↓		↓		↓		↓		↓		↓		↓		↓		↓		0	1	1	2	2	3	3	4	5	6	7	8	10	11	14	15	21	22	30	31
E	13	↓		↓		↓		↓		↓		↓		↓		↓		↓		↓		↓		↓		↓		↓		↓		0	1	1	2	2	3	3	4	5	6	7	8	10	11	14	15	21	22	30	31	44	45
F	20	↓		↓		↓		↓		↓		↓		↓		↓		↓		↓		↓		↓		↓		↓		0	1	1	2	2	3	3	4	5	6	7	8	10	11	14	15	21	22	30	31	44	45	↑	
G	32	↓		↓		↓		↓		↓		↓		↓		↓		↓		↓		↓		↓		↓		0	1	1	2	2	3	3	4	5	6	7	8	10	11	14	15	21	22	30	31	44	45	↑		↑	
H	50	↓		↓		↓		↓		↓		↓		↓		↓		↓		↓		↓		↓		0	1	1	2	2	3	3	4	5	6	7	8	10	11	14	15	21	22	30	31	44	45	↑		↑		↑	
J	80	↓		↓		↓		↓		↓		↓		↓		↓		↓		↓		↓		0	1	1	2	2	3	3	4	5	6	7	8	10	11	14	15	21	22	30	31	44	45	↑		↑		↑		↑	
K	125	↓		↓		↓		↓		↓		↓		↓		↓		↓		↓		0	1	1	2	2	3	3	4	5	6	7	8	10	11	14	15	21	22	30	31	44	45	↑		↑		↑		↑		↑	
L	200	↓		↓		↓		↓		↓		↓		↓		↓		↓		0	1	1	2	2	3	3	4	5	6	7	8	10	11	14	15	21	22	30	31	44	45	↑		↑		↑		↑		↑		↑	
M	315	↓		↓		↓		↓		↓		↓		↓		↓		0	1	1	2	2	3	3	4	5	6	7	8	10	11	14	15	21	22	30	31	44	45	↑		↑		↑		↑		↑		↑		↑	
N	500	↓		↓		↓		↓		↓		↓		↓		0	1	1	2	2	3	3	4	5	6	7	8	10	11	14	15	21	22	30	31	44	45	↑		↑		↑		↑		↑		↑		↑		↑	
P	800	↓		↓		↓		↓		↓		↓		0	1	1	2	2	3	3	4	5	6	7	8	10	11	14	15	21	22	30	31	44	45	↑		↑		↑		↑		↑		↑		↑		↑		↑	
Q	1250	↓		↓		↓		↓		↓		0	1	1	2	2	3	3	4	5	6	7	8	10	11	14	15	21	22	30	31	44	45	↑		↑		↑		↑		↑		↑		↑		↑		↑		↑	
R	2000	↓		↓		↓		↓		0	1	1	2	2	3	3	4	5	6	7	8	10	11	14	15	21	22	30	31	44	45	↑		↑		↑		↑		↑		↑		↑		↑		↑		↑		↑	

⇩ = Use first sampling plan below arrow. If sample size equals, or exceeds, lot or batch size, do 100 percent inspection.

⇧ = Use first sampling plan above arrow.

Ac = Acceptance number.

Re = Rejection number.

371

TABLE 12.3B
Single-Sampling Plans for Tightened Inspection (MIL-STD-105D)

Acceptable Quality Levels (tightened inspection)

Each data cell shows **Ac Re** (Ac = Acceptance number, Re = Rejection number). ↓ = Use first sampling plan below arrow. ↑ = Use first sampling plan above arrow.

Sample size code letter	Sample size	0.010	0.015	0.025	0.040	0.065	0.10	0.15	0.25	0.40	0.65	1.0	1.5	2.5	4.0	6.5	10	15	25	40	65	100	150	250	400	650	1000
A	2	↓	↓	↓	↓	↓	↓	↓	↓	↓	↓	↓	↓	↓	↓	↓	↓	0 1	1 2	2 3	3 4	5 6	8 9	12 13	18 19	27 28	41 42
B	3	↓	↓	↓	↓	↓	↓	↓	↓	↓	↓	↓	↓	↓	↓	↓	0 1	1 2	2 3	3 4	5 6	8 9	12 13	18 19	27 28	41 42	↑
C	5	↓	↓	↓	↓	↓	↓	↓	↓	↓	↓	↓	↓	↓	↓	0 1	1 2	2 3	3 4	5 6	8 9	12 13	18 19	27 28	41 42	↑	↑
D	8	↓	↓	↓	↓	↓	↓	↓	↓	↓	↓	↓	↓	↓	0 1	1 2	2 3	3 4	5 6	8 9	12 13	18 19	27 28	41 42	↑	↑	↑
E	13	↓	↓	↓	↓	↓	↓	↓	↓	↓	↓	↓	↓	0 1	1 2	2 3	3 4	5 6	8 9	12 13	18 19	27 28	41 42	↑	↑	↑	↑
F	20	↓	↓	↓	↓	↓	↓	↓	↓	↓	↓	↓	0 1	1 2	2 3	3 4	5 6	8 9	12 13	18 19	27 28	41 42	↑	↑	↑	↑	↑
G	32	↓	↓	↓	↓	↓	↓	↓	↓	↓	↓	0 1	1 2	2 3	3 4	5 6	8 9	12 13	18 19	27 28	41 42	↑	↑	↑	↑	↑	↑
H	50	↓	↓	↓	↓	↓	↓	↓	↓	↓	0 1	1 2	2 3	3 4	5 6	8 9	12 13	18 19	27 28	41 42	↑	↑	↑	↑	↑	↑	↑
J	80	↓	↓	↓	↓	↓	↓	↓	↓	0 1	1 2	2 3	3 4	5 6	8 9	12 13	18 19	27 28	41 42	↑	↑	↑	↑	↑	↑	↑	↑
K	125	↓	↓	↓	↓	↓	↓	↓	0 1	1 2	2 3	3 4	5 6	8 9	12 13	18 19	27 28	41 42	↑	↑	↑	↑	↑	↑	↑	↑	↑
L	200	↓	↓	↓	↓	↓	↓	0 1	1 2	2 3	3 4	5 6	8 9	12 13	18 19	27 28	41 42	↑	↑	↑	↑	↑	↑	↑	↑	↑	↑
M	315	↓	↓	↓	↓	↓	0 1	1 2	2 3	3 4	5 6	8 9	12 13	18 19	27 28	41 42	↑	↑	↑	↑	↑	↑	↑	↑	↑	↑	↑
N	500	↓	↓	↓	↓	0 1	1 2	2 3	3 4	5 6	8 9	12 13	18 19	27 28	41 42	↑	↑	↑	↑	↑	↑	↑	↑	↑	↑	↑	↑
P	800	↓	↓	↓	0 1	1 2	2 3	3 4	5 6	8 9	12 13	18 19	27 28	41 42	↑	↑	↑	↑	↑	↑	↑	↑	↑	↑	↑	↑	↑
Q	1250	↓	↓	0 1	1 2	2 3	3 4	5 6	8 9	12 13	18 19	27 28	41 42	↑	↑	↑	↑	↑	↑	↑	↑	↑	↑	↑	↑	↑	↑
R	2000	↓	0 1	1 2	2 3	3 4	5 6	8 9	12 13	18 19	27 28	41 42	↑	↑	↑	↑	↑	↑	↑	↑	↑	↑	↑	↑	↑	↑	↑
S	3150	0 1	1 2	2 3	3 4	5 6	8 9	12 13	18 19	27 28	41 42	↑	↑	↑	↑	↑	↑	↑	↑	↑	↑	↑	↑	↑	↑	↑	↑

↓ = Use first sampling plan below arrow. If sample size equals or exceeds lot or batch size, do 100 percent inspection.
↑ = Use first sampling plan above arrow.

Ac = Acceptance number.
Re = Rejection number.

372

TABLE 12.3C
Single Sampling Plans for Reduced Inspection (Table II-C of MIL-STD-105D)

Acceptable Quality Levels (reduced inspection)†

Each cell shows the acceptance number (Ac) and rejection number (Re); ↓ = use first sampling plan below arrow; ↑ = use first sampling plan above arrow.

Code	Sample size	0.010	0.015	0.025	0.040	0.065	0.10	0.15	0.25	0.40	0.65	1.0	1.5	2.5	4.0	6.5	10	15	25	40	65	100	150	250	400	650	1000
A	2	↓	↓	↓	↓	↓	↓	↓	↓	↓	↓	↓	↓	↓	↓	↓	↓	↓	1 2	2 3	3 4	5 6	7 8	10 11	14 15	21 22	30 31
B	2	↓	↓	↓	↓	↓	↓	↓	↓	↓	↓	↓	↓	↓	↓	↓	↓	0 2	1 3	2 4	3 5	5 6	7 8	10 11	14 15	21 22	30 31
C	2	↓	↓	↓	↓	↓	↓	↓	↓	↓	↓	↓	↓	↓	↓	↓	0 1	0 2	1 4	2 5	3 6	5 8	7 10	10 13	14 17	21 24	↑
D	3	↓	↓	↓	↓	↓	↓	↓	↓	↓	↓	↓	↓	↓	↓	0 1	0 2	1 4	2 5	3 6	5 8	7 10	10 13	14 17	21 24	↑	↑
E	5	↓	↓	↓	↓	↓	↓	↓	↓	↓	↓	↓	↓	↓	0 1	0 2	1 4	2 5	3 6	5 8	7 10	10 13	14 17	21 24	↑	↑	↑
F	8	↓	↓	↓	↓	↓	↓	↓	↓	↓	↓	↓	↓	0 1	0 2	1 4	2 5	3 6	5 8	7 10	10 13	14 17	21 24	↑	↑	↑	↑
G	13	↓	↓	↓	↓	↓	↓	↓	↓	↓	↓	↓	0 1	0 2	1 4	2 5	3 6	5 8	7 10	10 13	14 17	21 24	↑	↑	↑	↑	↑
H	20	↓	↓	↓	↓	↓	↓	↓	↓	↓	↓	0 1	0 2	1 4	2 5	3 6	5 8	7 10	10 13	14 17	21 24	↑	↑	↑	↑	↑	↑
J	32	↓	↓	↓	↓	↓	↓	↓	↓	↓	0 1	0 2	1 4	2 5	3 6	5 8	7 10	10 13	14 17	21 24	↑	↑	↑	↑	↑	↑	↑
K	50	↓	↓	↓	↓	↓	↓	↓	↓	0 1	0 2	1 4	2 5	3 6	5 8	7 10	10 13	14 17	21 24	↑	↑	↑	↑	↑	↑	↑	↑
L	80	↓	↓	↓	↓	↓	↓	↓	0 1	0 2	1 4	2 5	3 6	5 8	7 10	10 13	14 17	21 24	↑	↑	↑	↑	↑	↑	↑	↑	↑
M	125	↓	↓	↓	↓	↓	↓	0 1	0 2	1 4	2 5	3 6	5 8	7 10	10 13	14 17	21 24	↑	↑	↑	↑	↑	↑	↑	↑	↑	↑
N	200	↓	↓	↓	↓	↓	0 1	0 2	1 4	2 5	3 6	5 8	7 10	10 13	14 17	21 24	↑	↑	↑	↑	↑	↑	↑	↑	↑	↑	↑
P	315	↓	↓	↓	↓	0 1	0 2	1 4	2 5	3 6	5 8	7 10	10 13	14 17	21 24	↑	↑	↑	↑	↑	↑	↑	↑	↑	↑	↑	↑
Q	500	↓	↓	↓	0 1	0 2	1 4	2 5	3 6	5 8	7 10	10 13	14 17	21 24	↑	↑	↑	↑	↑	↑	↑	↑	↑	↑	↑	↑	↑
R	800	↓	↓	0 1	0 2	1 4	2 5	3 6	5 8	7 10	10 13	14 17	21 24	↑	↑	↑	↑	↑	↑	↑	↑	↑	↑	↑	↑	↑	↑

⇕ = Use first sampling plan below arrow. If sample size equals or exceeds lot or batch size, do 100 percent inspection.
↑ = Use first sampling plan above arrow.
Ac = Acceptance number.
Re = Rejection number.
† = If the acceptance number has been exceeded, but the rejection number has not been reached, accept the lot, but reinstate normal inspection.

373

TABLE 12.4A
Double Sampling Plans for Normal Inspection (Table III-A of MIL-STD-105D)

Acceptable Quality Levels (normal inspection) — values shown as Ac Re.

Sample size code letter	Sample	Sample size	Cumulative sample size	0.010	0.015	0.025	0.040	0.065	0.10	0.15	0.25	0.40	0.65	1.0	1.5	2.5	4.0	6.5	10	15	25	40	65	100	150	250	400	650	1000
A				↓	↓	↓	↓	↓	↓	↓	↓	↓	↓	↓	↓	↓	↓	↓	↓	↓	↓	↓	↓	↓	↓	↓	↓	↓	↓
B	First	2	2	↓	↓	↓	↓	↓	↓	↓	↓	↓	↓	↓	↓	↓	↓	↓	0 2	*	*	*	*	*	*	*	*	*	*
B	Second	2	4																1 2										
C	First	3	3	↓	↓	↓	↓	↓	↓	↓	↓	↓	↓	↓	↓	↓	↓	0 2	0 3	*	*	*	*	*	*	*	*	*	*
C	Second	3	6															1 2	3 4										
D	First	5	5	↓	↓	↓	↓	↓	↓	↓	↓	↓	↓	↓	↓	↓	0 2	0 3	1 4	2 5	*	*	*	*	*	*	*	*	*
D	Second	5	10														1 2	3 4	4 5	6 7									
E	First	8	8	↓	↓	↓	↓	↓	↓	↓	↓	↓	↓	↓	↓	0 2	0 3	1 4	2 5	3 7	*	*	*	*	*	*	*	*	*
E	Second	8	16													1 2	3 4	4 5	6 7	8 9									
F	First	13	13	↓	↓	↓	↓	↓	↓	↓	↓	↓	↓	↓	0 2	0 3	1 4	2 5	3 7	5 9	7 11	*	*	*	*	*	*	*	*
F	Second	13	26												1 2	3 4	4 5	6 7	8 9	12 13	18 19								
G	First	20	20	↓	↓	↓	↓	↓	↓	↓	↓	↓	↓	0 2	0 3	1 4	2 5	3 7	5 9	7 11	11 16	*	*	*	*	*	*	*	*
G	Second	20	40											1 2	3 4	4 5	6 7	8 9	12 13	18 19	26 27								
H	First	32	32	↓	↓	↓	↓	↓	↓	↓	↓	↓	0 2	0 3	1 4	2 5	3 7	5 9	7 11	11 16	17 22	25 31	↑	↑	↑	↑	↑	↑	↑
H	Second	32	64										1 2	3 4	4 5	6 7	8 9	12 13	18 19	26 27	37 38	56 57							
J	First	50	50	↓	↓	↓	↓	↓	↓	↓	↓	0 2	0 3	1 4	2 5	3 7	5 9	7 11	11 16	17 22	25 31	↑	↑	↑	↑	↑	↑	↑	↑
J	Second	50	100									1 2	3 4	4 5	6 7	8 9	12 13	18 19	26 27	37 38	56 57								
K	First	80	80	↓	↓	↓	↓	↓	↓	↓	0 2	0 3	1 4	2 5	3 7	5 9	7 11	11 16	17 22	25 31	↑	↑	↑	↑	↑	↑	↑	↑	↑
K	Second	80	160								1 2	3 4	4 5	6 7	8 9	12 13	18 19	26 27	37 38	56 57									
L	First	125	125	↓	↓	↓	↓	↓	↓	0 2	0 3	1 4	2 5	3 7	5 9	7 11	11 16	17 22	25 31	↑	↑	↑	↑	↑	↑	↑	↑	↑	↑
L	Second	125	250							1 2	3 4	4 5	6 7	8 9	12 13	18 19	26 27	37 38	56 57										
M	First	200	200	↓	↓	↓	↓	↓	0 2	0 3	1 4	2 5	3 7	5 9	7 11	11 16	17 22	25 31	↑	↑	↑	↑	↑	↑	↑	↑	↑	↑	↑
M	Second	200	400						1 2	3 4	4 5	6 7	8 9	12 13	18 19	26 27	37 38	56 57											
N	First	315	315	↓	↓	↓	↓	0 2	0 3	1 4	2 5	3 7	5 9	7 11	11 16	17 22	25 31	↑	↑	↑	↑	↑	↑	↑	↑	↑	↑	↑	↑
N	Second	315	630					1 2	3 4	4 5	6 7	8 9	12 13	18 19	26 27	37 38	56 57												
P	First	500	500	↓	↓	↓	0 2	0 3	1 4	2 5	3 7	5 9	7 11	11 16	17 22	25 31	↑	↑	↑	↑	↑	↑	↑	↑	↑	↑	↑	↑	↑
P	Second	500	1000				1 2	3 4	4 5	6 7	8 9	12 13	18 19	26 27	37 38	56 57													
Q	First	800	800	↓	↓	0 2	0 3	1 4	2 5	3 7	5 9	7 11	11 16	17 22	25 31	↑	↑	↑	↑	↑	↑	↑	↑	↑	↑	↑	↑	↑	↑
Q	Second	800	1600			1 2	3 4	4 5	6 7	8 9	12 13	18 19	26 27	37 38	56 57														
R	First	1250	1250	↓	0 2	0 3	1 4	2 5	3 7	5 9	7 11	11 16	17 22	25 31	↑	↑	↑	↑	↑	↑	↑	↑	↑	↑	↑	↑	↑	↑	↑
R	Second	1250	2500		1 2	3 4	4 5	6 7	8 9	12 13	18 19	26 27	37 38	56 57															

↓ = Use first sampling plan below arrow. If sample size equals or exceeds lot or batch size do 100 percent inspection.

↑ = Use first sampling plan above arrow.

Ac = Acceptance number.

Re = Rejection number.

* = Use corresponding single sampling plan (or alternatively, use double sampling plan below, where available).

TABLE 12.4B
Double-Sampling Plans for Tightened Inspection (MIL-STD-105D)

Acceptable Quality Levels (tightened inspection)

Sample size code letter	Sample	Sample size	Cumulative sample size	0.010		0.015		0.025		0.040		0.065		0.10		0.15		0.25		0.40		0.65		1.0		1.5		2.5		4.0		6.5		10		15		25		40		65		100		150		250		400		650		1000	
				Ac	Re	Ac	Re	Ac	Re	Ac	Re	Ac	Re	Ac	Re	Ac	Re	Ac	Re	Ac	Re	Ac	Re	Ac	Re	Ac	Re	Ac	Re	Ac	Re	Ac	Re	Ac	Re	Ac	Re	Ac	Re	Ac	Re	Ac	Re	Ac	Re	Ac	Re	Ac	Re	Ac	Re	Ac	Re
A																																																							
B	First	2	2																																				0	2	0	3	1	4	2	5	6	10	9	14	15	20	20	35	23	24	·	
	Second	2	4																																				1	2	3	4	4	5	6	7	15	16	23	24	34	35	52	53	52	53		
C	First	3	3																																		0	2	0	3	1	4	2	5	3	7	6	10	9	14	15	20	23	24				
	Second	3	6																																		1	2	3	4	4	5	6	7	11	12	15	16	23	24	34	35	52	53				
D	First	5	5																																0	2	0	3	1	4	2	5	3	7	6	10	9	14	15	20	23	24						
	Second	5	10																																1	2	3	4	4	5	6	7	11	12	15	16	23	24	34	35	52	53						
E	First	8	8																														0	2	0	3	1	4	2	5	3	7	6	10	9	14	15	20	23	24								
	Second	8	16																														1	2	3	4	4	5	6	7	11	12	15	16	23	24	34	35	52	53								
F	First	13	13																												0	2	0	3	1	4	2	5	3	7	6	10	9	14	15	20	23	24										
	Second	13	26																												1	2	3	4	4	5	6	7	11	12	15	16	23	24	34	35	52	53										
G	First	20	20																										0	2	0	3	1	4	2	5	3	7	6	10	9	14	15	20	23	24												
	Second	20	40																										1	2	3	4	4	5	6	7	11	12	15	16	23	24	34	35	52	53												
H	First	32	32																								0	2	0	3	1	4	2	5	3	7	6	10	9	14	15	20	23	24														
	Second	32	64																								1	2	3	4	4	5	6	7	11	12	15	16	23	24	34	35	52	53														
J	First	50	50																						0	2	0	3	1	4	2	5	3	7	6	10	9	14	15	20	23	24																
	Second	50	100																						1	2	3	4	4	5	6	7	11	12	15	16	23	24	34	35	52	53																
K	First	80	80																				0	2	0	3	1	4	2	5	3	7	6	10	9	14	15	20	23	24																		
	Second	80	160																				1	2	3	4	4	5	6	7	11	12	15	16	23	24	34	35	52	53																		
L	First	125	125																		0	2	0	3	1	4	2	5	3	7	6	10	9	14	15	20	23	24																				
	Second	125	250																		1	2	3	4	4	5	6	7	11	12	15	16	23	24	34	35	52	53																				
M	First	200	200															0	2	0	3	1	4	2	5	3	7	6	10	9	14	15	20	23	24																						
	Second	200	400															1	2	3	4	4	5	6	7	11	12	15	16	23	24	34	35	52	53																						
N	First	315	315													0	2	0	3	1	4	2	5	3	7	6	10	9	14	15	20	23	24																								
	Second	315	630													1	2	3	4	4	5	6	7	11	12	15	16	23	24	34	35	52	53																								
P	First	500	500											0	2	0	3	1	4	2	5	3	7	6	10	9	14	15	20	23	24																										
	Second	500	1000											1	2	3	4	4	5	6	7	11	12	15	16	23	24	34	35	52	53																										
Q	First	800	800								0	2	0	3	1	4	2	5	3	7	6	10	9	14	15	20	23	24																														
	Second	800	1600								1	2	3	4	4	5	6	7	11	12	15	16	23	24	34	35	52	53																														
R	First	1250	1250							0	2	0	3	1	4	2	5	3	7	6	10	9	14	15	20	23	24																																
	Second	1250	2500							1	2	3	4	4	5	6	7	11	12	15	16	23	24	34	35	52	53																																
S	First	2000	2000			0	2																																																		
	Second	2000	4000			1	2																																																			

⇩ = Use first sampling plan below arrow. If sample size equals or exceeds lot or batch size, do 100 percent inspection.

⇧ = Use first sampling plan above arrow.

Ac = Acceptance number.

Re = Rejection number.

· = Use corresponding single sampling plan (or alternatively, use double sampling plan below, where available).

375

TABLE 12.4C
Double-Sampling Plans for Reduced Inspection (MIL-STD-105D)

Each data cell shows **Ac Re** (Acceptance number / Rejection number). ↓ = use first sampling plan below arrow; ↑ = use first sampling plan above arrow. Arrow-region cells are marked ↓ or ↑.

Sample size code letter	Sample	Sample size	Cumulative sample size	0.010	0.015	0.025	0.040	0.065	0.10	0.15	0.25	0.40	0.65	1.0	1.5	2.5	4.0	6.5	10	15	25	40	65	100	150	250	400	650	1000
A				·	·	·	·	·	·	·	·	·	·	·	·	·	·	·	↓										
B				↓																									
C				↓																									
D	First	2	2	↓	↓	↓	↓	↓	↓	↓	↓	↓	↓	↓	↓	↓	↓	↓	0 2	0 3	0 4	0 4	1 5	1 6	2 7	3 8	5 10	7 12	11 17
D	Second	2	4																0 2	0 4	1 5	3 6	4 7	4 8	6 9	8 12	12 16	18 22	26 30
E	First	3	3	↓	↓	↓	↓	↓	↓	↓	↓	↓	↓	↓	↓	↓	↓	0 2	0 3	0 4	0 4	1 5	1 6	2 7	3 8	5 10	7 12	11 17	↑
E	Second	3	6															0 2	0 4	1 5	3 6	4 7	4 8	6 9	8 12	12 16	18 22	26 30	
F	First	5	5	↓	↓	↓	↓	↓	↓	↓	↓	↓	↓	↓	↓	↓	0 2	0 3	0 4	0 4	1 5	1 6	2 7	3 8	5 10	7 12	11 17	↑	↑
F	Second	5	10														0 2	0 4	1 5	3 6	4 7	4 8	6 9	8 12	12 16	18 22	26 30		
G	First	8	8	↓	↓	↓	↓	↓	↓	↓	↓	↓	↓	↓	↓	0 2	0 3	0 4	0 4	1 5	1 6	2 7	3 8	5 10	7 12	11 17	↑	↑	↑
G	Second	8	16													0 2	0 4	1 5	3 6	4 7	4 8	6 9	8 12	12 16	18 22	26 30			
H	First	13	13	↓	↓	↓	↓	↓	↓	↓	↓	↓	↓	↓	0 2	0 3	0 4	0 4	1 5	1 6	2 7	3 8	5 10	7 12	11 17	↑	↑	↑	↑
H	Second	13	26												0 2	0 4	1 5	3 6	4 7	4 8	6 9	8 12	12 16	18 22	26 30				
J	First	20	20	↓	↓	↓	↓	↓	↓	↓	↓	↓	↓	0 2	0 3	0 4	0 4	1 5	1 6	2 7	3 8	5 10	7 12	11 17	↑	↑	↑	↑	↑
J	Second	20	40											0 2	0 4	1 5	3 6	4 7	4 8	6 9	8 12	12 16	18 22	26 30					
K	First	32	32	↓	↓	↓	↓	↓	↓	↓	↓	↓	0 2	0 3	0 4	0 4	1 5	1 6	2 7	3 8	5 10	7 12	11 17	↑	↑	↑	↑	↑	↑
K	Second	32	64										0 2	0 4	1 5	3 6	4 7	4 8	6 9	8 12	12 16	18 22	26 30						
L	First	50	50	↓	↓	↓	↓	↓	↓	↓	↓	0 2	0 3	0 4	0 4	1 5	1 6	2 7	3 8	5 10	7 12	11 17	↑	↑	↑	↑	↑	↑	↑
L	Second	50	100									0 2	0 4	1 5	3 6	4 7	4 8	6 9	8 12	12 16	18 22	26 30							
M	First	80	80	↓	↓	↓	↓	↓	↓	↓	0 2	0 3	0 4	0 4	1 5	1 6	2 7	3 8	5 10	7 12	11 17	↑	↑	↑	↑	↑	↑	↑	↑
M	Second	80	160								0 2	0 4	1 5	3 6	4 7	4 8	6 9	8 12	12 16	18 22	26 30								
N	First	125	125	↓	↓	↓	↓	↓	↓	0 2	0 3	0 4	0 4	1 5	1 6	2 7	3 8	5 10	7 12	11 17	↑	↑	↑	↑	↑	↑	↑	↑	↑
N	Second	125	250							0 2	0 4	1 5	3 6	4 7	4 8	6 9	8 12	12 16	18 22	26 30									
P	First	200	200	↓	↓	↓	↓	↓	0 2	0 3	0 4	0 4	1 5	1 6	2 7	3 8	5 10	7 12	11 17	↑	↑	↑	↑	↑	↑	↑	↑	↑	↑
P	Second	200	400						0 2	0 4	1 5	3 6	4 7	4 8	6 9	8 12	12 16	18 22	26 30										
Q	First	315	315	↓	↓	↓	↓	0 2	0 3	0 4	0 4	1 5	1 6	2 7	3 8	5 10	7 12	11 17	↑	↑	↑	↑	↑	↑	↑	↑	↑	↑	↑
Q	Second	315	630					0 2	0 4	1 5	3 6	4 7	4 8	6 9	8 12	12 16	18 22	26 30											
R	First	500	500	↓	↓	↓	0 2	0 3	0 4	0 4	1 5	1 6	2 7	3 8	5 10	7 12	11 17	↑	↑	↑	↑	↑	↑	↑	↑	↑	↑	↑	↑
R	Second	500	1000				0 2	0 4	1 5	3 6	4 7	4 8	6 9	8 12	12 16	18 22	26 30												

↓ = Use first sampling plan below arrow. If sample size equals or exceeds lot or batch size, do 100 percent inspection.

↑ = Use first sampling plan above arrow.

Ac = Acceptance number

Re = Rejection number

* = Use corresponding single sampling plan (or alternatively, use double sampling plan below, when available).

† = If, after the second sample, the acceptance number has been exceeded, but the rejection number has not been reached, accept the lot, but reinstate normal inspection (see 10.14).

376

TABLE 12.5
Multiple-Sampling Plans for Normal Inspection (MIL-STD-105D)

Acceptance Quality Levels (normal inspection)

Sample size code letter	Sample	Sample size	Cumulative sample size
A			
B			
C			
D	First	2	2
	Second	2	4
	Third	2	6
	Fourth	2	8
	Fifth	2	10
	Sixth	2	12
	Seventh	2	14
E	First	3	3
	Second	3	6
	Third	3	9
	Fourth	3	12
	Fifth	3	15
	Sixth	3	18
	Seventh	3	21
F	First	5	5
	Second	5	10
	Third	5	15
	Fourth	5	20
	Fifth	5	25
	Sixth	5	30
	Seventh	5	35
G	First	8	8
	Second	8	16
	Third	8	24
	Fourth	8	32
	Fifth	8	40
	Sixth	8	48
	Seventh	8	56
H	First	13	13
	Second	13	26
	Third	13	39
	Fourth	13	52
	Fifth	13	65
	Sixth	13	78
	Seventh	13	91
J	First	20	20
	Second	20	40
	Third	20	60
	Fourth	20	80
	Fifth	20	100
	Sixth	20	120
	Seventh	20	140

AQL columns (each with Ac = Acceptance number, Re = Rejection number): 0.010, 0.015, 0.025, 0.040, 0.065, 0.10, 0.15, 0.25, 0.40, 0.65, 1.0, 1.5, 2.5, 4.0, 6.5, 10, 15, 25, 40, 65, 100, 150, 250, 400, 650, 1000

↓ = Use first sampling plan below arrow (refer to continuation of table on following page, when necessary). If sample size equals or exceeds lot or batch size, do 100 percent inspection.

↑ = Use first sampling plan above arrow.

Ac = Acceptance number.

Re = Rejection number.

⇊ = Use corresponding single sampling plan (or alternatively, use multiple sampling plan below, where available).

⇈ = Use corresponding double sampling plan (or alternatively, use multiple sampling plan below, where available).

† = Use corresponding single sampling plan (or alternatively, use multiple sampling plan below, where available).

• = Acceptance not permitted at this sample size.

377

with their tabulated values (for single, double, and multiple schemes) are provided, the average sample size curves vary considerably in different plans. Therefore, MIL-STD-105D also provides tables for double and multiple sampling plans similar to the single sampling plan discussed above (Tables 12.4A, B, and C, and Table 12.5).

In double sampling plans, two sample sizes per sample-size code letter are given, with the corresponding acceptance and rejection numbers. Use of these tables is similar to the technique described under single sampling plans. Inspection starts with the first sample size for the particular code letter and the decision is made applicable to accept/reject criteria found in the table. If, however, the number of defectives found in the first sample falls between the first acceptance number and the rejection number, a second sample of the size given in the plan is inspected and a decision is made. The sample comparisons among normal, tightened, and reduced inspection that were given earlier for single sampling plans are applicable to double sampling plans. Similarly, a change in the sample size code letter and the sample size itself whenever an arrow is encountered has the same interpretation. Where an asterisk (*) is used in double sampling, it means that the corresponding single sampling plan is applicable because the relationship between the acceptance number and the sample size is unrealistic.

The multiple sampling plans provide for up to seven sample sizes which may sometimes be required to reach an accept/reject decision. Because of the similarity in the technique used, a detailed description of the use of these tables is not provided. Information for normal inspection using multiple sampling plans can be found in Table 12.5. The reader is advised to consult the standards for more details and explanations.

Lot Tolerance Percent Defective Plans

The lot tolerance percent defective (LPTD) plans governing the acceptance or rejection of the lots are based on the aspect of consumers' risk. When it is desirable to protect the consumers, one would be interested in selecting OC curves that are fairly similar at the tail portion (see e.g., Figure 12.9), signifying a low probability of acceptance. To engineers and

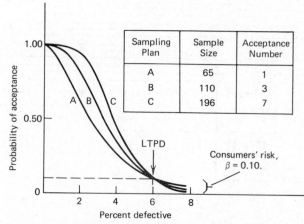

FIGURE 12.9   OC curves for some frequently used sampling plans showing similarity at the tail portions for a typical LTPD of 6%.

receiving inspectors, this means the percent defective that will regularly be rejected; that is, the percent defective for which the probability of acceptance would be very low.

Lot tolerance percent defective is defined in the Dodge-Romig Tables as an allowable percent defective, a figure which may be considered as the borderline of distinction between a satisfactory lot and an unsatisfactory one. The MIL-STD-105D standard contains LTPD plans with 5% and 10% acceptance criteria. The term *limiting quality* (LQ), which is used in the standard, is synonymous with LTPD. Commonly, the LTPD is the value specified for a product quality that has a 10% chance of acceptance, (consumers' risk, $\beta = 0.10$). The technique for determining LTPD values or choosing a sampling plan based on the given LTPD (in percent defective) with the help of the tables of MIL-STD-105D is similar to that described in connection with AQL plans.

Average Outgoing Quality Limit Plans

The sampling plans based on average outgoing quality limit (AOQL) values ensure that regardless of the percent defective of the product submitted for inspection, the quality of the outgoing product will never exceed the specified AOQL value on the average. The AOQL sampling plans can therefore be defined as the worst average quality that can exist, in the long run, in the accepted products, after all the rejected lots have been 100% inspected and all the defective items have been replaced by nondefective items.

Earlier, when considering AQL and LTPD plans, we looked into the similarity of OC curves either at the top portions or at the bottom tails. Suppose we wish to group together OC curves which are neither alike at the top shoulder portion nor anywhere similar at the tail. Such a plan would be a mixture of OC curves taken from Figures 12.8 and 12.9. A common feature of these curves is that when they are used with 100% inspection of all the rejected lots, the outgoing products will not, on the average, be worse than their AOQLs. Therefore, when a plan is specified for an AOQL, one has to select an OC curve with an AOQL equal to the specified value.

However, the difficulty remains, since a large number of sampling plans (with quite different n and Ac values) can be given the same AOQL value when used for one lot size, but reject very different amounts of defective products, as shown in Table 12.6 (compiled from the Dodge-Romig 2% AOQL Single Sampling Table).[8]

To make it easier to select the right sampling plan, Dodge and Romig have created tables of sampling plans that are classified according to their AOQLs. There are also tables for

TABLE 12.6
Lot size, $N = 1000$

2% AOQL Sampling Plan		Percent of Product rejected (approx.)	Necessary Level of Quality (%) to Assure Regular Acceptance about 98% of the Time
n	Ac		
90	3	11	1.1
65	2	14	0.9
40	1	19	0.5
18	0	30	0.1

sampling plans classified according to LTPDs. Both of these classifications will now be discussed.

Dodge–Romig Tables

The inspection tables developed by Dodge and Romig take into account both the LTPD and AOQL sampling plans. There are two sets of tables for each of these plans: single sampling and double sampling. No provision is made for multiple sampling. The plans differ from those in MIL-STD-105D in that they assume 100% inspection of all rejected lots with defectives replaced by acceptable items.

LTPD Tables. Lot tolerance percent defective (LTPD) tables are based on the probability of acceptance of a particular lot that has a percent defective equal to the LTPD. These plans are thus associated with consumers' risk, β. Table 12.7 shows a typical Dodge–Romig Table for a single sampling plan, when LTPD = 1.0%, (β = 0.10 or 10%). Other tables have LTPD values ranging from 0.5 to 10.0% defective.

TABLE 12.7
Dodge–Romig Single Sampling Lot Inspection Table, Based on Lot Tolerance Percent Defective[a]
LTPD = 1.0%

Process Average (%)	0–0.010			0.011–0.10			0.11–0.20			0.21–0.30			0.31–0.40			0.41–0.50		
			AOQL			AOQL			AOQL			AOQL			AOQL			AOQL
Lot Size	n	c	(%)	n	c	(%)	n	c	(%)	n	c	(%)	n	c	(%)	n	c	(%)
1–120	All	0	0	All	0	0	All	0	0	All	0	0	All	0	0	All	0	0
121–150	120	0	0.06	120	0	0.06	120	0	0.06	120	0	0.06	120	0	0.06	120	0	0.06
151–200	140	0	0.08	140	0	0.08	140	0	0.08	140	0	0.08	140	0	0.08	140	0	0.08
201–300	165	0	0.10	165	0	0.10	165	0	0.10	165	0	0.10	165	0	0.10	165	0	0.10
301–400	175	0	0.12	175	0	0.12	175	0	0.12	175	0	0.12	175	0	0.12	175	0	0.12
401–500	180	0	0.13	180	0	0.13	180	0	0.13	180	0	0.13	180	0	0.13	180	0	0.13
501–600	190	0	0.13	190	0	0.13	190	0	0.13	190	0	0.13	190	0	0.13	305	1	0.14
601–800	200	0	0.14	200	0	0.14	200	0	0.14	330	1	0.15	330	1	0.15	330	1	0.15
801–1,000	205	0	0.14	205	0	0.14	205	0	0.14	335	1	0.17	335	1	0.17	335	1	0.17
1,001–2,000	220	0	0.15	220	0	0.15	360	1	0.19	490	2	0.21	490	2	0.21	610	3	0.22
2,001–3,000	220	0	0.15	375	1	0.20	505	2	0.23	630	3	0.24	745	4	0.26	870	5	0.26
3,001–4,000	225	0	0.15	380	1	0.20	510	2	0.24	645	3	0.25	880	5	0.28	1,000	6	0.29
4,001–5,000	225	0	0.16	380	1	0.20	520	2	0.24	770	4	0.28	895	5	0.29	1,120	7	0.31
5,001–7,000	230	0	0.16	385	1	0.21	655	3	0.27	780	4	0.29	1,020	6	0.32	1,260	8	0.34
7,001–10,000	230	0	0.16	520	2	0.25	660	3	0.28	910	5	0.32	1,150	7	0.34	1,500	10	0.37
10,001–20,000	390	1	0.21	525	2	0.26	785	4	0.31	1,040	6	0.35	1,400	9	0.39	1,980	14	0.43
20,001–50,000	390	1	0.21	530	2	0.26	920	5	0.34	1,300	8	0.39	1,890	13	0.44	2,570	19	0.48
50,001–100,000	390	1	0.21	670	3	0.29	1,040	6	0.36	1,420	9	0.41	2,120	15	0.47	3,150	23	0.50

[a]n, size of sample; entry of "All" indicates that each piece in lot is to be inspected. c, allowable defect number for sample. AOQL, average outgoing quality limit.

Source: Reproduced by permission from H. F. Dodge and H. G. Romig, *Sampling Inspection Tables—Single and Double Sampling,* 2nd ed., New York: John Wiley & Sons, Inc., 1959, and AT&T Bell Laboratories.

The principal advantage of these tables is their economy. If the quality manager selects a plan under the correct process average (the figures indicated on the first top row of the table), he will minimize the total number of pieces that must be looked at (thus reducing costs) both for sampling and for 100% inspection. At the same time while setting a fixed maximum AOQL, he can achieve budgeting for the combination of sampling and 100% inspection.

To use the Dodge–Romig Table (e.g., Table 12.7), one must decide first whether it will be used for single sampling or double sampling. Next, the usual lot size and the *process average* are determined. The process average figure given in the table is the percent defective at which the product normally runs. The sample size is given in column n under the applicable process average, and the acceptance number (denoted by c instead of Ac) as given in column c. Notice that in addition to a stated AOQL, the plans also have a particular value of LTPD. The word ''All'' indicates that each piece of the lot is to be inspected. For example, if $N = 1001$ and process average $= 0.22$, the required single sample plan for LTPD of 1.0% is found from Table 12.7 as: $n = 490$, Ac (or c) $= 2$, AOQL $= 0.21\%$. Any other plan for a given LTPD, using either single sampling or double sampling, can be obtained in a similar manner.

Three important economical features of the Dodge-Romig tabulations are to be noted.

1. The process average has been tabulated only up to a maximum number that is one-half of the LTPD. An extension beyond this is unnecessary since, from a cost point of view, 100% inspection control then would be more economical.

2. The amount inspected increases with the increase in process averages. This means that an improvement in the process results in fewer inspections and hence lower costs.

3. The inspection costs are increased by choosing larger lot sizes, since an increase in lot size decreases the relative sample size.

AOQL Tables: The AOQL tables were developed to be used for assurance that after all sampling and 100% inspection of rejected lots, the average quality over many lots will not exceed the AOQL. The AOQL plans thus limit the amount of poor outgoing quality on an average basis, but give no assurance on individual lots. The AOQL values given in the tables range from 0.1 to 10.0. Taking single and double sampling together, there are a total of 26 tables in all. A typical Dodge-Romig Table for single sampling is Table 12.8 for an AOQL of 3.0%.

The technique for the use of Dodge-Romig AOQL Tables is similar to that explained in the case of LTPD tables.

12.8 ACCEPTANCE SAMPLING PLANS BY VARIABLES

It is fairly common to find situations where a product's acceptance depends on one or two isolated quality characteristics, for example, the excitation current in transformers, the amount of sulphur and phosphorous present in specialty steels, the fat content in high-grade meat, and so forth. All of these have to do with isolated problems where the operators in the inspection areas are encouraged to be more careful in a given matter, or the engineering team is asked to devote more effort towards a particular difficulty having to do with engineering specifications or process control parameters.

TABLE 12.8
Dodge–Romig Single Sampling Lot Inspection Table, Based on Average Outgoing Quality Limit[a]
AOQL = 3.0%

Process Average (%)	0–0.06			0.07–0.60			0.61–1.20			1.21–1.80			1.81–2.40			2.41–3.00		
			LTPD			LTPD			LTPD			LTPD			LTPD			LTPD
Lot Size	n	c	(%)	n	c	(%)	n	c	(%)	n	c	(%)	n	c	(%)	n	c	(%)
1–10	All	0	—	All	0	—	All	0	—	All	0	—	All	0	—	All	0	—
11–50	10	0	19.0	10	0	19.0	10	0	19.0	10	0	19.0	10	0	19.0	10	0	19.0
51–100	11	0	18.0	11	0	18.0	11	0	18.0	11	0	18.0	11	0	18.0	22	1	16.4
101–200	12	0	17.0	12	0	17.0	12	0	17.0	25	1	15.1	25	1	15.1	25	1	15.1
201–300	12	0	17.0	12	0	17.0	26	1	14.6	26	1	14.6	26	1	14.6	40	2	12.8
301–400	12	0	17.1	12	0	17.1	26	1	14.7	26	1	14.7	41	2	12.7	41	2	12.7
401–500	12	0	17.2	27	1	14.1	27	1	14.1	42	2	12.4	42	2	12.4	42	2	12.4
501–600	12	0	17.3	27	1	14.2	27	1	14.2	42	2	12.4	42	2	12.4	60	3	10.8
601–800	12	0	17.3	27	1	14.2	27	1	14.2	43	2	12.1	60	3	10.9	60	3	10.9
801–1,000	12	0	17.4	27	1	14.2	44	2	11.8	44	2	11.8	60	3	11.0	80	4	9.8
1,001–2,000	12	0	17.5	28	1	13.8	45	2	11.7	65	3	10.2	80	4	9.8	100	5	9.1
2,001–3,000	12	0	17.5	28	1	13.8	45	2	11.7	65	3	10.2	100	5	9.1	140	7	8.2
3,001–4,000	12	0	17.5	28	1	13.8	65	3	10.3	85	4	9.5	125	6	8.4	165	8	7.8
4,001–5,000	28	1	13.8	28	1	13.8	65	3	10.3	85	4	9.5	125	6	8.4	210	10	7.4
5,001–7,000	28	1	13.8	45	2	11.8	65	3	10.3	105	5	8.8	145	7	8.1	235	11	7.1
7,001–10,000	28	1	13.9	46	2	11.6	65	3	10.3	105	5	8.8	170	8	7.6	280	13	6.8
10,001–20,000	28	1	13.9	46	2	11.7	85	4	9.5	125	6	8.4	215	10	7.2	380	17	6.2
20,001–50,000	28	1	13.9	65	3	10.3	105	5	8.8	170	8	7.6	310	14	6.5	560	24	5.7
50,001–100,000	28	1	13.9	65	3	10.3	125	6	8.4	215	10	7.2	385	17	6.2	690	29	5.4

[a]n, size of sample; entry of "All" indicates that each piece in lot is to be inspected. c, allowable defect number for sample. LTPD, lot tolerance percent defective corresponding to a consumer's risk (β) = 0.10.

Source: Reproduced by permission from H. F. Dodge and H. G. Romig, *Sampling Inspection Tables—Single and Double Sampling*, 2nd ed., New York: John Wiley & Sons, Inc., 1959, and AT&T Bell Laboratories.

The sampling plans based on variables apply to a single quality characteristic for which continuous measurements can be made. In the production of steel, for example, the percentage of deleterious elements present in the molten metal bath is regularly checked with the help of a precise instrument and records are kept on the individual measurements. For any heat poured out of the furnace, as few as two or three readings suffice to arrive at an accept or reject decision. This information is of great help in assessing how good or bad the mechanical properties of the steel are going to be. In such similar situations where the characteristic inspected can be judged on a variable scale, the variable sampling plans may be used. Of course, the entire focus here is on one specified critical characteristic which must be monitored for quality protection.

MIL-STD-414

The most commonly used variable sampling plan is the one provided by MIL-STD-414[9] *Sampling Procedures and Tables for Inspection by Variables for Percent Defective.*

The format and terminology of this standard are very similar to those in MIL-STD-105D. Tables for sample-size code letters, OC curves, levels of inspection, AQLs and lot acceptability criteria are all provided with sufficient description. The basic approach used in MIL-STD-414 is that it assumes normal distribution for each of the independent variable measurements involved in the acceptance decisions.

Although the standard makes provision for nine different procedures that can be used to evaluate a lot for acceptance or rejection, sampling plans based on unknown variability are the ones used most frequently. There are two alternative methods given for sampling plans based on unknown variability: the standard deviation method given in section B and the range method, given in section C. Procedures and examples for known variability are given in section D of the standard.

The sample-size code letters are based on the lot size and the inspection level as shown in Table 12.9. Of the five inspection levels shown, unless otherwise specified, inspection level IV is most commonly used. To familiarize readers with the techniques of use, an example for unknown variability is given in the following example that makes use of two other tables published in MIL-STD-414 (Table 12.10 and Table 12.11). The reader is referred to the standard itself for more details on other procedures and tables.

Example: (Standard deviation method, single specification limit) A farm equipment manufacturing company has ordered a lot of 70 special grade ductile iron cast axle tubes from

TABLE 12.9
Sample-Size Code Letters[a] (MIL-STD-414)

Lot Size		Inspection Levels				
		I	II	III	IV	V
3 to	8	B	B	B	B	C
9 to	15	B	B	B	B	D
16 to	25	B	B	B	C	E
26 to	40	B	B	B	D	F
41 to	65	B	B	C	E	G
66 to	110	B	B	D	F	H
111 to	180	B	C	E	G	I
181 to	300	B	D	F	H	J
301 to	500	C	E	G	I	K
501 to	800	D	F	H	J	L
801 to	1,300	E	G	I	K	L
1,301 to	3,200	F	H	J	L	M
3,201 to	8,000	G	I	L	M	N
8,001 to	22,000	H	J	M	N	O
22,001 to	110,000	I	K	N	O	P
110,001 to	550,000	I	K	O	P	Q
550,001 and over		I	K	P	Q	Q

[a]Sample size code letters given in body of table are applicable when the indicated inspection levels are to be used.

TABLE 12.10
AQL Conversion Table (MIL-STD-414)

For Specified AQL Values Falling within These Ranges	Use This AQL Value
— to 0.049	0.04
0.050 to 0.069	0.065
0.070 to 0.109	0.10
0.110 to 0.164	0.15
0.165 to 0.279	0.25
0.280 to 0.439	0.40
0.440 to 0.699	0.65
0.700 to 1.09	1.0
1.10 to 1.64	1.5
1.65 to 2.79	2.5
2.80 to 4.39	4.0
4.40 to 6.99	6.5
7.00 to 10.9	10.0
11.00 to 16.4	15.0

a foundry. After delivery, a sampling inspection by variables to an AQL of 2% is to be carried out. To ensure toughness, a good metallurgical process control is necessary. The minimum elongation of a tensile test bar cut from the samples supplied is limited to 10%.

The engineer is asked to draw the necessary sample size, have the samples tested in the laboratory, and make an acceptance or rejection decision for the lot of 70 axle tubes supplied.

Solution

Table 12.10 is used to find the corresponding AQL value of the plan for the given AQL of 2%. The given AQL of 2% falls within the range 1.65 to 2.79, which means an AQL equal to 2.5 is to be used.

The engineer now proceeds to determine the sample size. Since no specific inspection level was given, he or she chooses inspection level IV. From Table 12.9, for a lot size of 70 and inspection level IV, the sample size code letter F is obtained. Using Table 12.11, the sample size n is determined, corresponding to sample-size code letter F, which is 10. The constant k (acceptability criteria) for the applicable AQL is 1.41.

Choosing 10 random samples for tensile testing, let us assume that the engineer finds the percentage elongation values to be 8%, 9%, 11%, 12%, 8%, 7%, 13%, 11%, 9%, and 13%.

Next, the mean, $\overline{X}$ and the standard deviation are calculated.

$$\overline{X} = \frac{\Sigma X}{n}$$

$$\overline{X} = \frac{8 + 9 + 11 + 12 + 8 + 7 + 13 + 11 + 9 + 13}{10} = 10.10$$

$$s = \sqrt{\frac{\Sigma(X - \overline{X})^2}{n - 1}}$$

TABLE 12.11
(MIL-STD-414) Master Table for Normal and Tightened Inspection for Plans Based on Variability Unknown, Standard Deviation Method (Single-Specification Limit, Form 1)

Sample size code letter	Sample size	Acceptable Quality Levels (normal inspection)													
		.04	.065	.10	.15	.25	.40	.65	1.00	1.50	2.50	4.00	6.50	10.00	15.00
		k	k	k	k	k	k	k	k	k	k	k	k	k	k
B	3	↓	↓	↓	↓	↓	↓	↓	▶	▶	1.12	.958	.765	.566	.341
C	4	↓	↓	↓	↓	↓	↓	↓	1.45	1.34	1.17	1.01	.814	.617	.393
D	5	↓	↓	↓	↓	↓	↓	1.65	1.53	1.40	1.24	1.07	.874	.675	.455
E	7	↓	↓	↓	↓	2.00	1.88	1.75	1.62	1.50	1.33	1.15	.955	.755	.536
F	10	↓	↓	↓	2.24	2.11	1.98	1.84	1.72	1.58	1.41	1.23	1.03	.828	.611
G	15	2.64	2.53	2.42	2.32	2.20	2.06	1.91	1.79	1.65	1.47	1.30	1.09	.886	.664
H	20	2.69	2.58	2.47	2.36	2.24	2.11	1.96	1.82	1.69	1.51	1.33	1.12	.917	.695
I	25	2.72	2.61	2.50	2.40	2.26	2.14	1.98	1.85	1.72	1.53	1.35	1.14	.936	.712
J	30	2.73	2.61	2.51	2.41	2.28	2.15	2.00	1.86	1.73	1.55	1.36	1.15	.946	.723
K	35	2.77	2.65	2.54	2.45	2.31	2.18	2.03	1.89	1.76	1.57	1.39	1.18	.969	.745
L	40	2.77	2.66	2.55	2.44	2.31	2.18	2.03	1.89	1.76	1.58	1.39	1.18	.971	.746
M	50	2.83	2.71	2.60	2.50	2.35	2.22	2.08	1.93	1.80	1.61	1.42	1.21	1.00	.774
N	75	2.90	2.77	2.66	2.55	2.41	2.27	2.12	1.98	1.84	1.65	1.46	1.24	1.03	.804
O	100	2.92	2.80	2.69	2.58	2.43	2.29	2.14	2.00	1.86	1.67	1.48	1.26	1.05	.819
P	150	2.96	2.84	2.73	2.61	2.47	2.33	2.18	2.03	1.89	1.70	1.51	1.29	1.07	.841
Q	200	2.97	2.85	2.73	2.62	2.47	2.33	2.18	2.04	1.89	1.70	1.51	1.29	1.07	.845
		.065	.10	.15	.25	.40	.65	1.00	1.50	2.50	4.00	6.50	10.00	15.00	

Acceptable Quality Levels (tightened inspection)

All AQL values are in percent defective.
↓Use first sampling plan below arrow, that is, both sample size as well as k value. When sample size equals or exceeds lot size, every item in the lot must be inspected.

X	$X - \overline{X}$	$(X - \overline{X})^2$
8	−2.1	4.41
9	−1.1	1.21
11	0.9	0.81
12	1.9	3.61
8	−2.1	4.41
7	−3.1	9.61
13	2.9	8.41
11	0.9	0.81
9	−1.1	1.21
13	2.9	8.41
		$\Sigma(X - \overline{X})^2 = 42.9$

$$s = \sqrt{\frac{42.9}{9}} = \sqrt{4.77} = 2.18$$

Since it is the minimum value of characteristic, L (limit of elongation) that is of importance, Q_L (lower quality limit) is calculated as:

$$Q_L = \frac{\overline{X} - L}{s} = \frac{10.10 - 10}{2.18} = 0.05$$

Comparing this value of $Q_L = 0.05$ with the factor k ($= 1.41$), the lot meets the criteria of $Q_L \geq k$. However, since 0.05 is less than 1.41, the engineer rejects the lot.

General Methodology in the Use of MIL-STD-414 Tables

An application of the standard deviation method with a single specification limit when variability is unknown was demonstrated in the previous example. If the variability of the process is known, the known variability plan is the most economical to use. To provide flexibility in application, however, MIL-STD-414 provides a number of alternative procedures. Using either the standard deviation method or the range method, the acceptability criteria for single and double specification is decided by using forms 1 and 2. The following is a summary of a step-by-step procedure to be followed for deciding lot acceptability by using MIL-STD-414 tables.

1. Determine the sample-size code letters (use Table 12.9) corresponding to the lot size and inspection level.

2. Obtain the sample size, n and acceptability constant, that is, factor k or M (from MIL-STD-414 tables, which are indicated in the column of the table corresponding to the applicable AQL value).

3. Select a random sample of n units from the given lot; inspect or test for the specified quality characteristic and record the measurements for each.

4. Calculate the sample mean, $\overline{X}$ and either the standard deviation, s or the average range, $\overline{R}$ as the case may be.

5. Calculate the quality indexes, using whichever of the following tabulated formulas is applicable in the particular situation. In the case of a double specification limit, determine the lot percent defective, p from sample size tables (not given in this text).

6. Determine the acceptability of given lots. The acceptability criteria for the two situations are given in the table below as: single specification limit: accept, if Q_u or $Q_L > k$ (or negative); double specification limit: accept, if $p \leq M$.

Situation	Form Number	Acceptability Factor	Quality Indexes (upper and lower limits)
(a) Standard deviation method, single specification limit (variability unknown)	No. 1	k	$Q_U = \dfrac{U - \overline{X}}{s}$ $Q_L = \dfrac{\overline{X} - L}{s}$ (accept if; Q_U or $Q_L \geq k$)
(b) Standard deviation method, double specification limit (variability unknown)	No. 2	M	$p = p_U + p_L$ where, p_U and p_L determined from sample size table in MIL-STD-414
(c) Range method, single specification limit (variability unknown)	No. 1	k	(accept if; $p \leq M$)
(d) Range method, double specification limit (variability unknown)	No. 2	M	

Shainin Lot Plot Plan

The Shainin *lot plot plan*[10] is a variable sampling plan that uses a plotted frequency distribution to evaluate a sample for decisions concerning the acceptance or rejection of a lot. The decision is based on a comparison of the actual histograms (called lot plot) obtained against 11 different types of standard lot plots, each having its own rough estimates of the extremes (lower limits and upper limits; see the section on histograms in Chapter 5) as their limits. A standard sample size of 50 observations is to be maintained. See Grant and Leavenworth[11] for details.

12.9   OTHER SAMPLING PLANS

There are many other special sampling plans, both for attribute and variable schemes, available for special situations that have appeared in the literature in recent years. To include even a brief description of the more common ones would require many pages. Out of these, only the chain sampling plans used for continuous production, will briefly be mentioned here.

Continuous Sampling Plans

The concept of sampling for continuous production was first devised by H. F. Dodge[12] in 1943 with a plan which has been commonly known as CSP-1 (continuous sampling plan).

Later, this plan was modified to CSP-2 and CSP-3 plans, followed by other multilevel continuous sampling plans. In the case of manufacturing operations where production takes place by a continuous process on a conveyor or other straight-line system, CSP plans have become quite popular. The technique of inspection can be summarized as follows.

1. To start with, inspect 100% of the units and continue inspection until i units in succession are found without defects.

2. When i units have been found free of defects, stop 100% inspection and inspect only a fraction of the units by selecting them one at a time from the flow of the production.

3. If a defective unit is found, revert to 100% inspection and continue until i units are found without defects.

CSP-2 and CSP-3 Plans

Two modifications of CSP-1 plan have appeared, called CSP-2 and CSP-3. Continuous sampling plan 1 requires a return to 100% inspection whenever a defect is found during the sampling inspection. Continuous sampling plan 2 calls for 100% inspection only when a second defect occurs in the next k or fewer sample units. The procedure offered by CSP-3 is a further refinement of CSP-2, and gives more protection against a sudden run of poor quality.

Much of the early work on the continuous sampling plans based on Dodge's presentation was incorporated into MIL-STD-1235 (ORD) which was superseded by MIL-STD-1235A (MU) in 1974. Inspection according to this standard is done by attributes for defects or defectives using the three classes of severity, (i.e., critical defects, major defects, and minor defects). It is indexed by AQLs and based on the AOQLs, comparable to the system used in MIL-STD-105D.

CSP-T Plans

The multilevel continuous sampling procedure provided by this plan applies to alternating sequences of 100% inspection and sampling inspection. The special provision includes reducing the sampling frequency upon demonstration of superior quality.

CSP-V Plans

This plan provides a single-level continuous sampling procedure. A return to 100% inspection is required whenever a defect is discovered during the inspection of the first i sample units. From there on, however, the clearance number, $i,$ is reduced by two-thirds. More details of this type of plan can be found in the standard itself.

Sequential Sampling Plans

The concept of item-by-item sequential sampling, developed by A. Wald,[13] is similar to the multiple sampling schemes, except that sequential sampling can theoretically continue indefinitely. However, in practice, sampling is terminated after the number inspected equals three times the number inspected by a corresponding single sampling plan. The plan is suitable for costly inspections or destructive tests. The general methodology is similar to the sequential reliability testing technique described earlier in Chapter 8. Edward Schilling gives much of the details on the subject of acceptance sampling.[14]

12.10 KEY TERMS AND CONCEPTS

Acceptance sampling is statistically deciding whether or not a lot representing a statistical population should be accepted (or rejected) after the sample has been inspected.

An **acceptance sampling plan** states for a given lot size the sample size (n) and the acceptance number (Ac) for the largest number of defectives allowed in the sample. Acceptance sampling plans can be prepared for single, double, and multiple or sequential sampling. They are also designed for attributes and variables.

The **operating characteristic curve** shows, for each acceptance sampling plan, the probability of acceptance for a given level of quality (i.e., percent defective) in the lot (population). Acceptance on the basis of sampling results and inferences are subject to error however; so-called producer and consumer risks. Users of acceptance sampling plans must specify the acceptable quality level (AQL), the associated producer's risk, and the lot tolerance percent defective (LTPD) with the consumers' risk. Each acceptance sampling plan also has an associated average outgoing quality (AOQ) curve with an average outgoing quality limit (AOQL). See the text of this chapter for detailed definitions.

The **producer's risk** is the probability of rejecting a lot based on sampling results, even though the lot quality, expressed as AQL, is, in fact, acceptable. This alpha-risk is a normal statistical error.

The **consumer's risk** is the probability of accepting a lot, even though the actual quality of the lot, expressed in terms of LTPD would warrant a rejection. The risk is also known as the beta-risk.

Sampling takes the forms of simple random sampling, stratified random sampling, systematic sampling, or cluster sampling for purposes of statistical inferences, depending upon the properties of the respective statistical population. Sampling must be unbiased in order to subject it to error with known probability laws.

The **MIL-STD-105D** is a widely applied standard that describes sampling procedures and tables for inspection by attributes. The scheme allows for differentiation in normal, tightened, and reduced inspection. The MIL-STD-414 is the similar standard for variables.

The **Shainin lot plot plan** is a variable sampling plan that uses a plotted frequency distribution to evaluate a sample in conjunction with inspection.

Chain sampling plans (CSPs) are designed for inspection of continuous production.

12.11 DISCUSSION AND REVIEW QUESTIONS

(12–1) In what way does statistical quality control represent a significant departure from the older philosophy of inspection?

(12–2) What are the types of acceptance control? For any person who makes purchases from stores in a shopping mall which type of acceptance control would be preferred from a quality assurance point of view, and why?

(12–3) Explain the meaning of the following terms.
   a. Consumers' risk
   b. Producer's risk
   c. Probability of acceptance
   d. Acceptable quality level, (AQL)
   e. Lot tolerance percent defective, (LTPD)
   f. Average outgoing quality limit, (AOQL)

(12–4) *The real problem in most acceptance sampling is to design a satisfactory acceptance sampling system or, more commonly, to select such a system from a number of possible systems that incorporate risk analysis.* Explain.

(12–5) *An important advantage of modern acceptance sampling systems is that they exert more effective pressure for quality improvement than is possible with 100% inspection.* Give comments with reasons. Under what circumstances must one resort to 100% inspection?

(12–6) Discuss the needs of top management involvement in a sampling inspection program.

(12–7) Name the types of sampling plans with examples of the areas of their applicability.

(12–8)   Name the three levels of inspection severity, and discuss the criteria of shifting from one level to the other.

(12–9)   Are rule-of-thumb sampling plans always wrong? Explain.

12.12   PROBLEMS

(12–1)   Draw an OC curve that has ideal discriminating power and then superimpose one that yields very poor discrimination.

(12–2)   A single sampling plan uses a sample size (n) of 10, and an acceptance number (Ac) of 1. Using a. hypergeometric, and b. the Poisson probabilities, compute the respective probabilities of acceptance when a lot of 60 is known to be 2%, 5%, 10%, and 20% defective. Explain the symbols in the formula you use.

(12–3)   Draw different series of OC curves that signify:
a.   effect of varying sample size.
b.   effect of varying lot size.
Discuss the protection provided by a sampling plan where n and N vary.

(12–4)   Draw the OC curves for a single sampling plan with the following conditions: acceptance number = 0, and sample size = 2, 5, 10, and 20.

(12–5)   Calculate and draw a family of OC curves for a single sampling plan with the following conditions: sample size = 30, and acceptance number = 0, 1, 5, and 10.

(12–6)   A foundry is required by one of its customers to do destructive tests on an expensive 100 pounds of ductile iron casting by cutting a piece out for tensile testing. Assume that five castings comprise a lot. If the test is successful, the foundry tests one and ships four. If the test is not successful, the remaining four are scrapped.
a.   Specify the characteristics of this single sampling plan and derive the OC curve.

b.   Give your comments regarding double sampling scheme in this case to save costs to the foundry.

(12–7)   Construct the AOQ curve for a sampling plan with lot size, $N = 200$, sample size, $n = 10$ by using the OC curve data given in the following table. Determine the value of AOQL from the plot.

Probability of Acceptance	Percent Defective
0.06	40
0.08	35
0.13	30
0.22	25
0.37	20
0.55	15
0.77	10
0.90	5

(12–8)   A manufacturer submits a lot of 1200 pieces for inspection by attribute to an AQL = 2.5%. Using MIL-STD-105D, for single sampling plan and normal inspection, determine:
a.   the sample size, n.
b.   the accept/reject criteria.
c.   the accept/reject criteria for tightened inspection.
d.   the accept/reject criteria for reduced inspection.

(12–9)   A production lot of 3000 units is subjected to sampling by attributes. For the scheme of single sampling, normal inspection, and an AQL of 1.0%, determine the following as per MIL-STD-105D.
a.   The probability that the above chosen plan will reject the production lot if it contains 2% defectives
b.   The minimum number of defectives that will result in rejection of the production lot
c.   The LTPD value for the plan

(12–10) For a production lot of 3000 units, determine the following when AQL = 1%, and double sampling plan, normal inspection applies as per MIL-STD-105D.
    a. The size of the first sample
    b. The accept/reject criteria for the lot, if the first sampling results in zero defectives, one defective, two defectives and three defectives
    c. The number of defectives to be found in the first sample to necessitate drawing of a second sample
    d. The size of the second sample

(12–11) Inspection by variables (as per MIL-STD-414) is applied to a production lot of 100 high-strength low-alloy steels. Variability is not known, AQL = 1.5% and the single specification limit applies. The measure of percent of elongation of this steel is considered the critical characteristic and must not be less than 15%. Tensile tests were performed on a random sample of 10 and the following readings were recorded.

Sample Number	1	2	3	4	5	6	7	8	9	10
Percent of Elongation	22	18	17	15	16	23	20	19	17	22

Using the standard deviation method, determine the acceptability of the lot.

12.13 NOTES

1. ANSI/ASQC A2-1978, *Terms, Symbols and Definitions for Acceptance Sampling,* American Society for Quality Control, Milwaukee.

2. See Cameron, J. M., "Tables for Constructing and for Computing the Operating Characteristics of Single Sampling Plans," *Industrial Quality Control,* July 1952, p. 37; also, Schilling, Edward, *Acceptance Sampling in Quality Control,* Marcel Dekker, New York, 1982.

3. Hamaker, H. C.; in *Philips Technical Review,* Dec. 1949, March 1950, and June 1950 issues.

4. More examples of probability calculations for double and multiple sampling OC curves can be found in, Grant, Eugene L. and Leavenworth, R. S., *Statistical Quality Control,* 4th ed., McGraw-Hill Book Company, New York, 1972; Besterfield, Dale H., *Quality Control—A Practical Approach,* Prentice-Hall Inc., Englewood Cliffs, New Jersey, 1979.

5. Puri, Subhash C., *Statistical Aspects of Food Quality Assurance,* Publication No. 5140, available from Information Services, Agriculture Canada, Ottawa, K1A OC7, Ministry of Supply and Services, 1981.

6. Copies of the MIL-STD-105D standard may be obtained by directing requests to the Commanding Officer, United States Naval Supply Depot, Attn: Code DMD, 5801 Tabor Avenue, Philadelphia, PA. 19120. This information is also available as ANSI/ASQC Standard Z1-4-1980, *Sampling Procedures and Tables for Inspection by Attributes,* published by the American Society for Quality Control, Milwaukee. The Canadian Government version is available as CG5B-105-GP-1 and the International Organization for Standardization (ISO) version is number 2859.

7. The AQLs are in a geometric progression, each being approximately 1.585 times the preceding ones.

8. Dodge, H. F. and Romig, H. G. *Sampling Inspection Tables,* 2nd ed., John Wiley & Sons, Inc., New York, 1959.

9. Copies of MIL-STD-414 may be obtained by directing requests to the Commanding Officer, U.S. Naval Supply Depot, Attn: Code DMD, 5801 Tabor Ave., Philadelphia, PA 19120. MIL-STD-414 is now available as ANSI/ASQC Standard Z1.9, 1980, *Sampling Procedures and Tables for Inspection by Variables for Percent Nonconforming.*

10. Shainin, Dorian, "The Hamilton Standard Lot Plot Method of Acceptance Sampling by Variables," *Industrial Quality Control,* Vol. 7, No. 1, July 1950, p. 15, and "Recent Lot Plot Experiences Around the Country," *Industrial Quality Control,* Vol. 8, No. 5, March 1952, p. 22.

11. Grant, E. L. and Leavenworth, R. S., *Statistical Quality Control,* 4th ed., McGraw-Hill Book Company, New York, 1972.

12. Dodge, H. F., "A Sampling Inspection Plan for Continuous Production," *Annals of Mathematical Statistics,* vol. 14, September 1943, p. 264.

13. Wald, A., *Sequential Analysis,* John Wiley & Sons, Inc., New York, 1947.

14. Schilling, Edward G., *Acceptance Sampling in Quality Control,* Marcel Dekker, New York, 1982.

12.14 SELECTED BIBLIOGRAPHY

American National Standards Institute/American Society for Quality Control, ANSI/ASQC Standard A2-1978: *Terms, Symbols & Definitions for Acceptance Sampling,* American Society for Quality Control, Milwaukee.

Beainy, Ilham and Case, Kenneth E., "A Wide Variety of AOQ at ATI Performance Measures With and Without Inspection Error," *Journal of Quality Technology,* January 1981, p. 1.

Burr, Irving W., "Management Needs to Know Statistics," *Quality Progress,* July 1984, pp. 26–30.

Canadian Standards Association, *Introduction to Sampling Procedures for Materials and Manufactured Product,* CSA Special Publication Z90, Rexdale, Ontario, 1975.

Craig, Cecil C., "What Statistical Tool is Best?" *Quality Progress,* January 1975, p. 21.

Dodge, Harold F. and Romig, Harry G., *Sampling Inspection Tables: Single and Double Sampling,* 2nd ed., John Wiley & Sons, Inc., New York, 1959.

Dodge, Harold F., "A Sampling Inspection Plan for Continuous Production," *Journal of Quality Technology,* July 1977, p. 104.

Dodge, Harold F., "Keep It Simple," *Journal of Quality Technology,* July 1977, p. 102.

Miller, L. W., "How Many Should You Check?" *Quality Progress,* March 1975, p. 15.

Schilling, Edward, *Acceptance Sampling in Quality Control,* Marcel Dekker, New York, 1982.

Schilling, E. G. and Johnson, L. I., "Tables for the Construction of Matched Single, Double and Multiple Sampling Plans with Applications to MIL-STD-105D," *Journal of Quality Technology,* Vol. 12, No. 4, 1980, pp. 220–9.

Sommers, D. J., "Two-Point Double Variables Sampling Plans," *Journal of Quality Technology,* Vol. 13, No. 1, 1981, pp. 25–30.

Stephens, Kenneth S., *How to Perform Skip-Lot and Chain Sampling,* ASQC Publication, 1982.

Weber, E. C., "Getting Buyer and Producer Together," *Quality Progress,* July 1975, p. 18.

CHAPTER 13

QUALITY OF PERFORMANCE: CUSTOMER RELATIONS AND LIABILITY

Both the vendor and buyer (vendee) behave as a quality assurer when they meet certain obligations during the buying process, as well as during the performance phase of a product. The customer, after having ordered and bought an item or service, expects that what the vendor delivers will be in accordance with the specified quality, so that its performance will be accordingly adequate. However, whether the *quality of performance* will truly turn out to be adequate remains somewhat uncertain at the point of sale.

The example of the purchase of a car illustrates the situation fairly clearly. The customer usually arrives at the decision to buy after carefully weighing many factors that give some indication about the probable future quality of performance of the car. Warranties help to reduce the fear and repercussions of malperformance occurring within a short period after the purchase of a new or used car. Still, it is understood that any car will sooner or later end up on a scrap yard; it has a life. During its lifespan the quality of its performance will vary. Repair services will be required and the quality of these will be an important factor in determining the further quality of performance of the car.

Similar human contacts that determine quality of performance can also be observed in cases of a service to be rendered, such as a haircut. The customer and the barber usually agree before ''production,'' on the general kind of haircut to be given and then remain in direct personal contact (through mirrors) while the haircut is carried out. Quality of performance in this case, means that the hair will be cut in the manner desired by the customer. Sooner or later haircuts grow out and will normally be repeated. One can see that quality of performance derives from good quality of design and conformance and is the crucial result of these factors. It is the quality which the customer expects and which quality assurance is to establish and render. *Performance* circumscribes the material or immaterial use of an item or service over a timespan as perceived by the customer or owner. The performance is futhermore the actual functioning or servicing as an observable fact.

In this chapter we will discuss the closing of a production cycle with regard to a particular item or service. The cycle started with the designing and specifying of quality. We recall that this was done with the customer in mind and the customer participating either directly or indirectly through market surveys or other gauges of customer requirements. For the custom-made item or project, the customer maintained direct contractual contacts during the subsequent production phase, just like the customer receiving a haircut. Production and inspection planning and its implementations through operators and inspectors were done to assure quality of conformance. Now we return to the customer for delivery, installation, and servicing, of the product. Quality of performance can begin and will hopefully lead to high customer satisfaction. A new life phase of the product or service starts a new cycle here.

In this phase, the cycle of performance continues to involve quality assurance activities. These include a final check of the item before delivery, careful handling until it is safely transferred to the customer, assuring that the customer is well informed, and possibly even training the purchaser in using and maintaining the item. All of these contacts and defect-preventive measures are to assure quality of performance. The actual performance expectations during and after the possible warranty period will vary, depending on the subsequent owners and the actual performance history of the item. The kind of product and the respective quality characteristics and performance expectations are, of course, decisive for the actual quality of performance. Food will usually be consumed and digested within a short timespan, perhaps with some exceptions. Cars normally have a somewhat longer lifespan. Many factors, foreseen and unforeseen, have a continuing impact on the quality of performance, until the useful life of the item or service comes to an end.

In the following we shall trace this cycle of quality of performance through the early stages of customer–vendor contacts and then discuss reliability as measured by performance, preventive maintenance, customer obligations for quality assurance, customer feedback, and product liabilities. Marketing by the producer is a closely related function but we shall again focus on quality assurance without ignoring the many other surrounding decisions and activities. Similarly, the wide field of modern *consumerism* will be discussed only briefly, although it has many relationships with modern quality assurance. After this chapter we shall review all quality assurance activities so far covered with reference to a particular item, and bring them into Part Three, where we discuss perspectives of overall quality management and the establishment of a *company wide quality assurance program*.

13.1 CONTRACT REVIEW AND DELIVERY

Before the producer or vendor ship and deliver the contracted quality, the contractual obligations should be reviewed. This is to ensure conformance in the final production stages and customer satisfaction at the crucial delivery and acceptance point. Customers should receive precisely what was advertised, negotiated for, and put forth, either as a sample or prototype as understood in the contract. Moreover, the item should be well-packed and preserved during transportation and other handling. Vendor–vendee interfacial structure is shown in Figure 13.1 in order to relate many of these factors.

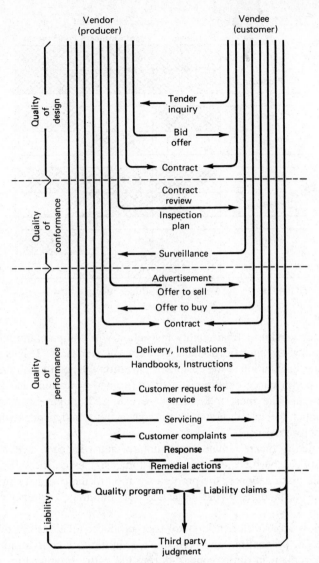

FIGURE 13.1 Vendor-vendee interfacial structure with particular reference to the life-cycle approach to quality.

A contract review by both the producer and the customer should be done to verify that quality specifications and standards are adequately achieved in the final product. This review is more important when producers and customers have not already cooperated during the product design and production phases. During the design and production period, mutual trust and confidence regarding quality assurance can develop. A customer merely ordering an article and then passively and often patiently waiting for delivery, is usually more tense and uncertain when the item arrives and must be inspected for acceptance.

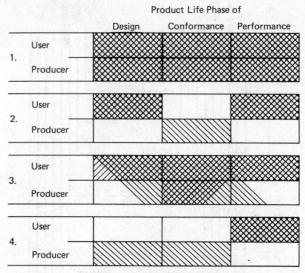

FIGURE 13.2   Types of user–producer relationships.

A producer's (or vendor's) final quality inspection will compare and test critical quality characteristics against the specification and performance standards. This is carried out either in the plant or after installation at the customer's location. Depending on the item, final tests under the actual performance conditions at the customer's location, and possibly with the customer present, are the most reliable and revealing. It might be more advantageous, however, to conduct the test in-house or through a third, independent party, in order to remedy defects before delivery.

The valuable quality image remains more untarnished with such a *predelivery test*. It is even better if tests can be repeated until the customer agrees on the satisfactory quality of performance. This clarification of the proper execution of the contract can avoid misunderstanding and also mishandling of the item, and thus minimize customer complaints in the future.

Figure 13.2 shows four different kinds of user–producer relationships during the product life stages of design, production, and performance. In the *first* configuration, both the user and the producer remain in direct contact during the entire product life, as, for example, with computers and airplanes. In the *second* configuration, the user designs the product, has it subcontracted to the producer, and then receives the item without further contact between them. In the *third* configuration, both parties cooperate for some overlapping period during intermediate stages. Finally, in the *fourth* configuration, the producer designs and produces a good with the only user contact being that established through market research. There is normally no further contact after the item is sold, except on those occasions when customer-initiated contacts occur in the form of complaints, returns, compliments, and so forth.

The contract documents, along with all other major quality assurance records and reports, have to be reviewed as well. Quality assurance extends beyond meeting the quality specifications of the product and/or service. It includes other obligations, such as delivering at the right time to the contracted destination. The packaging and mode of transportation are usually also contracted and understood to be safe in accordance with legal and conventional norms. The inspection plan usually stipulates delivery conditions and respective verifications. In case of future customer inquiry or complaints about quality of performance, the documented history of major items must be available. This means that items, once in the hands of customers, need to carry identifications that correspond with quality control documents. This closed-loop information system will be discussed in more detail in a later chapter.

Predelivery Contract Reviews

A widely applied technique for final predelivery contract reviews are checklists. This is in addition to functional testing. The salesperson, being the company's most likely representative, and hopefully a trusted one, could possibly conduct this final check, as has been said, in the presence of the customer. Naturally, upon proper confirmation of what one could call *quality at delivery* the customer should sign a document affirming the satisfactory finding. Often the fact that the price has been paid is, in itself, an implication of the acceptance of adequate quality.

Cases where either the company's representative or the customer finds that the delivered quality deviates from the contracted quality specifications will usually result in rejection. The situation is similar, but not identical, when a customer complains after having originally accepted the item without registering a protest. Rejection by a customer because of claimed nonconformance should be taken as a relatively serious incident. Future business relationships and the quality image of the products and services are at stake. Final predelivery checks are done to avoid this situation. Rectification depends on the seriousness of the defect and the management's commitment to attaining adequate quality of performance in a reasonable amount of time. Consumer protection laws have been established to help both the vendor and the buyer to act in a fair and correct manner.

Figure 13.3 is based on the life-cycle concept of possible *user's contact points*. It shows the various user contacts in a system view with the product moving from left to right. The arrow that points from right to left is the feedback of information into the product design stage. In other words, the user-producer contact describes a cycle that accompanies the quality assurance cycle and respective flow of activities from design to production to performance, and back to design review.

13.2 RELIABILITY AND LIABILITY

Quality of performance is most closely related to reliability, understood as fitness for use and failure free performance for a certain period under stated conditions. As a major quality characteristic, reliability is expected by customers for practically all items and services. It is sometimes referred to in similar terms, such as dependability, freshness, and so forth. During the product design phase, reliability was already considered, mainly in connection with a failure mode and effects analysis (FMEA) technique. Failure and reliability have inverse relationships, since minimizing causes for failures maximizes reliability. Once the item has been delivered to the customer, a failure or breakdown in performance will normally induce a formal and systematic *field failure analysis.*

This repeated analysis and investigation of the causes of failure provides important feedback for a review of the study of failure modes and reliability undertaken during the design phase,

1. Quality of design
    1.1 Market research
    1.2 Design assurance
    1.3 Contract review
    1.4 Inspection and test planning
    1.5 Suppliers

2. Quality of conformance
    2.1 Inspection
    2.2 Workmanship

3. Quality of performance
    3.1 Producer initiated contact
    3.2 Consumer iniated contact
    3.3 Product liability
    3.4 Product reliability

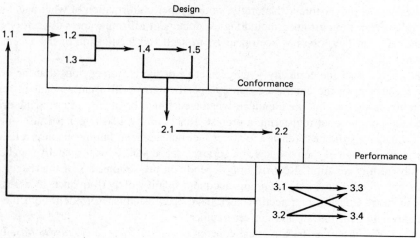

FIGURE 13.3   User contact points.

and lead to correction of causes, prevention in the future, and, thus, to more reliable products and also to better services.

Example: The following details derived from a newspaper article illustrate customer reactions to unreliable and defective deliveries of a major order of transit buses.

> . . . Transit is withholding payment on 200 trolley buses ordered from Bus Industries Ltd. until design and mechanical problems are ironed out . . . . [t]here are a large number of minor and aggravating problems, including faults with doors, windows, and the propulsion system . . .
> Many of the difficulties were not entirely unexpected. These buses have modern state-of-the-art control systems and a lot of problems wouldn't show up until they are actually put into service . . . Most of the blame for the problems lies with outside suppliers. We are beefing up our quality control program. The company recognizes the need for solid quality control prior to shipment of orders . . . .

Thorough design and quality control during production are indeed essential for adequate reliability in performance. A bus door opening due to failure of the control mechanism is unreliable and hazardous. Appendix E of this book contains a list of product safety checkpoints which can be used in establishing a product safety program.

It was shown in an earlier chapter how the reliability of an assembled item is determined. Once the random failure rate for a particular component and part at a specific time in the life and use of the item is estimated, a reliability coefficient can be calculated. This coefficient describes the probability of a component and/or product functioning properly. These probabilities are fractions ranging from zero to one; 0.5 means 50% probability of an event occurring. The difference between one and the failure rate, which describes the probability of the item not functioning properly, is the measure of the reliability.

Assuring Reliability

Providing safe, reliable products and services that meet the needs of consumers is the basic purpose of all quality oriented procedures. Acceptance of this responsibility throughout the company is essential to preserve the confidence of customers, and thus retain their business and compete successfully in the marketplace. Moreover, meeting this responsibility supports the basic objective of providing products and services that improve the quality of life in home, work, and leisure environments.

The intent of this policy is to emphasize the importance of continuing attention to customer needs and expectations concerning product integrity (product safety, reliability, quality, and environmental effects) and to define objectives and responsibilities for the design, manufacture, testing, transportation, installation, maintenance, and disposal of products that meet or exceed appropriate standards.

To achieve high reliability, it is necessary to have a broad-scope reliability program as part of the company-wide comprehensive quality assurance program. A variety of methods have been developed for quantitative prediction of reliability, including the factors of maintainability and availability. All of these constitute what is now well recognized as the separate field of reliability engineering, the details of which were given in Chapter 8.

13.3 PRODUCT LIABILITY

Reliability and liability are closely related. With high product reliability and quality of performance, the risk of liability action against the producer decreases. Unfortunately, the risk of liability can never be entirely eliminated.

The method of reliability analysis based on failure rate distribution has already been presented. Once this is understood, the practical problem of its application is one of data compilation and analysis.

Product liability is the obligation, responsibility, and the risk of having to compensate for any harm and costs caused by a significant lack of quality of performance. The producer is liable for gross negligence. Liability is an important legal term; so is gross negligence. A person who is injured because of a faulty product can take legal action against the producer. This is due to established acceptance of the idea of product liability.

The producer's risk of being held liable has increased in recent years due to relatively greater hazards with many modern products, to consumers' rights and protection movements and legislation, and to changes in the principles of buyer beware and *privity of contracts*. Under privity of contract the user of a product and the manufacturer have a contractual relationship.

A quality assurance program and evidence of adequate quality planning and control, along with proper information to the user regarding the product and its possible hazards, reduce the risk of being held liable when problems do arise. The *user* can also be held liable for *gross negligence* and for *failing to comply* with his or her obligations and responsibilities for quality of performance.

A product will normally be judged *legally defective* when the following conditions exist.

1. When the product carries a significant risk to consumers that manufacturers do not anticipate or guard against.

2. The degree of risk is sufficiently great that foreseeable use poses a danger to the user. (In some cases even misuse of the product is included.)

3. When it can be shown that a vendor did not follow the state of the art or did not know about the risks even when marketing the product and generating a use for it.

4. When product distributors are not properly trained for representation of the product to the end user. Distributors should be qualified to advise the end user on safe use of the product, and constantly keep up-to-date by the manufacturer. Service personnel must follow prescribed safety controls.

5. The manufacturer fails to provide clear, detailed, written instructions on the proper use and administration of the product.

6. When manufacturers do not maintain accident analysis records and records of failures, and when they fail to investigate the reasons for failures in order to establish future product recall or modification improvement programs.

All of the above items are considered a part of the product loss control process.

We pointed out earlier that the best and strongest protection against liability claims and their possibly disastrous impact on producers is quality assurance and its respective procedures and documentations. When a major customer has placed a liability claim and the *public image* of the producer is tarnished, the introduction of a quality control program, as was decided in the example of the bus company, might come too late. Prevention of liability claims is the best policy, especially where high reliability in performance must be expected by the customer and the general public. Before we turn to the discussion of managing comprehensive quality control programs in Part Three, appropriate contact and communication between vendor and vendee has to be discussed.

Contacts and Communications for Quality of Performance

The various contacts and communications between producers, vendors, and customers, possibly with governments or other third parties participating, are shown in Figure 13.1. The contract, with explicitly or implicitly agreed upon quality of performance, is the focal point and base, not only in the legal sense, but also in any information exchange between the parties concerned. Any change in contractual stipulations need to be as thoroughly negotiated, decided upon, documented, and controlled, as the original agreement.

Contracted quality of custom-made items before actual production has begun is particularly prone to review, change, and misunderstanding. Because of less standardized and more individualized quality specifications, contact between producer and customer must be more frequent and the communication more detailed and personal. However, once the item has been delivered and accepted, communication and cooperation regarding quality of performance become even more sensitive. In the following we distinguish between *producer-initiated* communications and those that are *customer-initiated*. We disregard those of the third parties because they usually play a secondary role of inconsequential importance.

Producer-initiated Contacts

Potential vendors are contacted either through bidding on a tender or through market research. The tender documents direct the bidding procedures. Market research is the systematic gathering, recording, and analyzing of data concerning the distribution and sale of products. Quality assurance personnel should participate in the planning of such research projects to ascertain that the expected quality of performance is known.

Market research as conducted by company staff or independent agencies, uses various survey and interview methods. The most frequent instrument used is the *questionnaire*. Quality assurance staff should check the questions being asked, and might also contribute to establishing statistical validity and reliability of the results.

Quality is always an issue before market contacts are made. Market research serves the purpose of finding answers to these quality related questions. This gathering of information helps in determining the deciding on quality of design conformance and performance. Quality assurance specialists should participate in all these customer approaches directly, or indirectly through staff training and review of quality specifications and assurance claims.

Truth in Advertising

Truth in advertising can serve as an example of the need for quality assurance checks. Various forms of advertising are used to create an interest in the product or service, and ultimately, to acquire a sales contract. When quality assurance viewpoints and safeguards are observed, realistic, fair, and true quality related perceptions and expectations develop in the mind of the public.

A quality assurance department is, in itself, a most important information source for potential customers and, by definition, one that implies integrity of the information. Unreasonable or exaggerated claims should not be handled lightly, as doing so will provide easy evidence to plaintiffs in liability controversies.

In planning and conducting valid and effective advertising and other product or service promotions, feedback from customers through market surveys plays an important role. Through well-planned and tested questionnaires, interviews, and other contacts, experiences and disappointments, suggestions for improvements regarding quality, and the associated quality assurance procedures necessary can be compiled and analyzed and possibly used for promotion. However, these inquiries often concentrate on negative experiences, field failures, and customers dissatisfaction. Customers will not always convey these messages through complaints, liability claims, or by discontinuing business relationships. Distributors and sales staff, once trained and informed, can become instrumental as the liaison between customers and quality assurance staff.

User-initiated Contacts

User-initiated contacts mostly take the form of inquiries and complaints. Such instances offer valuable data on actual quality of performance and should be treated as a first step towards a potential product liability action. The producer should have a well-prepared data recording scheme through which quality of performance information can be ascertained. Often, but not always, a user does know best what he or she needs.

Figure 13.4 is a schematic diagram of a complaint handling procedure. This diagram might help the person receiving the complaint to understand further procedures and decision points. Naturally, quality assurance should become involved in all complaints. Procedures will have to be written in such a way that the right person can make the right decision in a touchy matter, such as a complaint. It is a good policy to assume that a complaint is the first step in a product liability action. Settlement of a complaint should be fair, consistent, and legal. Figure 13.5 is an example of a field report used for analyzing customer's complaints. A somewhat elaborate consumer affairs case can be summarized in the manner of that shown by the format of Figure 13.6.

Channels for user feedback information, according to the ANSI Z 1.15 Guide, are the following: product return and failure analysis, product failure early warning system, complaints, product acceptance survey, user review panels, product safety performance, and other external feedback. Channels with user-initiated contacts are product return, failure warning and reporting, and complaints.

This feedback signals negative experiences with quality of performance. Positive experiences are also useful in showing strengths rather than weaknesses in quality. Both kinds of information must be carefully monitored, analyzed, and reported or stored for product

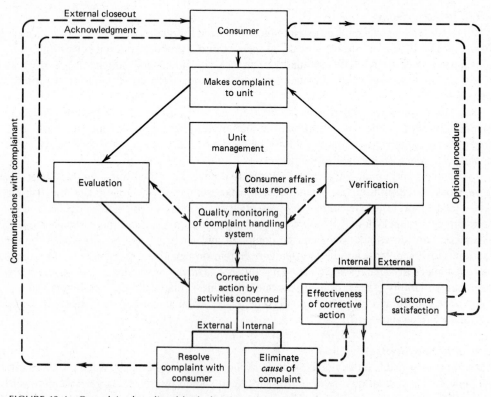

FIGURE 13.4   Complaint handing block diagram.

Field Report Mail to:	Product: Service Order: Charge to: Cust. ☐ Service ☐ Cost (estimate): Copy to:	
Customer:	Location (Plant):	Contact:
Equipment was originally sold to:		
Warranty claimed: Yes ☐ No ☐	Warranty expiration date:	Date trouble reported:
Date Work done:	Customer Complaint:	
Type of work done:		
	Complaint resolved: Yes ☐ No ☐ if "No", Action suggested:	
Repair complete: Yes ☐ No ☐		
From:	Date:	

FIGURE 13.5   Field report form for customer complaint.

design purposes, for improvements in production and conformance, and for improvement in customer services. Another major reason for such information processing is *product liability protection*.

However, contract review always entails a meeting of minds and, therefore, direct contact between parties. Negotiation and clarification, with communication in both directions, should result in a clear understanding of the obligations and rights of both parties. This will avoid future misunderstanding.

The Canadian Standard Association (CSA) Standard Z 299.1[1] for a quality assurance program outlines requirements for contract review. The ability of the producer to comply with tender requirements should be established before bidding. Later on, the contractual requirements must be compared with those in the original tender documents. Particular attention should be given to the quality specifications, applicable codes and standards, governmental controls, and administrative arrangements that might have an impact on quality.

A quality-minded producer will observe quality assurance viewpoints and obligations in the product and service, as well as in associated communications. This prudence will help to create a positive quality image in the marketplace and, at the same time, will provide protection against unreasonable and unscrupulous customers. Safety and product liability problems always grow—right from the conception of a product—and become bigger and bigger—up to the end use by the consumer—because of an increasing number of problems.

Complainant: _____		File No.: _____	
Unit(s) Involved: _____		_____	
Addressed to: _____		Date Rec'd: _____	

Summary of Complaint:

A   Acknowledgement Date: _____   By: _____

Consumer Affairs Action Taken:

Action Requested From: _____   Date: _____

Analysis of the Case:

B

Actions Taken re Claimant:

Closeout Letter Dates: Unit: _____ by _____   C.A.: _____ by _____

Internal Corrective Actions Required:

C

Verification: Date: _____   By: _____

Case Closeout: ☐ A   ☐ B   Date: _____   By: _____   ☐ C

FIGURE 13.6 Consumer affairs case summary.

Management responsibilities are diffused throughout the life cycle, no matter what the effective approaches are.

Who Is to Be Held Liable?

A woman bothered by bugs bought a can of insecticide, but instead of spraying it on the furniture or in the air, she sprayed it on herself. She had a violent allergic reaction and sued and collected from the manufacturer.

In another case, a man who was injured in an accident while riding in a 13-year-old car sued the manufacturer. When he was thrown against the gear shift, the knob on top of it broke apart, and he was impaled on the shaft. The knob was made of white plastic that had oxidized over the 13 years and developed hairline cracks. The man sued the company, claiming that the company had made the car with faulty materials, and he won his suit. So did the women who plugged a 115-volt vacuum cleaner into a 220-volt plug, causing the sweeper to blow up. It is hard to see how a motor burning out could blow up, but somehow the woman got hurt in the process.

Since the consumer has no facility to alter the final product, the *manufacturer's responsibility* should be all precautions in design construction commensurate with the state of the art when the product is built, as well as compliance will all product safety statutes, administrative laws, and, in particular, full conformance with the customs, practices, and standards of the industry.

All of the following allegations can be applied in accusations of product liability.

1. Defective construction of materials
2. Failure to comply with codes
3. Failure to investigate the sciences and comply with state of the art
4. Failure to properly warn the user of hazards
5. Failure of the product to perform as advertised
6. Improper design
7. Fault of two or more manufacturers

Product liability is quite a fascinating facet of quality assurance. The readers are no doubt aware that a record number of lawsuits are being initiated every day as a result of injury, death, or property damages allegedly resulting from faulty products, inferior services, or poor workmanship. Such horror stories are prevalent in the news. We can't turn a deaf ear to them. Quite a large number of lawsuits have already driven many manufacturers into bankruptcy. One law journal conservatively estimates that the number of lawsuits in North America alone will increase beyond the million mark by the end of the 1980s. Appendix F contains a list of articles regarding particular issues and aspects of product liability.

The *caveat emptor* (buyer beware) rule of the eighteenth century does not completely apply any longer, if at all. Instead, it is *caveat vendictor* (let the seller beware). This stands to reason, given the various frightening statistics that tell us how consumers lose lives, become permanently injured or disabled, and get deprived of properties or loved ones, on account of inferior quality aspects in one form or another. Indeed, behind the rush of product cases lies more than half a century of conflict over liability, among producers, consumers, and legislators. Who is to be held liable and who should pay for the casualties or inconveniences of our consumer society, is the question being asked by all of us.[2]

13.4 CONSUMER BEHAVIOR AND DESIGN ASSURANCE

Production managers have traditionally preferred a design that allows wide-ranging standardization of processes, work assignments, and supplies. Such stability allows production of relatively large lot sizes, if the market warrants it, as well as infrequent changeover and continuous line production. Quality assurance also benefits, because little change in product variety not only minimizes design activities and design reviews, but also reduces the number of defects. When a design allows this kind of production, it is simultaneously tested more and debugged by both producer and consumer.

During the period from 1960 to 1970 consumer behavior in many industries changed as more variety and integrity in products and services were demanded. Starr[3] pointed out that because of a shift towards greater *quality of difference* by consumers, production mangers were faced with a challenge to adapt. Quality assurance was also called upon to assure that designs adequately met what the consumer wanted. In fact, quality assurance evolved to include design during the early 1960s under these market pressures, because, before that time, quality control departments merely accepted designs and assured conformance to design specifications.

Escalating Consumer Demands

The demand for a greater variety in products and services was not limited to automobiles. Homes were sold with fireplaces, family rooms, multiple plumbing, and other extras instead of the simpler and more uniform homes previously designed for middle-income families. Brands of cereals, beverages, wines, and other food products multiplied, with only a few leaving the market. New types of insurance policies were designed to cater to individual circumstances and needs. Fashions, although always changing with the seasons, also exploded in terms of variety. Shelves with pharmaceuticals and drugs overflowed, making it difficult to select the best suitable for a given situation.

This escalating variety in products and product designs can be attributed to a high degree of market saturation, consumption oriented lifestyles, available production capacities, investable funds, and emphasis on marketing, rather than on production details. During that period of economic growth and full employment, consumers demanded custom-made designs which expressed their individuality and made visible their wealth and success. Marketing, of course, induced much of this demand for ever-changing designs, new models, and innovative products and services, in a struggle to claim a market share and outsell the competition.

This, however, does not mean that consumers have become less demanding, as far as designs are concerned. The opposite is the case. The design must still be pleasing and cater to individuality and self-assurance. To appear rich is still important for most. However, in addition, health, higher demands for education, entertainment, safety, and general quality of work and life, have had great impact on modern design tasks. Here, in particular, quality assurance is called upon to act as the communication link between consumers and designers. While traditional marketing, as a function and discipline, is expected to attract and satisfy customers in the traditional fashion, quality assurance must apply and develop a different kind of expertise.

Customers, when stressing more quality of performance and conformance, rather than quality of difference expect of producers a different kind of assurance, a more technologically demanding one. Consumers have become sceptical of changes in brand names and different forms of packaging.

Managerial Responsibilities

The quality assurance function acts as an intermediary between consumers and designers when adapting to changing consumer behavior. As illustrated in Table 13.1, quality assurance functions have to adapt to basic consumer behavior changes over time, along with cyclical fluctuations. In order to cooperate with other functions in an organization, such as design and production engineering, marketing, financing, personnel, and purchasing, at the corporate and plant level, scientific decision-making methods must be mastered and applied. For

TABLE 13.1
Consumer Behavior versus Design Assurance

Situation	Consumer Behavior	Design Assurance
Growth period	Demand for greater variety, uniqueness, quality of difference, higher quality level, and extras	Design assurance increasingly complex; shorter time available; increasing participation of quality assurance and marketing departments
Market saturation	More quality and services demanded, selective from among competitors; greater reaction to defects	Design stresses improvements in visible quality characteristics; marketing, service, quality control become important; modular designs are favored
Recession (inflation, unemployment)	Demand for more utility and performance, conformance to standards, less change and minimum variety	Design assurance stresses the use of new technology for improving basic quality of performance; strong quality assurance influence
Underdeveloped technology	Simple living standard justifies low cost but good quality demanded with minimal services and extras	"Robust" designs with quality assurance involvement; more quality control emphasis
Accident	Greater awareness for safety and product integrity	Review of design in industry concerned; involvement of public and government; quality assurance role expands
Breakthrough (innovation)	Demand for design change and sharing of innovation depending on economic situation; consumer pressure for higher quality level	Design assurance appraises potential; cautious design adaptation; quality assurance sceptical until innovation proven and matured
Supervisory methods and management	More direct participation in design assurance, self-assembly, self-inspection, "do it yourself" type of attitude	Use of management science methods in design; more consumer and quality assurance input; cooperative design assurance
Competitor entry	Preference for successful competitor, consumers expect other producers to follow suit	Competitive benchmarking through design review; major attempt of design improvement or substitution
Life Cycle Early phase	Participation in field testing to seek improvements depending on individual preference	Analysis of old and similar designs; collecting of ideas; little quality assurance involvement

TABLE 13.1 (Continued)

Situation	Consumer Behavior	Design Assurance
Start-up	Expect benefits from quality of difference; exercise individuality; relatively highly defect conscious	Stress on infancy defect prevention, debugging; quality assurance increasingly involved; interest in consumer feedback
Steady state	Increasing demand for variety, becoming more selective; demand more extras and improved service	Design reappraisal; competitive benchmarking; stress quality assurance and quality control input; compliance with new standards
Termination	Switching to new designs; more defect conscious	Little new design effort on old model; quality assurance more protective against liability claims; phasing-out of design assurance and substituting design
Antique	Demand visibility for demonstrating wealth, education, and culture, defect accepted if not appreciated	Design assurance exploits antiques for new ideas, quality assurance scarcely involved

instance, network analysis methods help to schedule and delineate responsibilities and actions in design projects. Simulation can test different combinations of modular designs and production, impact of decentralization of inspection, and impact of statistical techniques, FMEA, and so forth. Heuristic programs, using modern computer facilities, test designs for consumer reactions, under the influence of different stochastic and assumed conditions. Again, forecasting plays an important role in predicting major turning points in consumer behavior and hopefully, allowing quality assurance personnel a head start.

Predictions of changes in consumer behavior concerning a new product during the early phase of product planning mainly affect cosmetic quality characteristics, such as the trim of cars, packaging, display, color, and other visible attributes, and the associated services. Demand for variety normally did not induce major pragmatic design alterations, such as in car engines or food processing methods. Production, rather than marketing, becomes more involved in order to cope with subsequent and more basic design changes, such as for new electronic devices, front-wheel-drive in cars, new building materials for homes, new and different home appliances, and new innovative commodities in the sports and leisure goods industry.

Modular designs have been developed in which the process of transformation and assembly are separated. "It is the essence of the modular concept to design, develop, and produce those parts which can be combined in the maximum number of ways."[3] Computers, along with *computer-assisted design* (CAD) facilities make it possible to master extreme combinational possibilities and to combine a myriad of different parts into final assemblies of an ever greater variety.

One can imagine that quality assurance has to play an important role as assistant to production management and design engineering. When quality is what consumers expect, and are told to expect through the various media, then design engineers cannot work in complete isolation on the drawing board. Design engineering has always benefited from consumer demands for variety, because gradually the demand shifted from mere outwardly visible variety to demands for more meaningful inherent quality and technical improvements. This development in consumerism and demands for product integrity and adequate performance moved the emphasis from marketing, to production, to the design and associated research and development function around the time of the oil crisis of the 1970s. With inflation and subsequent unemployment, consumer behavior had to change again from a throw-away pattern to a more neoconservative one, demanding real utility and performance.

Consumer behavior can, of course, be expected to change in a cyclical pattern, somewhat correlated with that of income and material living standards. Perhaps before it was the designer's task to please by the pleasant appearance of the design; now technological integrity is to be achieved. Governments also have stepped up and seen to it that quality assurance provides benefits for all concerned. This information should not end with salespersons, marketing managers, distributors, or even with production and plant managers. As has been demonstrated, once turning points in consumer behavior have been monitored, the quest is for major changes in product and production design for better quality.

Quality assurance offers, in well organized and functioning systems, the additional communication link. The opportunity, for instance, to deal with customer complaints lies in monitoring customer dissatisfaction. While during the early phases of such monitoring emphasis is on more or less previously overlooked features that the customer needs and expects, these complaints, during later analysis, reveal more basic design faults. In short, customer complaints, normally handed over from the marketing and sales departments to quality assurance personnel are the single most important opportunity to adapt design and subsequent production to suit customers. With a closer link between consumer behavior changes and design, quality assurance has to play an essential role. In fact, this has been the main force in the recent evolution of quality assurance management. Table 13.1 assesses various items of design assurance in respect to consumer behavior.

Consumer Loyalty

Communication regarding quality, as with other business issues, should contract business partners, producer and customer, in a long-term, satisfying relationship with mutual benefits. Loyal customers return once they have parted with one item, a car for instance, because they were satisfied. Quality of performance as perceived and experienced by customers leads in the continuum of time and as communicated to the producer to affirmation or change of the quality of design. "After the Sale is Over"[4] is an interesting article in which the author suggests how sellers must be alert and sensitive to customer needs to keep an effective "relationship management" working for the company.

13.5 LEGAL ASPECTS OF PRODUCT LIABILITY

The new philosophy in product liability is toward *strict* liability. New laws are making it easier for the plaintiff to obtain a winning judgment. The emphasis has been changing from privity of contract to breach of warranty and negligence.

In 1965, the American Law Institute issued its *Restatement of the Law of Torts* declaring that:

> If a product because of a defect becomes unreasonably dangerous and causes an injury, it is defectively made. A manufacturer, who makes and sells that defective product, has committed a fault. It is implied that he was negligent and, therefore, strictly liable to the injured party.[5]

The Consumer Product Safety Act (CPSA) of 1972 became a significant part of legislative law for consumer safety with the main intent of preventing hazardous or defective products from reaching the customer. The other hardest-fought of the major safety laws were the National Traffic and Motor Vehicle Safety Act of 1966, and its companion, the Highway Safety Act of 1966. This brought together the enforcement of all licensed vehicles, including trailers and motorcycles, under federal regulations. The CPSA legislation also established the Consumer Product Safety Commission (CPSC) with a given mandate and broad power to cover an estimated 10,000 products. Its regulatory armament includes a wide range of penalties aimed at unsafe products and their manufacturers and distributors.

The Warranty Act (known as Magnuson–Moss Warranty Act) became a law in 1975 with powers granted under the Federal Trade Commission that apply to written warranties on tangible personal property—property normally used for household, personal, or family purposes. The subject matter of warranties forms a part under the Uniform Commercial Code which also includes various exemptions, the designations of *full* or *limited* warranties, the duties of *warranty protection,* and differences in *written warranty* and *service contract,* including their definitions and interpretations.

In the areas of food and drugs, the Food and Drug Act created laws that are far more restrictive in nature, since no new drug may be marketed without prior approval of the Food and Drug Administration (FDA). Instead of leaving the control of quality to industry, more recently the FDA has promulgated standards of ''good manufacturing practices'' based upon the prior work of industry and formulated in collaboration with industry committees.

It is pertinent to note that product liability laws vary somewhat from jurisdiction to jurisdiction and in all likelihood the lawsuit is tried in the state or province where the injury occurs.

As a result of a lawsuit, regardless of who wins it, the defendant always incurs massive legal expenses. The result is unfortunate for most of the smaller asset companies, since it not only affects their public image and competitiveness in the market, but many times leads to bankruptcy.[6] Appendix G describes in brief the outcome of some of the precedent-setting cases concerning product liability.

13.6   QUALITY ASSURANCE OF SERVICES

It is a fallacy to consider services as less amenable for quality assurance than products. Whenever customers pay a price for a service received, either directly in cash or indirectly through taxes, they do have a perception of the quality of such services and whether or not their need has been satisfied and whether or not the organization is able to provide better services. Avoidance of negative perceptions and possible liability claims is a major concern and objective in all service organizations since the present proportion of the labor force that is engaged in service in North America is simply too large (85:15). It is obvious that improvement in our standard of living is highly dependent on better quality services.

The Nature of the Quality of Services

Service is understood as rendering aid, convenience, and utility directly from person to person. It is either the customer personally who receives that service, such as advice from a physician or a lawyer, or a customer's property that is serviced such as with repairs to a home or maintenance of a car. Close customer contact with the service institution and its staff is an essential feature of any service, having considerable bearing on the quality and quality assurance of the service rendered.

Basically, utility and quality received from a product or a service are not very different. Product performance and customer service are closely linked in any quality program; the greater the attention to product quality in production, the fewer the demands on the customer service operation to correct subsequent problems. This is true for the automobile industry as much as for a hotel or restaurant establishment. The latter, however, is considered a typical service institution, while the automobile industry is not. In a restaurant, excellent food and wine would not necessarily satisfy the customer when the service lacks proper atmosphere, promptness, courtesy, friendliness, and so forth.

While many services are simple conveniences not requiring much skill or training, others are normally rendered by a professionally trained person with extensive education and experience. However, as in workmanship of physical products, *personal ability* and *aptitude* are important inherent elements of the person providing the service.

The nature of services and the provision of these have somewhat changed in the postindustrial era towards more impersonal service facilities. Transportation relies extensively on the quality of respective facilities, because travelers need reliable and safe transportation vehicles along with courteous and capable personnel. Quality of medical services, particularly in hospitals, rests as much on the equipment available, as on the skill and professional integrity of the medical staff. Computer and modern communication devices have replaced or augmented services by human operators in ever greater measure. Service in these cases becomes increasingly impersonal, more formalized and standardized. Unfortunately, however, besides the unstated assumption that procedures exist and people in service organizations follow them to satisfy customers, the methods and techniques of quality assurance are very poorly known or practiced in service organizations, except perhaps in a few. Ask anyone these days about his or her one trouble (better termed "horror story") regarding quality of service and chances are you will hear plenty.

It is amazing to learn that, in spite of so many laws and consumer protection acts, and in spite of customer-satisfaction-guaranteed advertisements, the buyer beware situation still remains. It is somewhat true that the service industries tend to have relatively captive markets and are not exposed to foreign competition. This does not necessarily mean that there is no need for quality assurance of services. A system of quality assurance with the identical life-cycle approach as that for products is much needed in service organizations. The effects of malpractice and mistakes in a profession or service institution are extremely costly, just as those in a defective product sold to a consumer. Without repeating much from previous chapters, we shall point out some salient features of quality assurance in services.

Principles of Quality Assurance in Services

The entire field of quality control in service operations needs attention. The organizational function usually does not exist in any formal sense as it does in manufacturing. Standards for quality do not exist, perhaps because the dimensions of quality have not even been identified.[7]

As we pointed out, the difference between quality assurance of products and services may not be as pronounced as one might assume. In practically all previous chapters we have given examples from traditional service industries, indicating that quality assurance principles apply wherever a price is paid and utility is contracted.

The life-cycle approach of quality assurance planning and the three phases of it, namely, the quality of design, quality of conformance, and the quality of performance still apply, whether considering provision of services or product manufacturing. Some principles of quality assurance that seem to be particularly important when services are rendered are the following.

1. Essential quality criteria for the service should be specified with the consumer in mind and then emphasized in the design, operation, and delivery modes.

2. For services with predominantly interpersonal contact, the person rendering such service must be adequately trained, selected, and supervised. Often personal integrity is of relatively great importance.

3. Consumers need to be well-informed about the services they can expect to receive, so that the definition and delineation of the service is well understood and communicated.

4. When interpersonal contact is minimal in service operations, formal and standardized procedures help to avoid errors. In these cases, computerized processes are particularly effective, as, for instance, in modern banking operations involving a large volume of transactions.

5. Capacity to service must be adaptable to customer demand in terms of unique desires, requirements, timing, and place. With proper planning, control of demand, and readiness for service quality can be considerably enhanced, for example, by efficient procedures for appointment scheduling, handling and rehandling of huge numbers of small items, and other details.

6. Customers in close-contact services might participate in quality assurance through clarification of their needs.

7. Services can enhance the quality image and quality of design, production, and performance of products, and vice versa. For instance, warrantees assure certain specified services after the point of sale; teaching and learning in a course is supported by lesson notes and textbooks.

8. *Self-service* is an attractive alternative in low risk and time saving instances, such as when buying gasoline. However, proper customer aids and supervision becomes an important low-contact service and prerequisite. Computer facilities increasingly provide low cost services, normally with greater speed and convenience than the same service facility without computer aids. *Time* is an extremely important factor to control for.

9. Management of quality assurance is widely delegated to those staff members in contact with customers. Supervisory management maintains performance control of service staff through operational procedures and inspection and/or training.

10. Statistical quality control concepts and techniques should be applied when service criteria and operational processes can be quantified; for example, with errors in banking operations, and, generally, with records of customer complaints. Direct customer quality surveys and statistical analysis offer important information for improved quality and quality assurance.

11. Emphasis must be placed on prevention, rather than on correction of poor service. Poorly delivered services normally cannot be easily repaired or reworked, as, for example, when the wrong tooth is pulled by a dentist, when letters are delivered to the wrong address, or when names are typed incorrectly on a driver's license.

12. Modern planning and control methods for quality assurance of services should be adopted, developed, and applied. This calls for a blue-print of the procedures, clearly written down in step-by-step form on a flowchart.[8]

Approach and Methods

A service to be rendered starts with the assessment of the need and desire of individuals and groups. This design activity determines major qualitative attributes of what the customer expects in performance, and what the server and the service institution is, in fact, to provide.

Once the service is specified, mode of delivery, type of server and service facility, and timing and location for the service is also largely clarified. The service institution will now have to develop, plan, and provide the personnel and respective physical facilities. Service capacity that satisfactorily meets the major individual demands of people is not an easy task because of the great variety of personal idiosyncrasies and expectations. Service staff must be trained and certified in accordance with the risks of providing the service. A waiter in a restaurant has fewer and less serious risks to avoid than a surgeon or an airline pilot.

As soon as the service has been rendered, the individual server, their organization, and the employer should confirm the perception of the quality of the service. Questionnaires are an important medium for such communication. A satisfied customer is still the most effective advertisement, because he or she will normally help to attract other customers. Competition in service industries is often very strong in people-dominated rather than in facilities-dominated services, such as with restaurants.

Methods to be applied in the planning and control of services during the design, operation and postoperation stages are of a qualitative and quantitative kind. Surveys and personal inquiries still play the most important role in gathering service data and in communication and prediction. Cooperation with marketing function helps to apply an arsenal of well-proven methods.

Statistical process control techniques are directly applicable to service evaluation, because they separate random and nonrandom deviation from nominal value, such as with accounting error, waiting time, frequency of customer complaints, and so on. Waiting line models, simulation techniques, and computer assisted planning and operational control, are aids in attaining optimal service-system design under given conditions and trends. Some of the applications of these decision-making methods for assuring quality of services have been demonstrated earlier in this book, for example, in the simulation of the child-care and adoption process.

Customer complaints and comments about services received, or not received, call for careful analysis by supervisors and operators, using, for instance, cause–effect diagrams. Operational procedures, codes of ethics, and performance standards serve as guides and means for personal conduct during service operations. Such aids facilitate performance evaluation and auditing of compliance.

Discussing quality assurance of services had led to review of much of what has been covered on quality assurance in previous chapters. Throughout this book we, as authors, intend to induce and direct the learning of our readers and students. Therefore, is this book a service or a product? We think it is both, simply because we have written it with you, the student, in mind.

In Part Three of this book we shall broaden our view on quality assurance from the individual product or service to the organization and institution. We shall see, once more, that it is not only the manufacturing establishment that requires a rational, systematic, quality assurance program, and that such programs are as much required in the various service industries as well. The reader should keep this in mind as we proceed, because service industries do adopt traditional and modern quality assurance concept and methods.

The quality cycle for one particular item, product line, or type of service, does not end with the item itself. It extends through new designs, either by the same producer, or a competitor, along with general social, economic, and technological factors. One must, therefore, recognize quality assurance in a wider context and not only at the level of one product. In this part, "Planning and Controlling Quality," discussion centered on the particular singular unit, the product, project, or service. In the next part, "Managing Quality Assurance," a wider perspective will be adopted, with the quality assurance program as the focal point. We shall see in the next part those concepts and practices necessary for a comprehensive approach to quality assurance programs.

13.7   KEY TERMS AND CONCEPTS

Quality of performance is determined by the actual use, application, and consumption of an item by the customer and the respective perception of satisfaction or dissatisfaction with the product supplied or service rendered. It refers to the major *life phase* of the item.

Customer relations are the various contacts and communications between the consumer and the producer during the life phase of a product, contract, or service. Harmonious and constructive interaction can support attainment of satisfactory quality and quality assurance. Customers should participate in quality assurance for their own benefit, under the principle of *buyer beware.*

Contract review, as used in this chapter, is a final verification and confirmation of conformance with quality specifications and quality assurance obligations, before delivery and contract completion.

Reliability relates to quality of performance as failure-free performance for a certain period under stated conditions. Maintainability and availability are similar product-related features that contribute to customer satisfaction and general benefit.

Liability is the obligation of producers, and, to a certain extent, also of the consumer, to comply with principles and laws and to assure quality in the broadest sense. Nonnegligence is of crucial importance in providing a safe product by virtue of respective quality assurance procedures and activities. Product liability is a risk against which protection through prudent measures is necessary. It has become an intricate and important legal term and concept.

Customer complaints express dissatisfaction and need to be carefully analyzed in order to prevent serious liability litigation and avoidable recurrences.

Consumer behavior as a psychological phenomenon, has become an important element for planning and controlling quality assurance effectively, for creating maximum satisfaction of consumers, and for meeting the competition. Consumer behavior concerning desired quality of products and services varies with general economic trends and developments and general living standards.

13.8   DISCUSSION AND REVIEW QUESTIONS

(13–1)   *Quality of performance describes responsibilities for both producer and consumer.* Explain and discuss.

(13–2)   *The basic form and content of producer–customer relationships change with the life phases of the product.* Explain.

(13–3) *Customer relations means, foremost, customer communications.* Explain and discuss.

(13–4) *"Reliability" and "Liability" are closely related terms, particularly with reference to quality assurance.* Explain.

(13–5) *Consumer protection laws have raised the product liability of producers.* Explain.

(13–6) *Quality assurance demonstrates a policy of producer beware, but what about the traditional policy of buyer beware?* Explain and discuss.

(13–7) *Advertisements can support quality assurance and can also diminish it.* Explain and discuss.

(13–8) *Producers are as much responsible for their dealers and distributors as they are for their suppliers.* Discuss.

(13–9) *Consumer behavior concerning quality of design varies when compared by periods of economic growth and of recession.* Explain.

(13–10) *There are major determinants of consumer behavior which must be considered when validating the design.* Explain.

(13–11) *The customer is always right; almost always.* Discuss from the viewpoint of perception of defects in service operations where every person sees quality slightly differently.

13.9 PROBLEMS

(13–1) Producers communicate with customers through advertisements, sales promotion brochures, user manuals, and other similar means.
  a. Select an advertisement and comment on it, using information concerning quality assurance of the product or service.
  b. Select a sales promotion booklet and assess it for its quality assurance information to the customer.
  c. Select a user manual and assess it from a quality assurance viewpoint; that is, its usefulness as a guide to the customer for

maintenance of quality of performance, such as information concerning repairs, complaints, and other related issues.

(13–2) You are assigned the task of analyzing data from warranty claims for quality assurance purposes. Select any product or service and:
  a. state the purpose and objectives of such an analysis.
  b. prepare a form to be used by the warranty claim department for compiling data related to quality assurance.
  c. describe the method of analysis of this data and the report on a standardized form of your own.

(13–3) Prepare a simple questionnaire for inquiry on customer satisfaction for a:
  a. restaurant.
  b. shoe store.
  c. garment manufacturer.

(13–4) You are assigned the task of preparing a procedure for handling customer complaints in a resort hotel.
  a. Outline the plan for writing the procedure.
  b. Draft the procedure and respective report forms.
  c. Prepare a plan for implementing and reviewing the procedure.

(13–5) Products and services are closely interlinked from the quality assurance point of view. Explain this by means of an example with particular reference to statistical control technique.

13.10 NOTES

1. Canadian Standard Association publication, *Quality Program Standards*, Z 299.1, 1979, 178 Rexdale Blvd., Rexdale, Ontario, Canada, M9W 1R3.

2. For a revealing description of the outcome and evolution of court decisions on liability see, Maslow, Jonathan E., "Product Liability Comes of Age," *Juris Doctor* (1975), MBA Communications, Inc. Other articles and publications of interest are: Seagraves, D. W., "Product Liability Problems—Growing Towards Crisis," *Quality Progress*, Vol.

10, No. 1, January 1977, p. 16. Piehler, H. R., Twerski, A. D., Weinstein, A. S. and Donaher, W. A. "Product Liability and the Technical Expert," *Science*, Vol. 186, No. 4170, December 1974, p. 1089. Nelson, P. C., "Current Impact of Product Liability," *Quality Progress*, Vol. 9, No. 8, p. 20, August 1976. A book by John Kolb and Steven Ross, *Product Safety and Liability—A Desk Reference*, McGraw-Hill Book Company, New York, 1980, is of special interest and covers many topics related to management of product safety and liability.

3. Starr, Martin, "Modular Production—A New Concept," *Harvard Business Review*, November–December, 1965.

4. Levitt, Theodore, "After the Sale is Over," *Harvard Business Review*, September–October 1983, p. 87.

5. This is based on common law and on legal decisions of then recent cases (called *case law*).

6. The reader will benefit from various articles published in the field of product liability, such as: Clancy, W. W., "The Knot-Hole in the Solution to Product Liability," Part 3, *Quality Management and Engineering*, Vol. 11, No. 6, June 1971, p. 22; Nelson, P. C., "Current Impact of Product Liability," *Quality Progress*, August 1976, pp. 20–7; Seagraves, D. W., "Product Liability Problems—Growing Toward Crisis," *Quality Progress*, January 1977, pp. 16–18; Koch, W. H., *Products Liability Risk Control*, Technical Paper IQ75-538, Society of Manufacturing Engineers, Dearborn, Michigan, pp. 5–12.

7. Buffa, E. S., "Research in Operations Management," *Journal of Operations Management*, Vol. 1, No. 1., August 1980, pp. 1–7.

8. Shostack, G. L., "Designing Services that Deliver," *Harvard Business Review*, January–February, 1984, pp. 133–9.

13.11 SELECTED BIBLIOGRAPHY

American Society for Quality Control, Product Safety & Liability Prevention Technical Committee, *Product Recall Planning Guide*, 1981, American Society for Quality Control, Milwaukee.

Bajaria, Hans J., (editor), *Quality Assurance: Methods, Management and Motivation*, Society for Manufacturing Engineers, 1981.

Carrutha, Eugene R., *Assuring Product Integrity*, Lexington Books, Lexington, Mass., 1975.

Eginton, Warren W., "Minimizing Product Liability Exposure," *Quality Progress*, January 1973.

Feigenbaum, Donald S., "Liability Prevention—Quality Control, New Partnership," *Quality Progress*, 1973, p. 14.

Kytle Jr., Rayford P., "Evaluating Product Liability before Marketing," *Quality Progress*, February 1974, p. 16.

Levitt, Theodore, "After the Sale is Over," *Harvard Business Review*, September–October 1983, p. 87.

Powderly, Daniel D., "Arbitration–A Way to Resolve Quality Disputes," *Quality Progress*, June 1980, p. 26.

Troxell, Joseph R., "Service Time Quality Standards," *Quality Progress*, September 1981, p. 35.

PART III

MANAGING QUALITY ASSURANCE

In this part we shall broaden the view and discussion of quality assurance to the level of the entire company or organization. The understanding of product and service oriented quality assurance concepts and techniques, as they were covered in Part Two, is a prerequisite for the pursuit of this part. Much of the information presented earlier will be reviewed and applied for designing and implementing *company-wide quality assurance programs.*

Planning and controlling decisions and activities for accomplishing goals of quality and quality assurance in an enterprise describe the management functions. Managing quality assurance will not be restricted to the sphere of a quality assurance manager, but will involve the entire staff. This is consistent with the underlying philosophy in this book, described as total quality assurance.

The logic of chapter topics in this part follows the cycle of decision making. Chapter 14, Objectives and Strategies, lays out the long-term operational framework in which the organization, discussed in Chapter 15, becomes active. Managers initiate, lead, and support on the basis of sound information and decisions, as covered in Chapter 16. Quality cost accounting, described in Chapter 17, is particularly valuable for performance planning and control at the top management level. However, all individual decisions, procedures, and activities need to be coordinated and integrated through a quality assurance program, which is discussed in Chapter 18. Given the environment of competition, technological, economic, and social changes must be reinforced in quality assurance programs and their implementation must be examined through audits, which are covered in Chapter 19.

Part IV will complete the study of *The Management of Quality Assurance,* with a survey of community-wide involvement with quality assurance management.

417

CHAPTER 14

OBJECTIVES AND STRATEGIES

"'Would you tell me, please, which way I ought to go from here?' asked Alice. 'That depends a good deal on where you want to get to,' said the cat. 'I don't much care,' said Alice. 'Then it doesn't matter which way you go,' said the cat.'"[1]

This excerpt from *Alice in Wonderland* can be considered as a parable that bears much truth about management determining where to go and how to get there. Quality assurance, as we have seen, is both an objective and a strategy. Satisfied customers are obtained and retained through quality of products and services. This objective demands systematic plans and energetic actions, well conceived and communicated. In order to bring about quality assurance, everybody in the company should know the objectives and strategies decided upon for quality assurance. Otherwise, decisions and actions lack direction and thus principles of good workmanship and high quality of output do not become implemented in the company and at the individual workplace. Managing for quality assurance means, first and foremost, setting clear, meaningful, and challenging objectives, and planning ways and means for achieving these. In an interesting article Feigenbaum[2] writes: ". . . Effective international competition today is a combination of competition in its highly visible and traditional form—product versus product—together with a less outwardly visible, but equally powerful, competition involving companies' effectiveness in quality and productivity management. . . ." In order to better understand this particular responsibility and task for top managers in large organizations, and owner-managers of small enterprises, a sample from the "in-basket" of such a person might be instructive. Perhaps we should imagine the desk of the president of the bus manufacturer that we used as an example in a previous chapter, as he or she encounters major defects in a $44 million project.

The note on the rejection of 200 buses is, of course, a bombshell. As an outsider, not knowing the details and possible mitigating circumstances, it is easy to condemn the senior management of the manufacturer. It appears, however, that the president did not previously realize the importance of a comprehensive quality control program, because this is what the president is now proposing, though, of course, fairly late.

Other notes in the president's basket at this time are probably related to major quality problems: questions by departmental managers and supervisors regarding what kind of quality program standard he or she has in mind; what kind of advertisement to use to recruit the required staff; and where to publish it for the notice of qualified quality personnel.

There is a letter from one of the major suppliers demanding new guidelines and an explanation of why the supplier's company was mentioned negatively in the newspaper. There is also a request for an interview by the local broadcasting company, and a note by the union representative suggesting a meeting on the matter.

Naturally, a senior manager will delegate much of the work to aids and departmental managers. Yet, what remains for the often very lonely person at the top, particularly before and after such bombshells are dropped, are the decisions on major issues and problems that touch on the long-term existence, and success or failure, of the company. In the positive sense, through this person the long-term needs and opportunities must be monitored and analyzed. Responses to major changes in the marketplace must be worked out and approved, and the wishes of the owners and the customers addressed and satisfied. Senior corporate management cannot involve itself in daily and current problems in ongoing operations. However, these lower level decisions and actions need to be directed and controlled through top management. Such directives and decision aids for lower echelons basically take the form of statements of objectives and policies, or strategies.

"The leader conceives the things for which he strives as things already attained" said the poet Goethe. In the mind of the president and of others, the objective and intention for introducing a quality control program in the company is still a desired state of affairs at some future time. The strategy to get there still has to be worked out. Again, the president, as chief executive officer, will have to articulate and communicate his understanding of approaches (i.e., strategies and policies) for achieving the objective of an effective quality control program. Many executive officers find themselves in a quandary when establishing objectives and strategies for quality assurance. Awareness achieved only through bombshells is a fairly common status quo on all levels of some organizations. The fact remains that quality leadership, in large and small companies and organizations alike, must emanate from senior management. Quality control, conceived mainly as inspections, and delegated as such to lower supervisory levels, does not substitute for such leadership in the long run.

14.1   THE PROCESS OF DECISION MAKING

Before we discuss the setting of objectives and respective strategies for quality assurance in more detail, a brief review of decision-making systems will help in understanding the functions and functioning of top management.

A *system,* as we have seen earlier, is designed to achieve specific objectives and purposes. The purpose of the productive system is to create products and services, using available resources and technological processes that are *fit for the use.* Decision systems are analogous to any other productive system, in that decisions are to be produced that in turn initiate and direct actions. Statements of quality assurance objectives and strategies are outputs from the corporate decision system, partly personified through the president. The inputs into these decision-making systems are information, perceptions, communications of opportunities, and developments and problems external or internal to the organization.

In the example of the bus manufacturing company, when the chief executive officer has decided that a quality control program is to be introduced, this objective needs to be translated by senior management into intermediary goals and milestones. Furthermore, as shown in Figure 14.1, goals need to be translated again into specific tasks that can be assigned to lower level operational staff. Policies or objectives, as well as tasks, establish the desired outcomes in a hierarchical order. Before we consider this goal setting in quality assurance, related determination of ways and means for arriving at these destinations will have to be mentioned.

In our understanding, *policies* and *strategies* describe approaches and principles for the attainment of objectives and subordinate goals (Figure 14.1). Establishing a quality control program is clearly an objective for the company in question. Does this mean it is also a policy? Some would claim that yes, it is, because the program is for the purpose of attaining a higher degree of quality in the future and thus having more satisfied customers. One could call this a policy of ultimate ends, in which all objectives are streamlined down to individual tasks for a final justification. We shall not pursue this largely academic discussion, but the reader should realize that such interrelationships do exist. What matters, paraphrasing the cat's advice again, is the destination and the way that leads to it.

In the following, we shall first focus on establishing such goals and destinations, and then describe respective approaches to them. The *hierarchical structure* will help us to delineate corporate from operational management. As opposed to the focus of Part Two on the quality of the product and thus on the middle management perspective, we will now take a *company-wide* view. We can not exhaustively deal with *management by objective,* or any other related more general topics of management. Concentration will remain on establishing company-wide quality assurance that embraces all individual products, projects, orders, and subsystems in a productive system.

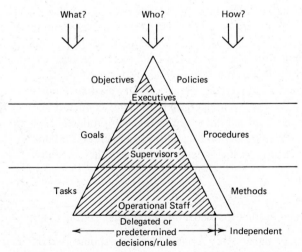

FIGURE 14.1   Elements of decision making and planning in the hierarchical order.

14.2 SETTING OF OBJECTIVES FOR QUALITY ASSURANCE

Objectives serve as a guide for the decisions and actions necessary for their accomplishment. They are further described in the form of subgoals and subtasks, as shown in Figure 14.1. At the same time, objectives themselves are decisions arrived at after careful consideration of the need, desirability, and feasibility for them in the given context. When the president of the bus manufacturing company decided to establish a quality control program, he or she obviously did this in response to a pressing need for dramatic defect-prevention measures. The objective of a suitable and effective quality control program will have to be further explained and clarified in more detail.

Some questions to be answered are the following.

1.   What should the elements in such a program be?

2.   What published quality program standard should be complied with?

3.   Should the program include design assurance or should it just be a defect-corrective inspection system?

4.   Should the program involve a radical change in current control practices and procedure (a breakthrough), or a more gradual change and improvement?

5.   What individual projects and project goals would lead to development and implementation of such a quality control program?

6.   Who should be in charge of such projects and who should participate?

7.   What deadlines should be set for the accomplishment of the quality control program and for the individual milestones leading up to it?

All these questions, and many more, require answers that will lead to the formulation of instrumental goals and task assignments. The quality related hierarchy of objectives, goals, and tasks will follow the sequence shown in Figure 14.1

This kind of structuring provides for systematic delegation of responsibilities from the top, down to the operational functions in a company, and thus for wide participation in determining tasks and responsibilities, and for constructive communication and rational decision making. Objectives for better quality and quality assurance can readily be understood by the company staff, particularly when the need (as in the example of the bus manufacturing company) is obvious and the possibility of blaming others no longer exists.

Corporate and senior management can use quality assurance objectives for the purpose of achieving general improvement in operations and staff cooperation. The new and innovative quality control program will require major changes in production planning and control, new purchasing procedures, introduction of quality- and operation-related audits, and other measures that will have impact on the general work life in the company.

Major Principles

There are many principles that should be observed when setting quality assurance objectives.

1.   The need for greater quality assurance efforts should be convincingly demonstrated and analyzed. Otherwise, objectives will not be perceived as important challenges and the probability for optimal attainment will be reduced.

2. Objectives must be realistic in view of the financial and human capacity of the company. Ends and means need to be tested, as conflicts between them usually create frustration and disharmony. In working for quality assurance, such adversity can very quickly become counterproductive.

3. Objectives must be clear, acceptable, and aligned with policy statements. Visible management input and approval must exist at all times together and in compliance with:

  a. Existing codes and standards
  b. Facilitation of wide participation of all
  c. Allowance for independent decisions and partial goal setting
  d. Coordination of objectives, goals, and individual tasks for quality assurance
  e. Translation of goals and tasks into fair and workable performance standards
  f. Visible and meaningful recognition for goal achievement
  g. Fair and sufficient support in case of difficulties
  h. Possible revision of tasks and goals; and
  i. Adaptability

Rules and Procedures

Most of the rules for sound goal setting seem to be common sense, although their violation frequently leads to just those conditions in a company that breed poor workmanship and poor quality. The style of goal setting for quality assurance, as well as for other outcomes and achievements, depends on senior management policies and the personalities involved. A chief executive officer must see to it that laws, codes, regulations, and directives from government sources are complied with. On the other hand, they must also represent the interests of the company and thus actively contribute to quality assurance, not only internally, but also externally. For instance, many major customers impose compliance with published quality assurance standards, such as MIL-Q-9858A, CSA-Z 299.1, and so on. Corporate officers have ample opportunity to participate in the writing of these standards and to participate in setting quality assurance objectives in their industry.

Methods and Practice

Methods for goal setting range from independent conception, formulation, and communication by the boss to more participatory approaches. The latter type of approach involves the operational staff by the use of quality circles or the more conventional project teams. Through such dynamic goal setting at the grass roots level, many problems obstructing proper task achievement and workmanship can be overcome without direct senior management involvement. At the same time, more serious and general problems and opportunities for improvement in current quality assurance can be monitored and brought to the attention of supervisory management. Active and comprehensive goal setting for better quality assurance should proceed from the top to the bottom and, to be realistic, also in the reverse order.

The recently developed and most frequently applied institutional arrangement is the formation of project teams, each having been assigned specific goals and tasks. Such project teams allow direct input and participation of senior, as well as other, managerial and operational levels of staff. Depending upon the need and the environment, the formation and execution of critical projects become a significant milestone in the improvement of quality.

In many cases, special tailor-made planning is needed for each key project. The exact form of such a project is decided upon by a combination of the following.

1. Assessing the status quo.

2. Analyzing customers' complaints.

3. Analysis of major failures and defects, using Pareto Analysis, histograms, and other similar methods.

4. Base line audits to determine strengths, weaknesses, and voids in current programs.

5. Comparing the existing program with generic standards.

6. Deriving goals from existing corporate objectives and policies.

7. Considering the setting of tasks and methods at the operator level.

In the example of the bus manufacturer and the declared objective of establishing a quality control program, aside from deriving and communicating subordinated goals, the president might very well arrange for various ad hoc projects, in order to get things done quickly and systematically. As has been pointed out earlier, objectives by themselves represent benchmarks and hopefully challenges for those who have to achieve them. Objectives, by definition, do not normally describe approaches without clear-cut strategies and policies. Such guidelines also have to be formulated by chief executive officers to complement the objectives. Table 14.1 gives a detailed example of an objective where policy, resource allocation, departmental goals, task assignments, and deadline are clearly stated.

14.3 STRATEGIES FOR ACHIEVING OBJECTIVES

A *policy* is a general guide for decisions and actions; a *strategy* is a more direct and explicit guide towards achieving specified objectives. According to Massie, "A policy is an understanding by members of a group that makes the action of each member more predictable to other members."[3]

Policies usually originate at the executive level of a company and are influenced by many factors, such as need, laws and codes, standards and convention, appeals and suggestions by staff, customers, suppliers, audit reports and other information requiring corrective action and general directives from senior and supervisory management. Generally speaking, both policies and strategies express a point of view and represent established or intended ways and approaches that are to be complied with by all those concerned.

A policy and its strategy should be clearly related to the objective they purport to attain, because otherwise they lack rationality and direction. Major principles to be observed are the following:

1. As a guide, for further decisions and actions it must be formulated in written, understandable, and acceptable terms. A policy is by nature not a command but designed to support operational managers and staff.

2. A policy and its strategy must be general enough to allow delegation of responsibility, and specific enough for consistent and goal oriented decisions and actions.

3. It must anticipate future developments and difficulties and plan resolutions to possible counteractions. This will reduce uncertainty and render policies more realistic and adaptable.

TABLE 14.1
An Example of Quality Assurance Strategy

Objective	Improved quality of supplies (incoming material)
Policy	Clear specification concerning product and quality assurance of supplier (applying CSA standard)
Resource	Vice president in charge of production, special budget allocation
Quality Assurance Department Goals	Determination of defective and returned supplies during a given year, showing item, supplier, defect, percent defective, time of delivery, purchase order, and any other relevant information.
	Procedure: Listings from respective documents, format to be approved; statistical analysis and graphs. Deadline June 30, of the following year.
	Responsibility: Quality Assurance and Purchasing Departments. Other related goals: determination of specification, vendors rating special quality audits; comparative analysis of price and quality; improved receiving procedures, acceptance sampling, cooperation with suppliers; employing of consultants; change of company's requirements; change in purchasing contracts; appointment of an ad hoc committee and any similar measures needed.
Task Assignment	Allocation of purchasing and receiving documents; initial assessment and reporting to quality assurance manager.
	Method: Meeting with accounting department; sampling inspection of entries in documents; written report completed by April 30.
	Responsibility: Mr. X, Inspector, quality assurance.

4. It should be used by senior management to create the atmosphere of confidence and cooperation that is instrumental for good workmanship and quality assurance.

The declared objective of the bus manufacturer in our example is to establish a quality control program. The chief executive officers will have to weigh their decisions in light of many of the guidelines taken from the available standards. For instance:

Because any quality program must receive complete support of management to be effective, management must demonstrate this support by clearly defining the policies and objectives of their quality program.[4]

Effective management for quality shall be clearly prescribed by the contractor. Personnel performing quality functions shall have sufficient, well-defined responsibility, authority and the organizational freedom to identify and evaluate quality problems and to initiate, recommend or provide solutions.[5]

The contractor is responsible for planning and developing a program which assures that all his management, design and technical responsibilities for quality are integrated and executed effectively.[6]

The quality control system includes receiving, identifying, stocking and issuing parts and material; all manufacturing processes, packing, storing and shipping.

The system is designed to ensure customer satisfaction through quality control management of supplies made and services performed here, and by our suppliers at their facilities. It is designed to spot processing problems early so we can correct them before we've produced a lot of faulty items. Written inspection and test procedures will be prepared to supplement drawings and other specifications, as necessary. . . .[7]

All of the above statements imply certain objectives and strategies; for example, in order to ensure satisfaction of customers, quality problems are to be spotted as early as possible in order to prevent defects. This expresses essentially the objective and strategy of a quality control system or program that must be part of the company's policy.

Typical Policy Statements

In its most common form, any statement of policy that top management makes remains vague to lower levels of management. However, it should be remembered that a policy statement is normally prepared to provide a general guideline for planning the quality assurance program that the company has set out to create. Consider, for example, the following typical policy statement.

It is the policy of XYZ company to provide products and services that:

1. Satisfy customers' needs and perform their required function safely and reliably for the intended use.

2. Meet all applicable codes and regulations of the federal, provincial, and country-wide industrial standards.

3. Are accurate per the company's advertisement and sales promotion.

Many different versions of the above statements can reflect the messages and quality mission. It is also possible to specifically state quality policies quite separately from overall company policies, for the specific managers responsible for achieving the objectives of such policies by, for example:

1. Establishing and implementing written procedures appropriate to the business and taking other necessary actions to provide products and services that are safe, reliable, and that preserve or improve the present environment.

2. Reviewing and appraising legislation (enacted and pending) related to the safety, reliability, quality, and environmental effects of the company's products and services, and taking appropriate action.

3. Reviewing and appraising all situations and claims involving personal injury, property damage, and environmental damage arising from the use of the company's products and services, and being aware of all pertinent actions taken.

4. Periodically reviewing and appraising the effectiveness of the division's product-integrity efforts and initiating whatever changes are necessary to ensure achievement of their product-integrity objectives.

These departmental policy statements are somewhat more direct and also consider on-line managerial responsibilities. On the surface, everyone may seem to be in favor of quality; but the concern of top management has to be genuine. In order to ensure that practice does not differ from pronouncements, one approach that is actually in use is to divide policy planning according to subsystems, and then to make statements for each of the elements in each subsystem.[8] Table 14.2 is an example of policy statements suitable for a *receiving/ incoming material control* subsystem.

14.4 POLICIES, PROCEDURES AND WORK ELEMENTS: THE INTERRELATIONSHIPS

Policies and strategies as general guidelines for goal achievement have limitations and need to be complemented by more specific procedures and performance instructions in order to become meaningful to all concerned. This interrelationship of policies, procedures, and work elements (including methods) at the individual job level can be drawn upon, as in Figure 14.2*a* and Figure 14.2*b*.

While objectives are translated in the form of goals, subordinated goals become results of specific procedures. For instance, in the example of the bus manufacturer, a decision was made to establish a quality control program to be achieved by adopting a particular quality program standard. One goal here would be to improve the quality of supplies (Table 14.1), and another would be to review and assure the quality of the incoming materials (Table 14.2). Many other goals would also be listed by the president. Improvement of supplies would be delegated to the purchasing agent or a project team and quality assurance of supplies would be specified through procedures. While doing this, the directives for assuring quality of supplies were already considered in conjunction with product or project-related quality assurance. The president would become involved mainly through approval of such procedures and controlling the compliance through audits.

Procedures should allow for independent decision making whenever and wherever possible, for more challenging and satisfactory jobs and work assignments. Yet, the operational staff might require more detailed *task assignments* and *performance standards,* particularly in the case of high risk operations and items. As Figure 14.2*a* illustrates, even these specific job-related directives and aids become coordinated only with overriding procedures and policies, and thus with respective goals and ultimate objectives. Therefore pinning down a formalized plan is not an easy task. Top management must consider a wide latitude in what it may take to materialize a quality plan in perspective of other business plans of the company (see Figure 14.2*b*).

However, the degree to which subordinate decisions and actions are governed by supervisory and top management staff via decisions in the form of policies or procedures, is increasing toward the lower levels of responsibility in the organization. The nonshaded area in Figure 14.1 indicates a proportion of delegated authority and responsibility for independent decision making and action. The bus manufacturer will probably have to rely heavily on the ability and cooperation of supervisors and operational staff to design and implement procedures and improved work methods. The quality control program is urgently needed and much depends on carrying out the new policy through staff initiative and competence. One procedure for quality program planning, which interlinks all levels of management, is described in Figure 14.3, together with steps leading to the quality program audit.

TABLE 14.2
Examples of Typical Policy Statements

Elements	Policy	Scope
Vendor capability survey	Vendor quality capabilities must be assessed periodically be reviewing vendor performance.	Assessment of vendors must take place by discussions with their representatives or visits to vendor sites.
Records and vendor feedback	The record-keeping requirements must be provided and feedback of information maintained to assure quality.	The requirements of such policies must cover the acceptance–rejection criteria based on incoming inspection control procedures, with means of communicating information to vendors and management.
Incoming material testing	Established procedures must be followed for testing of all incoming materials and records must be kept.	The procedures should indicate the method of testing and the quality characteristics being tested.
Purchasing requirements	Purchasing procedures must conform to the requirements of quality specifications, including sampling and testing procedures, in the manner of agreed-upon contracts.	The purchase documents must contain the specifications for each item. Any modification and/or acceptability criteria must be included in order to help the purchasing department select vendors.
Vendor capability survey	Vendor quality capabilities must be assessed periodically by reviewing vendor performance.	Assessment of vendors must take place by discussions with their representatives or visits to vendor sites.
Approval of samples	Established procedures must be followed for approving the samples in consultation with the engineering, design, and purchasing departments.	Sample testing must adhere to the prescribed sample size, sampling plan, average quality level, or any other prescribed scheme as laid out in the procedure manual.

If the president lacks the ability to formulate sound policies and directives, and if the more objective oriented strategy pertaining to the new quality control program is not sufficiently understood and accepted by the staff, improved quality assurance cannot be achieved quickly enough. The article ''Japan—where operations really are strategic'' in *Harvard Business Review*,[9] describes cases in which top management, through consistent setting of long-term quality assurance objectives and strategies, make short-term decisions more workable. As a result, in each case, lower-echelon manufacturing managers clearly understand the strategic significance of their day-to-day concern with operational detail. The writer of the article claims that in American companies such a coherence between strategic decisions and those at the operational level is usually not as strong. Quality assurance, among other objectives

Approach	Used By	Extent of Influence	Scope
Policy	Managers	Affects every unit in company	Very broad
Procedure	Middle-managers, supervisors	Specialized by function, decision, etc.	Specific, limited in overall application
Method	Lead hands, workers	Directs operation, job etc.	Job restricted, technical, definite

(b)

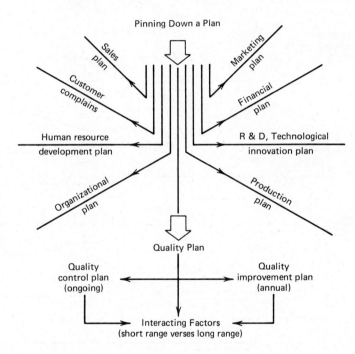

- Statement of what the company has set about to do.
- Business philosophy with respect to quality, understanding of what people buy and why.
- Means by which the company intends to do business, in terms of quality.
- Strength in budget to back up quality demands.
- Overall alignment and interdepartmental relationships.
- Administration and ability of the managers.
- Understanding of company's capabilities and shortcomings.
- Details of planning at both the abstract level and the operational level.
- Priority the company gives to built-in quality philosophy.
- Flexibility and adjustments in overall plan with respect to quality.

FIGURE 14.2 (a) Guides for decision and action. (b) Consideration of interacting factors for pinning down a quality action plan.

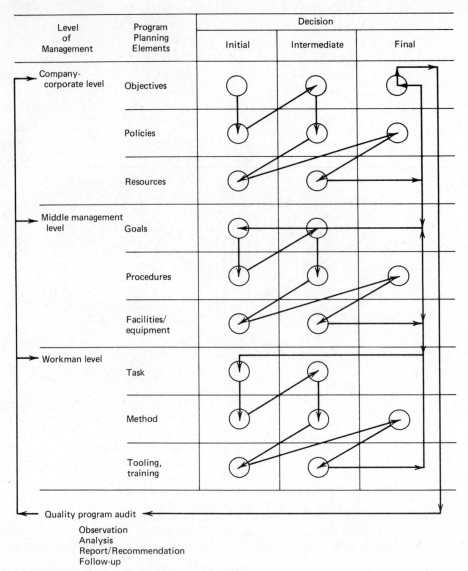

Level of Management	Program Planning Elements	Decision		
		Initial	Intermediate	Final
Company-corporate level	Objectives			
	Policies			
	Resources			
Middle management level	Goals			
	Procedures			
	Facilities/equipment			
Workman level	Task			
	Method			
	Tooling, training			

Quality program audit
Observation
Analysis
Report/Recommendation
Follow-up

FIGURE 14.3 Procedure for program planning.

of the company, in fact results from a quality program with continuous direct involvement of senior management. For this the policy statements are of crucial importance.

Some major principles for effective goal setting for quality assurance are also expressed by the following writers. It seems to be important that goals, policies, and action programs are well-coordinated in order to lead the company to desired achievements.

Eliminate use of goals and slogans posted for the work-force in an attempt to increase productivity. Zero defects is an example. Such slogans, in the absence of quality control, will be interpreted correctly by the work-force as management's hope for a lazy way out, and as indication that the management has abandoned the job, acknowledging their total inadequacy.[10]

In short, meeting higher quality and performance goals does not inevitably mean higher costs. . . . It is not necessary to exceed competitive quality by a wide margin, nor is it usually commercially rewarding. . . . When setting goals of zero defects, U.S. managers need to know what level of quality customers expect and what quality problems they will tolerate. It may not be necessary to achieve an absolute level of zero defects, but management should be aware of the exact standard necessary to meet competition—and then should continually measure actual performance against that standard. Our experience has been that making quality the organization's most important goal can make work more satisfying for all concerned.[11]

To give the leadership of the quality function to top management requires the creation of new management tools, and one of these is the annual quality program. The annual quality program includes the policies, objectives, plans, and other elements of a broad approach for defining what needs to be done and how to do it.[12]

Create a mechanism for establishing quality goals and reviews for every level. Such goals should be consistent with a hierarchy of goals, the highest of which are established to support ultimate company objectives by top management, which is fully responsible for quality. Hopefully, such goal-setting will proceed from high-level analysis of customer needs, competitive activities and company capabilities.[13]

14.5   A STRUCTURED APPROACH TO POLICY SETTING

Decisions on a quality assurance policy can be arrived systematically in at least five major steps, as follows.

1.   Determine the objective and draft the respective approach either as a general policy or a more specific strategy. Gather all relevant facts. Generate alternative courses of action; then analyze and select, possibly with participation of staff and outside assistants.

2.   Have policies approved and finalize the draft.

3.   Release, communicate, and interpret the policy to all those concerned. Those supervisory managers who are to establish procedures have to be adequately informed in order to be able to implement the policy correctly.

4.   Implement the policy when respective procedures and performance directives and standards have been finalized.

5.   Control adherence to policies through audits and corrective action; including a review of the policy.

Such a structured approach to policy setting follows a similar pattern for determination of objectives and can to a large extent be combined with this. In a small business environment, a description of the quality control policy might very well suffice. The bus manufacturer, however, does operate on a somewhat larger scale with a more diversified company and staff. The president, after having stipulated the quality control program for preventing major defects and loss of business, would most likely form project teams to work on the matter. Each team would be given terms of reference and goals, as was mentioned earlier when discussing goal setting. These teams would interpret the goals and associated policy statements by their procedures and task assignments.

Such a project team structure provides the president with ample opportunity to influence and follow up the establishment of the quality control program until it has been established in fact. Moreover, available staff resources and aids can constructively participate in improving quality assurance through various projects. Senior management will act as coordinator of these projects and thus provide leadership beyond merely setting objectives and strategies.

One of the main prerequisites for effective goal achievement in quality assurance projects is the existing organization, with its regular rules, work practices, personalities, and their interrelationships. Setting of objectives and the ways and means for achieving these is a continuous task for management, because dynamic developments, both inside the company and external to it, demand adaptation and positive responses. In the next chapter, the question of creating an organization and team for optimal quality assurance will be discussed. Objectives and policies need the backing of the people in the company. Such policies are prepared by and intended for the company as a whole and, in more specialized form, for the departments and individual functions. The management cycle of decisions and activities only starts with goal and policy making.

14.6 PROBLEM SOLVING THROUGH GOAL SETTING: A CASE EXAMPLE

At the present time, Transport Windair Limited is experiencing delays and suffering large costs in trying to make new navigational equipment available to the flying public.

When a new piece of equipment is installed it is given a proof of performance test before it is made available. In this test, engineering, operations and quality assurance personnel go over the site and equipment specifications to ensure that the equipment will perform its intended job and that it is maintainable. Once the test is completed, all deficiencies are noted and the group comes to a mutual decision as to whether the operations group will take it over from the engineering group and put the system in the air.

Presently, enough deficiencies are being found that operations will not take over the equipment and engineering is forced to redo many parts of the installation, causing increased installation and travel costs, as well as a delay in making the facility available. The changes then have to be checked again by quality assurance and operations. These facilities can be thousands of miles apart; so travel may be expensive. Management initiatives resulted in the following structured approach.

Objective: To reduce the time needed to make new navigation aids available, and thus save large engineering and travel costs as well as providing prompt service to the public.

Goals: To reduce the number of major deficiencies found in new installations during proof of performance testing.

Tasks:

1. Testing of equipment before it is installed.
2. Monitoring of private contractors.
3. Training of engineering installation staff.
4. Involvement of quality assurance and operations staff on initial site decision proposals.

The policy of the telecommunications branch of the company is that large costs and time spent in making new facilities available to the flying public can be reduced substantially by greater interdepartmental participation in site design, more training in good installation practices, and closer monitoring of site building contractors.

Work Elements:

1. Test all equipment before it is shipped and installed so that its capabilities are known before proof of performance testing.

2. Closely monitor building contractors to make sure they fulfill the contract and follow building and electrical codes.

3. Make all installation staff aware of Transport Windair standards for installation and workmanship.

4. Give all divisions input on initial site design proposals so objections can be dealt with before the site is built and completed.

Approval for Implementing the Proposal Would Involve:

1. Producing data on past installations showing how much delay was caused in getting new facilities commissioned and in the air by deficiencies uncovered in proof of performance tests.

2. Providing data on the total costs of:
    a. Correcting contractors' mistakes at site.
    b. Equipment reengineering at site.
    c. Labor and travel costs for installation personnel to correct deficiencies found during proof of performance tests.

3. Providing detailed listing of costs to implement all the changes proposed.

Implementation Plan

STEP 1: *Employee meeting number one (first week).* Initial meeting with operations, quality assurance and engineering staff conducted by quality assurance managers. Outline program and conduct question and answer session. One four-hour session at the regional office.

STEP 2: *Courses for engineering staff (second and third week).* Set up courses, conducted by quality assurance staff on:
a. Monitoring of private contractors, review of building, electrical codes, and where to go for information. Meeting to be conducted in two sessions of one week each in the engineering workshop.
b. Workmanship standards: what is expected of them to conform to Transport Windair standards and where to get information.

STEP 3: *Design coordination meeting (fourth week).* Conducted by managers to all supervisors in quality assurance, operations and engineering. The purpose will be to develop a policy to give all departments input to initial project proposals. This policy will then be used on the next project.

STEP 4: *Equipment inspection (first project).* Equipment to be installed in first project is removed from stores, set up at workshop, and tested thoroughly. A full report is made to the implementation team of all defects found.

STEP 5: *Employee meeting number two (middle of first project).* Meeting of all staff in all three divisions to keep everyone up to date and to get feedback on how the project is going. Query for suggestions. The meeting should be conducted by quality assurance implementation staff, for one half-day at the regional office.

STEP 6: *Proof of performance test.* Conducted by representatives of all divisions at the end of the project. A detailed analysis of the results is obtained, prepared by quality assurance and supplied to all divisions.

STEP 7: *Employee meeting number three (completion of project).* Meeting of all staff, conducted by the regional manager, to discuss results obtained in the first project. Meeting includes a question and answer session. Ask for suggestions for improvements. Meeting should be held for one half-day at the regional office.

STEP 8: *Cost of quality report (completion).* Detailed costs of the program up to completion of the first project, with a breakdown in costs for the project itself and the initial start-up costs. The report is supplied to all divisions.

STEP 9: *Audit program.* The program is audited by quality assurance evaluating the program results and the cost of quality. This report is supplied to all divisions.

Program Start Up:

a. Finalize implementation plan.
b. Obtain approval of regional management.
c. Have regional management introduce the program to operations, engineering and quality assurance.
d. Hold introductory meetings with all staff at the regional office to explain the purposes of the program and what is hoped to be accomplished.
e. Set up courses for engineering staff on contract monitoring and workmanship standards.
f. Conduct meetings between managers of all the divisions to implement a procedure for the divisions having input to the project design.
g. Monitor the results of the first project for results of equipment inspection and results of proof of performance testing.
h. Conduct regular meetings with staff involved, to promote awareness of results and to get feedback on programs and suggestions for improvements. Have an evaluation done by the quality assurance department on the results of the program and the final cost of quality.

14.7 DECISION MATRIX AND DECISION TREE

In the following, quantitative methods for management are demonstrated and reviewed with reference to quality assurance. Strategies and policies, as decided by senior management, can be analyzed by means of *decision matrix* and *decision tree* approaches. The matrix relates different expectations (states of nature) to alternative management-controlled paths of action. Demands in terms of quality vary in the market, depending on numerous forces that management has little or no control of. The demands are either relatively *high, moderate,* or *low.*

Quality could be defined as related to price level or as a degree of freedom from defectives. The alternatives for management can be expressed in terms of quality assurance effort, measured by program *intensity* and *comprehensiveness*. The body of the matrix, as shown by the table in the following example, has an estimated payoff for each pairing of the various states of nature with a quality assurance strategy, expressed in dollar terms.

Example: Consider the three quality program strategies and their demands based on the customer's specifications as shown in the following table.

Quality Assurance Program (strategy)	Demand of Quality Level (based on specification)		
	Low (0.2)	Moderate (0.5)	High (0.3)
	payoff		
Comprehensive	(6)	3	14
Intermediate	9	12	13
Final inspection	12	13	10

1. Determine the *maximin* decision, that is, the worst possible payoff decision for each alternative.

2. Determine the *maximax* decision, that is, the highest possible payoff decision for each alternative.

3. Determine the *maximum expected monetary value* decision.

Solutions

1. 10 (final inspection)

2. 14 (final inspection)

3. $-6(0.2) + 3(0.5) + 14(0.3) = -1.2 + 1.5 + 4.2 = 4.5$ (comprehensive)
$9(0.2) + 12(0.5) + 13(0.3) = 1.8 + 6 + 4.9 = 11.7$ (intermediate)
$12(0.2) + 13(0.5) + 10(0.3) = 2.4 + 6.5 + 3 = 11.9$ (final inspection)
Strategy: To select the final inspection program.

In the example just given, if, for instance, a relatively low quality demand will materialize and the company in expectation of such a state in the future, has implemented only end-item inspection, (that is, a modest and low-cost quality assurance program), the payoff will be relatively high. When competition and customer demands force the required quality upwards to a relatively high level, then this same company will have a lower payoff; it will be less competitive in the market.

As has been mentioned the payoffs are estimates and it depends on the experience and research effort of the analyst whether figures will be fairly valid. With the matrix established, management can now choose, usually from two basic decision-making approaches: to determine the worst possible payoff for each alternative strategy (quality assurance effort) and then select the relative highest payoff (*maximin principle*), or to determine the highest possible payoff for each (*maximax* principle). The selection of either of these two principles reflects and depends upon the expectation and attitude of management. Selection of the maximin connotes a more pessimistic outlook into the future. In the previous example the maximin principle leads to the cell (in the example table) with $10 million of low quality assurance effort and high quality demand. The maximax principle would yield $14 million payoff for high quality assurance effort and high quality demand.

The decision matrix can further be refined by attaching probabilities to each state of nature and then calculating the expected monetary values. The sum of these probabilities is one. If the low demand is associated with 0.2 probability, the moderate with 0.5, and the high with 0.3, the alternative (strategy) that will render the highest expected monetary value would have to be chosen.

Analysis Using a Decision Tree

A *decision tree* is similar to decision matrix, in that it helps to project strategies and outcomes into the future, using estimates. The tree-like schematic has branches that emanate either from decision-of-event-points or *nodes*. With the original node representing the present time, all other nodes represent future decisions and events as they are interrelated.

A tree for a similar case as used in the above decision matrix example, although abbreviated, is shown in Figure 14.4. In this example, management must decide whether to finance and establish a moderate and less expensive quality assurance program (QA1), such as a simple inspection program, or a comprehensive and more powerful one (QA2). The QA1 branches represent the policy of waiting for the outcome of the demand for quality; there is a 0.7 probability for low demand and a 0.3 probability for high demand. If, in fact, the high demand does materialize, then the choice would be to introduce a QA2. This will give an estimated payoff of $20 million. The payoff will be zero if nothing is done, and $40 million if the production facilities are reequipped in order to provide better quality.

The other branches must be interpreted similarly. In order to evaluate the entire decision tree and to arrive at a decision for the right-most decision node, the highest payoff branch in each future decision point is chosen and then calculated further back to the left and the original node. The decision must be to adopt QA1.

The payoff estimates in this example were taken in terms of present value. The three analyses can be further refined through introduction of time periods, costs, and revenues.

Management benefits from the use of decision matrixes and trees mainly because the models lead to a systematic assessment and inclusion of factors and forces that essentially influence success and failure of quality assurance strategies and policies. The *payoff criterion* then leads to the most promising path of action. These models help management in decision making in a general sense, and many other deliberations and methods are used before the final decision is made. Many practitioners also have to overcome the reluctance to settle on estimates, which these decision models require, realizing that nothing is more difficult to predict than the future. However, to do nothing and let fate decide, is not desirable.

Example 1: The Britannia Yacht Company will establish a new manufacturing facility. President Charles Prince must decide whether to opt for automated assembly, or for a manual assembly method.

Mr. Prince has researched the method and has come up with the following figures.

1. Automated method:
   a. Fixed cost of equipment is $X.

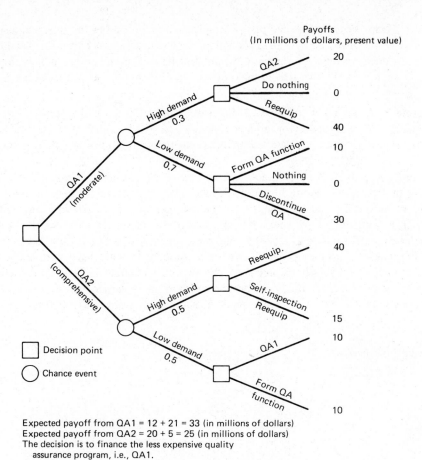

Payoffs
(In millions of dollars, present value)

Expected payoff from QA1 = 12 + 21 = 33 (in millions of dollars)
Expected payoff from QA2 = 20 + 5 = 25 (in millions of dollars)
The decision is to finance the less expensive quality
  assurance program, i.e., QA1.

FIGURE 14.4   Decision tree analysis of the example used.

b. There is an 80% chance of a smooth implementation resulting in few defects and high quality yachts.

c. If this occurs, market research predicts an 80% chance of sales of 2000 units, and a 20% chance of sales of 1500 units.

d. The automated process is new and relatively untried. There is a 20% chance of installation problems with defective installation of equipment impacting on the quality of the final product. This will affect sales, with an 80% chance of selling 500 units, and a 20% chance of selling 400 units.

2. Manual production:

a. Fixed cost of equipment Y.

b. There is a 70% chance of attaining good employee morale and quality production using quality circles. This will result in the following sales expectations: a 20% chance of selling 1000 units; a 50% chance of selling 1500 units; and a 30% chance of selling 2000 units.

   c. There is a 30% chance of low employee morale due to the overzealousness of the foreman, Mr. Bligh. Production will be curtailed at 800 units, according to a decision made at a general meeting of the executive bodies. Which alternative should be chosen so as to maximize expected output? For the answer, see Figure 14.5.

   Example 2: A company called B. J. Precision Limited is a small enterprise producing circuit boards especially made for large computers. The owner's plan of production strategy covers the next five years. A detailed analysis and forecast of existing opportunities is as follows.

   The company is to bid for a contract with the Federal Department of Supplies. The chance of getting the contract is about 50 percent. If the contract is acquired, no other jobs except for some repair work could be accepted by the company for the next two years. After that period, for the remaining three years, demand for the conventional circuit boards is the same as if the contract had not been attained. The prerequisite for bidding is to establish a comprehensive quality assurance program that is in compliance with the department's quality assurance standards. The net income from the two-year contract is $100,000. Income for the remaining three years will be in proportion to what is earned in five years with the contract not acquired. The cost of the quality assurance program is $10,000 per year.

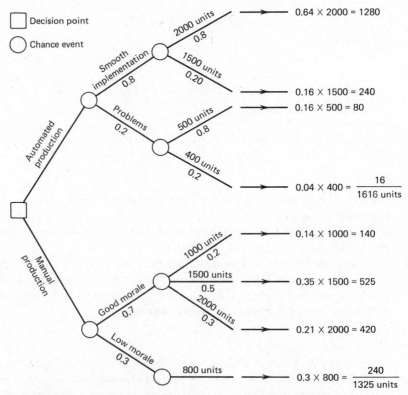

FIGURE 14.5 A decision tree analysis where automated production leads to the greater number of expected units.

The probability of experiencing a high demand for regular circuit boards if the bid was made but not received is .6, with moderate 0.2, and low 0.2. This is still more favorable than if the bid was not made, because the new quality assurance program has increased the probability for a high demand. The demand probability pattern for regular circuit boards is 0.3 for high, 0.2 for moderate, and 0.5 for low demand. A high demand is expected to furnish a net income over five years of $300,000, a moderate demand to furnish $250,000, and a low demand to net $150,000.

1. What should the decision be on the basis of *expected monetary value* (EMV) criterion? Draw the decision tree and justify the decision.

2. During negotiations with the federal department, B. J. Precision Limited found out that by also installing new test equipment along with the quality assurance program, the probability of receiving the contract would increase to 0.8. Should the above decision be revised? For the answer see Figure 14.6.

14.8   SCHEDULING

Scheduling is the planning against long, intermediate, and short-term horizons. Quality assurance programs along with capacity planning and location analysis are long-term decisions. In the intermediate planning, production and provision of services is determined for about a year. Assignment of work for the immediate upcoming short period, again has to be specific and final. Methods and models for such planning vary with the time horizon. Quality assurance is an objective and determinant in any planning over time.

Components for long-term planning that influence quality assurance are the general strategies in selecting products, markets, processes, and resource bases, with regard to quality level. Prices and costs, as well as competition for markets and resources, are interrelated with the policy for quality.

Production strategies for the intermediate range interpret long-range policies and further operationalize these. Forecasting methods help to estimate the demand in more or less aggregated terms. The demand is a function of many factors and can be influenced, positively or negatively, by the quality assurance policy of the company. Demand is for items of adequate quality, but output might include defective items.

For each month the demand will have to be predicted, because there may be *seasonal, cyclical,* and *trend* fluctuations. If the demand remains steady over this timespan, production planning is not as complicated as if the demand fluctuated. Nevertheless, the strategy for meeting demands is either to fluctuate production output in accordance with the demand, or to maintain output at or near a particular level. The least-cost strategy minimizes the total of fluctuation, inventory, and shortage costs. Level production tends to maximize inventory holding and shortage costs, keeping production fluctuation (that is, overtime and/or subcontracting) at a minimum. Linear programming and simulation can be used, among other methods, to determine the least-cost *aggregate schedule*.

Quality assurance aspects are usually subsumed in these models within the production costs. Aggregate schedules entirely based on quality costs, in addition to regular production planning, would inform about the implementation of long term quality assurance policies. Overtime and subcontracting can introduce disturbances with detrimental impact on quality costs. The same is true of fluctuating the workforce.

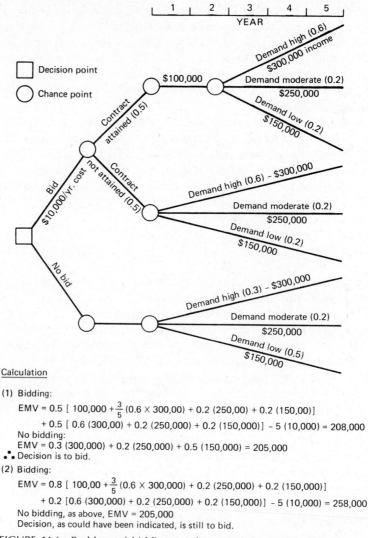

Decision point

Chance point

Calculation

(1) Bidding:

EMV = 0.5 [100,000 + $\frac{3}{5}$ (0.6 × 300,00) + 0.2 (250,00) + 0.2 (150,00)]

    + 0.5 [0.6 (300,00) + 0.2 (250,000) + 0.2 (150,000)] − 5 (10,000) = 208,000

No bidding:

EMV = 0.3 (300,000) + 0.2 (250,000) + 0.5 (150,000) = 205,000

∴ Decision is to bid.

(2) Bidding:

EMV = 0.8 [100,00 + $\frac{3}{5}$ (0.6 × 300,000) + 0.2 (250,000) + 0.2 (150,000)]

    + 0.2 [0.6 (300,000) + 0.2 (250,000) + 0.2 (150,000)] − 5 (10,000) = 258,000

No bidding, as above, EMV = 205,000

Decision, as could have been indicated, is still to bid.

FIGURE 14.6   Problem of bidding on the contract solved by using a decision tree analysis.

On the other hand, a strategy that maximizes inventory holding and back-orders carries respective risks of loss of quality. Aggregate scheduling with linear programming, simulation, and similar methods, allows for experimenting with quality cost data or estimates, in order to integrate the final plan with quality assurance objectives.

The aggregate schedule is operationalized by disaggregating demand, workforce, inventory, fluctuations, and back-orders into specifics. Once the quality-strong products, resources, workers, and other factors are known in conjunction with disaggregated quality cost compilation, respective priorities can be assigned. Production fluctuation and changeovers usually cause extra high quality costs, as can subcontracting. A steady workforce offers a better chance to build high quality performance and motivation.

Companies producing to order, as in a typical job shop situation, have less opportunity to standardize outputs and processes. In such companies, quality is directly specified in conjunction with the customer. Meeting deadlines is often the most crucial quality characteristic and proper scheduling is therefore of utmost importance. Orders that compete to be delivered on time must be assigned in a manner that optimizes meeting of delivery dates and utilization of available capacity. While management has little control over the detailed specification of the order, other than to accept or reject it, they can choose from various strategies of prioritizing orders and assigning them to work systems.

One example of a method to be used is the *assignment method* of linear programming; (see Section 14.9). The assumption that any work system can serve each order alike is a fairly limiting one. Other such scheduling methods have similar limitations.

The most versatile and practical method to use for selecting the optimal *sequencing rule* is *simulation*. Such rules to choose from are first come/first serve, longest or shortest operating time, remaining operating time, time remaining to deadline, and the like.

When orders are also categorized by degree of necessary quality assurance, sequencing rules can be set accordingly. Some orders can always be more easily rushed, with the risk of higher quality costs, than others. Moreover, one criterion of efficiency for the optimal sequence and schedule, other than the degree of meeting delivery dates or the utilization of capacity, would be the maximizing of quality assurance. Priority could be given to orders with either high or low quality assurance requirements, depending on the quality assurance policy. Actual or estimated quality costs could also serve as a criterion in ranking the priority of orders.

Scheduling of individual orders, their routes through the various processes, and the load of individual shops and work systems can be done using simple bar charts. Scheduling of inspection and testing is no different than for any other productive operation. For scheduling of large-scale orders or projects with many interdependent operation factors, modern network analysis and scheduling methods have been developed. These models, such as Program Evaluation and Review Technique (PERT) and Critical Path Method (CPM) facilitate computer usage during the project design and execution phase.

Figure 14.7 illustrates an example of a project in quality assurance. It is shown in the PERT with determistic time data. Completion is after 18 days, with the *critical path* indicated in the chart (Figure 14.7). A more complete demonstration of this method is reviewed in Chapter 18. In conjunction with Figure 14.7, consider the following elements of a start-up project plan for a quality audit.

A. The members of the audit team meets with company management to be informed of what is expected of them. The scope and objectives of the quality control system is laid out for the auditors.

B. The project manager develops an audit plan (or modifies an existing one) as a result of the meeting.

C. The auditors gain an operating knowledge of the parts of the system they are going to be auditing.

D. The auditors review all requirements of the existing quality control system as laid out in the company's policy manuals.

E. The auditors achieve physical familiarization with the plant.

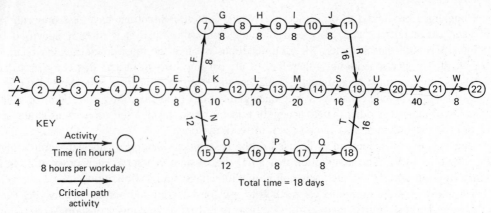

FIGURE 14.7   PERT network.

F.   Quality control procedures are audited to determine their adequacy. This is a comparison of the stated procedures with applicable standards.

G.   The auditors check for the existence and adequacy of a defects prevention program.

H.   Process and manufacturing controls are audited.

I.   Finished product inspection procedures are audited.

J.   Special processes are audited.

K.   Purchasing procedures are audited.

L.   Receiving department procedures are audited.

M.   Procedures of material control for production processes are audited.

N.   The record keeping department is audited.

O.   Drawing and change control procedures are audited.

P.   The metrology control system is audited.

Q.   Statistical quality control procedures are audited.

R.   The auditors prepare a report of activities 6 through 10.

S.   The auditors prepare a report of activities 11 through 13.

T.   The auditors prepare a report of activities 14 through 17.

U.   The audit team meets and discusses their findings.

V.   A final report is compiled.

W.   The project manager meets with company management to present findings.

14.9   LINEAR PROGRAMMING

In *linear programming,* an objective, such as the minimizing of quality costs, can be specified in terms of an *objective function.* For products with different quality costs per unit linear programming determines the quantity for each product that would minimize the total quality costs within a set of prevailing quantitative production constraints. On the basis of fairly restrictive assumptions, linear programming renders an optimum solution and its alternative. The quality cost, that is the sum of inspection, defect prevention, and failure costs, are minimized in the solution.

Minimization of quality costs is obviously not the only goal in a company. It competes with short-range profit maximization. Linear programming, and other mathematical programming techniques, serve as strategic planning tools. The question of what quantities should be produced for most efficient quality assurance is not an irrelevant one. The quality cost data, could be interrelated or incorporated with overriding costs, or profit data. Consequently, a linear program prepared for quality assurance can be handled as a module in more comprehensive linear programs.

Any such modules render restricted information in accordance with the assumption and selection of objectives and variables. In the light of outputs from the various linear programs that share the same setting and assumed conditions, trade-offs can be made and policies formulated. *Goal programming* might also be used to decide on the priority ranking of competing goals for the same available resource base.

In quality assurance, other circumstances that call for application of linear programming are where ones concerns are to:

1. Assign available production capacity, divided into individual resource constraints, to product lines, markets, or time periods so that an objective is minimally or maximally achieved.

2. Assign plant or department capacities to competing customer orders.

3. Assign inspection and testing capacities to competing customer orders and contracts.

4. Determine the best mix of suppliers and ingredients, in order to maximize quality; latter measured in terms of defect-free items.

The following is an example of linear programming with particular reference to quality assurance. In this application the simple *graphic* and *algebraic* linear programming techniques are used.

Example 1: The ABC Company produces two products, with quality costs of $1.50 (product A) and $2.00 (product B). The total quality costs are to be minimized, given the minimum work-time constraints imposed by union contract. Fabrication technique requires four hours for product A and six hours for product B. The other processes in this case are assembly and quality control tests. These have the following time requirements per product.

	Product	
	A	B
Assembly	5 hours	4 hours
Testing	4 hours	1 hour

According to the union contract agreement the minimum total work time per process for a two week period is 60 hours for fabrication, 60 hours for assembly, and 20 hours for testing. The workers are prepared to work overtime at the regular pay rate; whereas the company wishes to minimize quality costs and, at the same time, to operate at or above the minimum work level.

Given a set of restraints and variables, we wish to know the nonnegative values of these variables that will satisfy the constraints and maximize or minimize some linear function of the variables.

Solution

If the total quality costs (Z) are to be minimized, we can make a symbolical statement that $Z = 1.5A + 2.0B$, subject to: $4A + B \geq 20$, $4A + 6B \geq 60$, $5A + 4B \geq 60$, and $A,B \geq 0$.

Figure 14.8 shows the *restrictions* for the above limiting cases plotted graphically. It is seen that point C defines the combination of products A and B, for which we will obtain maximum benefit.

The desired solution is given by the minimum ($1.5A + 2B$) line, which touches the *feasible solution* area; that is, when $A = 9$ and $B = 4$.

Therefore, to minimize quality costs, The ABC Company should produce 9 units of A and 4 units of B.

The respective figures can now be calculated: quality cost = $(1.5)(9) + (2)(4) = \$21.50$, fabrication time = $4(9) + 6(4) = 60$ hours, assembly time = $5(9) + 4(4) = 61$ hours, and testing time = $4(9) + 1(4) = 40$ hours.

Example 2: The ABC Company operates three plants, all of which have the capacity of producing any of four products. However, the quality costs per unit vary, due to such factors as quality of supplies, workmanship, efficiency of quality control, quantitative capacities, and product and process specialization. The following matrix summarizes the prevailing conditions. Using this information, determine the total minimum quality costs.

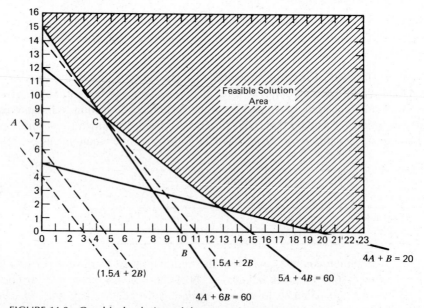

FIGURE 14.8 Graphical solution of the example used.

Plant	Product				Supply ('000)
	A	B	C	D	
I	3	5	7	5	30
II	6	2	3	5	50
III	4	3	6	2	40
Demand ('000)	20	40	30	30	

Solution

We will use the technique usually known as *Vogel's Approximation Method* (VAM) to reduce the amount of work required to develop a solution.

Vogel's Approximation Method[14] computes the initial solution by steps one through four below; steps five through seven generate the optimal solution.

1. Determine in each row and column the difference between the lowest and the next lowest cost.

2. Select the row or column, with the largest difference; if this is a tie, choose arbitrarily.

3. Allocate as many units as possible into the cell with the lowest cost and eliminate the row or column, from further assignment when supply is assigned or demand satisfied.

4. Repeat steps one through three until all assignments are made. If demand exceeds supply, or vice versa, a "dummy" row or column has to be inserted with zero cost before step one can be taken.

5. Test for degeneracy in which sum of loaded cells must equal number of rows and columns minus one. If not equal, add small assignments as load to an open unloaded cell in order to equalize.

6. Evaluate all open cells by moving one unit into the open from a loaded cell and make the counter move so that the total supply and demand in each row or column do not change.

7. When a move in step six allows for cost reduction, assign the maximum load possible to the formerly open cell and make the counter move. Repeat steps five and six until no cost savings through reassignment are possible.

Test for degeneracy: as stated in step five, rows + columns − one = loaded cells (cells with units assigned in table); or $3 + 4 - 1 = 6$. Hence no further assignment is necessary.

Evaluation of open cells: In accordance with step six above, the first open cell is Factory 1/Product C (termed as 'cell' 1/C). Add \$7 in 1/C and subtract \$3 from 2/C. Make counter move by adding \$2 in 2/B and subtracting \$5 from 1/B. Cost per unit would increase by \$1; $(7 - 3 + 2 - 5 = 1)$. This move would not be carried out.

The evaluation of the remaining open cells would also result in added cost as shown in the following calculations: $1/D = 5 - 2 + 3 - 5 = 1$, $2/A = 6 - 2 + 5 - 3 = 6$, $3/A = 4 - 3 + 5 - 3 = 3$, $3/C = 6 - 3 + 2 - 3 = 2$, $2/D = 5 - 2 + 3 - 2 = 4$.

		Product						Difference			
	A	B	C	D	Supply	1	2	3	4	5	
Factory 1	$3 20	$5 10	$7	$5	30	$2	2	2	2	(2)	
Factory 2	$6	$2 20	$3 30	$5	50	$1	1	(4)	—	—	
Factory 3	$4	$3 10	$6	$2 30	40	$1	1	1	1	—	
Demand	20	40	30	30	120						
1.	$1	$1	$3	($3)							
2.	1	1	(3)	—							
Difference 3.	1	1	—	—							
4.	1	(2)	—	—							
5.	1	2	—	—							

Total minimum quality cost: The total quality cost is the sum of the units assigned in each loaded cell multiplied by the cost per unit. For example, cell 1/A has a quality cost of 20 units × $3 = $60. The same calculation for the other cells yield: cell 1/B = 10 × 5 = 50; 2/B = 20 × 2 = 40, 3/B = 10 × 3 = 30, 2/C = 30 × 3 = 90, and 3/D = 30 × 2 = 60. The combined total cost is $330.

Example 3: A company producing pharmaceuticals has four inspectors with the same qualification but with different performance levels in terms of time conducting an inspection assignment. There are four inspection jobs with the table below showing inspector time requirements on the average, expressed in hours.

		Inspector		
Job	A	B	C	D
1	8	6	2	4
2	6	7	11	10
3	3	5	7	6
4	5	10	12	9

Assign the inspectors so that the total inspection time is minimized.

Solution

The *assignment method* of Linear Programming[14] will be used to solve this problem.

STEP 1: *Subtract the smallest number in each horizontal row from itself and every other number in the row, and enter into a new table.*

	A	B	C	D
1	6	4	0	2
2	0	1	5	4
3	0	2	4	3
4	0	5	7	4

STEP 2: *Subtract the smallest number in each vertical column from itself and every other number in the column, and enter the results in a new table.*

	A	B	C	D
1	6	3	0	0
2	0	0	5	2
3	0	1	4	1
4	0	4	7	2

STEP 3: *Determine the number of minimum lines needed to cover all zeros.*

	A	B	C	D
1	6	3	0	0
2	0	0	5	2
3	0	1	4	1
4	0	4	7	2

Since four lines (because of 4 inspectors) are needed we must proceed to step 4.

STEP 4: *Subtract the smallest uncovered number from every uncovered number and add it to the numbers at the intersection of covering lines, which results in the following.*

	A	B	C	D
1	7	3	0	0
2	1	0	5	2
3	0	0	3	0
4	0	3	6	1

STEP 5: *Determine again the minimum number of lines needed to cover all zeros. These are now equal to the number of inspection jobs (rows) and the optimum assignment can be made.*

	A	B	C	D
1	7	3	0	0
2	1	0	5	2
3	0	0	3	0
4	0	3	6	1

STEP 6: *Make assignments by starting with the row and column having only one zero: "x" is the assignment.*

	A	B	C	D
1			x	
2		x		
3				x
4	x			

Example 4: Jobs can be scheduled with inspection following the operation using the *Johnson Method.*[15]

Given the following jobs one through four, each with a particular time requirement for the machine operation and the subsequent inspection, determine the sequence of the jobs, so that the total time required is at a minimum.

Job	Machine Hours	Inspection Hours
1	4	7
2	9	5
3	3	2
4	7	6

Solution

STEP 1: *Select the job with the shortest time required for machine operation or inspection. It is Job three with a time of two hours for inspection.*

STEP 2: *If the shortest time is at the machining operation, then schedule this job first in the sequence. Otherwise schedule it last in the sequence. A figure of two hours is at the inspection operation and therefore job three is scheduled last.*
We now have:

position	1st	2nd	3rd	4th
job				3

STEP 3: *Select the next shortest time and repeat step two. It is job one with four hours at the machining operation. Job one will, therefore, be scheduled first in the sequence.*

STEP 4: *Repeat step three and proceed until all jobs are assigned. Next is job two with five hours for inspection. Therefore, it is to be assigned in the second to last position. Finally, the second position will be assigned to job four.*

We now have the following sequence:

position	1st	2nd	3rd	4th
job	1	4	2	3

The easiest way to determine the minimum time for machining and inspection of the four jobs is to construct a chart with a time scale starting at zero hours.

Time (hours) 0	4	11	20	23
Machining	Job 1	Job 4	Job 2	Job 3
Inspection	/////	Job 1	Job 4 /////	Job 2 Job 3
Time (hours) 0	4	11	17	20	25	27

 = Idle time.

The chart shows that the total time for the four jobs in this sequence will be 27 hours, including 7 hours of idle inspection time. Any other sequence would require more time.

14.10 KEY TERMS AND CONCEPTS

An **objective** is a desired state and outcome that is specified in terms of object, time, and place. It is interrelated with subordinate goals, targets, and missions, and also with other objectives of the same organization. An objective is a decision and serves as a challenge, guide, and performance criteria.

A **strategy** describes in some technical detail the method of attainment of objectives. Both objective and strategy are typical long-term decisions and action guides.

A **policy** outlines paths of action and rules in more general terms than a strategy. It often implies the objectives and goals.

Goal setting can be understood as a systematic procedure for establishing goals with participation of supervisory and operational staff. Satisfactory quality and respective quality assurance is considered a goal that is of concern to everyone in an organization. Goal setting must be coordinated with overriding objectives, and goal attainment must comply with policies and strategies.

A **procedure** outlines an activity in operational and technical detail, in a sequence of steps, in order to achieve the desired outcome of such activity. Procedures operationalize policies and the respective attainment of higher ranking objectives.

A **program,** such as a quality assurance program, is the composite of procedures and other decisions and information that purport to implement policies and to attain objectives. A program guides decisions and actions in a general and long-term sense, and thus differs from a project. A program can be designed and implemented by means of various specific interdependent projects. Programs and projects are basically defined by the objectives they are to achieve and the policies they are to implement.

Decision methods that are applied for the determination of objectives or policies are for the purpose of assuring rationality and factuality of the decision and subsequent action. These methods, such as *decision trees, network analysis/PERT* and *scheduling* can be applied for the attainment of sound quality assurance.

14.11 DISCUSSION AND REVIEW QUESTIONS

(14–1) *Senior managers cannot involve themselves in problems of current operations and business of the day.* Discuss.

(14–2) *Quality leadership must emanate in large and small companies from senior management.* Discuss.

(14–3) *An objective decided upon by senior management to establish a quality assurance program must be further translated into more specific goals, milestones, and tasks.* Explain.

(14–4) *Certain rules govern the effective setting of quality assurance objectives.* Explain.

(14–5) *There is normally also room for goal setting at the supervisory and worker level.* Explain and discuss.

(14–6) *Project teams are often very suitable for establishing goals and their attainment.* Explain and discuss.

(14–7) *A policy statement for quality assurance is, by necessity, fairly general.* Explain.

(14–8) *Policies, procedures, and work elements or tasks are closely interrelated.* Explain.

(14–9) *Policies should be established in accordance with certain procedures.* Explain and discuss.

(14–10) *Decision matrixes and trees can help in determining a suitable policy.* Explain.

(14–11) *In scheduling the implementation of policies and strategies, various methods can be applied.* Explain and discuss.

(14–12) *Linear programming appears to be applicable for planning and implementing a quality assurance policy.* Explain and discuss.

14.12 PROBLEMS

(14–1) The senior management of a distillery has established the following goal; quality costs are to be reduced by 5% within the next year and a more marketable, appealing bottle of whiskey is to be offered at the same time. The quality assurance manager has established the following goal-related tasks and general approach.

Tasks: One hundred percent visual line inspection and statistical end item inspection; incoming material inspection; organoleptic (taste) tests of whiskey; and pH, color, clarity, and strength tests.

Approach: Both the inspection costs and defects costs are to be reduced at all inspection points. A new bottle design with a more appealing label will be field tested. Current customer complaints will be analyzed. Packaging and handling will be improved to reduce breakage. Special inspectors will be trained. This plan is to be assessed concerning its effectiveness in attaining the objective. Recommend improvements on the plan.

(14–2) The following is a statement of a printing company. "The quality of printing that we do on the offset presses is the highest possible for the type of paper our customers order. We run free of all types of trash. We use the highest gloss and best scuff free of inks we can buy. We register and hold register throughout the run. We control both in the presses and the bindery. An okay is given on all runs by the foreman. Signatures are pulled each hour of the run and the time, date, and signature of operator is noted. These samples are inspected daily by the plant superintendent."

a. The new owner wants to establish challenging objectives for improving quality, because it is felt that the above statement is a slight exaggeration. Outline steps to be taken by the new owner for setting objectives.

b. Outline measures for attaining such objectives.

(14–3) A small-garment manufacturer wants to reduce customer complaints through a new inspection program. For this reason, quality costs are to be accounted for.

a. Formulate the objective for the quality cost accounting system.

b. Establish related goals and tasks for individual departments.

c. Describe measures for implementing and controlling these measures.

(14–4) A manufacturer for compressors to be used in refrigerators has found that the frequency of repairs and returns soon after sale is increasing disturbingly. As a prerequisite for taking remedial actions and for establishing goals for improvements, a consultant has been contacted.

a. Formulate the terms of reference for this consultant.

b. Describe your approach to assure that the consultant will actually provide the information necessary for setting the goals and the new strategy.

(14–5) With reference to problem (14–4), the relations with the supplier of this company are to be improved, as it is felt that supplied material and parts have been found defective too often. The company relies heavily on its own receiving inspection for quality assurance.

a. Formulate objectives and approaches for an improvement project.

b. Establish subgoals and milestones that can be used in preparing the project plan and schedule.

(14–6) A small metal plating shop with ten employees is required by its major customer to prepare a quality manual. The owner has declared that this document will be ready within two months. Individual employees are to be given an assignment for helping in preparing this manual. Describe the approach.

(14–7) The quality objectives of The Wellfit Garment Corporation are to provide customers with well-fitting, durable garments that are visually appealing and suitable for a desired end use at the lowest possible price. These objectives are met, as well as is deemed possible, without any formal quality assurance plan. The garment fit has been created through the use of some customer specifications, but basically through trial and error. Quality policies in existence are not recorded or documented. When a customer is dissatisfied with a product, the product is returned and either the customer's account is credited, or a new product is sent as a replacement.

   a.   Should this company formulate quality objectives and policies?

   b.   Outline steps for assessing the need for quality objectives and policies.

   c.   Write a letter to the owner/manager of Wellfit Garment Corporation, explaining your assessment and recommendations.

(14–8) Linear programming—transportation model: Three branch plants producing hydraulic equipment for electrical power generation plants require the continuous services of three independent inspection and consulting institutions. The following table describes the data. Costs per one item of inspection are entered in each cell.

Inspection institution	Plants			Inspection capacity
	A	B	C	
1	7	3	5	10
2	4	6	5	10
3	9	6	7	13
Demand for inspection	7	16	10	

Assign inspection capacity, based on least cost, to each plant.

(14–9) Scheduling of inspection jobs.

Develop a least-time assignment plan using the assignment method of scheduling, given the data below:

Job	Inspectors			
	A	B	C	D
1	4	5	9	8
2	6	4	8	3
3	7	3	10	4
4	5	2	5	5

(14–10) Given the following inspection times in the laboratory and operation time at the machines, develop the job sequence with the least flow time; apply the Johnson Method.

Job	Machine	Lab
1	7	2
2	15	8
3	4	8
4	9	10

14.13 NOTES

1. Carroll, Lewis, *Alice's Adventures in Wonderland,* New Junior Classics, Mabel Williams and Marcia Dalphin, editors, P. F. Collier and Son, Corp., New York, 1949, p. 51.

2. Feigenbaum, A. V., "Quality and Business Growth Today," *Quality Progress,* November 1982, p. 22.

3. Massie, Joseph A., *Essentials of Management,* Prentice-Hall, Inc., Englewood Cliffs, N. J., 1964.

4. *Quality Program Requirements for Contractors,* standard DND-1015, Department of National Defense, Ottawa, Canada.

5. Military Specification MIL-STD-Q9858A, *Quality Control System Requirements,* Government Printing Office, Washington, D.C., 1963.

6. Canadian Standard Association, CSA standard Z299.1, *Quality Assurance Program Requirements,* Rexdale, Ontario, Canada, 1978.

7. United States Small Business Administration, "The Management Aid for Small Manufacturers," MA 243, *Setting Up a Quality Control System,* Washington, D.C., 1979.

8. See, for example, Caplan, Frank, *The Quality System—A Source Book for Managers and Engineers,* Chilton Book Company, Radnor, Pennsylvania, 1980, in which the author identifies nine subsystems and 103 quality elements, comprising the total quality system.

9. Wheelwright, Steven C., "Japan—where operations are really strategic," *Harvard Business Review,* July–August, 1981, p. 67.

10. Deming, W. E., "What Top Management Must Do," *Business Week,* July 20, 1981, p. 19.

11. Reddy, J. and Berger, A., "Three essentials of product quality," *Harvard Business Review,* July–August 1983, pp. 153–9.

12. Juran, J. M. and Gryna Jr., F. M., *Quality Planning and Analysis,* 2nd ed., McGraw-Hill Book Company, New York, 1980, p. 550.

13. Dorsky, L. R., "Management Commitment—To Japanese Apple Pie," *Quality Progress,* February 1984, p. 18.

14. See, for example, Chase, R. D. and Aquilano, N. J., *Production and Operations Management—A Life Cycle Approach,* 3rd ed., Richard D. Irwin, Inc., Homewood, Illinois, 1981, p. 198.

15. See Meredith, J. R. and Gibbs, T. E., *The Management of Operations,* 2nd ed., John Wiley & Sons, Inc., New York, 1984, p. 367.

14.14 SELECTED BIBLIOGRAPHY

Chase, R. B., "Where Does the Customer Fit in a Service Operation?" *Harvard Business Review,* November–December 1978, pp. 137–42.

Chase, R. B. and Aquilano, A. J., *Production and Operations Management, A Life Cycle Approach,* 3rd ed., Richard D. Irwin, Inc., Homewood, Ill., 1981.

Hays, F., "Quality Control in Airline Operation," *ASQC Quality Congress Transactions,* Detroit, 1982, pp. 607–10.

Juran, J. M., "Quality Control of Services," *1974 Japanese Symposium, Quality Progress,* December 1981, p. 10.

King, C. A., "Quality Controls in Hospitality Service Operations," *ASQC Quality Congress Transactions,* Boston 1983, pp. 412–17.

Langevin, R. C., "Quality Control in Bank Operations," *ASQC Quality Congress Transactions,* Boston, 1983, pp. 131–5.

Main, J., "Toward Service Without a Snarl," *Fortune,* March 23, 1981, pp. 58–66.

Rosander, A. C., "Service Industry QC—Is the Challenge Being Met?" *Quality Progress,* September 1980, pp. 34–5.

Sasser, E., Olsen, R. P., and Wycoff, D., *Management of Service Operations: Text, Cases and Readings,* Allyn and Bacon, Boston, 1978.

Scanlon, F. and Hagan J. T., "Quality Management for the Service Industries," Part I and Part II, *Quality Progress,* May 1983, pp. 18–23 and June 1983, pp. 30–5.

Shostack, G. L., "Designing Services that Deliver," *Harvard Business Review,* January–February 1984, pp. 133–9.

Takauchi, H. H. and Quelch, J. A., "Quality is more than making a product," *Harvard Business Review,* July–August 1983, pp. 139–45.

CHAPTER 15

ORGANIZATION: DESIGN AND DEVELOPMENT

Where people work together to achieve certain purposes and objectives, tasks must be assigned to groups and individuals. Whether or not the people work together in an effective, organized fashion to achieve these tasks, goals, and general company-wide objectives depends primarily on management's ability to design a sound organizational structure, through which the staff becomes directed towards cooperation and goal achievement. In the previous chapter we described policies, procedures, and directives for individual job performance as guidelines for the company and its staff. We have seen that three hierarchical levels exist in the company: top management, middle or supervisory management, and operational staff. The triangular shape used in Figure 14.1 showed that the number of people at each level varies, with relatively few supervisors leading their subordinate teams or departments. These supervisors themselves are subordinates of senior managers to whom they report.

The main question in this chapter is what the organization's hierarchical structure should be, so that everybody can perform as expected and, in fact, achieve their tasks. Without seeking solutions to the many problems implied in this question, any policy or program for quality assurance of products and services can not be practically carried out.

A senior vice president of a major corporation stated: "More than forty years of practical experience and observation in both government and industry have convinced me that the most shocking waste of human spirit and capabilities occur through poor organizations...most defects in organization arise from the disregard of a few fundamental laws."[1] Obviously, two major principles are that everybody must be assigned a well-defined task that is well-delineated from tasks of related jobs, and that the person responsible is capable of achieving his or her task when certain prerequisites exist in the workplace.

To design and develop an organization that allows the best possible performance of the company's staff, both jointly and individually, is a major management task and responsibility. Its accomplishment ensures that people can observe rules for good workmanship and thus contribute to quality assurance. In fact, customer complaints most often are traced to some *organizational defect,* such as the following.

1. Poor work instructions.
2. Unsuitable staffing.
3. Poor supervision.
4. Lack of management control.
5. Lack of clear-cut objectives and policies.
6. Pressure of manufacturing quotas.
7. Unpleasant personal relationships in the company.

In this chapter we shall briefly define and describe organizations from the perspective of quality assurance. We will discuss the problem of designing organizations for quality assurance and describe some modern developments in the organizational field, as these directly involve quality assurance. We shall not and cannot present general organizational theory and practice, but refer the reader for a broader study of the subject to the references at the end of the chapter.

One fact that should be kept in mind, is that having set a clear and challenging objective of quality assurance, an important and new impetus for reorganizing has been created in an organization. We only need to remember the predicament of the president of the bus manufacturing company described in Chapter 14. Surely, before the quality control program design was announced as an objective, an organizational review of the company would be necessary. Such a review might have found that the president's "in-basket" had many notes and requests that revealed confusion about tasks, responsibilities, reporting lines, and so on. Moreover, the president might have realized that much of his time is spent on matters that should be the concern of others in the company. Friction between departments and individuals caused by the bombshell of major customer complaints indicate urgent need for organizational review and changes.

15.1   ORGANIZATION FOR QUALITY ASSURANCE

Organization is the relationship structure of work assignments and responsibilities to functions and persons in a company. Designing and developing such an organization involves dividing the work of achieving purposes and objectives into hierarchical levels of responsibilities and laterally-ordered individual functions.

Factors that influence the organization and the organizing activities are these.

1. The kind of products and services the organization is to provide.
2. The scope of operation.
3. Management and operational styles.
4. Complexity of applied technology and economic social environments.

5.   Competitive pressures.

6.   Availability of managerial, technical, and operational staff.

7.   The personalities of people in the company.

Because all these factors and their degree of influence on the management and operation of the organization change continuously, organizational *reviews* and *adaptations* are necessary. By design, each organization must be tailor-made to the unique conditions that prevail in each empirical case.

Quality assurance as a major task and responsibility of the organization and the company of people should be assigned and accepted in each unit function and job in explicit form. Where it is a strong tradition and understanding, sound work ethic, motivation, and performances ensure satisfactory quality of products and services almost automatically. This often occurs in small enterprises with intimate personal bonds between the people. In large organizations with many people employed in a great variety of specialized functions and tasks, a formal organizational structure is mandatory.

The more a company grows, the more special quality assurance functions must be designed and integrated with the existing organization. These functions need to be staffed with specialists in quality assurance, able to design policies, procedures, and operational systems in their fields. The relative soundness of the internal quality assurance organizational units has a decisive impact on the overall quality performance of the company and organization. Major customers tend to judge a company's potential for quality work to a large extent by the overall quality organization, its status, relationships, and personalities.

Consider the following two examples.

Example 1: Not long ago, the head technician in a quality assurance laboratory was on sick leave. Due to the fact that the organization is small (4 laboratory technicians and one supervisor reporting to the quality assurance manager) none of the other technicians was able to do the highly specialized work of the head technician, thus creating a backlog in the release of finished goods for sale.

This type of problem may occur in any organization, large or small, that does not have a full training program for their personnel.

The necessary training program should be as complete as possible, so that all the persons in one unit or section may perform any duty of that section. There should be procedures for the training of all the personnel in the section. The training should be documented for future audits.

Example 2: One local manufacturing firm has an interdepartmental communication problem with its supplier. The problem is that the machine shop rejects casting parts due to difficulty in machining. The supplier is requested to anneal the castings for easy machining. The original design requires high strength and hardness values in the material specification. Thus the supplier is caught between manufacturing and design engineering. This, of course, is due to the lack of interdepartmental communication within the firm, and creates organizational problems for quality assurance.

An organizational change should be made to establish a liaison engineer to perform the interdepartment communication duty between the designer and the production personnel, for quality assurance requirements.

The standards for quality assurance programs do require explicit accountability for organizational structure in quality assurance. The following excerpt from CSA document Z299.1 serves to illustrate this point.

> In the development of an organization and the assignment of responsibilities and authorities it should be recognized that the quality program is interdisciplinary and involves most of the organization. . . .Unless the responsibility for quality is fully acknowledged and understood by all, from the top executive to the shop worker, full compliance with the Standard cannot be achieved. Also persons assigned responsibility for assurance of quality should be aware of, but free from, the pressure of cost and production, and be given the necessary authority to perform their roles effectively. . . .[2]

This stipulation includes many of the fundamental rules and principles for organizing work. Some of those with particular relevance to quality assurance are the following.

1.   Every member in the organization must have a clear task assignment, associated with respective responsibilities, authorities, performance standards, and reporting lines. There should normally be one direct supervisor.

2.   Decisions should be delegated wherever and whenever feasible, but without negative impact on quality of performance. This principle of exception relieves the supervisors of unnecessary routine decisions and can render the work of subordinates more challenging and interesting.

3.   All functions and tasks must be assigned to qualified persons. This requires careful selection, training, supervision, recognition, and often changes in the job assignments and procedures of recruitment.

4.   An optimal match of function and tasks with appropriate people requires frequent organizational review, redesign, and adaptation. The need of the organization for adequate individual and joint work achievements and employment expectations must be optimally fulfilled by the organization.

5.   Human aspirations and creativity must be developed and utilized as a mainspring for quality of work and work life.

6.   While the internal organizational structure and changes must be based on objective analysis of the need and decision making, it must also become integrated with other organizations in the wider community. For instance, basic human rights and democratic principles of fairness must prevail in the organization.

In any dynamic and competitively strong organization, these stipulations are constantly in the mind of the manager. Rapid and frequent developments—internal and external to the organization—tend to disturb the working relationships and the prevailing harmony. In the example of the bus manufacturer, a review of the organization would start at the president's desk. The review would first formally assess the existing organizational layout, with regard to functional and task-related performances. Various possible formal arrangements such as line-staff structures, committees, project teams, and so on, provide basic structural patterns to choose from.[3]

15.2 DESIGNING FOR QUALITY ASSURANCE ORGANIZATIONS

Organizational Principles

Many organizational principles other than those described earlier would further help managers in charge of company design and development to build an organization in which quality assurance is attained at each and every workplace. Procedures and methods for organizing follow their own logic and can be fairly straightforward. Much of what has been outlined so far explains the fundamental approach. In principle, any new organizational structure or change in function is a decision by superior management. Before such a decision is made, the following questions should be answered.

1. What are the needs for the structure or function and the specific purpose and objective of the tasks in both the long-term and short-term perspective?

2. What are its relationships to other functions, vertically and laterally in the hierarchical structure?

3. What are the responsibilities, of the function, task, or job? How should the job description be documented, formulated, and made acceptable within set rules and directives?

4. What are the required qualifications for the person involved in the organizational change? What are the available or potential personnel resources? What training will be required?

5. What methods are to be employed to evaluate performance. What improvements need to be made, and how?

Based upon the above questions, the important principles leading to an effective and efficient organization can be summarized as follows.

1. Adequacy for the type of program and its objectives (scope, complexity, timing, resource availability).

2. Sufficient responsibility and authority.

3. Clarity in task (job) descriptions, reporting lines, supervisory powers, delineation from related jobs.

4. Formalized and properly communicated and approved tasks in sound relationship to the abilities of people, equipment, facilities, and working conditions.

5. Optimal functional division and specialization with regard to scope and complexity of program.

6. Optimal delegation of responsibility and authority for quality to operational jobs and related functions.

7. Adequate cooperation in accordance with the organizational arrangement. Fair personal advancement and promotion.

8. Frequent appraisal of the organization and its general and specific performance in achieving goals.

9. Adequate independence of quality control and inspection as a trustee of customer and corporate management (company) interest in quality.

10. Adequate utilization of existing talents and expertise for quality assurance.

11. Maximum quality assurance with regard to established standards and regulatory requirements.

12. Top management responsibility and visible commitment.

Functional Structures

Even the organization consisting of one person will show a pattern by which work is carried out. Certain predetermined tasks will be listed, and then ranked in priority and scheduled for execution. Many of these functions repeat themselves as regular tasks, while others come and go along with the daily operations. As the work grows, the number of people and special functions also increase. Once the organization takes on giant scope, it must be contained in independent, well-integrated subunits, such as plants or branches, depending upon one or more of the following factors.

1. Permanence (function oriented versus program oriented).

2. Organizational units, (line, line-staff, committee, matrix).

3. Scope of business, (small versus large business).

4. Evolutionary stage of quality assurance (self control, simple inspection).

5. Functional emphasis; (e.g., reliability, customer service, incoming inspection).

6. Organization for maintaining control, (status quo versus breakthrough).

7. Degree of centralization, (single plant versus multiplant operations).

Organizational Schemes

Figure 15.1 depicts a relatively well-known *line-staff* organizational chart (Chapter 3 has more examples). It is the one suggested for small manufacturing enterprises. The advantage here is that one staff member is explicitly and visibly assigned to look after quality assurance, and reports to the top management.

More complex and specialized organizational schemes take the form, illustrated in Figure 15.2, of the quality assurance department in a larger organization. These schemes are all fairly self explanatory, but they provide only a general overview. Functions are not identical with tasks, and such schemes mainly show the functions and intended combined integral authority.

Functions are regular activities and responsibilities to be performed on a long-term, continuous basis. The position for a quality manager, for instance, would be a function which has a job description and status as indicated in the *organizational chart* or diagram. *Tasks* are merely derived from specific goals, and are usually assigned in conjunction with committee work and project teams.

Standards for quality programs require that the quality control and inspection functions have sufficient independence. The primary reason for this is to remove the control over acceptance and rejection of manufactured goods from the very organization that is responsible for its production. The philosophy is to keep the quality control organization in a better position to remain unbiased and therefore, able to fulfill its functions more effectively. Several external standards which are required by major customers demand such independence in the

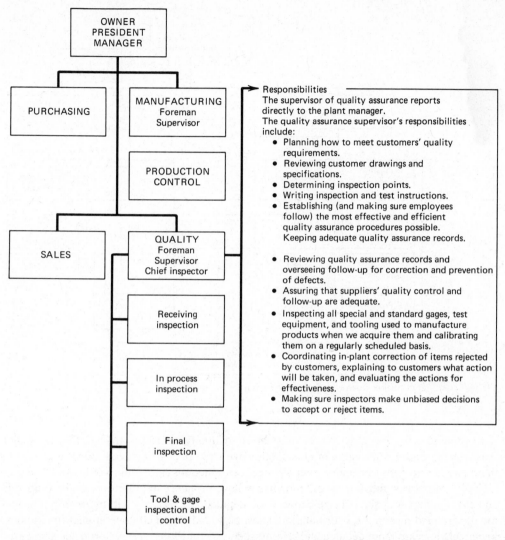

FIGURE 15.1   A simple form of organization chart showing quality control responsibilities. Adapted from *Management Aids* No. 243, U.S. Small Business Administration, Washington, D.C.

internal organizational schemes, based on the premise that production should be concerned with cost and schedule.

In Figure 15.1, a simple outline of an organization with more detailed quality assurance responsibilities is shown. Any well-written job description is derived from the official designation of the major job function, sometimes called the *position* or the *department*. Along with a general list of duties and qualifications, a job description might include specific performance standards, not only as quantitative and nonquantitative, but also in verifiable terms, like appraisal criteria and methods, assignments to committees and project groups, and procedures for temporary replacements and changes in the job description.

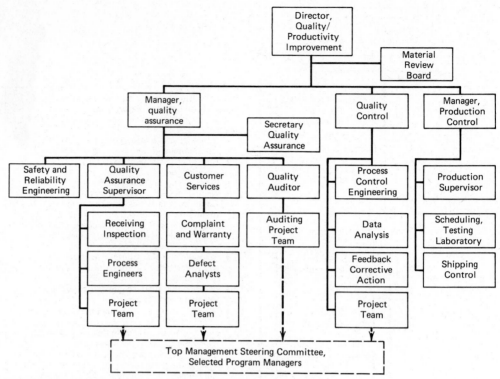

FIGURE 15.2   A complex form of organization chart.

In principle, the job descriptions must be individually approved and verified by senior management and its delegates in consultation with the person involved. This is to ensure that there are no gaps nor undue overlaps between functions in the unit vertically as well as laterally. Moreover, the holder of the job has to have influence in order to make it acceptable and challenging. A sound balance between too detailed and possibly stifling descriptions and too general and meaningless formulations must be found. Normally operational functions at lower echelons are more detailed and refer to specific procedures and performance standards; while more senior positions outline duties and authorities only.

The following is a brief example of a job description for the position of Director of Quality Assurance.

Scope: Planning and implementation of total quality assurance system.

Responsibilities: The Director of Quality Assurance is responsible for the quality assurance organization, including the formulation, implementation, coordination and promulgation of policies related to quality. He or she is also directly responsible for the management of the quality audit to maintain the effectiveness of the quality system. The director is the company's representative, authorized to resolve all quality-related conflicts to the satisfaction of its customers. He or she maintains a master copy of the quality manual, in reference to which all other departments must comply with published directives in all facets of the quality program.

Authority and relationships: Reporting directly to the Senior Vice-President, the director is the final authority on all matters related to quality. His or her relationships with other managers should aim to effect planning and coordination in sales, design engineering, new product development, manufacturing process control, and other departments, in order to meet customers' expectations.

Accountability: The director is accountable for all quality related programs that contribute to overall cost and profitability.

Matrix Management

Quality assurance, being basically a staff function with its own separate internal organization (at least in large organizations), has direct interrelationships with practically all other functions and departments. This interface can best be demonstrated in a *matrix*-like diagram, as in Figure 15.3. In order to maintain clarity of the presentation, each level in the hierarchical organization structure can be shown separately.

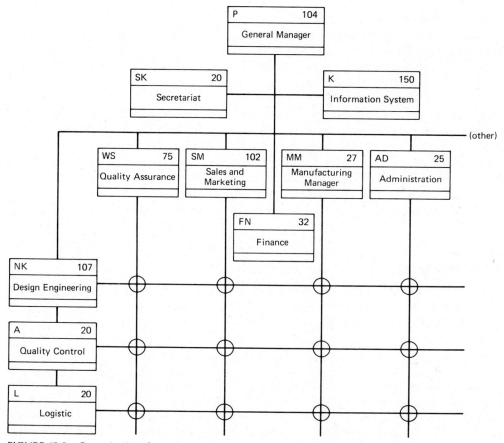

FIGURE 15.3 Organizational matrix structure with respect to quality assurance.

Staff functions, such as design engineering, quality control, logistics, and so on, can be organized into departments or as standing committees, with the latter providing more flexibility in membership and specific responsibilities. Such staff functions can also be augmented by other *task forces* and *project teams* as the need emerges. For instance, in the case of the bus manufacturer, top management might very well establish a *task force* for the intended quality program.

Concepts of *matrix management* suggest that many different forms of organizational interfaces coexist between special assignments, tasks, programs, and projects on one side of the matrix, and regular functions, departments, and individual positions on the other side. This relatively recent complementation of the traditional hierarchical structure, with its strict chain of command and departmentalization, has a positive impact on quality assurance. It allows for the introduction of quality programs that coordinate all respective duties and activities towards achievement of company-wide quality goals.

15.3   QUALITY PROGRAMS AND PROJECTS

Programs and projects have specific tasks and objectives to accomplish. This involves staff, and thus requires organizational planning and control. *Programs* here are understood as having long-term purposes and objectives, and as being comprised of many different activities that must be coordinated. A program often involves *projects* with special, complex, ad hoc tasks. A project in conjunction with a quality program can be to determine the applicable needs and standards and to design the quality program.

Design and implementation of quality programs will be the topic of another chapter. Here the question is how to *organize* for the establishment of quality programs and projects? Any number of departments, while being charged with regular long-term functions, can become involved actively or passively in quality programs and projects. Quality programs, as described in standards, coordinate individual quality assurance activities and procedures, and thus govern company-wide performances.

Here are some essential elements of programs for achieving quality assurance required by customers.

1.   General quality policy, objectives, and respective functional organization.

2.   Documentation of the quality program, for instance a quality manual.

3.   Design assurance, contract review, documentation procurement control, including receiving inspection, production control, installation, and test equipment control.

4.   Nonconformance control and corrective action with inspection status of work-in-progress and records.

5.   Information system and program maintenance, staff services and training.

These program elements demonstrate areas of activity with direct reference to the company's products and services. Production stages encompassing quality of design, conformance, and performance become clearly distinguishable. Figure 4.4 (Chapter 4) describes the accountability in matrix organization, in which program elements, in somewhat different designations, are assigned to functions of quality control manager and others.

The cooperation of these persons in various areas of the quality program is further clarified as primary or secondary responsibility. Other possible kinds of participation could be authority to approve, informed participation, chairing of meetings, and other stipulations for departmental and personal involvements. This organizational plan can effectively coordinate functions and staff activities for quality assurance, that otherwise could not be achieved by merely inserting such duties in the respective job descriptions.

Human interaction for the purpose of quality assurance also takes place in various projects that must be undertaken in conjunction with the more permanently established quality program. As already mentioned, a good example of a typical project would be the quality program design, writing of the quality manual, and carrying out of the program's actual implementation. In each case the complex tasks and project goals are assigned to a team of selected staff members, usually under the chair of a representative of the quality assurance department.

Organizational Aspects of Project Planning and Control

The particular organizational aspects of project planning and control would be a most interesting and important subject that, however, exceeds the scope of this book. We briefly discussed a new approach for quality oriented projects in Chapter 3. We then dealt with quality circles designed to promote staff participation in quality problem-solving at the operators level. At that time we explained quality circles as being somewhat the same as project teams. Both of these should become more closely intertwined with the regular project structures that operate under direct management control.

Projects can be staffed effectively only when their activities are adequately defined and delineated. Figure 15.4 illustrates a composite of activities for implementation of a quality program in fairly self-explanatory terms. Modern project planning and control systems that employ computers allow for easy deployment of staff, once the project structure is known in a network format. Assignment of project tasks in accordance with the schedule of starts and completions, and other related stipulations, often requires the further preparation of assignable work packages. Figure 15.5 relates a project network and the existing line-staff organization via an intermediate matrix of the familiar type.

The various traditional and modern organizational structures, as briefly demonstrated here, provide for formal insertion of quality assurance responsibilities and activities. (In fact, pressures for improved quality assurance, as exemplified in the bus manufacturer's case, led to an innovative pattern of matrix management.) We pointed out earlier that any organizational arrangement must meet the unique needs of dividing, assigning, and coordinating the work of people in the company, jointly and individually. Without this structure and system and the various job descriptions and procedures that are derived from the overall pattern, the company cannot perform, compete, and attain satisfactory quality assurance. Yet the organization as it exists on paper might conflict with the actual work performance and the human interaction.

In a competitive and turbulent business environment, unsuitable functional or procedural arrangements and incomplete, outdated, or unacceptable job descriptions and work assignments disturb, rather than promote and support performances. Quality assurance is particularly sensitive to organizational decisions and dynamic human interaction. Effective organization for quality assurance can create harmonious and vibrant teamwork. Neglect of quality as-

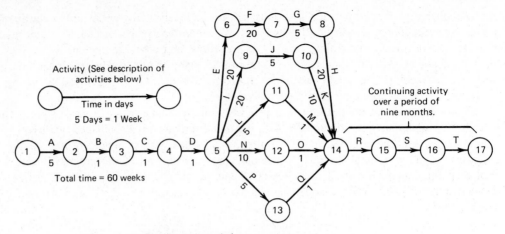

A. Finalize program design.
B. Establish progress criteria milestones and standard.
C. Attain superior management's approval and commitment.
D. Have superior management formally introduce the program.
E. Establish a library, with the cooperation of the engineering department, of all pertinent drawings, specifications and standards and catalog them for control of revisions and issues.
F. Purchase necessary calibration standards for test and acceptance tools and/or establish contracts for doing calibrations.
G. Establish control of all test and acceptance tool adjustment.
H. Calibrate and record all test and acceptance tool adjustment.
I. Establish an information recording and retrieval system for data from inspection so that reports may be compiled on vendor ratings, quality costs, etcetera.
J. Train office staff in the processing of data at I.
K. Record training and qualifications of office staff.
L. Train receiving inspection staff.
M. Record training and qualifications of receiving inspection staff.
N. Train in-process inspectors and production personnel.
O. Record training and qualifications of in-process inspectors and production personnel.
P. Train final inspection staff.
Q. Record training and qualifications of final inspection staff.
R. Monitor the program through progress reports measured against the criteria established at B and take corrective action as required.
S. Provide quarterly progress reports to superior management.
T. Have an independent audit of the system done 12 months from point 5 and submit final report to management.

FIGURE 15.4 PERT Network: Introduction of quality assurance program for ABC Tube Benders Co. Ltd.

surance, resulting in loss of customers and job security, disrupts the company. After having established a suitable organizational structure, organizational reviews must account for weaknesses due to changing internal and external forces and corrections must be induced. After all, in every organization there is a formal structure and an actual behavior mode of the people. These two factors must match positively.

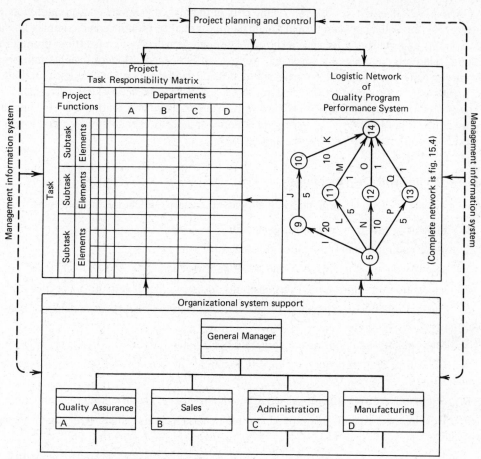

FIGURE 15.5 Project planning and control diagram involving line-staff organization via an intermediate matrix of system support.

15.4 ORGANIZATIONAL DEVELOPMENT

We have not separated the concept of formal organization, as it is documented and communicated, from the actual staffing and work of the groups and individuals in the company. We have considered organization and company as different sides of the same coin. The question then is, should the organizational pattern be devised first, and then the functions, tasks, and project teams, staffed and assigned to persons; or should the existing behavior of the persons, and their understanding of their roles be the mainspring for organizational structuring? This is somewhat a philosophical question that can not be pursued here, even on the surface. However, it has a very simple, practical answer. The formal organizational structure and its many elements and stipulations for guiding the company towards collective growth in the marketplace and achievement of set objectives, must help, challenge, and move the staff to the satisfaction of everyone as much as possible.

Obviously, "as much as possible" is the crux of the matter. Various managerial functions dictate that the match between formal organization for quality assurance and complex human motives and expectations must and can be made through such decisions as staffing, promoting, training, and reorganizing whenever and wherever it is needed. Modern insights into organizational behavior equip decision makers to structure and develop psychological and sociological forces in a company, (i.e., develop human assets for satisfactory work performances). Quality assurance and its many benefits for customers, owners, staff members, and the community, is both the result of a truly functioning organization and company, and a source of pride in workmanship and quality motivation.

Recent Trends and Developments

Recent developments in organizing for quality assurance are reflected in the following trends.

1. Increasing program oriented organization.

2. More committee and task force organization.

3. More frequent reorganization and adaption to needs.

4. Delegation of authority and responsibility to the worker level (self-inspection).

5. Clearer functional separation of quality engineering, quality management, quality auditing, and inspection.

6. More quality functions at the corporate management level.

7. Growth of staff functions for quality assurance at all levels (specialization, professionalization).

8. Stipulations of organizational requirements in generic quality program standards.

9. Development of information and communication systems, along with organizational structures.

Assuring optimal organizational change and adaptation to changing needs and conditions involves the following.

1. Testing and analyzing organizational defects for possible weaknesses and needs for corrective action.

2. Once the cause of nonconformances have been traced to inadequate workmanship or procedures, appraising remedial action in a nonpunitive manner.

3. Using such causes for the study of related jobs, in order to gain broad improvements, rather than only at the job which appears to have poor performance. Subjecting all procedures, rules, and supervisons to an appraisal for possible further improvements.

4. Securing necessary arrangements for corrective action, with wide participation of those concerned in the decision-making process.

5. Documenting the changes and communicating these to those concerned in such a manner that improved performance will actually be achieved. Possibly providing for training and other support for change implementation.

6. Not restricting remedial action to operational staff; extending changes to supervisors or even beyond.

7. Including the organizational change, new procedure, job description, realignment of responsibilities, and staffing changes in audits for evaluation.

8. Perform regular audits for follow-up control.

Training as a Means of Organizational Development

Needs for new knowledge and skills in a company demand training in quality, ever more frequently, at all levels of responsibility. The recent growing awareness of the importance of quality assurance requirements for the success and growth of companies, translates into strengthening the qualifications and motivation of all staff members, and not only those in quality assurance functions.

For staff members already employed with the company, various training courses help to update qualifications. Special training for improved quality assurance on the job and in the company as a whole should be explicity directed to well-defined needs and training objectives. Training must result in new and useful factual knowledge, skills, experiences, aptitudes, understandings, and important information that leads to better workmanship and thus, ultimately, pride in everyone's workmanship.

Principles for building a sound organization with quality assurance supported by the staff, individually and jointly, are primarily related to management's monitoring the need for organizational adjustments, and thus maintaining adequate contacts, communication, and meaningful participation in decision-making processes. Such exchange of information should move from management to the operational staff, as well as in the other direction. Modern organizational patterns directed to promote interhuman cooperation are closely related to information systems. The next chapter will deal with the topic of using information systems for achievement of quality.

15.5 KEY TERMS AND CONCEPTS

Organization is the relationship structure of work assignments and responsibilities to functions and persons in a company. An organization is people oriented by definition.

An **organization for quality assurance** is the work unit in charge of quality assurance procedures and programs. Quality assurance is also an inherent responsibility in all other work assignments in an organization.

Organizing for quality assurance denotes all activities and decisions that plan and control work assignments with regard to quality attainment. Organizing at the company level integrates individual assignments and procedures for quality attainment into a quality assurance program.

Organizational schemes show the division of work and functions in a company. Hierarchical line-staff

types of schemes vary from, or can be complemented by, matrix organization. The scheme reflects to a certain extent the management style in an organization. Flat hierarchical schemes and matrix organizations lend themselves to wide participation in quality assurance planning and control.

A **job description** is a person oriented outline of a work assignment that is interrelated with other job descriptions. The description entails a list of responsibilities, implied authorities, relationships, accountabilities, qualifications, and other related details of the employment contract. Workmanship standards and quality assurance responsibilities are implied, but should be explicitly defined.

Quality programs and projects describe goal and purpose oriented complexes of activities, approaches, and work packages, to be performed in an organization and by organizational units. Committees and project teams are assigned programs and projects. A program, as opposed to a project, has more long-term and general objectives and purposes. Standards describe features of quality programs.

Organizational development, with regard to quality assurance, means to design measures, programs, and projects, that will orient all work toward accomplishment of quality assurance. This implies an understanding, that the organization is more than a formal structure of work assignments and functional interrelations. An organization is designed by people, for people, so that they can properly cooperate and effectively work and compete as a group. Education and training for quality assurance fosters organizational harmony and effectiveness.

15.6 CASE: BILL'S DILEMMA

William Bosdorf is a student working towards a Master's degree in Business Administration, with some practical experience in hospital administration. He had originally enrolled in a program to become a physician like his father. After the first year he felt that because modern medicine has become more and more dominated by electronic equipment, studies in engineering might be more appropriate for him. He now holds an undergraduate degree in electrical engineering and computer science, and is in charge of maintenance of electronic equipment at the local hospital.

Many suppliers of this kind of special equipment are well-versed not only on technical matters, but also in methods of production and marketing. William soon found that graduate studies would provide essential knowledge and skill to work more effectively with suppliers. He was particularly interested in reviewing and improving the quality assurance of suppliers, before and after the sale of the equipment. At the same time, he wanted to integrate quality assurance management internally in the hospital administration.

One basic question continued to surface as he went about his studies What in fact is Management? He read much about it, but nothing seemed to be helpful with regard to his function in the hospital. The following reading[4] seemed to him the most realistic and convincing.

> . . .Too many American managers are not doing their jobs, and for one reason: They do not understand what their jobs are all about . . . [B]efore agreement can be reached as to what to do about US productivity, what is the job of a manager? . . . They think that the job is "understood." It is obvious from their answers that it is not . . . [S]everal professors of business administration do not do much better . . . [U]nder currently taught approaches to management, it is presumed that the relation between boss and worker is inherently adversarial. . . . [U]nless managements learn how to manage systems of machines and people, the productivity of their capital investment may not pay off . . . . At a very minimum, we should expect our schools of engineering and of business to be developing courses of instruction in productivity which are directly useful in this current need to catch up . . . . Engineers now leave schools essentially unaware of how statistics is to be related to productivity. Managers are untaught . . . . [E]ach State should inquire as to the image of management given by its business schools. These matters deserve debate. We need the tools of business management. But we need to have them wielded by people who understand their purpose in ways which are beneficial to our society. . . . What we are calling for is an intensive debate involving the schools of engineering, schools of business administration and business leaders, over what it means to be a manager. The revolutionary idea is that workers shall be genuine partners in not only doing the work but in making the workplace more effective.

Questions for Discussion

1. Do you share William's position that this definition of management is realistic and convincing? Explain your position.

2. What other definitions for management and, in particular, for management of quality assurance are there in general terms? What are their strengths and weaknesses? Be brief.

15.7 DISCUSSION AND REVIEW QUESTIONS

(15–1) *In a well managed organization, quality assurance is understood as inherent in workmanship and performance; there is no need for special programs.* Discuss.

(15–2) *Whenever there is a quality assurance function or department in an organization, it carries the full responsibility for quality.* What do you understand as "full responsibility"?

(15–3) *Many, if not most defects in quality can be traced to weaknesses in the organizational planning.* Do you agree? Explain.

(15–4) *The status of the quality assurance function in a company is indicative of the quality assurance policy and attitude of the management.* Explain.

(15–5) *In any company, including small enterprises, there should be at least one person in charge of quality assurance matters.* Discuss.

(15–6) *Everyone in an organization should participate in quality assurance planning and control.* Explain, and discuss how can this be achieved.

(15–7) *A quality inspector should report only to the quality manager.* Discuss various implications of this arrangement.

(15–8) *There are certain major determinants of an effective organization that in itself generates adequate quality assurance.* Explain and discuss.

(15–9) *A line-staff structure is not always the most adequate one for effectiveness of quality assurance in the organization.* Explain, and relate to other organizational patterns.

(15–10) *A job description for a manager of quality assurance must include certain information and directives.* Explain.

(15–11) *Quality assurance in an organization is usually performed by means of programs and projects.* Explain by giving examples of quality assurance programs and quality assurance projects.

15.8 PROBLEMS

(15–1) A common problem with a particular company is a situation like the following. The president is yelling at the production manager and calling him on the carpet for poor performance. The president tells him to reduce the scrap and improve the exactness of the cutting of the frost shields. The production manager in turn, yells at his cutters to have the machines set exactly to specification. However, the machines are so old that they are difficult to set exactly and they don't stay set for long. The president believes it is sloppy operators and careless maintenance people who are at fault. Consequently, he won't authorize the purchase of new cutting machines. Despite his frequent temper spasms, the quality has not improved, and meeting of delivery schedules is a problem. The main result of all this is bad feelings and very strained relations.

With reference to the subject matter of this chapter, discuss a solution you would like to offer.

(15–2) A major organizational problem in the quality assurance function is in reporting directly to the Corporate Director of Quality Control.

The idea itself is sound. The real problem comes from the many special projects being assigned to the Director of Quality Control that require him to be out of his office for weeks at a time. This leaves much of the necessary information required for efficient departmental operations bottlenecked on his desk. This situation has been in existence through the employment of many managers. Much time is spent trying to bootleg information that should be common knowledge. The obvious solution would be for the Quality Control Director to be more readily accessible to his subordinates. If this is not possible, then an assistant should be hired or an individual designated to disseminate information to the plant level on a timely basis.

Outline a solution to this problem.

(15–3) Effective work produces unemployment. If one understands willingness to work as the intention to achieve a particular objective, then working and performing well not only means to work more, but to share the work hours with others. Sharing the work can lead to better quality of performance. Stressing higher productivity and increasing the work load beyond a certain level of physical and psychological exertion will lead to a decline in quality of performance and resulting products and services. Discuss.

(15–4) Company Y is a small job shop specializing in precision machining and repair welding. Due to the company's size, the quality control manager is also the engineering manager who has a quality control supervisor reporting to him. Company Y has placed many new orders and is experiencing growth problems in the quality area. The quality control supervisor is assuming the quality control manager role on a daily basis but does not have the freedom to implement needed corrective actions, due to engineering and production pressure. The president of Company Y is receiving complaints about product reliability from his clients and decides to promote the quality control supervisor to quality control manager reporting directly to the president. The feedback cycle in quality assurance or quality control is extremely important. The company president identified this problem and saw the importance of establishing a direct line of communication with quality assurance and also, the importance of quality assurance assuming an unbiased attitude within the company organization.

Give your comments outlining a solution to the problem that company Y is faced with.

(15–5) To write a job description in a format suitable for inclusion in a quality program manual, you must be brief, unambiguous, and specific. Consider the two examples below. Both examples basically, say the same thing and we will leave it to you to decide which presentation is the most effective. Do not give critical attention to the words, we are concerned only with the format.

Example 1: *Job description, manager of quality assurance.* The manager of quality assurance is responsible for quality assurance within his or her range of activity. This includes analysis for improvement projects, quality planning, coordination, consultation, and troubleshooting. He or she shall identify quality problems, initiate or provide solutions to those problems, and control further work on deficient conditions until a proper disposition has been made. He or she has the responsibility and authority to stop operations when the quality requirements of the program have been violated. Sufficient background in research, engineering, and production is required to maintain a satisfactory level of proficiency. The manager of quality assurance will report to the president to provide this person with the appropriate avenue for implementing and enforcing the program.

Example 2: *Job description, manager of quality assurance.* The manager of quality assurance will:

1. Be responsible to the president for all aspects of the quality assurance program.

2. Initiate surveys to identify areas of improvement and provide program layout.

3. Coordinate all aspects of the quality assurance programs with other departments.

4. In consultation with engineering and production, cause work to be halted on any project that is not meeting the approved standard and to have the reason for the nonconformance rectified before work on the project continues.

(15–6) The satellite operation depicted in the lower block of this figure is geographically hundreds of miles from the head office of the company.

3. An article "Quality Organization in the Automotive Industries," *Quality*, 1969, No. 2, p. 39, deals with the subject of organization evaluation in the form of checklists in a quite elaborate form.

4. Excerpted from Tribus, M. and Holloman, H. J., "Productivity: Who is Responsible for Improving It?" *Agricultural Engineering*, Vol. 63, No. 7, pp. 10–20, 1982. By permission of the authors.

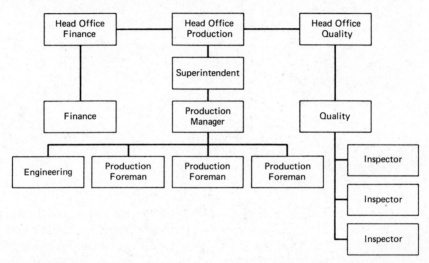

The problem presented illustrates a superintending manager who has limited information within his own operation. Production is the dominant feature, with poor communication between the production and quality departments.

What steps would you suggest to reorganize?

15.9 NOTES

1. Private communication to one of the authors.

2. CSA Standard, *Quality Assurance Program Requirements*, Z299.1-1978, Canadian Standards Association, Rexdale, Ontario, Canada, p. 31.

15.10 SELECTED BIBLIOGRAPHY

Bajaria, Hans J., (editor), *Quality Assurance: Methods, Management and Motivation*, Society of Manufacturing Engineering, Dearborn, Michigan, 1981.

Caplan, Frank, *The Quality System, A Sourcebook for Managers and Engineers*, Chilton Book Company, Radnor, Pennsylvania, 1980.

Fuchs, Jerome, H., *Administering the Quality Control Function*, Prentice-Hall Inc. Englewood Cliffs, N.J., 1979.

Harris, D. H. and Chaney, F. B., *Human Factors in Quality Assurance*, John Wiley & Sons, Inc., New York, 1969.

Mescon, M. W., Albert, M., and Khedouri, F., *Management: Individual and Organizational Effectiveness*, Harper & Row, New York, 1981.

16 CHAPTER

QUALITY INFORMATION SYSTEMS AND DECISIONS

The *management information system* is a well-planned and executed program, designed to collect facts and figures, analyze these, and then report the findings to the person and/or function that plans, controls, and possibly replans the production process. Management, in planning and controlling these activities, makes decisions based upon the results of such information.

Goals, procedures, organizational diagrams, and job descriptions exemplify information and decisions that direct people in the company when they receive them, or through which the same staff members activate and instruct others. Everybody in the organization and at the workplace receives messages on such issues as the daily work routine. The president of the bus manufacturer, for example, received the news of major quality problems about a set of buses that were delivered. If this message has caught the president by surprise, the internal information system most likely needs to be reviewed, along with the organizational setup. At least some warning should have alerted the chief executive officer to the problems. Metro Transit, being the major customer, was given an assurance of quality in one form or another. Such quality assurance is, by its very nature, important information extended and communicated by the manufacturer to the customer. This message, among other information, had contributed to informing Transit's decision to sign the contract with the bus manufacturer. Was the quality assurance a misinformation? Was the president misinformed by his or her own staff?

Many questions need to be answered. Data and facts on the major failures have to be investigated and compiled. The president will ask for an immediate report on the matter, and for this he will send out many specific instructions. One decision has already been communicated: a more suitable quality control program will have to be designed and implemented, in order to counteract recurrence of the failure.

One can fairly easily imagine what information flows into the "in-basket" and what decisions and information leave fairly swiftly from the "out-basket" on the president's desk. The influx of messages brings requests for decisions and responses to earlier messages, and requests for certain data and factual information. The president, as decision maker, will study and analyze the incoming communications, weigh the importance of these, and consider alternative paths of action for resolving the issues on hand. A decision will then have to be made in accordance with the chief executive officer's function and responsibility. This decision is the selection of the most promising alternative, considering all the known data at that time, which then will be sent out and communicated to subordinates. The information and decision-making system are closely interrelated, if not identical.

A management information system is designed to provide information when, where, and to whom it is required, in order to allow rational, well-informed and realistic decisions that in turn initiate action towards the achievement of goals and the solution of problems and obstacles. The information flow that continuously crisscrosses any organization needs to be planned and controlled, otherwise an organization can not practically function. With the growth of companies in scope and complexity and with the dynamic technological progress in the information and communication field itself, the quality of its information systems is of immense importance to the success of a company. Quality assurance results, to a large extent, from the staff having proper and sufficient information on matters of quality.

In this chapter we shall not be able to discuss the full scope of management information systems, but only their particular relationship to quality assurance. We shall inform the reader how the quality assurance organizations that we described in the previous chapter can now be further augmented and activated through an associated structured system of information exchange. First, we shall present a general picture of a *total quality information system*, followed by descriptions of data processing and quality reporting. The preparation of quality manuals and control procedures as two major kinds of documentation will be discussed as a separate topic. We shall learn that designing, implementing, and reviewing a quality management information system embraces a continuous cyclical activity in the life of an organization. This is a dynamic change that results from external and internal forces and developments.

16.1 FUNDAMENTALS OF A TOTAL QUALITY MANAGEMENT INFORMATION SYSTEM

Appointment to a function or job and the assignment of certain responsibilities and tasks, individually and in the organizational setting, implies basic needs for information. Under the totality concept, everybody contributes to the quality of products and services and thus participates actively in quality assurance. Consequently, there is no staff member isolated from the information flow specifically designed for quality assurance.

Quality Information System (QIS) as a Part of the Total Management Information System (MIS)

System is a widely used term. In the most general sense it can be understood as the phenomenon or idea of interacting interdependent elementary components that form a unified whole with a common goal and purpose. An information system, for instance, can be understood as the integrated, systematic allocation of data or information and systematic processing (arranging, interpreting, transforming) of these inputs. The output of the processed data serves its predetermined and predesigned purpose and objective. Wherever an unsystematic interaction of parts prevails, a common purpose or goal can only be achieved in a random fashion, rather than in an orderly and systematic approach. In such a chaotic situation, a system is nonexistent. In the area of information it would mean that data are compiled in a more or less unplanned, irregular fashion, and then processed, communicated, and used in a way that accomplishes its purpose of informing only by chance. The word "system", therefore, stands for plan, method, order, and arrangement.

The information system is often compared with the nervous system in the human body. It receives many external and internal impulses and brings forth proper reactions and actions; that is, in a healthy body. The organization, structured like an organic body with its flow of information, can be compared also with any other production system. As we have seen earlier in Chapter 2, a production system purports to satisfy demands for products and services.

Once the required quality is defined, processes and resources must be made available that permit the respective production. Previously we followed this sequence of quality of design, conformance, and performance, with regard to individual products and services. Many documentations and records were mentioned that accompany the efforts to assure quality. Now we are to consider quality assurance detached from the individual product, service, or project, and discuss company-wide functions and activities for quality assurance in a holistic manner.

Terms and Concepts

Information is knowledge given or acquired, in the form of facts that do help to form decisions and initiate actions. Data are facts that are collected, recorded, analyzed, and properly communicated as useful information. Data and information can be grouped into many different categories to make up reports describing a particular category. For instance, information can be defined as strategic, long-term planning information, or special product oriented design information used to write up a procedure, or be fit into many other categories. The controlled information system creates an information cycle that often leads to a review of the system design and plan, and then induces a new flow of information, decisions, actions, and controlled information.

Confidentiality implies that certain information and data are considered unaccessible to particular persons because of a potentially negative impact on the competitive strength of the company. Obtaining such information is called *corporate espionage*.

A *message* is the information sent and received. A message is not always based on data. For instance, if defects unexpectedly increase, the data might not reveal the cause. Yet a message might convey that the staff is poorly motivated!

Evidence is information that provides proof of an occurrence—a fact. Records and reports for instance, serve as evidence for decisions, actions, activities, or occurrences that are used in reviews, surveillance, audits, and particularly in cases of product liability actions.

Management reporting implies that the person responsible for planning and controlling quality receives information that is useful and which adheres to the principles of sound quality (fitness for use). Any redundant or missing information can inhibit proper decisions. Good coordination must exist in all management reports (see Figure 16.1).

Function	Records	Originator	Recipient	Frequency*
Quality management program	Quality manual	Management	As listed in the manual	
Product design	Order, contract blueprint etc.	Designer Customer	Production Quality Management	
Product specification				
Inspection/test plan				
Purchasing				
Receiving				
Work in progress inspection				
Nonconformance control				
Final inspection				
Inspection/test equipment				
Packing/shipping				
Customer service				

*D—daily M—monthly W—weekly BM—bimonthly Q—quarterly

FIGURE 16.1 Major elements of the quality management information system.

Charts, diagrams, and visual aids are often more effective means of presenting information than written words. The readers should recall various forms of such aids described in Chapter 5. Figure 16.2 shows a systematic diagram of a quality information system for current data reporting. Top management needs reports with information which is concise, analyzed, reliable, valid, and useful in condensed form, reporting only the most important messages.

Figure 16.3 is an example of a quality report intended to provide top executives with an idea of quality control activities. In addition to such regular quality reporting (monthly, quarterly, etc.), special reports on problems and corrective actions, including those of the costs (see Chapter 17) are also disseminated (see Figure 16.4). Without the visibility of current problems in real time, management remains uninformed and cannot make the right decision at the right time.

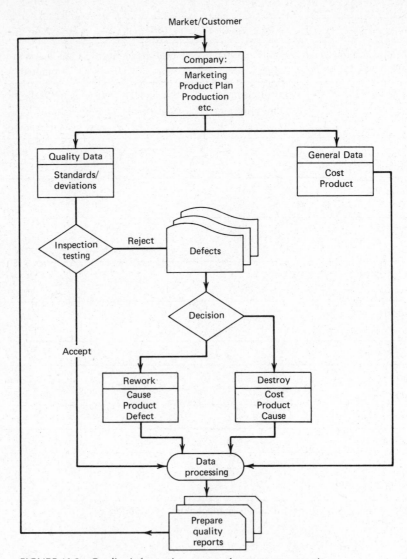

FIGURE 16.2   Quality information system for current operation.

Hierarchy of Information Flow

A total quality management information system embraces all information pertaining to the objective of quality assurance. As a system it has a special purpose and objective, through which it is related to higher ranking systems, and also to many subsystems. The familiar triangular schematic in Figure 16.5 shows the hierarchical structure and paths of information flow. It conveys the totality concept in that the information normally flows from the top downward in the hierarchy, in the form of decisions, instruction, procedures, approvals, and so on. However, in modern organizations the flow of messages, such as requests and recommendations, emanates from the lower echelons.

Rules governing delegated and independent decisions are communicated from the top; independent complementary decisions are made practically on each job, and reports on performances are transmitted back to supervisory positions. The intensity, extent, general optimization, and reliability of such hierarchical information exchange determines, in large measure, the quality of harmonious cooperation and workmanship at all levels of an organization.

Information System as a Productive System

Total quality information systems are also associated with quality assurance programs, projects, and specific productive systems. Figure 16.6 uses the familiar scheme of a productive system to illustrate the information system. The need for information, here pertaining to quality assurance, is satisfied through the respective compilation of data and other factual material, which is then processed through statistical and other analytical methods, until it is ready to be reported and communicated as information. One can see again the cyclical progression from need, to determining the process and resources, to the actual production of the information. An information system, being a production system in its own right, can produce information of different quality and usefulness. For instance, defective data put into a system cannot be fully compensated for through high quality processing. This is much the same as the fact that defective raw material will create waste through further processing.

Quality Report

Date/period	Receivers

Table of Contents

		Receivers
1.	Quality cost	
1.1	Total	1,2
1.2	Defect prevention costs	1,2
1.3	Inspection costs	1,2
1.4	Defect costs	1,2
1.5	Quality cost indices	1,2
2.	Defect compilation and analysis	
2.1	Defects by types, causes, places, and significance	
2.2	Defects as under 2.1 for product lines	1,2,3
2.3	Defects as under 2.1 for branches	1, 3
2.3	Defects as under 2.1 for other groupings	1, 3
3.	Comments, actions, results	
3.1	For senior management	1
3.2	For division managers	2
3.3	For others as named	3

Agenda For Receivers

1.	General manager
2.	Division managers
3.	Others as named

FIGURE 16.3   Quality report items of major interest.

Defect

Kind, type
Location
Cause
Frequency
Weight, criticality
Cost

Defect Listing

Defect Code No.	Description	Test	
103			

Defect Cause Listing

Defect Code No.	Causes (description)		
	Machine	Material	Other

Defect Frequency

Item:　　　　　　　　No. of Drawing　　　　　　　Date:
Inspection Point:　　　　　　　　　　Supplier:
Inspection Plan:　　　　　　　　　　Inspector:

Defect	Sample No.	Total	Weight	T × W

Defect Summation

Defect	Frequency* Weight Class	Total	Percent	Rank by Percent*

Defect Cost

Defect	Frequency* Weight Class	Total	Percent	Rank by Percent*

Pareto Analysis

100　　　　100%
% cumulative
weight or
  cost

*Defects
  ranked

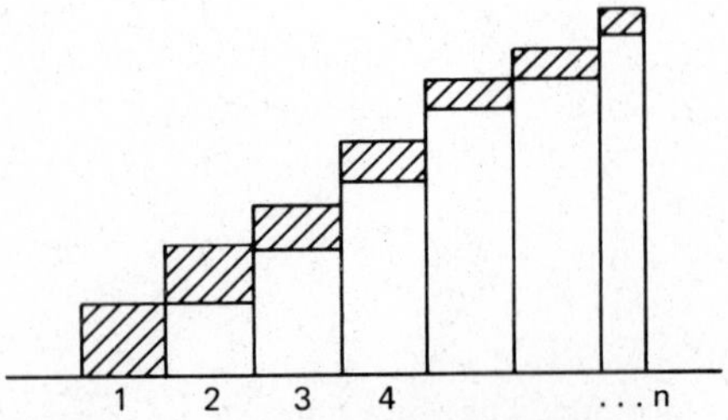

FIGURE 16.4 Example of set of records on defects, including Pareto Analysis.

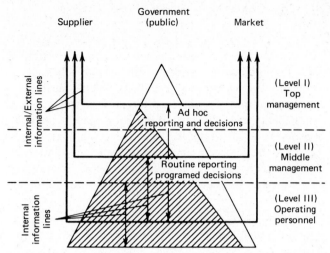

FIGURE 16.5   Hierarchical information flow (internal and external).

Quality of the Information System

Quality of information means that the information is useful to the receiver, or fit for use. The person in charge of an information system will design the system with the user of the information in mind. The user demands and needs information for deciding and acting, solving problems, selecting from alternative paths of action, and so on. Therefore, information must be relevant, reliable, current, sufficient, understandable, and optimal in every way, so that decisions are made in a timely, accurate, and effective manner. Given the hierarchy of managers within the company and the organization of reporting lines, top executives must receive condensed quality reports containing only the most important messages. Middle management requires more detailed information, and the operator needs to know elementary procedures and offered suggestions on how to perform the work properly.

Having explained the information system as a unique kind of production system, it should be clear that information systems have their own quality assurance aspects. The quality of design, conformance, and performance in information describes what an information system is actually to assure.

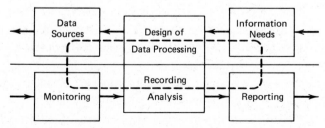

FIGURE 16.6   Information system flow.

16.2   MANAGING THE QUALITY INFORMATION SYSTEM

Depending on the importance, urgency, and complexity of the information, the degree of quality control of information will vary in order to maintain cost effectiveness. To always strike the right quality of information is extremely difficult, if it is possible at all. Any person requesting information has unique knowledge and intellectual capability. Therefore, it is important that the user of information influence the provision of information as much as possible. Maximizing quantity and frequency of reporting and dissemination of information has a known information-resistance impact. This is a psychological rejection of routine reporting that often leads to the neglect of important information and decisions. Many defects and failures in products and services are traceable to faults in the respective quality information system; the responsible persons were either misinformed or uninformed.

Structuring the Information System

Much of the information is well structured and routinized with regard to content, format, distribution, and data base. In such cases, the data processing and retrieval, that is, the entire informative production system, can be standardized. This should have a positive effect on the actual validity and reliability of the information and its associated decision making. Once standardized, computer facilities can be more fully utilized. However, the danger already mentioned is an unproductive overabundance of information.

The quality of ad hoc information should be assured through more careful design and processing, analogous to the production of custom-made high risk items. Realizing that information is providing potential power, some ad hoc intelligence information, and confidential data resources need to be protected. This is particularly important when computer and telecommunication facilities are used.

Receiving information is just one link in the continuous information exchange and flow. Once the information is used for a decision, follow-up is normally necessary. Factual data on the results of decisions and actions are needed, simply because uncertainty about the future always exists, even for the best-informed person. The better the feedback of data, the better is the subsequent information on the same matter. The assurance procedure in the three recognized areas (design, conformance, and performance) for a total quality management information system consists of the following measures.

1.   During the *design phase,* first determine the *need* for information in terms of type, quality, quantity, timing, destination, format, confidentiality, and distribution.

2.   Prepare an information specification, gain necessary approvals.

3.   Determine the *feasibility* for providing this information, including the availability of data, processing, staff capacity, and a comparison of benefits with costs.

4.   Finalize the information by design. *Interrelate* individual information and messages with the overriding reporting scheme. Eliminate inefficiencies, possible failure modes, and defects.

5.   Integrate information provision with other information exchanges regarding quantity, quality, timing, and flow direction; rationalize, systematize, simplify.

6.   Ensure that information needs and requests are satisfied optimally; avoid under or over informing.

7. During the *conformance phase,* retrieve, compile, monitor, acquire data, that is, pertinent facts and figures from the least costly and most reliable source.

8. Check the validity and reliability of the data.

9. Process the data in accordance with predetermined methods, facilities, and personnel (intermediate validity tests and checks might be necessary). Computerized, automated, and telecommunicated processes need particular safeguarding.

10. Conduct a final check against the original information request and its specifications.

11. Communicate and distribute information in accordance with procedures and decisions, obtaining documentation, receipts.

12. Retain and store documentation for future tracing and follow-up.

13. In the *performance phase,* the recipient of information should check the adequacy of the information regarding his or her need and the information specifications given with the request.

14. Ensure that information is sufficiently understood and constitutes actual valid additional knowledge, useful for arriving at a rational decision.

15. Reject any information or report that communicates unnecessary, unwanted information. Before rejection clarify the intent of those sending the message.

16. Provide feedback information in order to improve future quality of information.

A well-designed and implemented quality management information system demonstrates effective quality assurance, as it is instrumental for overall management. Its general and continuous purpose is to prevent defects and deficiencies in products and services. Moreover, the purpose of the information exchange is to support all functions, tasks, and persons in their quality assurance responsibilities and efforts, not only those employed in the quality assurance function. Figure 16.7 describes the closed control cycle between data monitoring, comparison with standards, and information feedback in case of significant deviations. The more detailed and useful the information feedback the better are the chances of detecting the causes for defects and failures, and consequently the more on-target remedial decisions and actions will be.

In the following we shall discuss some details of quality assurance data processing, reporting, and decision making. We do not pretend, however, to exhaust the topic. The references provided to the reader should be used as an information source for further study of this subject. We shall keep in mind that our interest is with the design and implementation of quality programs and restrict ourselves accordingly.

16.3 QUALITY DATA PROCESSING

As mentioned before, quality data are facts and figures that describe characteristics of products and services during all life phases, design, conformance, and performance. In addition, data must be compiled for preparation of decisions on matters of organization, quality programs, projects, and procedures. All decisions and actions require information, and, as we have seen, this information is generated from respective data. Therefore, the need for information, as it arises from problems, issues, goals, and any other objects of managerial decision making, determines the data required.

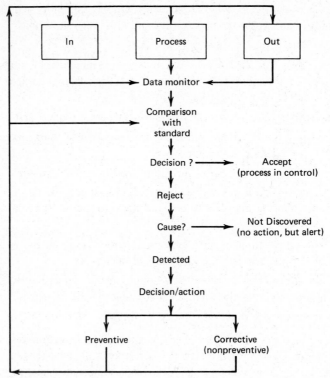

FIGURE 16.7 Close-loop process control monitoring information system.

Sources of Data

More specifically, the attributes of information, such as quantity, quality, timing, and whether or not it is decisive, direct the data collection and processing. The fact that potential problems and other matters can be anticipated but *happen* randomly is the reason why certain data must be monitored and compiled on a routine and regulated basis. The process control chart (see Chapter 11) for instance, helps to control processes by presenting process data. At the same time, the control chart informs the operator and/or the supervisor, about significant deviations from expected standards. This information leads to tracing of the causes of the problems, and if the causes are detected, to corrective decisions and actions. The chain of events for data collection and processing via control charts, for instance, is as follows.

1. A need arises for information on an out-of-control process.

2. A decision is made to use a particular type of control chart.

3. A method of analyzing the data is chosen.

4. A method of presenting the result is chosen, which emphasizes prevention and can be understood by the concerned party.

The main data for quality assurance are those acquired from the monitoring of defects and failures and any procedures and other measures for prevention. Defects and other nonconformances are monitored not only in order to take immediate corrective actions (such as rejecting the lot, changing process performances, or restoring required conditions); rather, tracing of defects to causes leads indirectly to prevention of reoccurrence, and thus to longer-term quality improvement. The more a defect reveals regarding possible causes, the more effective corrective and preventive measures can be. In some companies, a common practice is to post all the charts and graphs on defects and defectives in a common room where they are most visible to all the employees.[1]

Data Collection

Data collection takes the form of various recordings. Quality program standards, such as CSA Z299 state[2] that "records should contain documentary evidence of activities and data relevant to the quality of an item and testifying directly and indirectly that the item meets configuration and performance requirements and complies with jurisdictional and contractual requirements. . . ." Data are not only to satisfy the need for information of the decision maker, but are also to serve as material evidence. Yet *evidence* must be reliable, valid, and convincing information. This is the essence of quality assurance, namely that recordings provide documentary evidence that processes are controlled, procedures are implemented, and so on.

The following attributes of defects should be monitored whenever the benefits of having the information exceed the cost of data retrieval.

1. *Kind of defect:* Product or service related nonconformance with specifications of color, taste, measurement, and other factors at important design, conformance/production, or performance/usage phases. Indirectly, defects of products and services can be related to defective procedures, decisions, or other planning and control devices.

2. *Cause for defect:* Any defect, once reliably monitored and recorded, does have at least one assignable cause, with some known or unknown probability. A defective procedure, or noncompliance with a procedure, as well as poor workmanship, material, or process capacity, are among the major causes of defects. One or more such causes combined create defects that breed other defects, like an infectious disease.

3. *Location of defects detection:* Either the production phase or the inspection point. Both are important for tracing the defect cause and for assessing the relative importance of correction and prevention.

4. *Weight and relative significance of defect:* Predetermined ranks and classifications help to focus on the few relatively important defects and avoid wasting effort and time on many trivial defects.

5. *Evaluation of defect:* For some simple defects, correction by resorting to lower grading is necessary, while for the more potentially and actually costly defects, an intensive cause analysis and resulting prevention is advisable.

6. *Cost of defect:* Costs are the value of resources wasted due to the defect, and possibly opportunity costs in terms of lost sales.

The listing of defect attributes, nonconformances, and data thereof, concerns not only the products and services, but the underlying quality program, procedures, decisions, and quality assurance activities as well. In a total quality information system, defects are monitored and anticipated at all levels and phases of the organization. The rationale of this strategy and principle rests in the fact that most defects are caused by insufficient planning and control of work, that is, simply put, by poor quality management.

Essentials of Data Monitoring and Compilation

Predetermined data monitoring, compilation, and recording seldomly adequately informs the user. Raw data must be processed in order to convert them into effective and valuable information. The main technique used for this purpose is that of descriptive and inferential statistics (Chapters 5 and 6), because it condenses the wealth of individual data and extracts the revealing information. The chart for process control exemplifies the use of statistics as a data processing and information device. Other techniques used involve simple ranking, or just listing and comparison with related data. The well-known Pareto Analysis, the related Lorenz Analysis, or the ABC distribution, are some of the basic tools used for highlighting control of the relatively few but important defects. Figures 16.8a and 16.8b illustrate typical examples of a Pareto analysis and a systematic cause-effect analysis, applied to vendor analysis.

Advent of Computers

Computer facilities have revolutionized traditional manual data processing. Within integrated and computerized information systems, actual data processing links information requests and data retrieval when necessary (and cost effective) in the shortest possible timespans. Terminals and telecommunication devices serve for convenient data input and information receipt. Figure 16.9 illustrates such a comprehensive system of modern data processing with the advantage of using a computer for the sorting, processing, compiling, and communicating of data and information. In many companies, the quality assurance information system has now become closely intertwined with other specialized data processing systems, such as the financial, marketing, and personnel departments. In fact, most of the statistical techniques discussed in this book can be easily programmed using computers. A selected list of these programs is given in Table 16.1.

Modern advancements in microelectronics integrate actual production directly with so called *closed-loop* information and decision systems. It is important to realize the many fluid conditions that influence top executive decision making (see Figure 16.10).

16.4 COMPUTERS AND QUALITY ASSURANCE

Information systems are now increasingly computerized as total integrated systems and through isolated use of various types of microcomputers. An essential prerequisite for the optimal use of available hardware and software is the proper design of the data compilation, processing, and communication of such systems. In Figure 16.9, electronic data process was shown in the center position of the data and information flow network, between quality assurance and other functions.

Once a computer is adequately selected and programmed, it contributes considerably to obtaining fast and reliable information and thus, to decision making. This in itself augments quality assurance in performance and output of products. Examples of computer impact on

(a)

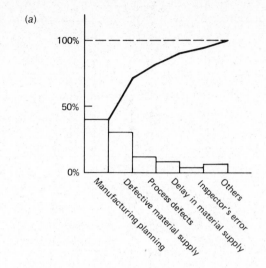

(b)

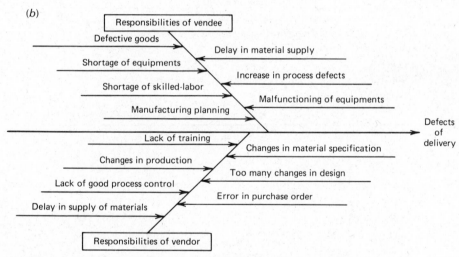

FIGURE 16.8   (a) Pareto chart for defects of delivery. (b) Cause-effect diagram for defects of delivery.

major quality assurance functions have been listed in Table 16.2. As personal computers become widely adopted, planning and controlling of quality will accordingly become more computer assisted. Direct involvement of all staff in assuring quality during the design, production/operation, and performance stages of a product will become more feasible and effective because of the tremendously improved information and communication facilities.

Computer assistance rests principly in:

1. Reliable monitoring of data.
2. Performing complex mathematical and statistical computations.
3. Reporting and graphing the user oriented results, as useful information.

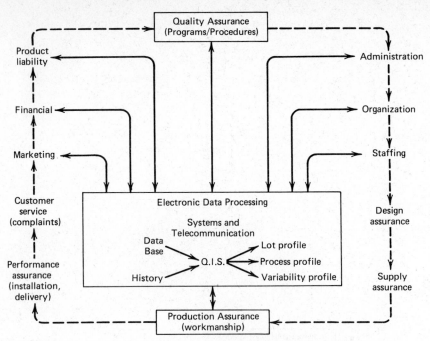

FIGURE 16.9 Integrated data processing for major functions in a quality assurance organization.

TABLE 16.1

Some Computer Programs Given in the *Journal of Quality Technology*[a]

Subject	Reference
Mean, median, frequency, histograms	January 1969, Vol. 1
	April 1972, Vol. 4
$\overline{X}$ and R charts	April 1969, Vol. 1
p and np charts	July 1969, Vol. 1
c and u charts	October 1969, Vol. 1
cusum charts	January 1970, Vol. 2
Algorithm for Scale selection in computer plots	April 1970, Vol. 2
Student t-test	October 1970, Vol. 2
Scatter diagram	January 1971, Vol. 3
Coefficient of correlation	April 1971, Vol. 3
Simple linear regression	July 1971, Vol. 3
Multiple linear regression	October 1971, Vol. 3
Sampling plans with AQL, LTPD	July 1972, Vol. 4
Double sampling plans	October 1972, Vol. 4
Multiple sampling plans	January 1973, Vol. 5
Probability charts	July 1973, Vol. 5
Dodge's continuous sampling plans	January 1975, Vol. 7
Statistical tolerancing	April 1976, Vol. 8

[a]Programs are in FORTRAN IV.

Source: Journal of Quality Technology, American Society for Quality Control, Milwaukee, Wisconsin.

486

TABLE 16.2
Computer Impact on Quality Assurance Functions

Function	Impact (examples)
Survey of customer quality demands	Larger sample size and variety; extensive analysis of data by means of modern, quantitative and statistical forecasting techniques.
Design specification and review	Computer-assisted design (CAD) permits a wide range of trials, tests, and alterations, in order to find the optimal design in the shortest possible time period for large-scope and complex items; convenient design changes are possible under controlled conditions.
Production and operations planning	Computer-assisted manufacturing (CAM) permits optimizing of processes with integration to design (CAD) and use of quantitative planning methods; any design change can be effectively translated into a process change; inspection plans can be properly coordinated and integrated.
Inspection planning	Standardized inspection plans for more critical and numerous inspection points, with automated inspection and process control.
Quality control	Inspection and testing is more reliable and consistent, scope and frequency can be adapted to changing needs; shorter reaction time in out-of-control situations.
Nonconformance handling	Nonconformances can be more thoroughly analyzed and remedial action taken without much delay; an alert can be communicated widely and quickly.
Performance control	Service information is communicated effectively and upon request to sales and service points and to customers; complaints are better processed and communicated, and handled.
Supervisor/operator performance	Front-line staff is better informed and can participate in quality assurance/inspection.
Senior management	Reports are more meaningful, allowing for more informed decisions and leadership.

4. Responding to information requests.

5. Communicating with other computers, operating systems, and individual files.

6. Maintaining and making available a complex data bank for the company.

When manual data and information transactions are automated, the scope for errors is reduced in the long run. Yet, with a greater number and variety of data processed in a shorter time with more complex systems, quality assurance planning and controlling decisions also become more complex. This leads to pressures and difficulties in negotiations, communications, and decision making. Only when certain essential prerequisites for the design and operation of computer assisted information and quality assurance systems are observed, can the feasible and desired quality of products and services be attained.

Major Prerequisites

The major prerequisites for an optimal computerized information system that is to enhance quality assurance are these.

1. All functions, jobs, and task assignments in an organization must be clearly and adequately defined, and information flow patterns must be delineated and specified. This includes the existence of operational procedure change systems, together with product and process specifications and standards.

2. Individual operating systems must be properly delineated and aligned. This concerns the interplay between personal and mainframe computer systems, and the separate design, operations, marketing functions, and so on, including their hierarchical levels of responsibility.

3. All systems should be coordinated and standardized for joint data compilation, processing, and storage.

4. Quality assurance of the system should be organization-wide and unified, extending over all phases of hardware and software engineering.

5. Users of computers should be properly trained and educated for reliable data input and recording, and meaningful information retrieval. With a policy that emphasizes assistance by computer, people, as producers and consumers, are to remain in a central role reaping the benefits of a higher quality of work.

6. All rational, reasonable, and largely predetermined kinds of information must be of satisfactory quality in terms of detail, quantity, time, and place. The quality assurance of such information extends to validity and reliability. Gaps and redundancy of information must be avoided.

7. The information system should be adequately integrated through design, flexibility and capability of adaption to need during operations, and through frequent independent audits of its effectiveness.

These prerequisites strongly suggest that quality assurance of hardware and software, in conjunction with the entire information system, is soundly instituted.

An important guide is the "Directory of Software for QA and QC."[3] Standards for quality assurance of software and computerized systems have also been published, some of which are listed in the bibliography at the end of this chapter.

Pitfalls

"There is now general recognition of the seriousness of the software problem, and formal programs have been developed to attack the problem. The main elements of the program draw upon some techniques used in controlling the quality and reliability of physical products."[4] One of the major results of controlled information systems is that people, as designers, producers and servers, can decide and act with quality assured information, that is, with knowledge and experience. Even when the machine is instructed via microprocessors, quality assurance is the ultimate safeguard that people are served.

16.5 REPORTS ON QUALITY

In any information system, data flow into the processing units and are either stored or directly transmitted for processing and then used for satisfying current information needs. Whether or not the receiver actually perceives and understands the messages and uses the information for sound decisions depends largely on the quality of reporting. The greater and more diverse the information needs are in a company, the more information systems must be formalized and computerized. This means that reports have to be standardized to transmit only the essential information.

Unrequired and unrequested routine reports have to be eliminated, so that important messages have a better chance of being received, perceived, and acted upon. Quality assurance reports about the causes of major defects must gain the attention of decision makers without undue delay. As defect monitoring is increasingly performed by on-line computers, critical defects can be communicated more quickly. Possibly, some predictable defects could be automatically corrected concurrently with recording and reporting.

Naturally, not all quality assurance activities and decisions can be supported through routine reports, no matter how fully automated the system is. Some reports are sent according to predetermined instructions to those persons and decision points that need to be informed and alerted; other reports are forwarded only upon request. Some information actively initiates and directs decisions, other communications merely keep persons informed. Some might lead to further inquiries, while others completely satisfy the information need.

Reports on quality assurance are interrelated in two directions: laterally at each level of responsibility in the organization and vertically within the organizational hierarchy. The integrated data processing system of Figure 16.9 implies lateral interdependent reporting.

Table 16.3 lists some typical reports that are instrumental in many functions of a quality control organization. Many of these, and other reports as well, entail a great amount of detailed information that is communicated to several departments, either routinely, or for special events and requests. The quality assurance program and its documentation in the quality manual will be discussed in one of the next chapters; so will the importance of quality cost reports. Most reports that are associated with the individual product, service, or contract cycle, starting with design assurance, have already been outlined in Part Two. The arrows in Figure 16.9 point in both directions, in order to indicate respective communication and interaction. There is not only a continuous influx of data into central processing but also continuous outflow of reports and information. Along with these continuous transmittals come decisions and actions and thus changes, dynamic adaptations and corrections for attainment of goals and quality assurance in every facet of the company.

In principal, all people need information in order to meet responsibilities and to recognize and understand the assigned tasks in the wider context of company operations and overall quality assurance. Motivation for good workmanship and meeting qualitative performance standards results to an ever-increasing degree from being properly informed and, at the same time, being able to inform others and to participate in the information exchange more meaningfully.

TABLE 16.3
Quality Assurance Functions and the Related Information Oriented Documentations

Function	Documentation
Quality assurance program	Quality manual
	Policy statement
	Assurance procedures
Administration	Office procedures
	Budgets/accounting
	Systems management, including management information system
Organization	Diagrams (line-staff), project networks, matrixes, guidelines, job descriptions
	Performance evaluation
Staffing	Listings of qualifications
	Performance appraisal
	Training/education outlines
	Motivational programs
Design assurance	Blue prints
	Quality specifications
	Test reports, failure mode analysis reports
Vendor assurance	Qualified suppliers list
	Purchase procedures
	Supplier surveillance report
	Supplier rating and performance
	Nonconformance report
Production assurance	Inspection reports
	Nonconformance reports
	Test equipment reports
	Internal audit reports
Performance assurance	Warranty reports, nonconformances
	Inspection reports, delivery reports
	Operators manual
Customer service	Complaints report
	Customer panel reports
	Marketing reports
Product liability	Investigation reports
	Court procedural reports
	Remedial action reports
	Quality assurance program appraisal

A well-structured and relatively open reporting scheme between the hierarchical levels of responsibility, that is, between executives, supervisors, and operational staff, has been indicated in Figure 16.5. The triangular, pyramid shape demonstrates the reporting principle of successive condensation of individual information and reports towards upper management. The underlying *broad* rule states that bosses require only analyzed data. Quality of information on quality assurance, as on any other matter, comes only from a convenient, timely, and understandable message. Too much irrelevant information acts as much as a communication barrier, as does information received too little or too late.

In general terms, executives are mainly interested in highly condensed information on quality assurance in all functional areas of a routine, and need to be informed about exceptional defects and occurrences of major company-wide significance. As shown in Figure 16.9, the quality assurance program and the functions at the top part of the scheme are more closely related with the executive management level. Towards the bottom of the organizational structure, department, branch, workplace and job-related reports convey more narrowly delineated and less condensed information. Workers, as opposed to bosses, need specific job-related instructions and data recordings.

16.6 DECISION MAKING AND INFORMATION SYSTEMS

The many filters and forms of information condensation along the reporting line often distort, unduly truncate, or unnecessarily augment reports as they move up and down the line. Quality assurance is particularly subject to unreliable and invalid information because of human desire to stress good news and remain silent or quiet about bad news. Modern automated and computerized processing and communication facilities can greatly improve quality of information, especially in the area of quality assurance of products and services. Once properly programmed and safeguarded, modern information systems provide reports with maximum perception of important messages when, where, and by whom the information is required.

Deciding Upon the Best Alternatives

By applying the systems concept, a decision can be considered as the output of interdependent and interrelated decision-making procedures that take the form of data processing. A decision is the selection of the best of several possible alternatives derived from factual information. As a decision commits certain available resources at a time when the final outcome and the degree of goal achievement is uncertain, the determination of an optimal decision in an environment of constant change and increasing risks is normally very crucial and complicated (see Figure 16.10). Thus, any information system needs to be controlled and updated regularly.

However, who can predict the reports and information required? Everyone would have to delineate reports, the information therein, the format, frequency, and other seemingly important attributes for the particular job, wherever it is in the hierarchy. Disinterest in quality assurance information would possibly have to be dealt with, when it is deemed unjustified and counterproductive. Neglecting quality assurance reports, a practice that might be revealed by major defects or product liability litigations, is a serious quality detractor, particularly when it occurs high up in the hierarchy. When sound quality assurance information systems

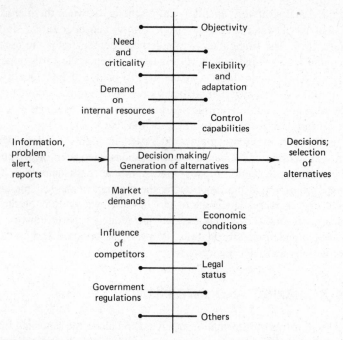

FIGURE 16.10   The process of managerial decisions and the factors that influence decision making.

and reporting remain ineffective because of the human element, corrective measures must be taken. Probably the most useful and generally acceptable approach is to initiate seminars, training, open discussions, and orientation in quality assurance.

Formal versus Informal Communication

In the age of computerized quality and management information systems, direct, personal, and informal communication is still necessary in many instances to make the system most effective. Perhaps the simple person to person reporting of trouble remains, for many reasons, an important quality assurance lever. Informality and direct communication survive not only in small enterprises, with their personal relationships. In larger companies, quality circles and project teams also allow for direct and effective information exchange and decision making.

As has been pointed out, the quality information system is the nervous system of the organization, designed to assure quality in a holistic fashion. From the senior management perspective, quality costs, quality programs, and quality audits are important information and documentations. These three topics lead us somewhat deeper into the information system design and operation and will be discussed in following chapters.

16.7   QUALITY MANUALS, STANDARDS, AND GUIDES, AS INFORMATION SOURCES

We pointed out that every company must assure quality and therefore, must practice quality assurance. Even for the most rudimentary program with simple procedures, a *quality manual* should be prepared and be made available. Customers, suppliers, employees, and other interested persons want and need to know what, in fact, the company does in order to assure quality. As we have seen, any assurance given must be verifiable. Therefore, companies, if not by their own policy, are required by others to document their quality assurance programs. This documentation might not always be called a manual, but serves the purpose of such a manual. A manual is a handbook that describes quality assurance activities as they are planned and implemented in a company.

As each company differs from most others in many respects, so must its manual reflect such differences. Even when the manual is written for a company with the identical program and standard, the manual should describe how the company has organized its own efforts in order to comply with the standard. Internal programs and procedures are the responsibility of the management of the company; other staff can only advise, approve, or reject.

A manual should be a complete and comprehensive outline of a company's quality assurance program. The program might have only a few procedures, or a great many. The *purchase order control procedure,* for example, stipulates that all orders must be approved, that the buyer sends the drawings and possible subsequent changes to the vendor, and finally, that all purchase orders are kept on file.

A procedure can be described as a set of instructions for the persons concerned in an endeavor. Here the information takes the form of directives for activities. The purchase order, in this case, is the data recording; so are design drawings. Instructions for updating the recordings and for safekeeping them are provided to maintain the validity and availability of such information.

The practice of preparing useful lists in the manuals, such as procedures, product specifications, and other pertinent information for the purchase order and associated documents, is the core element of any information system. Without such data and valid documentation, no report can be made. Recording (or records) and reports are interdependent.

Standards and guidelines are the main official information source for preparing a quality manual, and the associated recording and reporting system. The basic approaches are either to select a standard and then to design, implement, and control the quality information system accordingly, or to build upon the current documentation and information practices as they exist and then to use standards for improving the existing system. Probably a blend of both alternatives is the most practical. The writing of a manual, and a current review of recordings and reports, should lead to improved information flow. More details on this are provided in Chapter 18.

The Canadian Standards Association's (CSA) Series Z-299 lists quality program documents as being quality manuals, inspection and test plans, checklists, activity procedures, or descriptions. With four different CSA subprogram levels, the documentation ranges form simple end-item inspection recordings in the inspection program, to a comprehensive quality information system in the quality assurance program. This shows that depending upon its size and scope, a company might only need very simple recording of inspection results and information on defects to be able to control quality.

The more complex the product is technologically and the greater the hazards and risks of product liability, the earlier in the product life defects must be prevented. This means that much more information and data are required. For the simple inspection program, a simple procedure description suffices. For the large and complex information system in the quality assurance program (Z 299.1), a more comprehensive quality manual must be prepared that describes in some detail how all requirements of the program standard are to be met in the overall setting of the company.

As any company needs some evidence about its quality assurance activities in the broadest sense, the staff members should prepare a manual that informs about these activities and their respective procedures. We shall give some examples of such manuals in Chapter 18.

While manuals and procedures direct and describe activities, information on results of actual operations, and in particular on quality specifications, defects, and inspections performed, are recorded as data first. Standards describe the content and form of such quality records. The CSA Z-299.1 Standard emphasizes that records must provide "sufficient evidence" that activities have been carried out as required, that the defects have been monitored and analyzed with regard to causes, and that preventive action has been taken.

All records, recordings, and reporting must be outlined in the quality manual and this is done according to the program standard invoked. We shall study these program standards in more detail in Chapter 18.

The CSA Z-299 Standard is a typical example of its type. Any company that complies with this standard in program documentation and recording, meets other standards in general terms as well. The ANSI *Generic Guidelines for Quality Systems* (Z 1.15) suggests that the quality manual should specify the general quality policy, the major features and activities of the program, and the procedures. Therefore, the obvious purpose of the quality manual is to provide a general overview of the quality assurance effort of the company.

For control of the current operation, the guideline suggests that "supplier generated quality information on product acceptability may be used, where practical, to avoid unnecessary or redundant inspections and tests when all of the following conditions exist:

1. The type of information is defined and specified by the purchaser.

2. The information is a statement of fact or condition based on quantitative or qualitative observations.

3. The information can be verified by the purchaser.

4. The periodic verification program is established."[5]

If the purchaser is to rely on inspection and testing by the supplier, the respective information must be reliable. The same principle of verifying the quality of information is suggested for internally conducted inspection during production. The guidelines spell out in some detail the basic information for an effective quality system. These are: product identification, inspection and quality control procedures, inspection records, identification and recording of rejected items, scrap and spoilage reports, quality cost reports, and periodic summaries.

For each category of records and information, purposes and reasons are given. The guidelines describe the general principles and rules that must be followed in these recordings. Records must be complete, current, accurate, and traceable to the product. They must identify the staff involved and show sufficient information for tracing defects. These records must be retained and, finally, they must be analyzed for reports, necessary decisions, and corrective actions.

16.8 KEY TERMS AND CONCEPTS

A **management information system** is a program, planned and executed to collect facts and figures, to process and communicate these as information to aid decisions and actions. As a system, the program is formally structured.

Information is processed and communicated data, requested and designed to give knowledge (inform) to a person or machine and thus facilitate decisions or actions. Quality specifications and messages of deviation from standards are essential information for quality assurance.

Management reporting means to tailor the information flow for effective decision making by optimizing quantity, maximizing quality, and assuring proper timing and communication of information exchanges between appropriate destinations. Management reporting is the essence of a management information system. Special integrated management reports deal with information exchange on quality assurance matters.

A **quality information system** is the delineated and integrated section of a comprehensive management information system that collects, analyzes, and disseminates data and information concerning quality assurance. As a subsystem it can be manually operated or be computer-based.

Data and information on defects are an essential component in any quality information system. A defect analysis considers type of defect, cause, location, time, level of importance, consequences, preventive measures, and cost.

A **quality manual** describes policies, procedures, organization, and reporting schemes designed and maintained for quality assurance purposes. The document is often required to reflect compliance with applicable quality program standards and serves as a basis for quality audits.

16.9 DISCUSSION AND REVIEW QUESTIONS

(16–1) *Any quality assurance program is predominantly a management information system.* Explain.

(16–2) Define *quality of information.*

(16–3) *Effective management reporting depends on the observation of certain principles.* Explain.

(16–4) *A report on current quality assurance to senior management should include certain major points of information.* Explain.

(16–5) *Certain rules should be followed when compiling and processing quality data.* Explain.

(16–6) *The more and better informed employees are about quality assurance issues and developments, the higher their motivation for quality in performance.* Explain and discuss.

(16–7) *Defects and nonconformances must be carefully analyzed and the results widely communicated.* Explain and discuss.

(16–8) *Computers have a great impact on conventional quality records and reporting systems, including statistical control charts and the like.* Explain.

(19–9) *The need for information plays a similar role as the consumer demand in quality assurance.* Explain and discuss.

(16–10) *Each quality assurance function and special procedure should be associated with respective reports and reporting schemes.* Explain.

(16–11) *Reports and recordings should have a standardized format and content.* Explain, using an example.

(16–12) *The quality manual serves as a major information source for all concerned in quality assurance.* Explain and discuss.

(16–13) *Even when the quality assurance program standard does not require a quality manual, the company should prepare one.* Explain, and discuss the advantages.

16.10 PROBLEMS

(16–1) You are to prepare a quality report form for senior management.

a. List the steps to be taken, including a checklist of questions.

b. Prepare a record form that can be used for regular quality reporting.

(16–2) Select a record of inspection results and:

a. Critically evaluate the data recorded for further information processing.

b. Prepare a report of the analysis, presentation, and communication of the inspection results to senior management. Using this report, prepare a record form for inspection results.

(16–3) Write a procedure for handling self-inspection.

(16–4) List the purpose and major contents of a quality manual.

(16–5) Prepare a checklist for evaluating a quality manual.

(16–6) Write a procedure for evaluating the effectiveness of the quality assurance information system.

(16–7) Prepare a record form for defect reporting that can also be used for preparing a control chart for either percent of defectives or for number of defects.

(16–8) Draw a flowchart and an associated guide for acceptance sampling inspection.

(16–9) Prepare a flowchart for handling and analysis of items returned by customers because of dissatisfaction with the item.

16.11 NOTES

1. See for example, Leek, Jay W. and Riley, F. H., "Product Quality Improvement Through Visibility," 1978, *ASQC Annual Technical Conference Transaction,* p. 229.

2. CSA Standard *Quality Assurance Program Requirements,* Z299.1, Canadian Standards Association, Rexdale, Ontario, Canada, 1978, p. 49.

3. "Directory of Software for QA and QC," *Quality Progress,* March 1984, pp. 28–53.

4. Juran, J. M., and Gryna Jr., F. M., *Quality Planning and Analysis,* 2nd ed., McGraw-Hill Book Company, New York, 1980, p. 589.

5. American National Standard, *Generic Guidelines for Quality Systems,* ANSI/ASQC Z-1.15, clause 6.2 American Society for Quality Control, Milwaukee, Wisc., 1978.

16.12 SELECTED BIBLIOGRAPHY

American National Standard IEEE, *Standard for Software Quality Assurance Plans,* ANSI/IEEE Std 730–1981, The Institute of Electrical and Electronics Engineers, Inc., 345 East 47th Street, New York, N.Y. 10017

Baker, Frank H., "Improving Quality Information Management Systems," *Quality Progress,* July 1983, p. 42.

Bateman, V. W., "Software Quality—Key to Productivity," ASQC *Quality Congress Transactions,* Detroit, 1982, p. 767–76.

Berger, Roger W., "What Do You Look for When You're Looking for Software?" *Quality Progress,* March 1984, pp. 28–30.

Boehm, B. W., *Characteristics of Software Quality,* Vol. 1, TRW and North-Holland Publishing Company, New York, 1978.

Campanizzi, J. A., "Structured Software Testing," *Quality Progress,* May 1984, pp. 14–15.

Canadian Standards Association, *Software Quality Assurance Program,* Part I-CSA Q396.1-1982, Canadian Standards Association, 178 Rexdale Blvd., Rexdale, Ont., Canada, M9W 1R3

Carlsen, Robert D., Berber, JoAnn, and McHugh, James F., *Manual of Quality Assurance Procedures and Forms,* Prentice-Hall Inc., Englewood Cliffs, N.J. 1981.

Carpenter Jr., C. I. and Murine, G. E., "Measuring Software Product Quality," *Quality Progress,* May 1984, pp. 16–20.

Covino, Charles P. and Meghri, Angelo W., *Quality Assurance Manual,* Industrial Press Inc., New York, 1962.

Dunn, Robert and Ullman, Richard, *Quality Assurance for Computer Software,* McGraw-Hill Book Company, New York, 1982.

Durren, Michael J., "Computerized Inspection Instructions," *Quality Progress,* November 1982, p. 46.

Grinath, A. C. and Vess, P. H., "Making SQA Work: The Development of a Software Quality System," *Quality Progress,* July 1983, p. 18–23.

Kraft, Charles L. and Vincent, Albert W., "Inspector Instruction: Writing Them to the I-Point Level," *Quality Progress,* February 1982, p. 16.

Mendis, Kenneth S., "Quantifying Software Quality," *Quality Progress,* May 1982, p. 18.

Norquist, Warren E., "Quality: How Well Are You Communicating It?" *Quality Progress,* June 1981, p. 22.

North Atlantic Treaty Organization (NATO), *NATO Software Quality Control System Requirements* AQAP–13, 1981, NATO International Staff Defence Support Division, Material Management Systems Section, NATO Headquarters, 1110 Brussels, Belgium.

O'Reilly, C. A., "Variations in Decision Makers' Use of Informations Sources: The Impact of Quality and Accessibility of Information," *Academy of Management Journal,* Vol. 25, No. 4, December 1982, pp. 756–72.

Thomas, David W., "A Successful Management Information System," *Quality Progress,* January 1982, p. 24.

Williams, Harry E., "Communication and the Quality System," *Quality Progress,* September 1977, p. 34.

Williams, Theodore J., "Trends in the Development of Process Control Computer Systems," *Journal of Quality Technology,* April 1976, p. 63.

Wittaker, Della A., "Read It Aloud . . . To Improve Your Report," *Quality Progress,* April 1977, p. 34.

17 CHAPTER

QUALITY COST ACCOUNTING AND PERFORMANCE CONTROL

Defective products and services equal money lost for the producer, and often also for the consumer. Such nonconformances not only result in various costs but, in the extreme, major defects and failures with overriding liability suits can render a company bankrupt. A company's need to be competitive in terms of price, quality, and delivery, together with customers' expectations, legislation, and other requirements, underline the need to operate at optimum cost. Any measure, activity, procedure, or program that is designed to assure quality must save more than it costs. *Cost-effectiveness* of quality assurance is its main underlying principle. In short, quality assurance must contribute to profit, rather than be a cost center.

In order to achieve this objective, management needs respective cost information for the planning and controlling of quality assurance. The cost accounting system can be considered a special subsystem of the quality information system that was discussed in the previous chapter. Its purpose and objective is to inform about the cost and other financial aspects of quality assurance, and thus to support financially sound decision making. Moreover, quality cost accounting facilitates performance control of quality assurance functions, as well as that of other functions with regard to quality assurance activities and thus assures the quality of the products and services themselves.

In more specific terms, the objectives of quality cost accounting systems are these.

1. To serve as a guide to management, expressed in management language (i.e., profit, cost; in terms of dollars and cents).

2. To provide justification for quality improvement programs, since profits help gain management approval.

498

3. To measure success or failure in terms of money.

4. To aid decision making, allocation of resources, capital, current operating expenditures, budgeting, and so forth.

5. To aid in the evaluation of alternatives and the selection of the least costly inspection methods, tools, staff, and other necessities.

6. To help inform and motivate staff by showing them that costs are being reduced and profits are up.

A report on the bus manufacturer whose customer encountered design faults on the 200 buses delivered states, ". . . because of the problems experienced with the sale of 200 buses to Metro Transit, the company will not see a profit in its year-end financial statistics . . . ."

Much of the lack of interest in quality assurance among managers of some companies has been due to insufficient information on quality costs. Money, profits, and costs carry important messages for management, and therefore, quality cost accounting has an important role to play in inducing and motivating senior management to request cost information along with any proposal for quality assurance measures. The impetus for designing and implementing a quality assurance system must come from top management, and this in turn requires adequate information and evidence that such a system would be cost-effective.

In this chapter we shall discuss major concepts and practices in quality cost accounting. This will be followed with various special accounting techniques pertaining to data compilation and analysis, including cost reporting. We shall then briefly return to management decision making using quality cost information. The next chapter on quality assurance programs will show that the most important prerequisite for the design and implementation of such a program is that management can assess the financial cost implications. We therefore orient our discussion on quality cost accounting mainly to prepare for quality assurance programming.

17.1   QUALITY COSTS: CONCEPTS AND PRACTICES

Costs represent a sacrifice of values, which a business incurs for the purpose of attaining revenue. Not only assets, but also the service of staff, supplies, government support programs, are all resources that must be paid for directly or indirectly. Costs are distinguished from expenses and losses. *Expenses* are actual outflow of cash and thus costs applied against the revenue of a particular period. However, costs are a wider concept for the value of any resource acquired or used and measured in terms of money. *Losses* are a reduction in a firm's equity or in any resource for which respective revenue has not been attained.

The quality control function, aside from being responsible for reporting quality performance to management in terms of rejection and defective materials, must not overlook two important facts: all failure costs, no matter how they are caused, reduce profits; and preventive quality control, irrespective of when it is utilized, adds to cost savings. However, it should be kept in mind that the operation and effect of any measure costs money.

Quality Cost Categories

Quality costs are conventionally defined as the sum of defect *prevention costs, appraisal costs,* and *failure costs.* However, the definition and delineation of quality costs in individual cases is at times very difficult and complex. This is due to quality costs not being identical with costs incurred for creating quality of products and services. Such a definition would include all material and labor costs, as well as the overhead costs, and thus would be identical with general product costs. The following list contains some of the major elements in each category of quality costs

1. *Prevention* (Costs associated with the planning and controlling of a quality assurance system and program.)
   a. *Quality control administration.*
   b. *Quality systems planning and measurement.*
   c. *Vendor quality surveys;* here we look for salaries and expenses incurred by personnel outside of the quality control department, such as purchasing agents and design engineers who participate.
   d. *Quality data analysis and feedback;* also includes tabulating time.
   e. *Quality training;* formal programs and quality campaigns.
   f. In addition to those costs specially singled out above, there are the normal costs of quality control engineering planning for process control, appraisal planning, designing of test and inspection equipment, reliability studies, special process planning, and the other related factors of such a program.

2. *Appraisal* (Costs involved in the direct appraisal of quality, both in the field and in the plant.)
   a. All inspection and test costs no matter where performed or to whom the function reports. This does not include troubleshooting, repair, reinspection, and retesting, because these are failure costs.
   b. Laboratory acceptance testing costs related to evaluation of product.
   c. Setup costs for test and inspection.
   d. Product engineering review and shipping release.
   e. Field testing.
   f. Costs of materials used up in testing and evaluation.

3. *Internal failure* (Costs directly related to the occurrence of defective production within the plant.)
   a. Discrepant material review activity includes all expenditures by quality control engineering, purchasing, production control and design engineering attributable to evaluation, review, disposition and reprocurement of defective material.
   b. Rework; includes labor and material.
   c. Scrap; includes labor and material.
   d. Retest and reinspection, repair and troubleshooting.

4. *External failure* (Costs associated with the failure of a product or service in the field.)
   a. Returned material processing and handling.
   b. Returned material repair, including factory repair performed on in-warranty products.
   c. Field activity costs due to warranty replacement, including labor and material.
   d. Field activity due to engineering error, including cost of modifications.
   e. Field activity costs due to labor and material corrections in the field because of factory faults.

In a more general sense, two major categories of quality costs are distinguished. The first category consists of those cost elements which represent resources used for directly assuring quality of products and services, such as defect prevention and appraisal or inspection costs. One could call these the actively incurred quality assurance costs. These, of course, have the purpose of reducing, if not entirely avoiding and eliminating defects and failures of products and services. The second general category is the cost of defects and failures. Both of these cost groups need to be included in the total quality costs.

The delineation of individual cost elements in each category for computation of the sum and then for computing total quality costs, must be made from the definition as to what resource, or part thereof, was directly and explicitly spent, for instance, for defect prevention. Checklists of questions (see Table 17.1) can help one to arrive at a reliable and fair definition and delineation. If some training costs for operators are to be extracted as quality costs, then one would question what proportion of the course content was related directly to assure quality of workmanship. This would include, for instance, training in following quality assurance procedures, and training in self-inspection or testing. Many other practical examples for defining prevention costs could be cited. In general terms, any costs incurred for the design, implementation, and maintenance of a quality control program belong to the defect prevention costs.

Similarly, appraisal costs cover values of resources deployed not for preventing defects, but for inspecting and testing that which is already completed, either in part or entirely, for approval of products and services. While prevention costs focus on the quality control program and the general planning and control of quality assurance, appraisal of quality assumes availability of some completed work on the product or service, with its existing quality characteristics. Again, the productive work itself and the resources spent (even though work

TABLE 17.1
Checklist for Delineating Failure Costs

1. Are the sources of failures and their real causes identified and separated according to functional responsibilities? (Yes/No)
2. Are reports clear enough to distinguish the vital few costs from the trivial many for each of the responsible departments? (Yes/No)
3. Are the major cost contributors clearly identified to enable each department to act quickly to remedy the situation? (Yes/No)
4. Have all vendor-caused losses been reported to the receiving inspection department? (Yes/No)
5. Are the basic causes of defects found and corrected by all the concerned departments? (Yes/No)
6. Is nonconforming material salvaged economically, according to the accepted procedure? (Yes/No)
7. Are causes of high field failures identified and proper action taken on them? (Yes/No)
8. Are warranty changes audited appropriately and action taken to prevent recurrence? (Yes/No)
9. Does the quality improvement effort exist in harmony without an atmosphere of blame? (Yes/No)

might involve regular checks of the work by the worker) are not to be included in appraisal costs.

Appraisal must be understood as the verification that quality specifications for the item, supplies, processes, operator qualifications, and so on have been complied with. If a non-compliance has been detected by the person in charge of quality inspection, then from a certain point upwards, appraisal costs might very well result in defect prevention costs. This is the case when defects are further traced to their causes and preventive measures can be taken. The term "appraisal" must be literally and carefully applied for delineation of these active quality cost elements. This valid and reliable distinction sheds some light on the kind of quality assurance efforts needed. The less costly a defect is, the more simple will be the appraisal required. For a high cost defect preventive measures will be more cost-effective.

The defect and failure costs are also not always clearly definable in practice. Some scrap, for instance, is practically unavoidable because of technical reasons and natural variance in materials, production processes, and work performances. Consequently, only the avoidable scrap, that is, in general, any avoidable defect, should be included in this category.

The basic question to be asked is whether it would have been possible to prevent or eliminate a particular defect or failure if certain measures and controls had been carried out. A secondary question should be whether or not such a quality control device would have been reasonable, in that it would have saved more than it cost. The purpose of this question is to confirm the required cost-effectiveness of any quality assurance measure whether in a manufacturing organization or in the service industries. In principle, the net result must be a general and long-term reduction of quality costs.

The cost of a defect or mistake is hard to measure, especially if it only induces customer dissatisfaction. You order an item by phone, but the delivery person brings you the wrong thing; your luggage does not arrive at the airport when you do; you receive clothes back from the dry cleaner and the zippers don't work; monthly phone bill statements are wrong and it takes four months to correct them; the wedding invitation card arrives in the mail six months late; and so on and so forth. These are the factual happenings. Think of the direct costs involved in these defective services with added dissatisfaction. The cost of clearing up some of the mistakes in above examples may perhaps be small, but the cost of explaining how the errors occurred may be far greater.

17.2 QUALITY COSTS MODEL

Figure 17.1 is a common presentation of the interaction of the three main cost categories. The trade-off effect of more resources and efforts devoted to defect prevention and appraisal of actual defect and failure costs is indicated. Yet, the point where *total quality costs* are at a minimum demonstrates that there is always an optimum where these counteracting cost categories are in balance. Moreover, as the figure shows, by increasing expenditures on prevention costs, the costs of failure may be expected to fall. However, there is a point at which the total costs will be at their minimum.

Elements for the calculation of quality costs may differ from industry to industry. It is up to the top management to decide the portion of costs to be alloted to design control, manufacturing and production control, or to shipping control, for instance.

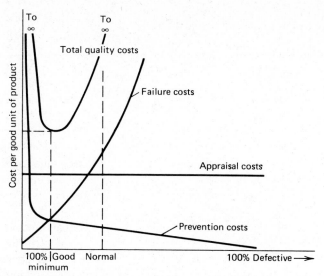

FIGURE 17.1   Quality cost model.

The Concept of Opportunity Costs

There are many unexploited opportunities for quality cost reductions in companies. The concept of *opportunity costs* is very meaningful for modern quality assurance practice. As the term indicates, even in a company with a suitable and well-implemented quality assurance program, there are always possibilities for further defect prevention and enhanced quality improvement. Management that is not kept informed through the quality cost accounting system cannot utilize such dormant opportunities. It is not surprising that from 15% to 40% of the manufacturer's cost of an average product is for hidden waste that the company tries to recover through higher prices passed on to consumers. Improving the systems of production and services will show tremendous opportunities for better quality and reduced prices.

Cost of Inspection Alternatives
Juran and Gryna, Jr.[1] have discussed the simple economics of inspection by a comparison of the total cost sampling inspection versus 100% inspection. Based upon the assumption that when no inspection errors occur and the cost to replace a defective item found in inspection is borne by the producer or is small compared to the damage or inconvenience caused by a defective, the total cost can be calculated as: NpA (no inspection), $[nI + (N - n) pAP_a + (N - n) (1 - P_a) I]$ for sampling inspection and NI for 100% inspection, where:

N = Number of items in lot.
n = Number of items in sample.
p = Proportion defective in a lot.
A = Damage cost incurred if a defective slips through inspection.
I = Inspection cost per item.
P_a = Probability that lot will be accepted by sampling plan.

TABLE 17.2
Important Cost Concepts with Application for Quality Assurance[a]

Concept	Feature	Examples
Variable (fixed)	Vary with production volume, material, labor	Inspection costs, 100% or sample size
Actual (estimated)	Historical or planned use of resources	Scrap, inspection, value training
Standard (normal)	Cost established for cost control, pricing, plans	Defect allowance, reduced quality cost targets
Incremental (sunk)	Costs affected by decision; selection of alternative	Determination of least-cost inspection plan
Capital (operational)	Costs for equipment, research projects, assets	Test equipment, quality program
Life cycle (production)	Costs compiled over the entire lifespan of a product, contract, or project	Design, production, performance phases; including prevention and maintenance
Short run	Costs per period with short-term showing different fluctuations than long-term trends	Prevention, appraisal, defect, and most other cost categories
Opportunity (income)	Revenue, profit contribution, not exploited but potential	Any possible cost reduction; less defects through cost-effective inspection
Traceable (unassignable)	Costs that can be traced to specific causes	Defect with identified cause
Controllable (uncontrollable)	Cost that can be influenced by decisions, actions, performances	Audits for examining quality assurance program
Direct (indirect)	Cost directly assignable to objects having used the value	Defect assigned to cost center having caused it; indirect cost would be assignable burden (overhead)
Hidden (known)	Costs that should be imputed; not recorded or reported	Customer lost because of poor quality; loss of good will; quality image amongst business partners
Out-of-pocket expenses	Cost actually paid during period	Warranty repair by service, laboratory expenses

[a]All concepts are interrelated in varying degrees. Moreover, total quality costs, the conventional categories of defect prevention, appraisal, and defect costs, and the individually defined cost elements in these categories, can be related to most of the concepts listed above.

Quality Costs Related to General Accounting Concepts

Managers, who are usually fairly well-versed in management cost accounting and decision making, expect quality costs to be related to other familiar concepts, which we can only mention here. These are listed in Table 17.2, together with a brief definition and typical references or examples of their relation to quality assurance. A comprehensive theory of quality cost accounting that could serve management as well as the existing management cost accounting system is still in the development stage.

Use of Index Numbers

An *index number* measures a percentage, a relative change, or the structure of a variable compared with some base. The *consumer price index* is probably the most widely known example of the index number. Similar indicators have been designed and used in quality assurance. The following are some examples of those applicable to quality cost analysis.

1. Monitoring changes in quality costs over time:

$$C = \frac{C_n}{C_0} \times 100 \qquad \begin{array}{l} C_n = \text{cost for a specific period} \\ C_0 = \text{cost for the base period} \end{array}$$

2. Summarizing quality costs over departments, product lines, and so forth:

$$C = \frac{\Sigma C_n}{\Sigma C_0} \times 100$$

C is a total of quality costs for different product lines, types of defects, quality assurance programs, departments, and so on.

3. Summarizing and weighing quality costs: the reason for weighing could be the fact that quality costs vary with the different life phases of a product over the period, or the relative importance of the costs in comparison with price, labor costs and other costs.

$$C = \frac{\Sigma C_n \times q}{\Sigma C_0 \times q} \times 100$$

q is a weight for the individual quality cost components. The basis for different weight must be known and justified, for instance, for product lines or defect categories that vary in importance.

Any index must have a reliable and valid base and a clearly defined purpose, adequately explained, understood, and communicated. Index numbers must be used in decision making with utmost caution. Table 17.3 gives an example of the calculation of index numbers with reference to quality cost reporting.

Example: A major auto manufacturer has a method of evaluating the products it produces, whether it be subassemblies or finished cars. This method is known as the *Quality Index Rating System*.[2] The rating system example for the trim quality was evaluated as follows.

When a reject is found during the audit procedure, a discrepancy evaluation is made. The discrepancy is put into one of three categories: one demerit, four demerits, or ten demerits.

A *one demerit* item is categorized as a "minor discrepancy." These are items that would not effect the build of a car, but would cause slight customer dissatisfaction. Examples of these are:

TABLE 17.3
Reporting Quality Cost Based Upon Index Numbers

Period	Total Cost	Reworked Cost	Quarterly Average		Quality Index Number Based on First Quarter	
			Total Cost	Reworked Cost	Total Cost	Reworked Cost
Jan	52	23				
Feb	70	41	54	27.3	$\dfrac{54}{54} = 1$	$\dfrac{27.3}{27.3} = 1$
Mar	40	18				
Apr	62	28				
May	87	53	63.3	32.0	$\dfrac{63.3}{54} = 1.172$	$\dfrac{32}{27.3} = 1.172$
June	41	15				
July	90	62				
Aug	72	42	81	54.7	$\dfrac{81}{54} = 1.50$	$\dfrac{54.7}{17.3} = 2.00$
Sept	81	60				
Oct	78	72				
Nov	90	60	88	68.3	$\dfrac{88}{54} = 1.63$	$\dfrac{68.3}{27.3} = 2.50$
Dec	96	73				

1. Soiled trim.
2. Long threads not trimmed.
3. Sealing strip staples missing.
4. Glue on face side.
5. Clip holes with excess glue.

A *four demerit* item is categorized as "moderate discrepancy." These are items that may be annoying to the assembly plant or the customer, but would not cause ultimate failure of the product. Examples of these are:

1. Sewn off notch (more than 1/4 in.).
2. Pleated or puckered trim.
3. Loose tension.
4. Wide seam.
5. Scuffed lace.
6. Narrow joint seam (between 1/4 in. and 1/8 in.).
7. Presew exposed.

8. Design sewn crooked or off location.

9. Sewn away from lace.

10. Defective embossing

A *ten demerit* item is categorized as a "major discrepancy." These are items that should not exist as they have been built, and that ultimately could end up causing major warranty claims. Examples of these are:

1. Sewing skips (joint or design sewing).

2. Cuts.

3. Needle holes.

4. Loose tension (major).

5. Open seams.

6. Narrow joint seams (less than 1/8 in.).

7. Sewn over seaming lace.

8. Wire pocket or bolster twisted, open, or missing.

9. Holes pierced off location.

Once the evaluation is completed, statistics are compiled.

Total Pieces Audited	Total Discovered	Total Demerit	Quality Index
100	2	14	140

The "quality index rating" (QIR) is then calculated by means of the following formula.

$$QIR = 145 - \frac{Discovered + Demerit}{Total\ Pieces\ Audited} \times 30$$

$$= 145 - \frac{2 + 14}{100} \times 30$$

$$= 140$$

The number 145 has been set arbitrarily by senior management for perfect performance; 30 was chosen for the trim quality assessment.

17.3 QUALITY COST ACCOUNTING: PRINCIPLES AND METHODS

Resources spent and invested for maintaining and improving quality must be accounted for in the same way as expenditures for any other business purposes. *Accounting* involves monitoring, recording, analyzing, and reporting financial and operational data in monetary terms. Only well-prepared and systematic accounting generates valid and reliable information. In cost accounting, and also in quality cost accounting, regular reports inform management and other users about short- and long-term developments in the general cost-effectiveness of quality assurance activities. At times special cost studies are undertaken for new product designs, for assessment of the financial justification of a quality control procedure, or even for entire quality control programs.

The following are some principles that relate to sound quality cost accounting.

1. *The purpose of the cost study needs to be clarified;* it directs data compilation, processing, and communication.

2. *As a management aid,* benefit and cost must be kept in sound proportion.

3. *Performance appraisal with cost data* should be done cautiously and in a constructive rather than a punitive manner.

4. *The quantitative basis for cost data,* such as defects or defectives needs to be approved, and respective recordings must be reliable.

5. *Cost data to be useful,* must be *available* without undue delay.

6. Only in exceptional and well-justified circumstances should quality cost data be kept *confidential.* Any objective and individual task and method has cost relationships. Once these are more widely and better understood, cost performance can improve.

7. *Cost must be defined* in a way that is generally acceptable and allows allocation of such data. Normally the user of such quality cost information decides, or at least contributes to decisions about data compilation, processing, and communication.

8. For proper interpretation, quality costs should be *compared on a suitable basis,* (e.g., as percent of sales and profits, etc.).

9. *Abnormal cost behavior* must be indicated and analyzed. Learning quality cost, for example, distorts normal cost during the program start-up phase.

10. Objectives, goals and tasks should be translated into costs.

11. *Causes for cost changes* not related to the quality program such as inflationary effects, must be neutralized.

12. *A quality cost data bank* should be gradually developed with the limitations of the data well signified. Initially, priority must be assigned on the basis of a Pareto Analysis. See the example in Figure 17.2 illustrating the relative costs of defects found in a ductile iron foundry. *Crude data is still better than none.*

13. Through close cooperation with accounting and other departments involved as suppliers or users of quality cost data, *quality cost information systems* must be integrated and coordinated for reasons of efficiency and effectiveness.

14. *Definition of quality costs* should not only deal with components, but also with the various types, such as opportunity costs, incremental, standard, normal, estimate, and historical costs, et cetera.

15. Only analyzed and approved quality cost information should be released. Broad principles (*bosses require only analyzed data*) must be observed.

16. Quality cost reduction normally should not impair quality. Whenever feasible, *consumer quality cost* should be included. In an optimal quality analysis, cost and usefulness are in sound relation to each other (e.g., degree of details, frequency, etc.).

17. Cost analysis must start with defining the purpose, and end with action and follow-up.

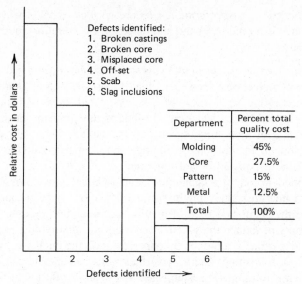

FIGURE 17.2   An example of ranking by application of Pareto Analysis for separating the highest contributor to cost figures.

Operating the Quality Costing System

The following basic accounting procedure leads to meaningful support of quality assurance planning and control.

1. *Responsibility for quality cost accounting to the quality assurance function:* The main reason for this is to establish communication between the financial management and quality control specialists regarding costs.

2. *The specific need, use, type, format, time, and distribution of quality cost information:* In case the scope and variance of such information exceeds the available accounting capacity, ranking by priority must be made, with top priority assigned by level of responsibility of management, relative importance of the project, product, or other subject to be studied, plus the weight of the potential defects. Pareto Analysis can be used.

3. *Content and form of the quality cost report:* To determine this, obtain approval from the user of the cost information, and verify feasibility of the included information.

4. *Assessment of the data:* Assess available data sources and other cost accounting reports in order to avoid duplication of cost accounting.

5. *New quality cost data compilation with the actual information needs in mind:* For this purpose, use data recording forms and techniques that result in valid and reliable data. Principles of sound data processing must be observed in order to assure quality of cost information.

6. *Further analysis and processing of the cost data:* If this is not done, sound raw data can be distorted through misleading comparison or relationships.

7. *Verification with the user of the information:* This is done to see whether or not the cost report meets the requirements and need. If it does not, corrective action should be taken.

8. *Assessing the use of cost information by the various users:* This is done in order to adapt to changing needs on the one hand and better cost accounting sources and processing capacities on the other.

This basic pattern describes an *accounting cycle* that can be coordinated with other ongoing cost accounting communications. However, quality cost accounting demands proper delineation and separation by the very nature of it being related to assessment of the cost effectiveness of the quality assurance program, and not of the company business and operation in general, nor of that of marketing, or any other function. After all, the purpose is not merely to inform, but to facilitate the planning, and, in particular, the control of quality assurance as a special responsibility and function.[3]

In cases of limited accounting capabilities and relatively low urgency of quality cost information, an ad hoc cost study might suffice. Later on, once management has gained deeper insight into the status and need of quality assurance in the company, more requests for cost information will induce the gradual expansion of quality cost accounting.

The accounting department, although it cooperates with the quality control department for operation of the overall costing system, nevertheless has a seperate role. Clear delineation of the two departments' roles must be established. For example, the quality control department may engage in the following activities.

1. Collection and analysis of all quality costs data.

2. Setting up quality cost objectives and interdepartmental policies on cost reduction and improvement.

3. Allocation of responsibilities for quality failure costs.

The accounts departments similarly can assist in the following endeavors.

1. Providing an operating report (or record) for the management.

2. Allocating appropriate funds as agreed upon.

3. Providing comparative data bases and projections for the quality department's improvement projects.

17.4 COSTS DATA: SOURCES, COMPILATION, AND REPORTING

Cost Data Sources

The sources of cost data in any company are normally so widely scattered that there are no established rules for searching out each and every data. Table 17.4 demonstrates a listing of cost data in the three main categories explained earlier and the possible or actual sources of these. In some cases estimates and forecasts will fill gaps left by *historical data*. It must be borne in mind that costs are values for resources used, or intended to be used; they are not actual or planned expenditures. Data recording forms with clear instructions and training for using these guarantee to a certain degree the validity, reliability, and traceability of the data. To decide whether or not a cost element falls under the heading of a quality cost is not always an easy or straightforward decision. Cost recording forms should therefore usually be accompanied by procedures and guides.

TABLE 17.4
Examples of Data Sources for Typical Quality Costs

Category	Department	Definition	Source of data
Prevention	Quality control	Engineering, technical, and supervisory costs of preventing the recurring defects	Payroll analyses
		Cost of investigation, analysis, and correction of causes of defects	Estimate based on an allowable percentage of costs incurred on engineering expense account
Appraisal	Design engineering Quality control Inspection Receiving	Costs of evaluating quality, and of identifying and segregating nonconforming parts and assemblies	Material review records, scrap reports, rework authorization reports
		Labor in sampling and sorting of purchased parts	Cost department productive labor records
		Direct labor of 100% inspection and 100% testing to separate defective from nondefective items	(Charges to account nos. 101, 107, etc,)
		Indirect labor in spot-check inspection or roving inspection to detect defective lots and to prevent production of defects	
		Labor and indirect costs of sampling, testing, and rating the outgoing product	
Failures	Quality control, servicing, and repair	Administrative cost of service to customers who receive defective items	Manufacturing expense reports, product cost information
		Material, labor, and burden of nonusable parts	Field repair, replacement and warranty cost reports
		Material and labor of fixing defective items (less the receipts from customers)	(Charges to account nos. 327, 120)

Where actual costs cannot be directly associated with specific elements, it may be necessary to make an expense allocation by *arbitration*. If such costs are significant, it is recommended that the necessary records be established for factual data. *Coding of data* permits consistency and ease in handling through computerized systems (coding practices used by account numbers are shown in Table 17.4).

Compilation of Cost Data

Defects are usually recorded by product, part, process, period, inspection point and work center, inspector, plant, supplier, customer, et cetera. Moreover, the type of defect, individual characteristic, relative weight and cause are also recorded in order to permit control of major defects and to facilitate cost detection and locating of causes for such defects. This division of defects into small classes can be used for dividing the respective costs as well.

For aligning the costs of defects and failures as closely and reasonably as possible with their causes, the respective prevention and appraisal costs also need to be known. Otherwise total quality costs cannot be compiled in sufficient detail. The degree to which quality costs are subdivided depends on the information request and need (see Table 17.4). Senior management is usually only interested in major grouping and developments. The middle and operational management levels holding responsibilities only in their assigned area have limited information needs, yet often need to have a comparison with other related cost centers.

Once data compilation is widely delegated, the cost analysis, and even the preparation of internal cost reports, can often be delegated. In centralized data processing with reports being returned to those providing the data, loss of information and relevancy often occurs. Therefore, a balance between centralized and decentralized cost reporting should be sought, particularly when the company grows and formality gains in importance.

Quality Cost Reporting

Regular and ad hoc quality cost reports communicate information by alerting the receiver to things they need to know or by satisfying an information request. *Standardized report forms* have the advantage of supplying a convenient grasp of the essential by a relatively large number of people. Ratios that relate quality costs to other compatible accounts for other functions, departments, branches, or plants, extract important messages for the reader. Figure 17.3 presents an example of a quality cost report.

The *usefulness* of quality cost reports from the user's point of view depends on the following.

1. Did the information add substantially to what was known already?

2. Was the breakdown of costs sufficient for undertaking further investigation and for planning corrective action?

3. Was the report timely?

4. Do the analysis and comparison show a definite trend?

5. Does the report allow me to appraise my own performances in the area in which I hold responsibilities?

6. Does this report and/or those forwarded to my superior clearly and convincingly

Report: Quality Costs

Period:
Company/Department:

1. Labor: quality engineering   _____
              inspection   _____
              other   _____   _____
2. Deduct adjustment   _____
3. Material/supplies   _____
4. Quality control/inspection in other departments
        4.1 _____   _____
        4.2 _____   _____
        4.3 _____   _____   _____
5. Total quality control cost (direct)   _____
6. Add overhead (indirect)   _____
7. Total quality control cost   _____
8. Internal failure cost (quality loss)
        8.1   Downtime   _____
        8.2   Rework   _____
        8.3   Downgrading   _____
        8.4   Scrap   _____
        8.5   Other   _____
9. Total internal quality costs   _____
10. External failure cost (quality loss)
        10.1   Adjustments   _____
        10.2   Warranty   _____
        10.3   Other   _____
11. Total quality costs   _____
12. Total manufacturing costs   _____
13. Sales Revenue   _____
14. Percent of 9 in 12.   _____
15. Percent of 10 in 13.   _____
16. Percent of 11 in 13.   _____

                      signed   _____

Note: Costs for the period listed above should be compared with budget and with previous periods for
      further analysis.

FIGURE 17.3   A summarized form of reporting quality costs.

demonstrate the profit contribution of quality assurance in general, and that of the quality assurance function in particular? Does this report convey my own meeting of performance standards?

7. Does the report on quality costs for individual products, services, processes, branches, et cetera, testify to impact of quality assurance and need for further strengthening of the related cost-effectiveness?

With management receiving quality cost information, decision making will normally be improved and the quality of products and services enhanced. This, after all, has been cited as the main objective for quality cost reporting.

17.5 QUALITY COSTS: MANAGEMENT DECISIONS AND PERFORMANCE CONTROL

Quality costs serve as important decision criteria at the operator, supervisor, and executive levels of responsibility. The operator, having been delegated inspection tasks, might be treated as a cost center and charged with quality costs, as defined above. In principle, the actual inspection costs and the respective costs of defects passed without such weeding-out of defects must be calculated and controlled for net savings. A well-trained and informed operator can carry out certain cost control duties once the recording forms support such controls and decision making. The operators should not only be told (and taught) what to do but what not to do as well. Both operator and inspector are interested in detecting and correcting any causes for defects as early as possible in order not to waste time and material on already defective items.

Figure 17.4 demonstrates the normally rapid increase in defect costs during the entire life-cycle phase of one particular product. A performance control system charging defect costs to the cost centers where they arose and where they should have been detected, motivates operators for proper quality control. While one cost center will be charged with defect costs, the other, having detected the defect and cause, might be credited. However, one must watch to ascertain that such cost controls for assuring quality do not undermine cooperation, particularly when penalties might be feared by the persons concerned.

While operators directly influence the methods for attaining quality specifications, supervisors can use quality cost information for assessing performances on the basis of standard costs, budgeted costs, and cost reduction goals. Standard costs are predetermined, similarly to any other performance standard.

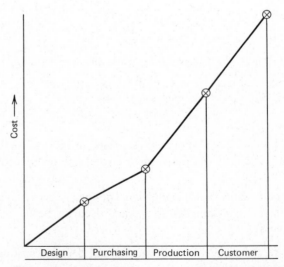

FIGURE 17.4   Cost increase per one major defect during the life cycle of one particular product. The earlier a defect is detected and prevented the lower are the costs.

Among the cost accounting approaches to be used are the valid and reliable estimates using statistical techniques based upon past satisfactory performances. *Comparison of actual and standard costs,* possibly applying statistical control chart techniques or Pareto Analysis, reveal significant deviations and noticeable results. Cost analysis can have the important advantage of an overall performance control, rather than control of specific quality characteristics. Standard costs allow sound budgets to be set for individual cost centers, departments, or plants.

17.6 PLANNING FOR QUALITY COST STUDIES

Critical areas in terms of high quality costs (cost centers) no matter where they are (i.e., with suppliers, engineering, production plans or products, projects, etc.) call for remedial measures mostly by supervisory management. In Figure 17.5 the sequential analysis applied to finding the most costly source is shown from the results obtained in a ductile iron foundry. A total of $15,035 of scrap in one week was sufficient to attract the attention of upper management.

Special cost reduction projects, possibly in conjunction with direct participation of operators, suppliers, or even the customers concerned allow for concerted and effective correction. These projects can be given a budget and appraised in their success in reducing specific and overall quality costs. Quality circles, similar to management-initiated regular task forces and projects, aid in discovery of costly waste of valuable human and material resources, including otherwise unused opportunities for improving quality and gaining market competitiveness.

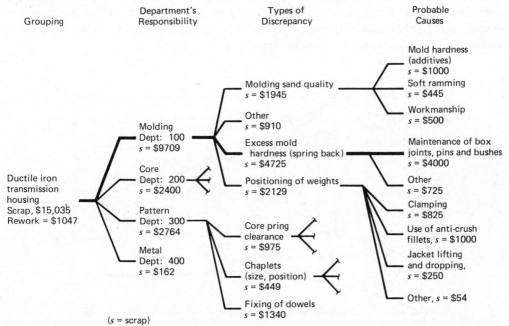

FIGURE 17.5 Sequential analysis applied to finding the most costly source of scrap; case example from a ductile iron foundry.

Planning for Cost Reduction Strategies

Consider the following table with examples of three different quality cost compositions indicating alternative quality assurance strategies.

	Case 1	Case 2	Case 3
Customer complaints	41.3%	21.3%	19.2%
Internal item defects	44.1%	26.3%	43.1%
Inspection	14.0%	50.3%	29.0%
Defect prevention	0.6%	2.1%	8.7%
	100%	100%	100%

In case 1 management leaves the detection of defects to the customer. Moreover, the small investment in inspection and defect prevention explains high defect frequency. The second policy relies heavily on inspection to minimize defects, while in the third case, more emphasis is placed on defect prevention. The cost structure might fluctuate and shift. For instance, with more inspection over a long period inspection costs will eventually be lowered and, as a result, the total cost will decrease. Higher defect prevention costs lower the need for item inspection in the long run, and thus have a positive effect for total cost reduction.

At the plant- or company-wide level, creation of a quality assurance program is a crucial decision, as exemplified by the bus manufacturer case mentioned in previous chapters. The chief executive officer finally decided to introduce a quality control program, implying of course that this would help to reduce liability costs and all quality costs. It is not only the relative emphasis to be placed on defect prevention and appraisal, but also the design and implementation of the program itself that minimizes quality costs in the long run and in overall terms. This approach for long-term quality cost reduction, with its associated levels and trends of quality costs, is shown in Figure 17.6.

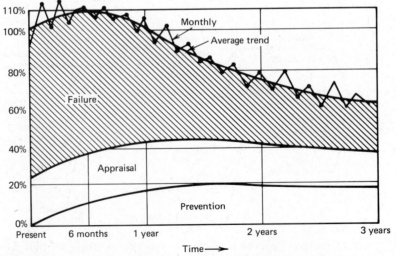

FIGURE 17.6 Expected trend in the long-term quality cost reduction project.

Quality program standards differentiate between simple inspection and other more comprehensive quality assurance programs. As will be discussed in the next chapter, a major decision in selecting the least costly quality program and its respective standard must be made after careful analysis of the prevailing conditions and cost relationships. In fact, while it is only realistic to assume that a zero-defect quality program is too costly in the beginning, it is not entirely unreasonable, since a reduction of the cost minimum, that is, the total quality cost curve (compare Figure 17.1) can very well be expected as the result of cost-effective quality assurance.

Quality program standards require planning and control of quality costs to a different degree. The following is an excerpt from the MIL-Q-9858A standard.

> The contractor shall maintain and use quality cost data as a management element of the quality program. These data shall serve the purpose of identifying the cost of both the prevention and correction of nonconforming supplies (e.g., labor and material involved in material spoilage caused by defective work and for quality control exercised by the contractor at subcontractor's or vendor's facilities). The specific quality cost data to be maintained and used will be determined by the contractor. These data shall, on request, be identified and made available for ''on site'' review by the government representative.[4]

The Canadian CSA Z299.0 standard and the ANSI/ASQC Z-1.15 guides, both describe features of quality costs for management decision making and quality performance planning and control. Audits, as we shall describe in one of the next chapters, inform management of the individual and overall quality cost deviations and possible improvement in the cost-effectiveness of quality assurance.

17.7 COST ANALYSIS IN DESIGN ASSURANCE

A new design or redesign must not only be technologically sound but also economically feasible. Quality assurance managers along with others involved in design activities can use various available analytical decision-making methods for arriving at final designs that meet customer expectations as well as imposed standards and regulations. Table 17.5 lists the conventional management methods and respective application for design assurance and reviews. These models are, of course, additional aids to the conventional checklist. Quality assurance managers seek to detect and prevent major design errors in the product or service as early as possible in the process. They furthermore need to prove that the design assurance programs they recommend are justified in terms of cost and payoff.

Financial analyses, using present value criteria, payoff periods or rates of return seems to generate useful information for upper management. The difficulty rests in obtaining good estimates of expected payoffs. Therefore, these models should only be used when defect costs can be fairly well estimated and then handled by the accounting specialists as opportunity costs and payoffs. If the resulting costs are extremely high, design assurance programs are usually not questioned in financial terms, but when they are doubtful these models become an important aid.

Cost Model: Break-even Calculations

A more modest cost model is the *break-even* calculation. The expected output volume justifies design assurance only when total costs remain below what the customer is willing to pay

TABLE 17.5
Quantitative Methods in Design Assurance

Method	Example of Application
Checklists, flowcharts	Interrelate major quality specifications with nonconformance, cause, and prevention method; cause-effect diagrams.
Statistical analysis	Market research, process capability studies, reliability testing.
Cost models	Interrelate costs, payoffs, and output for program approval.
Waiting line models	Determine testing capacity required or size of design review team.
Forecasting	Determining quality requirements and defect or failure modes.
Network analysis methods	Design, schedule design, and testing projects, design reviews.
Heuristic program	Determine layout of design and testing area.
Simulation	Determine product performance and reliability, design assurance capacity and size of team.

extra for in the price. The impact of design assurance on costs and revenue can be shown using the *break-even model* in different ways, as the following examples illustrate.

Example: A company producing household appliances considers redesigning a food mixer. The market seems to be fairly saturated at the time and competitors are expected to enter with an improved design. Further research indicates that consumers will not tolerate any major price increase. As shown in Figure 17.7a, costs for design assurance raise *fixed costs*

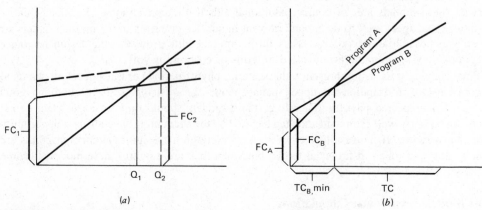

FIGURE 17.7   (a) Break-even analysis for impact of design assurance. (b) Break-even analysis for two design assurance programs.

(FC$_1$ to FC$_2$) and thus the *break even point* moves to a respectively higher volume (Q$_1$ to Q$_2$). The question that must be answered is whether the new design of the mixer will, in fact, meet better customer response and attract more sales. If not, a less intensive and less costly design assurance procedure, if any at all, seems to be justified.

However, management will have to realize that the better design (due to extra expenditure for design assurance) will yield a better payoff in terms of lower fixed costs. Special tests, including field tests, will prevent design errors that could lead to costly repair and recall actions. These hypothetical arguments might not be very convincing, but frequent recall action in the kitchen appliance industry demonstrates that defect preventive design assurance reduces rather than increases fixed costs in the long run.

Impact on the *variable costs* will most likely also be favorable, in that a good design will reduce defects and inspection costs during production. After all, the purpose of design assurance, is also to seek defect avoidance in all life phases of the design and the individual items.

Finally, if the new mixer finds the appeal that designers and producers hope for, a higher price might be even justified. Again, design assurance can be paid off. In many instances the relationships and interactions are hard to quantify and evaluate in dollar terms. Nevertheless, estimates can usually be made and the model can then be used for demonstrating the outcome. While the impact on price is particularly hard to estimate, two or more alternative design assurance programs with different *fixed* and *variable* cost calculations can be compared in terms of linear equations. Fig. 17.7*b* demonstrates this similar application of the model graphically.

Fixed costs are primarily the cost of the test equipment and research activities, while the variable materials and labor costs vary per item. Program A invests more in equipment and basic research for failure analysis, reliability testing, et cetera, while program B uses more conventional and manual testing. Again, the expected output volume in this model and the point of indecision are used as a decision criterion. Total cost (TC) for Program A (TC$_A$) are minimized to the left, and TC$_B$ to the right.

Investment Analysis

Any quality assurance programs, test and inspection facilities, supplier audits, or the assurance measures require investment and operational costs. These are expected to contribute to profit, at least in the long run. In the decision tree analysis the *net present value* (npv) of alternatives can be calculated by a rollback technique. The net present value is the discounted present value of future costs and/or revenues for set rates and periods.

The net present value criteria is often used for deciding on investment alternatives for expensive testing equipment. One can assume that the more advanced and expensive equipment will result in lower quality costs, if not immediately then at some future time. In addition to the initial investment and expected *salvage value* (SV) at the end of the equipment's use, expected lifespan quality costs account for the operational inspection or testing costs and for the costs of defects or failures.

In the following example the investment alternative with the highest net present value is to be determined and selected.

Example: Consider the two alternatives with the following investment figures.

Alternative 1
Initial cost:	$15,000
Economic life:	9 years
Salvage value:	$3000
Operational costs per year:	$6000

Alternative 2
Initial cost:	$25,000
Economic life:	6 years
Salvage value:	$8000
Operational costs per year:	$2000

Calculate net present values (interest rate is 14%) and decide on the best alternative.

Solution

Alternative 1: Initial Cost = $15,000. The present value (PV) of the salvage value (SV) of $3000, derived from tables giving present value of future amount, (see Appendix B, Table H) is calculated as PV of SV = $SV(PV_{sp}, i,n)$, where PV_{sp} = present value of single payment, i = interest rate, and n = number of years of economic life. Thus, the present value of salvage value = $3000(PV_{sp}, 14\%, 9) = 3000(0.3075) = \$922.50$.

Present value of operating costs (OpCost) derived from table giving present value of annuity (Table I, Appendix B) is calculated as PV of OpCost = $A(PV_a, i, n)$, where PV_a = present value of annuity. Thus the present, value of operating costs = $6000(PV_a, 14\%, 9) = 6000(4.9464) = \$29,678.40$.

Therefore the total cost of Alternative 1 = $15,000 + 29,678.40 - 922.50 = \$43,755.90$

Alternative 2: Initial cost = $25,000. The present value of salvage value = $SV(PV_{sp},i,n) = 8000(PV_{sp}, 14\%, 6) = 8000(.4556) = \$3644.80$. The present value of operating costs = $A(PV_a,i,n) = 2000(PV_a, 14\%, 6) = 2000(3,8887) = \$7777.40$. Therefore the total cost of Alternative 2 = $25,000 + 7777.40 - 3644.80 = \$29,132.60$.

Therefore based on the above cost considerations, Alternative 2 is selected.

Analysis Based on Internal Rate of Return

Other models for investment analysis often used are *payback periods* and *internal rate of return*. The payback model does not take cashflows beyond the payback period into account nor the salvage value. The simplicity of calculating the time when the new inspection program (or any other investment of funds for a special quality assurance purpose) has paid for itself, through lower costs and/or higher revenues, makes this practice attractive.

Similarly, computing the annual *rate of return* for an investment by relating investment and resulting returns (cashflows), estimating the lifespan, and then reading the rate off the table for present values annuities (Table I, Appendix B) is a fairly straightforward procedure. Naturally, this internal interest rate is merely an approximation, and usually leads to the same decision as the present value method. The next three examples are more illustrations for the payback or the annual rate of return investment criteria.

The break-even model is also often used to assess investments in quality assurance. It relates total costs to volume of output, with the costs divided into fixed and variable costs. More comprehensive quality assurance programs with costly test equipment require relatively more fixed costs than do simple end-item inspection programs that do not include any design assurance activities. The trade-off normally lies in the lower variable costs of highly automated equipment. Given a set profit contribution per unit and costs, a break-even point is defined by the model where the output volume is high enough to render profit.

The break-even model is limited through the assumption of linearity and mainly helps to demonstrate volume and revenue/cost relationships (Example 2). For the application in quality assurance the profit contribution per unit will be hard to determine in practice. However, if alternatives are to be compared with each other as linear functions based on volume output, break-even points serve to define rule decisions. Such an example for a test station with two alternatives as mentioned above, one with higher fixed and lower variable costs relative to the other, is shown in Example 3.

Example 1: Consider the data of the following table.

Payback Period	Alternative	
	1	2
Initial cost	$20,000	$15,000
Quality cost reduction		
Year 1	1500	2000
2	2800	6000
3	5000	8000
4	7000	6000
5	8000	

The standard interest rate is 12%. The new inspection procedure and equipment requires an initial investment (I) of $9000. The useful life is five years, and the quality cost reduction is estimated at $2000 per year.

Calculate the payback period and internal rate of return for the two alternatives.

Solution

Payback period

Alternative 1 Using the table, the cost reduction over the first four years $= \$16,300$. In year five it requires $3700 to reach the total original cost of $20,000. This balance is divided by the quality cost reduction for year five. Thus, $\dfrac{3700}{8000} = 0.4625$ of a year. Thus, the payback period $= 4.4625$ years.

Alternative 2 Original cost is recovered 7/8 way through the third year. That is, payback period $= 2\frac{7}{8}$ years.

Internal Rate of Return (IRR)

The IRR is the rate which discounts a project's cashflow (CF) to an npv of zero. Therefore:

$$0 = CF(PV_a, r, n) - I$$
$$= 2000 \, (PV_a, r, 5) - 9000$$
$$(PV_a, r, 5) = \frac{9000}{2000} = 4.5$$
$$(PV_a, 3\%, 5) = 4.5797$$
$$(PV_a, 4\%, 5) = 4.4518$$

or, IRR $= 3.6\%$.

Example 2: A manufacturing company produces product X at a variable cost of $8.00 per unit, plus fixed costs of $10,000.

A quality control program is implemented with a fixed cost of $2000, and with a resultant decrease (due to fewer defective units) of $1.60 in variable costs per unit. Determine the break-even point of the quality program and formulate a decision.

Solution

The break-even point is given by the point at which savings from implementing the quality program equals its cost. Hence: $1.60X = 2000$, or $X = 1250$ units.

Example 3: David Jones, president of the Acme Submarine Company is considering three methods of manufacturing a new line of minisubmarines. The methods and the costs associated with each are as follows.

		Manual	Semiautomatic	Automatic
Annual fixed costs for each method		$150,000	$350,000	$750,000
Variable cost per unit				
Materials		$5000	$5000	$5000
Direct labor		1000	700	400
Inspection cost		200	100	50
Defect cost		500	300	100
Launching cost (includes champagne)		103.70	83.70	53.70
Total variable cost per unit	=	$6803.70	$6183.70	$5603.70

The wholesale price will be $9200 per unit.

1. Construct a break-even chart for each of the three methods. If expected sales are 100 units per year, which manufacturing method should be chosen?

2. At a price of $8000 per unit, expected sales would be 1000 units per year. What method should be chosen, and how much profit would result?

Solutions

1. If expected sales are 100 units, the manual process should be chosen, since this choice will lead to a profit. The other choices have a break-even point of over 100 units (see Figures 17.8 *a, b,* and *c*).

Using manual production for 100 units the profit will be $= (100)(9200) - [150,000 + (100)(6803.70)] = 920,000 - 830370 = \$89,630$.

2. Expected sales are 1000 units at $8000.

Using the manual process, profit $= (8000)(1000) - [(150000) + (6803.70)(1000)] = 8,000,000 - 6,953,700 = \$1,046,300$.

Using the *semiautomatic process*, profit $= (8000)(1000) - [(350,000) + (6183.70)(1000)] = 8,000,000 = 6,533,700 = \$1,466,300$.

Using the *automatic process*, profit $= (8000)(1000) - [(750,000) + (5603.70)(1000)] = 8,000,000 = 6,353,700 = \$1,646,300$.

Therefore, to maximize profit at the production level of 1000 units, the decision would be to select the automatic process.

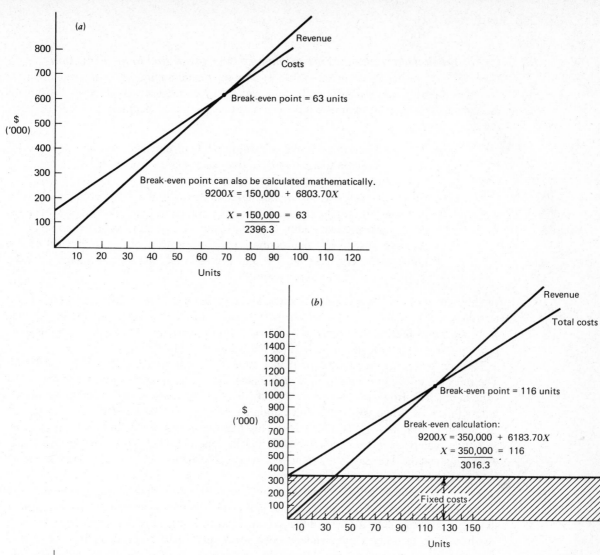

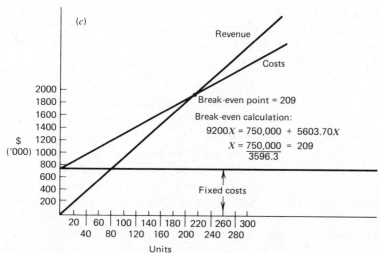

FIGURE 17.8 (a) Break-even chart using manual production system. (b) Break-even chart using semiautomatic production system. (c) Break-even chart for automated production.

Use of Break-even Model to Find Minima in the Cost of Incoming Inspection

Deming[5] has illustrated the break-even cost concept for minimum total cost evaluation for test of incoming material inspection in more detail. In a wide variety of practices, such plans tell the purchaser how to minimize the total cost of inspection of incoming materials or parts.

Let:

p = Average fraction defective in incoming lots of parts.

q = Average fraction that is not defective, such that $p + q = 1$.

k_1 = Cost to inspect (or test) one or first part.

k_2 = Cost to dismantle, repair, reassemble and test a part that fails because a defective part was put into the production line (i.e., the cost to correct a faulty assembly or the cost of downgrading a final product or scrapping the finished product).

k = Average cost to test one or more parts (as many more as necessary) until a nondefective one is found. The total average cost can be approximated as $\simeq k_1/k_2$ (for proof of this, see Deming[5]).

The ratio k_1/k_2 is the break-even point. The rules for *all-or-none* inspection are written as follows. If $p < k_1/k_2$, no inspection will be done (in this case, p or its distribution will lie to the left of break-even point). If $p > k_1/k_2$, 100% inspection will be done (in this case, p or its distribution will lie to the right of the break-even point).

The break-even quality is that particular value of p that satisfies the above conditions. The rules are effective even if a pure single-peak distribution of fraction defective does not exist in practice. However, it works well provided that the standard deviation of the mixture is less than 0.3. When the standard deviation of the means of the bimodal is 0.3 or higher, the situation becomes a state of chaos. This happens when the fraction defective in the incoming material becomes unpredictable, that is, one day being to the left of the break-even point and the next day to the right. Rules to deal with this situation have been developed by Joyce Orshini.[6]

According to these rules, a random sample of 200 parts is taken from each lot. If no defectives are found in the sample, the lot is accepted. If one or more defectives are found, defective parts are replaced with the good ones and the remainder is screened.

All-or-none inspection rules stated above for a single part being defective can also be extended to any number of parts. It should be recalled from Chapter 8 that the probability of failure would thus increase as the number of parts increases. The cost factor, k_2, likewise will increase. The real dilemma remains as to how much cost can be attributed to one particular defect (remember defects breed defects!).

17.8 KEY TERMS AND CONCEPTS

Quality costs are the sum of defect or failure, inspection and prevention costs. They are normally compiled and analyzed for products or contracts, product lines, plants or divisions, regions, time periods, et cetera, in absolute and relative terms. Generally, all cost concepts developed and used in accounting apply to quality costs. Costs are the value of resources used.

A relatively new concept is *quality life-cycle costs* that compiles and describes all costs incurred during the entire life of a product.

Failure or defect costs are avoidable losses of resources during design, production, and performance of a product or service. Internal failure or defect costs are normally relatively easier to control by quality assurance management than are externally incurred costs. Failure or defect costs tend to increase during production and early performance, suggesting the need for preventive quality assurance.

Inspection costs are incurred for testing and verifying compliance with quality standards, and for initiating corrective actions. Normally, but not necessarily, higher inspection costs will result in lower failure or defect costs.

Prevention costs are the value of resources deployed and invested for avoiding failures and defects during design, production and operation, and performance of a product or service. Training, participatory quality assurance, and quality programs and audits not only help to prevent failures and defects, but in the long run should also eliminate inspection costs.

Quality cost accounting is the discipline, functions, and practice of planning and controlling funds and properties, with explicit reference to quality assurance and control. The purpose of accounting is to safeguard, report, and inform, and to direct decisions and activities regarding prudent and effective resource deployment for the attainment of quality assurance.

Quality cost reports are produced by the *quality information system* in conjunction with the general accounting system. These reports facilitate managerial planning and control. The content and format of these reports should suit the information needs of the receiver.

Investment analysis is performed for selecting among various alternative means of financing productive capacity. With regard to quality assurance, investment analysis is to ensure and verify investments in quality assurance areas.

17.9 DISCUSSION AND REVIEW QUESTION

(17-1) Define *quality cost*, and comment on the statement: *Quality is free, but defects are costly.*

(17-2) *The purpose of accounting departments is not only to inform senior management about profit and loss, but also about the effectiveness of quality assurance.* Explain.

(17-3) *Costs of prevention and inspection are interrelated.* Explain.

(17-4) *The costs of internal and external failures should always be separately accounted for.* Explain why.

(17-5) *Costs must be divided into subgroupings in order to give meaningful insight into cost developments.* Explain.

(17-6) *Most, if not all, important cost concepts apply to quality assurance.* Discuss.

(17-7) *Costs should be related and expressed in index numbers to provide for effective cost control.* Explain and discuss.

(17-8) *Every quality assurance program or procedure should be translated into costs.* Discuss.

(17-9) *Cost analysis and accounting must start with defining the purpose and use of information.* Discuss.

(17-10) *Statistical quality control techniques can be applied to quality cost control.* Explain.

(17-11) *Quality assurance program standards do not mention quality cost accounting sufficiently, if at all.* Discuss.

(17-12) *Cost and financial investment models are also applicable to quality assurance.* Explain and discuss.

(17-13) *Development and implementation of a quality cost accounting system should start with a checklist.* Explain.

17.10 PROBLEMS

(17-1) Prepare a checklist for listing specific kinds of costs under:
   a. prevention costs.
   b. appraisal or inspection costs.
   c. failure and defect costs.

(17-2) Management of Igbal Woodworks Ltd. has set a goal of reducing quality costs by 5% within the next year. Describe measures for implementing this goal and prepare a cost control record.

(17–3) Inspection costs are to be forecasted for the coming year. Prepare an index number and the forecasting method for this purpose.

(17–4) Two inspection methods can be used for inspecting circuit boards at a particular inspection station. The fixed cost for method A is $3000, and for method B is $6000. The variable inspection costs and labor and material costs are $8 per unit for A and $4 per unit for B. What must be the inspection load in term of units in order to justify the method B, with its most costly equipment?

(17–5) A quality assurance manager wants to replace the current test equipment with a more modern and more effective design. There are two new designs on the market; the costs components for each are as follows.

Alternative 1: initial costs = $20,000, economic life = 10 years, salvage value = $4000, and operational costs = $7000 per year.

Alternative 2: initial costs = $30,000, economic life = 7 years, salvage value = $6000, and operational costs = $2500 per year.

Which equipment should be bought on the basis of the net present value with an interest rate of 12%?

(17–6) The following are to be designated as prevention, appraisal, or failure costs (if not sure which, explain).

a. Warranty costs
b. Vendor survey
c. Design of test equipment
d. Rework done by distributor
e. Set-up for tests
f. Writing of quality manual
g. Printing of cost reporting form
h. Expenses of inspector attending ASQC convention
i. Replacement of spare parts at dealer

In the above examples, spare part replacement might not always be related to a defect. Outline a procedure for assigning such costs to failure costs.

(17–7) Pistons for diesel engines produced in one part of a locomotive works are used later in the assembly area. One hundred percent inspection is performed, and an inspector can check pistons at an average rate of five minutes each. The total cost for the inspector, including fringe benefits, general burden, and salary is $9 per hour. The production output is 96% nondefective. If a defective piston is installed in an engine, it requires a mechanic one hour to replace it, at a cost of $15.50 per hour.

Is this inspection justified? What is the net savings or loss from performing the inspection?

17.11 NOTES

1. Juran, J. M. and Gryna, Jr., F. M., *Quality Planning and Analysis,* 2nd ed., McGraw-Hill Book Company, New York, 1980, p. 408.

2. Dowson, A. M., "Quality Emphasis Circles at the General Motors Trim Plant," Forum 27 Conference Transaction, American Society for Quality Control, Toronto Section, March 7, 1981.

3. Quality cost ideas and application is the subject of the April 1983 issue of *Quality Progress.*

4. Military Specification, *Quality Program Requirements, MIL-Q-9858A,* clause 3.6 Government Printing Office, Washington D.C., 1963.

5. Deming, W. Edwards, *Quality, Productivity and Competitive Positions,* Massachusetts Institute of Technology Center for Advanced Engineering Study, Cambridge, Mass., 1982, p. 267.

6. Orshini, Joyce, "Simple rules to reduce total cost of inspection and correction of products in state of chaos," Doctoral dissertation, New York University, Graduate School of Business Administration, 1982.

17.12 SELECTED BIBLIOGRAPHY

American Society for Quality Control, (ASQC) Quality Cost Technical Committee, *Bibliography of Articles Relating to Quality Cost Concepts and Improvement,* American Society for Quality Control, Milwaukee, January 1981.

ASQC Quality Cost Committee, *Guide for Managing Vendor Quality Costs,* American Society for Quality Control, Milwaukee, 1980.

ASQC Quality Cost Technical Committee, *Guide for Reducing Quality Costs,* American Society for Quality Control, Milwaukee, 1979.

Blank, Lee and Solorzano, Jorge, "Using Quality Cost Analysis for Management Improvement," *Industrial Engineering,* February 1978, p. 46.

Campanella, Jack and Corcoran, Frank J., "Principles of Quality Costs," *Quality Progress,* April 1983, p. 16. (This entire issue of *Quality Progress* deals with quality costs.)

Cheng, Philip, "A Conceptual Analysis of Quality Control Cost for Industrial Management Decision Making," *Industrial Management,* March–April 1976, p. 28.

Dawes, Edgar W., "Quality Costs—A Tool for Improving Profits," *Quality Progress,* September 1975, p. 12.

Groocock, J. M., *The Cost of Quality,* Pitman Publishing Co., London, 1974.

Gryna, Jr., Frank M., "Quality Costs: User vs. Manufacturer," *Quality Progress,* June 1977, p. 10.

Hagan, J. T., *Quality Costs,* International Telephone and Telegraph, New York, 1981.

Lambertson, W. A., "Internal Scrap Costs and Lunacy in a Job Shop," *Quality Progress,* December 1981, p. 22.

Lester, R. H., Enrick, N. L. and Mottley, H. E., *Quality Control for Profit,* Industrial Press, New York, 1977.

Meske, Albin F., "Economic Inspection Intervals: New Tool for Inspection Cost Management," *Cost Engineering,* November–December 1978, p. 219.

Moseley, Zach R., "Component Failure Cost," *Quality Progress,* January 1980, p. 34.

Pyzdek, Tom, "Impact of Quality Cost Reduction of Profits," *Quality Progress,* November 1976, p. 14.

Veen, B., "Investing in Preventive Quality Costs," *Quality Progress,* April 1978, p. 12.

Zabecki, David T., "Contribution Margin Analysis of Quality Costs," *Quality Progress,* October 1977, p. 34.

18 CHAPTER

QUALITY PROGRAM: DESIGN AND IMPLEMENTATION

In preceding chapters we discussed the setting of quality objectives, the organization, information system, and quality cost accounting including performance control as special areas for management attention and involvement. It stands to reason that any quality assurance system, as an integral part of the general productive system, must itself measure up to high standards of effectiveness in design and implementation. Company staff at all levels of responsibility want sound procedures, job designs, inspection resources and quality standards to guide and support them in their efforts for quality assurance and good workmanship. We have seen that staff members not only need to be properly informed, but also to be given some scope for active and independent participation in the management of quality assurance. Suppliers as well as customers have an active role to play in concerted efforts for attainment of quality.

Customers, particularly for high-risk items and major comprehensive projects, require of their suppliers a well-designed, documented, and implemented quality assurance program. For this purpose they invoke certain *quality program standards* and make compliance mandatory.

This shift to relying on the quality assurance program when evaluating a potential supplier's quality capabilities not only reduces overall costs, but at the same time promotes more independent planning and control of quality by the supplier and motivates them to prevent costly mistakes as early as possible. It is therefore in the mutual interest of both customer and supplier, to establish a sound and cost-effective quality assurance program. To wait until a major defect or a recall action occurs, (as in the case of the bus manufacturer), has become a practically untenable policy. We might recall that the president's immediate reaction and conclusion was to establish a quality control program, in order to prevent the reoccurence of such a disaster.

General benefits of quality assurance through sound management programs are the following.

1. Less defects and rework.

2. Less inspection and defect prevention costs in the long run and consequently, less total quality costs.

3. Greater *confidence* in the quality assurance given by the supplier, since the quality program is documented and can be verified.

4. Greater uniformity of quality assurance efforts throughout companies and industries when quality programs comply with generic program standards.

5. More *visibility* of concerted quality assurance efforts and success in the company.

6. More lucrative contracts with major customers for whom quality programs are major prerequisites of qualifying as a supplier.

7. Stronger *product liability protection,* since the quality program serves as convincing proof of adequate care taken in quality assurance by the management.

In this chapter we shall discuss some fundamentals of explicit quality assurance programming. It can be assumed that any company with a good quality record follows some program of action. However, the visibility of such a program to customers is often lacking. If a quality manual documenting such activities does not exist, the company forgoes the important benefits listed above. Thus, the opportunity costs are high until a proper quality program has, in fact, been established.

We shall first describe the concept of the quality program, using *standards* as the main *management aid.* The assessment of the actual need for a program and the selection of the appropriate standard will be done before details of the program are designed and outlined in the quality manual. Finally, the essential step is to implement the quality program into the life of the company, and thus achieve adequate quality of products and services continuously and consistently.

18.1 QUALITY PROGRAM: CONCEPTS AND STANDARDS

A *quality program* is the designed, documented, and effectively implemented totality of quality assurance policies, organization, and detailed procedures. Its purpose is to attain objectives of quality assurance and major benefits for all concerned and participating. With this explicit orientation to an objective, and the respective management approach and resource deployment, any quality program constitutes a real system. As a subsystem, it is influenced by its associated productive systems and other company systems, including the customers. Figure 18.1 demonstrates the customer–supplier relationship, from inspection and communication to customer reliance on the supplier's quality assurance program and even to the registration of such program by an independent third party.

In the introductory part of this book we explained the main features of a quality assurance system schematically in Figures 2.3 and 2.4. In accordance with this underlying systematic structure, essential components of any quality program include:

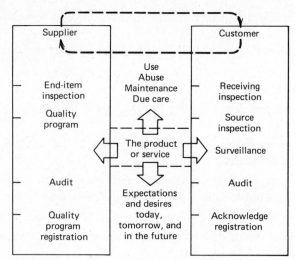

FIGURE 18.1 Elements of customer–supplier relationship management.

1. A purpose and objectives to be achieved.
2. Policies and basic principles for achieving the purpose and objectives.
3. Organization and major human resources.
4. Procedures and workmanship standards.
5. Documentation and communication.
6. Surveillance and program maintenance.

Except for these basic components, individual quality programs in private companies or public organizations vary considerably in order to suit their unique set of determinants, such as the scope of the operation, the kind of product or service, the technology, the people, management style, tradition, and existing needs. While it is senior management's responsibility to initiate and approve the proper quality program, it is usually a fairly difficult task. This is mainly due to the fact that often customers will have to approve the program. When more than one customer is involved management must then attempt to meet more than one set of different specifications and requirements. Under these circumstances, programs can become out of control monsters.

Program standards that describe features and stipulations in general (*generic*) terms have become a necessity for solving the otherwise conflicting and uncertain demands of customers. These standards facilitate a clear understanding and compliance with quality assurance requirements and commitments. They are prepared and published either by major procurement branches or by special independent and proficient standard-writing bodies.

Quality Program Standards

Table 18.1 lists the most widely applied international quality program standards and their major program components. These published documents have the following features.

TABLE 18.1
Stipulations of the Major International Quality Program Standards

PROGRAM COMPONENT	Fig. 18.2[a] Program Levels I	II	III	CANADA CSA Z-299.1 1979	CANADA CSA Z-299.2 1979	UK BS 5179 Part 3 1974	MIL-Q 9858A 1963	ASME NA-4000 1971	USA ANSI/ASQC Z-1.15 1979	USA ANSI/ASME NQA-1 1979	INTERNATIONAL ISO DIS 6215	INTERNATIONAL NATO AQAP-1 1972
1. Quality system/program	X	X	X	3.1/3.4	3.1/3.3	4	3.1/1.2	41-10,20,40	4	2	5	201
2. Quality assurance function	X			3.2	3.2	3	3.1	4200/4300	4.6/10	1	6	101,102,103 105,202
3. Design assurance/quality plan	X	X		3.4.2 3.5.2	3.3.2	8	4.1	4410	5	3	7	207
4. Supply/subcontract assurance	X	X		3.5.1 3.5.5	3.4.3	11.1 11.2	5.1 7.1	4430	6.1.2/6.1.3	4	8	210A,B
5. Quality manual/procedures/instructions	X	X		3.4.5 3.5.3	3.4.1	5	5.2 6.2	4441/4140	5.3.3	5	9	203,204
6. Document control	X	X		3.5.3	3.4.1	9	4.1	4420/4430	8.5 7.1.8	6	10	208
7. Receiving inspection	X	X		3.5.6	3.4.4	11.3	5.1 6.1	4441	6	7	11	210C,D 212,214
8. Production/process inspection	X	X	X	3.5.13 3.5.7	3.4.11 3.4.5	12.1 12.2	6.2 6.3	4451	7.1.4 7	9,10	13,14	211B 211A,214
9. Final inspection	X	X	X	3.5.8	3.4.6	14	6.3	4520 4530	7.2	11	15	213,214
10. Storing, Handling, Shipping	X	X		3.5.11 3.5.14	3.4.9 3.4.12	18	6.4	4460	7.3	13	17	217
11. Inspection/test equipment	X	X	X	3.5.4	3.4.2	10	4.2,4.3, 4.4	4530 4600	7.1.8	12	16	209+ AQAP 6
12. Product/status identification	X	X		3.5.10 3.5.9	3.4.8 3.4.7	18 17	6.1 6.7	4442 4452	7.1.9, 7.4 7.1.5	8,14	12,18	217,216
13. Nonconformance	X	X		3.5.16	3.4.14	16	6.5	4550	6.1.7 7.1.7	15	19	215
14. Corrective action	X	X		3.5.18	3.4.16	7	3.5	4800	8.4,8.6,9	16	20	206
15. Quality records/reports	X	X		3.5.15	3.4.13	6	3.4	4900	7.2.2, 7.5	17	21	205
16. Quality audits	X	X		3.3,4.1 4.2	—	20	—	4700	4.6.4	18	22	104
17. Training/motivation	X											
18. User contact/performance control	X											

[a] Level I = Inspection Program, Level II = Quality Control Program, Level III = Quality Planning and Control Program.

1.   They describe quality program features in general terms; scope, application, and definitions are included.

2.   They are mandatory only upon contract; codes become mandatory by law within a jurisdiction if adopted by regulation.

3.   They reflect the state of the art.

4.   They describe what prudent and proficient management would decide upon and implement for assuring quality without explicitly complying with such standards.

5.   They stipulate minimum requirements and allow for adaptation to existing conditions and circumstances.

6.   These standards can be used in conjunction with other related technical standards, such as standards for statistical control methods, audits, and other types of procedures.

7.   Special associated guidelines and schematic aids complement the standards for convenient application.

8.   As a management aid they contribute to minimizing quality assurance costs and to general rationalization (internal and external) of the company.

9.   These standards facilitate modern quality assurance management in all enterprises, industries, and communities, without undue interference from governments and major customers.

10.   Adequate compliance with these standards and the resulting effective quality assurance can be visibly recognized through generally approved quality program registration (e.g., registration by the Canadian Standards Association, the American Society of Mechanical Engineers, the Nuclear Regulatory Commission, and others).

Although it may appear desirable to reduce the current multitude of quality program standards, they nevertheless provide for useful alignment with special needs. Basically, quality programs, as the optimal composite of quality oriented decisions and procedures, range in type from simple end-item inspection programs to comprehensive quality assurance programs covering design, production, and item performance. Most standards divide into interdependent tiers of substandards, as demonstrated in Figure 18.2, with requirements or stipulations as shown in Table 18.1. This further permits the matching of the standard to different quality assurance needs.

As quality assurance concepts and practices evolve and improve, particularly in nonconventional areas of application, these program standards will play an important role as catalysts of change and improvement. Currently they are largely restricted to procurement documents and do not offer management the support in quality programming that is possible. Training and education, organizational behavior, production management, project planning and control, cost accounting, and modern computerization in communication and operation, are areas still not adequately addressed in these standards. With more general management awareness of quality assurance, an active involvement with quality standards and programming will be forthcoming.

Quality assurance programming in the sense of writing control procedures and coordinating activities is being performed by many, if not most, modern companies. Standards and guidelines are the main aids to this process. The responsibility for designing, documenting, and implementing a suitable and cost-effective quality program is assigned to the quality assurance function or carried by senior management directly.

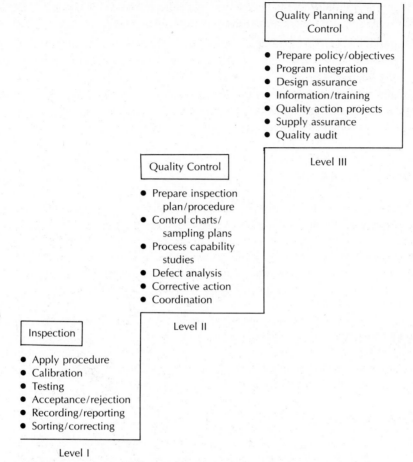

Quality Planning and Control
- Prepare policy/objectives
- Program integration
- Design assurance
- Information/training
- Quality action projects
- Supply assurance
- Quality audit

Level III

Quality Control
- Prepare inspection plan/procedure
- Control charts/sampling plans
- Process capability studies
- Defect analysis
- Corrective action
- Coordination

Level II

Inspection
- Apply procedure
- Calibration
- Testing
- Acceptance/rejection
- Recording/reporting
- Sorting/correcting

Level I

FIGURE 18.2   Interdependent tiers of substandards with important quality program elements. See Table 18.1 for a comparison of requirements with other generic and special standards.

Principles for Establishing a Sound Quality Assurance Program

Some important principles for the establishment of a sound quality program are given here.

1.   The need for the quality program must be surveyed and convincingly demonstrated.

2.   Standards and other management aids should be used, in order to minimize effort and resource requirement.

3.   Senior management must initiate and visibly support and identify the quality programing.

4.   Programing should be assigned and carried out by sufficiently knowledgeable and experienced staff.

5.   Programing should be conducted in the form of projects and with the active participation of staff in functions other than quality assurance.

6. Programing should be conducted systematically in phases of initiation, design, program documentation, implementation, and program maintenance.

7. Programing for quality assurance should be integrated with other ongoing management programs and should be handled as a continuous rather than as an ad hoc task and project.

8. Quality assurance program should be distinguished from annual quality plans, ad hoc types of quality assurance projects, or quality circles. A quality program constitutes the long-term, regular set of activities for quality assurance, and can thus serve as the master plan and framework for other quality assurance tasks and activities.

18.2 PROGRAM DESIGN

Each company and its management, in establishing a quality program, faces a situation that is different from that of other companies in many respects. Moreover, this situation changes continuously in a dynamic and competitive business environment. Therefore, designing the best program is a never-ending task. Initially, senior management, committed quality professionals and specialists, major customers, or just the competition might lead to the designing of a quality program. Later, regular audits of the quality program usually lead to changes and further improvements.

Programs are designed for solving problems and/or for bringing about major changes and breakthroughs towards a higher level of performance and improvement. In both cases analysis of the need is the first step, followed by goal setting, determination of approaches, and budgeting of resources. Programing describes the normal decision-making process, and the resulting program is therefore a decision of the management.

The Program Design Cycle

A quality program on a company-wide scope, over a certain time period consists of many sub-programs, such as writing the manual, revising receiving inspection, reducing critical defects within a given timespan, implementing statistical methodology, and so on. Each subprogram has a particular objective. The objective in turn determines both the resource requirement and the ways and means to accomplish the task. Of course, all these programs must remain part of the overall quality program and quality assurance effort. The design cycle follows the following sequence and steps.

STEP 1: *Appointment of a person.* Appointment of a person in charge of quality programming, possibly as chairman of a project or task-force.

STEP 2: *Survey of actual need.* A survey of the actual need for a quality program must be made in some detail. Such a study should checklist all pertinent factors and determinants, such as customer requirements, major quality defects, potential prevention and improvement projects, coordination of responsibilities, decisions and activities, comparison with quality programs of competitors and existing quality policies, and senior management position and strategy concerning quality assurance. The capacity of the company and its staff for gaining all the benefits of a larger market share through better quality assurance, must be clearly defined by the quality program. Needs should also be differentiated with regard to short-

and long-term developments, to special product lines, projects, customers and so on. They should be quantified as potential income in monetary terms.

STEP 3: *Assessing the management aids.* Once the need for the quality program is accepted and the design approved, available *management aids* should be assessed in order to prevent "reinventing the wheel." Quality program standards, guidelines, and published articles help to simplify the design process and to utilize the expertise and experience of others. Courses, conventions, and consultants in the field are sources of information.

STEP 4: *Selecting the appropriate quality program standard.* Selecting the appropriate quality program standard early during the design project is not only prudent but informs senior management, customers, staff, and other persons about the general attributes of the quality program. The selection can be performed by checklisting all factors, such as relative complexity of the product and service, maturity of the applied technology, costs of defects, major risks for liability, potential for additional business and contracts, and program requirements of the competition. For valid and reliable selection of the best alternative, factors could be ranked and weighted. Quality programs require a considerable investment of time and money, and selection of the most costly and comprehensive program standard and alternative is often not justifiable nor advantageous in the interest of customer satisfaction. A survey of quality costs in conjunction with program selection is therefore advised.

STEP 5: *Comparing the quality program with the requirements of the standard.* Comparing all existing components of the quality program with the requirements of the standard, and in particular with the required assurance procedures, is a wise step. This examination might reveal both gaps in the program and redundancies in it as well. Inviting representative of major customers to aid in this process, or applying for an external quality audit against the selected standard is always helpful. Customers stipulating compliance with standards can provide some guidance in the program design, partly as preaudit activities.

STEP 6: *Rationalizing and formalizing.* A review, adjustment, or realignment of quality assurance procedures becomes necessary to rationalize and formalize quality assurance decisions and activities.

STEP 7: *Preparing a draft outline.* Preparing a draft outline of the quality program and augmenting it with outlines on ad hoc projects, annual improvement plans, and implementation methods should be the next step.

STEP 8: *Participation by all concerned.* Present and discuss the complete draft with other supervisors and senior management, and make changes where required. Review of the draft is a learning exercise, since it prepares staff for the implementation of the program. Wide participation in the designing of the quality program motivates staff for quality assurance.

STEP 9: *Attaining formal approval.* Attain formal approval for the quality program design from supervisory management before submitting it to senior management for acceptance.

STEP 10: *Preparing the quality manual.* Once the quality program design has been approved and accepted as a draft, detailed documentation of policies, organization, procedures, in the form of a quality manual, can commence. This step is followed by the actual start-up of the new quality program.

In Chapter 15, (Figure 15.4) an example of a network plan for introducing a quality assurance program was shown, with a listing of typical activities. In view of the present discussion, typical features of the quality program in time perspective can be elaborated as in Figure 18.3.

18.3   PROGRAM DOCUMENTATION

Reasons for the formal documentation of any quality assurance program are that:

1.   Procedures, as the main components of the program, must be communicated in writing in order to direct decisions and practices correctly and consistently.

2.   Any program by definition can integrate and direct actions towards goal achievement only through formal directives and other pertinent information.

3.   Customers and government agencies need program documentation for independent and reliable verification and auditing.

4.   Quality program standards require and describe documentation.

5.   Such factors as performance control, corrective action, program implementation, program review and change, senior management's visible identification with the program, and program integration with other management programs for proper documentation of individual program components and of the quality program as a whole.

Long-range 5 years	Forecast of market and technological quality trends
	Establish quality objectives and strategies at the corporate level.
	Conduct feasibility study, including organizational and resource requirements.
	Obtain management approval.
	Design long-term quality programs with subprograms for the intermediate, immediately upcoming, and current year.
Intermediate-range 2 years	Determine quality needs for existing markets, processes and resources.
	Compare these needs with long-range objectives and strategies.
	Establish goals and action programs subdivided by priorities for first and second year stages.
	Obtain approval.
Next Year	Prepare implementation of first stage of two-year program.
	● Establish targets and translate these into tasks.
	● Formulate projects, organize, assign responsibilities and authorities.
	● Equip, inform and train.
	● Establish program and project management systems.
	● Integrate, coordinate and initiate.
Current Year	Implement various subprograms and projects.
	● Monitor progress, direct and correct.
	● Analyze progress, appraise, and reconsider long-term program.

FIGURE 18.3   Quality programs in time perspective.

The Importance of Written Procedures

Quality assurance programs are documented in the form of individual procedures and associated records and reports that are usually included, or referred to, in quality manuals.

Written quality assurance procedures are important because they direct and formalize the practices of decision making and work performance; as such they clarify management decisions. Figure 18.4 gives a typical example of the format of a quality assurance manual.

Principles of Effective Procedure Writing

The main and obvious principle of effective procedure writing is that the person concerned understands and can comply with the procedure. The following steps help in effective procedure writing.

XYZ Quality Control Manual	Title: Controlled Document Procedure	Approved by:		DP-105-4
		Copy No:	Review Date:	Page _____ of _____

1.0 Scope:

This procedure booklet consists of controlled documents issued to XYZ customers, individual departments, and/or groups relating to XYZ (unless, issued copy is stamped "FOR INFORMATION ONLY"). In this case, it would *not* be considered a controlled document.

1.1 This procedure will apply to Exhibit A, page 2, of this document.

2.0 Responsibility:

2.1 The Director of Quality Assurance shall be responsible for the overall control of revisions, issuance, and approval.

2.2 The QA Secretary, working together with the Quality Engineering Group, will keep an up-to-date QA Document Issuance Control Book, in which, all issuance of the above procedures will be controlled numerically.

3.0 Procedures:

3.1 The following procedures are in reference to a *request for a revision*. If an individual or customer is only requesting a copy of a controlled or uncontrolled procedure, steps D and E will apply.

    A. Request has been obtained for a revision.
    B. Upon request, a rough draft is typed and approved.
    C. Original is typed and approved.
    D. Copies are made.
    E. All controlled and uncontrolled copies will be issued from the QA Department. The controlled issuance will be immediately recorded in the QA Document Issuance Control Book and each will be distributed.

4.0 References:

4.1 QA Manual 101-1981 (Rev. 4/1/81), Section F.

FIGURE 18.4 Typical example of a possible page format. A good caption layout invites reading and contributes to usability of the manual.

1.   Determine the expected outcome of the procedure (e.g., inspector's reliability in making acceptance or rejection decisions).

2.   Define the object of the procedure (e.g., item, process, quality, characteristic, supplied lot, service maintenance).

3.   Select and appoint the procedure writer(s) and define their qualifications and authority.

4.   Establish the procedure writing approach and method (i.e., establish the current practice, gather data, determine the test, prepare the recording, checking test facilities and capacities, decision criteria, recording communication, etc.)

5.   Establish the format of the procedure for document writing and for communication purposes.

6.   Draft the procedure.

7.   Review, and attain approval.

8.   Implement tentatively; train, debug.

9.   Finalize procedure and attain approval.

10.   Review and/or audit.

Any quality program and its procedures have to contribute in the long run to the success of the enterprise. Senior management can only approve quality programs that promise cost-effectiveness. Moreover, the need for smooth implementation of the program requires the designer to involve those already employed in an area targeted for improvement with the designing and to make the program acceptable to them.

The practicality of the program design to a large extent determines the success of the program and the accomplishment of the objectives.

18.4   THE QUALITY MANUAL

One of the main purposes of the quality assurance program is to coordinate and integrate all decisions and actions towards quality goal achievement. In practice, this means to avoid any inconsistencies, conflicts, and undue overlaps or gaps through review of all procedures. Furthermore, procedures have to be augmented by documentation of other program components, such as the outline of policies, organization, recording and reporting forms, references to standards, program reviews, and change controls. Quality manuals describing the quality assurance program constitute the main program documentation.

In principle, any company explicitly claiming quality assurance should have a quality manual as formal documentation. This manual can create, in an appealing and convincing manner, an element of trust in the quality assurance, and confidence in the actual quality of products and services. However, certain principles must be observed in the preparation of quality manuals; the major one being that the manual must describe the company program as it truthfully exists. It is insufficient and misleading if the manual only demonstrates compliance with program standards. Principles governing quality manuals are these.

1. The manual must provide employees, customers, auditors, and other authorized persons with sufficient information about the quality program of the company, including policy, organization, procedures, and other program aspects. It should not include confidential matters, as these can be cross-referenced and recorded in other company documents.

2. The manual should be made as an official document, and not just as an information brochure. It must comply with requirements of in the standard. The manual should translate the standard's generic clauses into the procedures that suit the conditions of each work division.

3. The quality program documented in the manual must reflect the actual practices of the company. An auditor will first compare the manual with the standard and then compare the practices with the manual.

4. Any change in the standard, and/or the company quality program (policy, organization, procedures, etc.) must lead to an updating of the manual. Also, changes in actual practices, once approved, must be recorded.

5. A chart showing records and communication forms might be included in the manual.

There are numerous internal advantages for the company in having a written quality manual, such as:

1. The facilitating of communications among employees and outside customers, suppliers, and supervisory agencies.

2. The reinforcement of supervision.

3. The clarification and definition of individual duties, responsibilities, and authorities.

4. The formalization of quality control operations and company policies and objectives that help resolve arguments and give a sense of confidence to subordinates.

5. The fact that only written procedures and policies can be considered permanent features of the company, despite the mobility of personnel.

6. The manual as a good source of training and education for new personnel.

When it comes to designing and developing a written quality manual, many managers might continue to think of it as a necessary evil. There are many benefits that a manual has to offer, in addition to those outlined above. However, any quality manual becomes an effective working tool (capable of saving time and money) only when it is properly developed and designed, and kept flexible enough for easy updating.

Quality manuals are prepared and reviewed in a similar way to any procedure manuals. In fact, procedures should be written before they are entered, abstracted, or referred to in the manual. Program standards stipulate the form and content of quality manuals to a certain degree, particularly those for comprehensive programs. However, while some uniformity in manuals is advisable, the unique conditions of the company always have influence on the program and the manual as well.

The *United States Small Business Administration* has published a sample quality manual, *Management Aid 243,* to which we have already frequently referred. Figure 18.5 shows the style and contents of the first few beginning pages of the aid, which provides an overview of the manual page design.

TITLE PAGE

Company Name

QUALITY MANUAL

Approved by

Name   Title

Date

INTRODUCTION

This manual is issued to describe the quality assurance system to be employed at (Company Name) to attain compliance with the intent of the general inspection system requirements of the major government procurement agencies when specified in customer purchase orders, contracts, and subcontracts. The policy of (Company Name) is to apply the system to articles and materials received by (Company Name), as well as to articles produced by (Company Name) or its suppliers for end use in most governmental but especially in military and/or aerospace products.

The manual provides personnel and customers of (Company Name) with a description of company policy for maintaining an effective and economical quality assurance system planned and developed in conjunction with other planning functions.

Written procedures for implementing the policy described herein shall be established as dictated by complexity of the product design, manufacturing techniques employed, and customer requirements.

No changes in the manual or supplementary quality assurance procedures are valid until approved by the Plant Manager, or his assignee.

TABLE OF CONTENTS

Description	Section
Scope	1.0
Responsibilities	2.0
Purchase Order Control	3.0
Drawing and Specification	4.0
Control	5.0
Receiving Inspection	6.0
Raw Material Control	7.0
In Process Inspection	8.0
Assembly Inspection and/or	9.0
Functional Test	10.0
Final Inspection Test	11.0
Nonconforming Material	12.0
Control	13.0
Tool & Gage Control	14.0
Overrun Stock Control	Organization Chart
Packaging and Shipping	Purchase Order Form
Identification	Inspection Data Form
Appendix A	Identification Tags
Appendix B	Travel Card
Appendix C	
Appendix D	
Appendix E	

FIGURE 18.5 Arrangement of beginning pages of the quality manual. Selecting materials for inclusion in the manual starts with the development of an appropriate table of contents. Adapted from, *Management Aids*, No. 243, Small Business Administration, Washington, D.C.

Design and Development of The Quality Manual

Just as there is always a convincing need for a written quality manual, proper standardization (design and development) of it remains equally essential if it is to be used as an effective working tool. A proper design is determined by three factors: *format, layout,* and *writing style*. In general terms, these three factors must be kept in mind, so that a manual user (not the manual writer) can retrieve and understand information with minimum difficulty. Manuals are seldom permanently-bound volumes; rather, the manual is usually maintained as a loose-leaf book (with revisable printed pages), to allow for its easy updating and to keep costs within reason. Some companies however, prefer to document quality programs in a separate bound volume and procedures in a loose-leaf form.

Writing of the Quality Manual

The purpose of the manual is to help ensure quality through documented policies, organization, procedures, and forms for information dissemination. The manual informs management and other employees, as well as outsiders, about the quality assurance program of the company. It also coordinates activities and decisions, helps reduce product liability risks, and serves as a reference for audits.

The following principles should be observed in the writing of the manual.

1.   The manual must reflect the actual state of quality assurance in the company. It must be valid and true.

2.   Management must have officially approved the writing of it.

3.   The documentation must efficiently and effectively inform the reader to whom it is addressed. It must be clear, unambiguous, concise, understandable, and properly designed to be compatible with other criteria for proper management reporting.

4.   Recent changes must be indicated in the document.

5.   Auditors must be able to use the manual for the purpose of auditing the quality control program.

6.   Confidential matters and secret technical matters and processes should be cross-referenced only, not spelled out.

7.   The standard with which the quality assurance program is to comply should be mentioned in the document.

Beginning to write the manual is not an easy task. Either we start with the standard for the quality assurance program, or with the existing practices of the company. When using the standard, its requirements for the quality manual directly determine the approach. The major advantage of this approach, is that writing the manual first, and then experimenting with implementing it runs the risk of producing an invalid document and this possibility is avoided when the standard directs the writing of the manual. The manual must interpret the general standard for the specific setting and conditions of the company. The use of *prototype manuals,* prepared for a particular standard, can mislead the writer. The writer, in this case, may tend to ignore the existing reality of the company or practice.

Standard and prototype manuals must be adapted to company conditions. The manual must reflect the actual activities taking place, and the activities themselves must comply with the general requirements of the standard.

Once prepared, a useful and valid quality manual should have the following features.

1.   It should be *informative and understandable,* clear in purpose and objective.

2.   It should be *concise and complete,* with a valid interpretation of the standard or stated quality assurance policy.

3.   It should be *systematic* in layout.

4. It should *reflect actual quality assurance practices* and be amendable in accordance with set procedures. Its distribution should be known.

5. It should be subject to *periodic review*.

6. It should be prepared carefully as *an official document,* with wide participation of persons concerned. The manual writing should be handled as a special quality assurance improvement project with direct support by senior management.

7. The manual should be *coordinated and consistent* with other operations manuals of the company.

The following are suggested steps for a quality manual writing project.

1. Obtain approval for the manual and general directives for it from senior management.

2. Form a project team and procedure writing subcommittees.

3. Take an inventory of existing material for the content of the manual; this means, in particular, existing procedures, forms, et cetera. Compile the material.

4. Possibly prepare a project plan (network) showing milestones, resources, and costs.

5. Inform participating functions and persons about the project in order to gain acceptance, participation, and legitimacy.

6. Inform the members of the writing subcommittees about the subject matter and train them in procedure writing.

7. Prepare a draft of the manual.

8. Review the draft with the functions and persons concerned.

9. Review the draft with senior management, along with associated documentations, such as project-related reports, annual reports, and so on.

10. Determine with senior management, the distribution list for the manual.

11. Have management formally and visibly announce the launching of the program in accordance with the quality policy of the company.

12. Arrange for the auditing and updating of the quality manual.

13. Acknowledge the work of the project team and use this event and others to promote quality assurance in the company.

The writing of a quality manual should be executed as a project in order to apply the most suitable approach and to arrive at realistic and acceptable documentation. Along with the development or review of the manual, records and reports should be finalized in order that the respective forms can be displayed in the manual. Procedures usually involve activities for the recording of data and/or for the preparation of data reports.

The manual and the associated system of records and reports must be continuously adapted to changing conditions. Without such flexibility, effectiveness of quality assurance policies would become doubtful. Uninformed or poorly informed management and operational staff, are liable to make wrong decisions or no decisions at all.

Format, Layout, and Writing Style

The topical layout, the contents, and, in particular, the degree of detailing, depends on the size of the company, the type of product, contract, process, and industry, and the quality program standard invoked.

The manual published by the United States Small Business Administration might suit a small company with low liability risks and no particular standard to comply with. At the other end of the spectrum are manuals as described in the CSA Standard Z 299.1, the Canadian Department of National Defense DND 1015, MIL-Q-9858A and the like. The contents of any manual should be arranged in logical order. Normally, in the first part, the quality policy, organization, and functional responsibilities describe the long-term foundation of the program. In the next part, major procedures are listed, such as for quality control of purchases, production, and test equipment. In the CSA Standard these main functions are called *system functions*. These procedures must describe all related responsibilities, for example, who is to do what, when, how, where, and with what materials. The steps and the chronological sequence of each activity is given. Terms, symbols, and visual aids, such as flow diagrams, hold tags, et cetera, are explained. The procedures must further describe those records to be kept and reports to be made. This might include directions for data processing. The procedures might individually vary with regard to these elements, depending on the kind of quality assurance activity that the procedure is to regulate.

Format and the *styling* of the sheets (pages) of the manual's text should be chosen to include proper positioning of the constants and variables. Subtitles normally stay at the upper portion of each sheet. For example, if a format such as the one shown in Figure 18.4 is used, the *constant factor* (company's name: XYZ with a logo) would appear at the top left corner. Since the top right corner is the most visible corner of any page, one would normally put the document or subsection control number there, as a link between the manual and its *index*. Alternatively, as with the layout of Figure 18.4, the top right corner can be made up with a number of boxes to include other *variable factors*. In the format shown, a section title and a subject title (appearing in the middle portion) have been merged into one; they could be noted separately as well, and then each title would be given its code numbers in the main index.

The main point, is that all effort should be made to make the format inviting to read by giving it an uncluttered appearance. If more variables have to appear (such as page numbers, subsections, procedure numbers, etc.) the space at the bottom portion of the page can be utilized.

Layout concerns the written material for the text of the manual. There are many ways (styles) one can choose for the layout, for example conventional paragraphs, a playscript format, matrixes, illustrations (with or without flow charts), or decision tables.

No matter which style the writer selects, the manual text must be easy to read, interpret, and comprehend. Captions in the text with subsequent numbering of main items help the reader locate the desired point of emphasis clearly and easily. Remember, the manual is for the benefit of the user, not the writer!

Numbering System for the Manual Materials

An appropriate numbering system for the manual is another important requirement from the viewpoint of ease of referencing, flexibility, and easy expandability so that the pages can be added or revised without disrupting the established portions. There are several methods of numbering systems in use.

1. In *numeric sequence*, the subject matters are simply numbered "1,2,3. . ." in the sequence of their writing.

2. In the *alphabetic and numeric* system, a combination of letters and numbers are used. The letter designations can be used to identify the major headings and the numbers for identification of the specific subjects under that particular heading. For voluminous manuals this sytem offers a wide choice of arrangement of subject matter. There are many ways to arrange the letters and numbers, such as:

A1, A2, A3, . . . ; B1, B2, B3, . . . ; C1, C2, C3, . . .
A1–1, A1–2, . . . ; B1–1, B1–2, . . . ; C1–1, C1–2, . . .
A1.1, A1.2, . . . ; B1.1, B1.2, . . . ; C1.1, C1.2, . . .

3. A *numeric and alphabetic* system is the converse of the above alphabetic and numeric system. Here, two or three digit numbers are assigned to denote the major subject matters and followed by an alphabetic suffix for the subsequent matters. For example:

101, 101A, 101B, 101C, 101D, . . .
201, 202, 202A, 202B, 203, 204A . . .

4. A *decimal system* employs grouping of the various subject matters. Starting with for example, 1.01, the first master number 1 is followed by either a two or a three digit number with a decimal point separating the major (section) heading and minor (subject) heading(s).

5. The *MCA Digit System* is a special design developed by the Manuals Corporation of America, a New York consulting firm. This coding system allows up to 970,000 subject titles, as it is designed having three cycles of 99 numbers. The manual divider starts with 0000 and extends to 9999, with the natural ascending sequence, that is, 0001, 0002, ... Each change in number can create a new section title that can be further expanded to include a section number. For example:

0001–1, 0001–2, 0002–1, 0002–2, . . . , 0099–99
0101–1, 0101–2, 0102–1, 0102–2, . . . , 0199–99
0201–1, 0201–2, 0202–1, 0202–2, . . . , 0299–99

. . .

9901–1, 9901–2, 9902–1, 9902–2, . . . , 9999–99

Manual Writing as a Project

The quality assurance function takes charge of the manual writing and maintenance project. Only in a small organization will the owner–manager or the senior delegate, supervise the project personally.

A quality assurance program standard, such as the ANSI/ASQC *Generic Guidelines for Quality Systems, (Z-1.15)*, offers the most applicable, general, and authentic resource material. The following gives a step-by-step approach for a quality manual writing project.

Listing of Tasks

1. *Project initiation* by quality assurance staff, customers, consultants, and so on.

2. *Senior management decision;* formation of a steering committee with wide representation, including quality circles.

2.1 *Steering Committee conducts meetings for:*

2.1.1 Defining tasks, terms of reference, operational procedures; subject to approval by senior management.

2.1.2 Assessing the need for and purpose of the quality manual; data are customer requirements, quality problems, imposed standards, and competitors' actions in quality assurance.

2.1.3 Assessment of available writing aids: standards, guides, proto type manuals, internal manual writing expertise, and "How To" publications.

2.1.4 Assessment of internal manual writing resources and capacity: personnel, budgets, available time, and potential project team members.

2.1.5 Assessment of external resources: government support, consultants, and summer students.

2.1.6 Submission of a feasibility study to senior management to attain approval.

2.1.7 Nomination of a project manager and attaining approval.

2.1.8 Formation of a steering arm as an advisory committee to the project manager, and as an intermediary between the project manager and senior management.

2.2 *Senior management initiates project:*

2.2.1 Formulates and announces official quality assurance policy, possibly restates existing one with reference to quality manual.

2.2.2 Appoints and instructs project manager, makes respective announcements.

2.2.3 Informs staff about project goal and approach; requests support and participation, project permits quality assurance program improvement and enhances quality motivation.

3. *Quality manual writing project:*

3.1 *Design of project:*

3.1.1 Formation of project team, appointments by steering committee and nominations by departments or section members.

3.1.2 Team establishes manual-writing procedures and principles, including decision-making procedure.

3.1.3 Selection, compilation, and study of quality manual writing aids, such as standards, guides, "how-to" publications, prototype manuals, existing internal procedures and quality assurance documentation, assistance from experts (see, for example Appendix H, which lists useful military standards and specifications).

3.1.4 Training of team members in quality manual and procedure writing, drafting a prototype procedure.

3.1.5 Preparation of a quality procedure-writing guide or handbook.

3.1.6 Determination of general format and content layout.

3.1.7 Formulation of preface, stating purpose, objective, application, distribution, basic writing procedure, and implementation and review procedures.

3.1.8 Listing of major writing tasks and nomination of writers group.

3.1.9 Preparing a writing project plan and schedule: PERT network, milestones, budgets, and deadlines.

3.2 *Submission of project plan to steering committee, attaining approval.*

3.3 *Project team forms subwriting groups*

3.3.1 Listing of all subwriting groups activities; tasks assigned, members, deadlines, supervision, and support.

3.3.2 Commencement of procedure writing, concurrently or sequentially, by subgroups as follows (see writing procedure).

    a.   Quality manual amendment and review procedures

    b.   Customer relation; contract review, customer complaints

    c.   Design assurance and review

    d.   Configuration control

    e.   Supplier relationship

    f.   Inspection planning

    g.   Receiving inspection

    h.   Work-in-progress inspection

    i.   End-item inspection

    j.   Process control

    k.   Statistical quality control

    l.   Nonconformance control

    m.   Corrective/preventive action

    n.   Quality records and information system

    o.   Quality cost accounting

    p.   Quality motivation and training

    q.   Organization and coordination

    r.   Quality audit

3.3.3   Project team receives draft procedures from writing groups.

3.4 *Draft of quality manual:*

    3.4.1   Compilation of procedures.

    3.4.2   Attaining approval from supervisory management for their quality assurance procedures.

    3.4.3   Submission of draft manual to steering committee.

3.5 *Steering committee assessment of quality manual draft:*

    3.5.1   Verifies acceptance of quality manual by major customer.

    3.5.2   Submits manual to senior management to attain approval.

3.6 *Senior management assessment of manual:*

    3.6.1   Verifies compliance with original quality policy.

    3.6.2   Announces quality manual as official document, signing the manual.

    3.6.3   Recognizes the work of manual writers; dismisses project team.

    3.6.4   Reforms steering committee as quality audit committee.

4. *Quality manual implementation:*

4.1 Senior management announces procedure for quality manual implementation.

4.2 Quality assurance function or department takes charge of implementation, manual amendment, and review.

4.3 Quality audit committee plans and conducts audit for effective quality manual implementation; reports to senior management; possibly corrective action taken.

4.4 Senior management, supported by quality assurance staff, seeks formal audits of quality assurance program by external auditors.

4.5 Quality audit committee cooperates with external auditors, and conducts repeated quality audits.

A master network of the above step-by-step approach is shown in Figure 18.6. Standards stipulate information that manuals should entail, but do not specify the specific format and individual procedures presentation. The manual translates requirements of the quality assur-

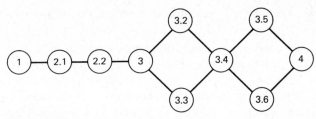

FIGURE 18.6   Master network of a quality manual writing project. See text for details.

ance program standard into the prevailing conditions of an organization. Auditors are then able to evaluate compliance with the standard. After having assessed the quality manual, they can then also determine the actual implementation of the program in the organization.

18.5   PROGRAM IMPLEMENTATION

Any program design and documentation is as useful as it is effectively implemented. According to Feigenbaum:[1]

> Certainly the characteristics of an effective control program for quality have become quite clear from the work of many quality professionals over the past years and these parameters can be readily stated. The demand today is for attitudes, knowledge and skills for implementing such control programs . . .

Program implementation is understood as:

1.   Planning and control of the start-up of the designed and approved program.

2.   Motivating, informing, and directing people participating in the program.

3.   Introducing report forms and schemes (information systems) and a quality manual; compiling and processing of data.

Little research has been done about the management of a transition phase (start-up). The responsibility for this usually rests with the quality control function that designed and initiated the program. Implementation, however, usually also involves company-wide personnel, and particularly the top management.

The degree of difficulty in implementation depends on the scope of the program, the intended change (improvement, status quo, or breakthrough), the quality of the program design, the qualifications of the people participating, support by superior management, and the quality of actual preparation for the implementation itself.

The one major rule for successful program implementation is the visible and competent leadership of management. If management is not sufficiently and convincingly committed to the program goals, no one in the company will be. An experienced quality manager, when asked what he would watch for in program implementation, suggested the following.[2]

1.   Provide employees with regular reports on the effectiveness of the program as it is implemented (i.e., decreases in defectives, decreases in cost, attainment of a higher quality product, etc.). This will provide an incentive to continue with the program and make further improvements.

2.   Establish a procedure for feedback from employees. In addition to the fact that this could be a major factor in defect prevention, it could also be a direct line to an employee who, having implemented a portion of the program, realizes how further improvements could be made. This could provide new information not previously considered when the program was prepared or might be only a ''looking ahead'' on the part of the employee.

3.   The program might have to be halted at any phase of its implementation. Consideration should be given to the method used to do this with the least disruption to the work force. This halt might be temporary or permanent, and would be the result of a justifiable, unusual or unforeseen circumstance.

Basic Principles of Program Implementation

We can summarize the important basic principles of program implementation as follows.

1.   Inform staff and management of the scope and objectives of the quality action program and approaches (changes) in general terms.

2.   Obtain visible management support of the program.

3.   Introduce participants to systems, reports, and tasks that each will be responsible for. Allow for learning. Setup project teams.

4.   Monitor and evaluate implementation of the plan.

5.   Modify the plan if it becomes necessary due to previously unforeseen difficulties or shortcomings in the plan.

6.   Modify the plan if it becomes necessary due to technological changes.

7.   Create a feedback mechanism to the staff to allow them to evaluate their personal performance in the program.

8.   Audit the quality control procedures and determine the adequacy of the stated procedures. Adjust as necessary.

9.   Once the system is operational, have workers and supervisors jointly set goals for quality control within their areas.

Securing a Positive Attitude Towards the Program

In order to attain a positive and cooperative attitude towards program implementation, the following ''*AIDA*'' approach has proven useful (A for attention, I for interest, D for desire, and A for action).

STEP 1: *Secure attention.* Kick off the campaign with a breakfast meeting of all supervisors. Invent a slogan (e.g., for insulated polyethylene pipe one manufacturer used: ''Quality: Our Pipeline to Success''). Concern for quality can be described by factory supervisors and sales personnel (practical-experience viewpoint) as well as by management. Determine a technique for measuring performance, enabling easy checking of accomplishments versus goals. Make management's support of the campaign visible to all employees.

STEP 2: *Get the interest of employees.* Convince employees that quality is important to their own well-being (job security, benefit sharing, pride in workmanship). Avoid too much attention getting by way of propaganda, slogans, and the like. Arrange for group sessions (employees with management). Invent methods of getting the employee to convince him- or herself that quality is important (e.g., the old suggestion-box routine).

STEP 3: *Create a desire.* Convince the employee that there is something he or she can do to make a real difference. Investigate processes and the operators on a one-to-one basis. Make the operator feel that he or she is in control of a situation (train them in the basics of quality control techniques). Give the operator more insight into the reasons for demanding such high precision.

STEP 4: *Prompt and support action.* Invite the operator to participate in analysis to discover causes of defects (quality circle approach). Training and development programs may be necessary, but the benefits must outweigh the costs (again, everybody must be trained to understand quality, not just the workers).

The following plan describes the implementation of a new quality assurance program by a small manufacturer. A project-type planning and control approach is applied. This implementation project was preceded by a similar one for designing the quality assurance program. It is to comply with a certain program standard.

The project team leader and the membership, having completed the program design, thoroughly won the support of senior management, and, in this case, having also sought a major customer's approval for the design of the manual, now carry on with the implementation. Naturally, the continuation of the project team, possibly with some changes in the membership and even the chairmanship, best utilizes newly acquired knowledge and experience in quality assurance programming. The major implementation project can be further augmented by quality circles that are like project teams. In this strategy the quality assurance program becomes a *quality action program,* and as such, a permanent dynamic and active element, force, and quality assurance mover and motivator in the company.

Layout of Implementation Plan

Having obtained final approval by the management committee for the new quality action program, the following implementation plan will commence immediately and it requires full visible support and cooperation of all management groups, supervisory and operational staff.

In order to achieve the goals for completion of implementation by week 52 (December 21, 1981) a strict compliance with established target dates must be assured by all involved parties. (See the time schedule in Figure 18.7.)

1. *Employee QC awareness meeting number one (week 29):* Initial employee QC awareness and introduction to action committee. Action program meetings, including question and answer period are to be conducted by QC managers (production and engineering manager) for all manufacturing division personnel in two and one-half to three hours. Group sessions (by department) are to be held in the lunchroom.

Start Aug. 1, 1981 ——————————— Finish Dec. 31, 1981

Week

No.	Steps of Implementation	29	30	31	32	33	34	35	36	37	38	39	40	41	42	43	44	45	46	47	48	49	50	51	52
1	Employee Q.C. Awareness meeting #1	▮																							
2	Supplier Q.C. Awareness		▮																						
3	Measuring & Testing Equip calibration and Identification			▮					▮				▮				▮				▮				
4	Quarantine lock-up					▮																			
5	Purchasing				▮																				
6	Incoming Inspection											Continuous →													
7	In process inspection										Continuous →														
8	Final inspection									Continuous →															
9	Material Review Board (MRB)							▮						▮				▮				▮			
10	Quality Measurement reports						▮				▮						▮					▮			
11	Cost of Quality Reports						▮				▮					▮				▮					
12	Corrective action										▮ ——— As required ———														
13	Employee Q.C. Awareness Meeting #2													▮											
14	Self Audit #1												▮												
15	Quality Program Review #1																▮								
16	Changes & Improvements																	▮							
17	Quality Program Review #2																			▮					
18	Self Audit #2																				▮				
19	Employee Q.C. Awareness Meeting #3																						▮		
20	Professional Audit																							▮	

FIGURE 18.7 Quality action program implementation plan.

2. *Supplier QC awareness (weeks 29–30):* A letter of introduction regarding QC action program will be sent to all major suppliers and customers by the plant manager, informing them in terms of how the program may affect them.

3. *Measuring and testing equipment (weeks 29–32):* Senior inspectors are to perform initial inspection, calibration, and identification of all measuring and testing equipment, jigs, fixtures, and special tools used by inspection departments and production personnel. An identification and inspection record is established for each item and is to be updated after every monthly inspection.

4. *Quarantine locker (weeks 29–30):* Maintenance will construct locker room space in the component warehouse to hold all quarantine items. The senior inspector is to verify completion and retain one key. The second key is to be handed to the QC manager (see engineering drawing QL-100-II). The material review board should be functional at this time, week 33.

5. *Purchasing (weeks 29–34):* The buyer to compile all established quality standards and procedures for existing purchased products and send copies to respective suppliers to assure supplier awareness of strict incoming inspection and supplier surveillance. This process will continue on all new future products or upcoming changes on existing products. At this point a supplier rating system might commence.

6. *Incoming inspection (week 31):* Receiving inspector is to perform incoming inspection duties on all received items and raw materials before acceptance of shipments as outlined and acquired by established standard and test procedures. The sampling plan is in place.

7. *In-process inspection (week 32):* In-process inspectors are to perform all in-process inspections and required tests and applicable procedures as outlined in detail on the process control card.

8. *Final inspection (week 33):* The chief inspector is to perform all final inspections and required tests as outlined in detail on the process control card and applicable procedures.

9. *Material review board, (MRB) (week 33):* The MRB to hold the first of their scheduled biweekly meetings.

10. *Quality measurement reports (week 34):* Q.C. managers are to compile and evaluate quality measurement reports (reject, rework, and scrap reports) on a monthly basis and send copies to the accounting department for analyzing.

11. *Cost of quality reports (week 35):* The accounting department is to provide monthly cost reports by analyzing and applying dollar values to quality measurement reports.

12. *Corrective action (week 36):* Plant managers, production and engineering managers QC managers, the production superintendent, and the senior inspector will discuss and analyze on a monthly basis quality measurement and the cost of quality, and decide on corrective actions to be taken in order to decrease the cost of quality. Corrective actions will be issued as written directives, closely monitored by assigned inspection and production personnel, and evaluated in the second and following corrective action meetings.

13. *Employee QC awareness meeting number two (week 39):* The second employee QC awareness meeting to be held by QC managers with review and highlights on achievements during the first three months of program implementation and discussion of problem areas and corrective measures followed by a question and answer period. (A memorandum is to be issued to all employees two weeks prior to the meetings advising meeting date and time.)

Note: A writer must be present at each meeting to take notes on comments and suggestions for improvements made by employees that will be evaluated by the management and supervisory team and considered for corrective action.

14. *Self-audit number one (week 41):* A first self-audit to be conducted by the vice president or his or her appointee or a project team leader, based on the audit checklists established by the QC program planning committee. A written detailed report on findings should be prepared and issued no later than week 43 to the plant manager, who will chair the program review committee.

15. *Program review number one (week 44):* The program review committee, consisting of QC and production and engineering managers, production superintendent, senior inspector, buyer, and plant manager (Chairperson) will review, analyze, and evaluate the total QC program based on the audit report, decide plan changes or improvements if necessary, and submit a written report on recommended changes to the management committee for approval. The auditor should be involved.

16. *Changes and improvements (weeks 45–47):* Implement changes and improvements if necessary after approval by management committee and assign inspection and supervisory personnel to monitor and report back in writing on implemented changes and improvements to the program review committee no later than week 47. The auditor should be involved.

17. *Program review number two (week 48):* The program review committee to review the QC program with emphasis on the latest changes and improvements and submit a written report on progress, value analysis, and cost to the management committee.

18. *Self-audit number two (week 49):* The vice president or his or her appointee is to conduct the second self-audit based on the audit checklists established by the QC program planning committee and prepare a written report to the program review committee and the management committee.

Note: Providing this report is satisfactory, the implementation of the quality action program is completed. Audit and certification are to follow.

19. *Employee QC awareness meeting number three (week 51):* The third and last employee QC awareness meeting during the program implementation period will be held by the plant managers for all employees for a review of the total quality action program after implementation. This will be followed by a 20 to 30 minute question and answer period. The meeting will take one and a half to two hours and should be held in the lunchroom. (A memorandum must be issued to all employees one week prior to the meeting, advising time, date, and location).

20. *Professional audit (week 52):* QC managers are to invite and schedule professional auditors to review the program.

A second example of an implementation plan that can be prepared would be similar to the project-type of quality program described in Chapter 15. The project plan for the audit will be shown in the next chapter.

In addition to a comprehensive implementation plan that covers the entire establishment of a quality assurance program, a more narrowly conceived start-up project initiates the implementation of the program. Whenever the quality program is to bring about major changes and breakthroughs, the start-up phase should be thoroughly planned.

Quality programs and start-up projects offer management a unique opportunity for creating a more harmonious and productive work environment and stronger cooperation. Everybody can understand the need for improved quality and thus, the purpose of a well-planned and implemented quality assurance program. Benefits for the individual and for the group and company are readily recognized.

Therefore, implementation of a quality program should marshall the dormant potential among the employees for active and constructive participation in program activities and goal achievement. Management must be aware that not utilizing this potential means opportunity costs. These costs are normally much higher than the expenditure required to finance a quality program and its subprograms. A cost analysis and budget might help to demonstrate the profitability of well-implemented and designed quality programs.

Implementation, as opposed to the start-up project, remains a continuous issue and problem in the life of the quality assurance program. Auditing, as will be discussed in the next chapter, is the major method for program maintenance.

18.6 APPLICATION OF PERT IN QUALITY PLANNING

In quality assurance planning the major purpose is building quality into the product rather than inspecting it in. This means a shift of effort to the preproduction activities of design and process control preparation. New product designs and major product reviews involve not only the quality assurance function, but practically every other function in a company. Therefore, in this project coordination becomes a main task, and project planning and control techniques, such as PERT, should be applied.

The advantages of *network analysis* methods for project planning and control rest mainly in preparing and maintaining an up-to-date schedule by means of a computerized information system. The *Project Analysis and Control System* (PROJACS) of IBM is one of these systems with great versatility. In a hierarchy of interdependent networks the interfacing areas of quality assurance with an overall product design or review can be conveniently handled as subnetworks.

The following material illustrates the interrelationships of a product design network and the interrelationships with a quality assurance and a cost accounting subnetwork. Because PROJACS does not process probabilistic networks, this simple model is partially analyzed manually. This is, however, only for the purpose of demonstrating the applicability of PERT. In practice, one should still use PROJACS, or other computerized models. Convenient updating of time estimates compensates for a PERT-like analysis.

Tables 18.2, 18.3, and 18.4 list the activities, predecessor relationships, and time data for the resulting planning information. The total project, shown in the *master network* (Figure 18.8 in conjunction with Tables 18.5 and 18.6) is estimated to require 65.5 weeks for completion with a standard deviation of 2.2 weeks.

Assuming a fit with a normal distribution, one can now calculate various probabilities (compare, chapter 6). For instance, what is the probability that the project will be completed in a time period of less than or equal to 63 weeks, rather than in 65.5 weeks? The answer

TABLE 18.2
Predecessor Relationships with Task Symbols

Task	Basic Predecessor	Quality Assurance	Cost Control
A	—	Aq	Ac
B	A		
C	B	Cq	Cc
D	C	Dq	
E	D		Ec
F	E		Fc
G	F		Gc
H	G	Hq	Hc
I	G,H	Iq	Ic
J	I	Jq	Jc
K	I		Kc
L	J,K	Lq	Lc
M	L	Mq	
N	M	Nq	
O	N	Oq	Oc

is a 13% probability; $Z = \dfrac{63 - 65.5}{2.2} = -1.13 = .13$ (from Table D of Appendix B).

Activities of the subnetworks have independent time estimates, and only if the estimated time exceeds that in the master network would the latter have to be revised. If the estimate for A_q were 9.1 instead of 2.1, the respective master time estimate would have to be raised from 8 to 9.1. Normally it is important that activities in subnets meet the late finish time of

TABLE 18.3
Quality Assurance Subnetwork[a]

Task	Optimistic	Time (weeks) Most Likely	Pessimistic	Estimate	Variance	ES	LS	EF	LF
Aq	2	2	3	2.16	.03		5.9		8
Cq	3	4	5	4	.11		11.5		15.5
Dq	2	2	4	2.3	.11		20.0		22.3
Hq	1	2	3	2	.11		38.2		40.2
Iq	1	2	2	1.8	.03		43.7		45.5
Jq	1	2	3	2	.11		49.3		51.3
Lq	2	2	2	2	0		53.3		55.3
Mq	1	2	2	1.8	.03		56.5		58.3
Nq	1	2	2	1.8	.03		60.5		62.3
Oq	1	1	1	1	0		64.5		65.5

[a]ES = early start, LS = late start, EF = early finish, LF = late finish.

TABLE 18.4
Cost Subnetwork[a]

Task	Time (weeks)			Estimate	Variance	ES	LS	EF	LF
	Optimistic	Most Likely	Pessimistic						
Ac	3	4	6	4.2	.25		3.8		8
Cc	1	2	2	1.8	.03		13.7		15.5
Ec	1	1	2	1.2	.03		24.3		25.5
Fc	2	3	3	2.8	.03		27.7		30.5
Gc	1	1	2	1.2	.03		33.3		34.5
Hc	1	2	3	2	.11		38.2		40.2
Ic	2	3	3	2.8	.03		42.7		45.5
Jc	1	2	2	1.8	.03		49.5		51.3
Kc	2	2	2	2	0		49.3		51.3

[a]ES = early start, LS = late start, EF = early finish, LF = late finish.

TABLE 18.5
Meaning of the Task-related Symbols Used in Figure 18.8

Symbol	Description of Tasks
A:	Preliminary design and basic product planning
B:	Study of prototype manufacture by top management
C:	Preliminary design
D:	Prototype manufacture
E:	Review and final decision
F:	Final design
G:	Final drawings
H:	Establishment of inspection standards
I:	Production line and process planning
J:	Production line and process design
K:	Procurement
L:	Arrange and adjust the production line
M:	Preproduction manufacture
N:	Review and readjustment of production line and each process
O:	Full manufacturing release
	Tasks for Quality Assurance
A:	Establishment of quality standard
C,D:	Evaluate the prototype product by productivity, reliability, serviceability, and conformity to market requirement
J,L:	Assessment of manufacturing processes
H, I, M–O:	Check conformity with the quality standard
M:	Assessment of quality of product
	Tasks for Cost Management
A:	Determine projected cost
C:	Compare projected cost with estimated cost
E:	Estimation of cost for full manufacturing and comparison of the cost with projected cost
F–K:	Collect actual cost data and fill the gap between actual and projected cost

TABLE 18.6
Master Network

Task	Predecessor	Time (weeks)					Event Time[a]			
		Optimistic	Most Likely	Pessimistic	Estimate	Variance	ES	LS	EF	LF
A	—	6	8	10	8	0.44	0	0	8	8
B	A	2	2	4	2.3	0.11	8	8	10.3	10.3
C	B	4	5	7	5.2	0.25	10.3	10.3	15.5	15.5
D	C	5	7	8	6.8	0.25	15.5	15.5	22.3	22.3
E	D	2	3	5	3.2	0.25	22.3	22.3	25.5	25.5
F	E	3	5	7	5	0.44	25.5	25.5	30.5	30.5
G	F	3	4	5	4	0.11	30.5	30.5	34.5	34.5
H	G	3	6	7	5.7	0.44	34.5	34.5	40.2	40.2
I	G,H	4	5	8	5.3	0.44	40.2	40.2	45.5	45.5
J	I	2	5	6	4.7	0.44	45.5	46.6	50.2	51.3
K	I	4	6	7	5.8	0.25	45.5	45.5	51.3	51.3
L	J,K	2	4	6	4	0.44	51.3	51.3	55.3	55.3
M	L	1	3	5	3	0.44	55.3	55.3	58.3	58.3
N	M	3	4	5	4	0.11	58.3	58.3	62.3	62.3
O	N	1	3	6	3.2	0.69	62.3	62.3	65.5	65.5

[a]ES = early start, LS = late start, EF = early finish, LF = late finish.
Time = 65.5 weeks.
$\sigma^2 = 4.66$, $\sigma = 2.16$.

the master network. Therefore, in the example just used, the late finish in the master network is the basis from which the time estimate is deducted; for activity A_q + 8 weeks minus 2.1 weeks is late start after 5.9 weeks. Possibly the quality assurance activity of A_q could be started earlier and is then also completed earlier, or "stretched."

Using the PERT model for processing a system of probabilistic subnetworks is practically impossible for manual data processing. Doing so is only useful for rough and initial estimates. More modern computerized systems would have to be used. In PROJACS, various modes of interrelationship of activities in the same network or related master network can be considered. The assumption that the succeeding activity starts after the predecessors are completed is at times unrealistic. It might have to start earlier, later, or at the same time, with related activities in the same network or in the master network. Furthermore, planners are interested in actual dates, rather than cumulated times that start at zero.

There should be no doubt that modern computerized project planning and control systems facilitate more intensive and reliable preparation of production, and in that context also facilitate quality assurance. The development of Program Evaluation and Review Technique in conjunction with Total Quality Control (PERT/TQC) to the stage of total quality assurance must still be achieved.

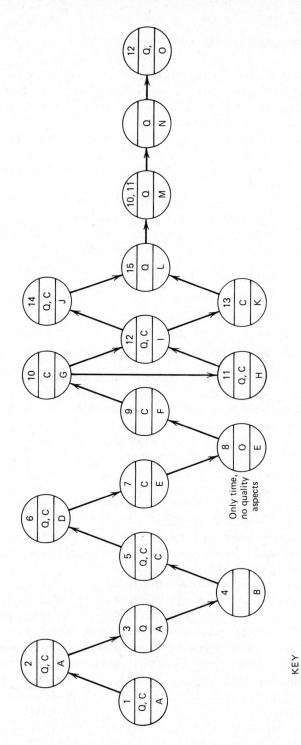

557

FIGURE 18.8 Network diagram for PERT/TQC example.

KEY

N = Number of activity

Q, C = Quality (Q) and cost (C) aspects

A = Code of activity

18.7   SOFTWARE QUALITY ASSURANCE PROGRAM

Software is a product which, like any other, must meet information requirements of people or machines in terms of quality, time, and place. With the rapidly increasing application and reliance on software, the need for special software quality standards and quality assurance programs has emerged as an important issue.

Repercussions from defective software can be disastrous. "The software cannot be allowed to fail. Consider, for example, nuclear power station control. Should a portion of software get into malfunction or inactivity, significant damages, casualties, or negative social consequences might happen."[3]

A software quality assurance program carries all the attributes described in this chapter. As a management system it reflects a disciplined approach to software development aimed at utmost defect prevention from the beginning. "In very general terms, the process of producing a programing product is comprised of the following five steps[4]:

1. Statement of requirements or objectives

2. Feasibility proof or prototype

3. Production of the final program

4. Testing

5. Service after release of the program to end users."

For defining general features of a software quality assurance program, generic quality program standards do apply. The exception to this is low risk software, for which simple end-item inspection standards suffice. All other software demands defect prevention procedures typically employed for high risk products.

Software quality assurance programs describe procedures and tests for the development process, starting with the specification of requirements. Standards supporting the production process are, for example, *MIL-STD-1679,* and the Canadian Standard *CSA Q 396.* Definition of software quality embraces numerous attributes, such as usability, reusability, testability, flexibility, portability, correctness, reliability, and efficiency. Software quality metrics measure each attribute by countable occurrences.[5]

Quality can also be measured in terms of program defects normalized by size over time. A quality index measures valid defects found by the end user over supported lines of code present in the field (in thousands), multiplied by years.[6]

The computer program can be divided into components and modules, each to be processed and tested separately. Network analysis methods offer controlled development and progress monitoring. The network for each project and subproject shows points of inspection and defect correction. These points can also be considered as major intermediate milestones and checkpoints for software integration and user clearance and approval. Inspection and reviews are performed by independent teams with broad representation. Formal test plans are prepared in accordance with procedures approved by supervisory management and customers. Test plans are to remove defects after general initial system structure, definition and delineation of components and modules, module logic, and module code completion. Finally, the software must be tested in the environment and under the conditions in which it will actually be used.

Attainment of perfect software quality requires usually many years of continuous reviews and formal auditing of the actual effectiveness.

18.8   KEY TERMS AND CONCEPTS

A **quality program,** also called a *quality assurance program,* describes organizational features and procedures designed to attain quality and quality assurance. The concept of a program also indicates the actual implementation of a *quality system* that is the designed interaction in an organization for goal achievement.

Quality program standards are published documents that authentically describe desired and required features of a quality program. *Generic standards* describes documents that are formulated in general terms so that the standard applies in all organizations. Program standards are augmented by guidelines that interpret the individual clauses and support compliance.

A **quality manual** documents the quality program of an organization and describes policies and procedures for quality assurance. Auditors use the quality manual in order to evaluate compliance with applicable quality program standards.

A **quality assurance procedure** is a formal and documented directive and guide to be followed in assurance activity and decision making. It differs from a *practice*, which merely expresses what is actually done and observed being done. A *policy* describes a general approach and sometimes an objective, and is specified for implementation by a procedure.

Procedure writing is carefully planned and controlled in order to assure actual application and attainment of purpose. Auditors, along with operators and supervisors, participate in the writing, in order to assure validity and effectiveness of the procedure and its adaptation to changed conditions.

Quality projects are special work packages in conjunction with the quality program, and with procedure design and control. A project has a more specific objective and more emphasis on a team approach than a program. Quality program design and maintenance can best be performed through a project approach.

Quality program implementation is a crucial phase in the life of a quality program, between the completion of the design and satisfactory attainment of actual program establishment (steady state).

Quality audits are the formal and planned evaluation of quality programs in all their phases and individual elements. Quality program standards and quality manuals provide the basis for auditing.

PERT/TQC stands for *program evaluation and review technique* applied in conjunction with total quality control. Quality oriented projects can apply this and other related methods. The method can be refined through computerized project management information systems and respective software.

A **software quality assurance program** is basically identical with generic quality assurance programs, but is described in separate standards with some special technical terminology.

18.9   CASE: ROXSHAR APPAREL CORPORATION

The quality objectives of the Roxshar Apparel Corporation are to provide customers with well-fitting, durable garments that are visually appealing and suitable for a desired end use at the lowest possible price. Elements included in these factors are fabric type, construction, and the degree or amount of manufacturing time needed before a delivery date.

These quality objectives are presently being met as well as is deemed possible without any formal quality assurance plan. The garment fit has been created through the use of customer specifications, and basically through trial and error. Construction and durability is created through the materials used in various garments, as well as the types of machines used in production. Various costs are controlled through mechanization that allows certain pieces to be sewn by automatic machines, rather than by individual sewing machine operators.

Existing quality policies are not recorded or documented. Basically, if a customer is dissatisfied with a product, the product is returned and either the customer's account is credited, or a new product is sent as a replacement. The amount of returns or the type of problems existing with the returns does not seem to get beyond the receiving department or the credit accounts. People in the design and engineering area learn of the problems only if they happen to read a return letter. This is due to a lack of knowledge in regard to quality control, and also due to the fact that there is not a person employed who has the authority to follow through on researching why the problems exist. In addition, no one has enough time to take away from their present job to ensure that a program begins. A person to specifically undertake this position is definitely needed.

Policies that *do* exist in an unwritten fashion are these. When fabric is received, the bolts are remeasured in an attempt to ensure the yardage as stated on the receiving documents. If there is a discrepancy the supplier is contacted. The fabric is also scanned to check for flaws and defects. If any are noticed the company owner is called and makes a decision whether to keep the goods and get some type of rebate from the supplier, or to send the entire shipment back. This decision is often based upon how the owner "feels," rather than on what would be best under present company circumstances, such as the need for the material, the current inventory, and other relevent factors.

If garments are found to be defective through the production stage they are repaired. This often includes having to recut portions of the garment which can itself result in many problems, including expensive lost production time, as well as the possibility that the color or dyelots may not be matchable, thus producing another second quality good. If the defects are found during shipping they are termed "seconds," and if they are not noticed, the defective garments are shipped to customers who, in turn, send them back.

The factory spends a fair amount of time and money recutting garment pieces to correct errors that should not happen and probably would not happen if a quality assurance plan existed. Instead of doing things right the first time, many goods are being produced and reproduced by trial and error. This is because there is not a well-documented system on how garments are to be made. People assume that others know what has to be done, but many times they do not.

There does not appear to be a documented quality plan anywhere in the company, except perhaps in the minds of various people, thus being only "documented" by what they *believe* is right, instead of what should be known to be right. There are, however, some specifications for garment seam construction, collar, pocket, and cuff construction, and button positioning, as well as specifications that state the seam size, finished garment measurements, and stitch length. These exist because of a quality program that was used a few years ago for western shirts, but now the factory is producing sportswear pants and outerwear, as well as shirts, so specifications need to be made for these products.

A quality program would be very beneficial, in that the system functions to follow would be kept within an adequate cost range so that the inspection and appraisal costs would not exceed the company's needs.

Questions for Discussion

1. List weaknesses and strengths of the company's quality assurance practice.

2. Formulate quality assurance objectives for the company and outline their implementation and expected approach.

18.10 DISCUSSION AND REVIEW QUESTIONS

(18–1) *There are many benefits of a quality assurance management program.* Explain.

(18–2) *A quality management program has fairly specific purposes and objectives.* Explain and discuss.

(18–3) *Any quality assurance management program has certain basic contents.* Explain.

(18–4) *Generic program standards have an important function.* Explain.

(18–5) *Quality program standards are often divided into substandards and have associated guides and handbooks.* Explain, and discuss further improvement of these standards as a management aid.

(18–6) *When establishing a quality assurance program in a company, certain principles and rules must be observed.* Explain and discuss.

(18–7) *A program design moves through several steps before it suits the unique conditions of a company.* Explain and discuss.

(18–8) *There are several reasons for the proper documentation of a company's quality assurance program.* Explain.

(18–9) *A quality assurance program is further complemented by long-term and short-term action programs directed towards quality improvements.* Explain and discuss.

(18–10) *As the quality assurance program includes various procedures of quality assurance, principles of effective procedure writing must be observed.* Explain.

(18–11) *Every company should have a quality manual.* Discuss.

(18–12) *Writing a quality manual should be performed in logical steps and arranged as a project.* Explain and discuss.

(18–13) *A quality manual is only a valid and realistic presentation after the program it describes is properly implemented.* Discuss, including the steps of implementation.

(18–14) *Program implementation can be enhanced through audits.* Explain.

(18–15) *The PERT model for project scheduling can be used for a combined production, quality, and cost plan.* Explain.

(18–16) *In addition to, or instead of PERT/TQC, any other computerized project planning system is applicable for design, implementation, and maintenance of quality assurance programs.* Discuss.

18.11 PROBLEMS

(18–1) Southend Meat Packing Ltd. has been asked by major customers for Wieners to have an assured average outgoing quality level of 2%. In order to meet this new requirement, you are assigned to establish necessary procedures of final inspection before shipment. The company does not have any formal quality assurance program, as it has formerly relied on normal government inspection.

a. Prepare a plan for meeting the customer's stipulation and write a memo to senior management outlining your recommendation for an action program.

b. As an AOQL of 2% requires respective acceptance sampling plans, write an inspection procedure, including all necessary records facilitating this inspection.

(18–2) For a company of your choice, assume you are to design a suitable quality assurance program.

a. Prepare a plan for the design of the program; the principle is to involve company staff and not outsiders, and to use project planning methods (PERT, etc.)

b. Determine aids for program design, such as applicable and suitable standards and guides, and explain their application for your task.

c. Write a memo to senior management recommending that the company proceed with your plan and giving justification for the plan.

(18–3) With reference to problem (18–2), assume management has approved your plan, but wants to know your approach for implementation. Prepare a project plan for implementing the quality assurance program.

(18–4) The owner of Harrislee Bakery Ltd. wants to qualify as a supplier of rolls to the fast food chain McPhee Hamburgers. McPhee auditors have asked for the quality manual that describes the quality assurance program. Harrislee does not have one as yet, although they have always stressed good quality of whatever they produce, and have a high quality reputation.

a. You are assigned to prepare this manual within one month with the help of company staff, including the owner. Outline your approach.

b. Prepare a project plan (PERT) for manual writing.

c. Prepare for the first meeting of the project team, including outlining tasks, goals, etcetera, so that team members can adequately perform their assignments. You are to chair the project team.

(18–5) Quality action programs as opposed to quality assurance programs have specific short-range goals to achieve. Project management principles and methods are therefore applicable and should be widely used.

a. Assuming that all projects have the ultimate goal of improving the quality of products and services, and that management of the project itself uses quality assurance principles for effective goal achievement, list for the project phases of planning, organizing, implementing, and control all points that must be considered for sound project performance.

b. List possible quality action projects for a company that is faced with typical quality assurance problems.

(18–6) Write a critique on one of the generic quality program standards from the viewpoint of management, and recommend improvements.

18.12 NOTES

1. Feigenbaum, D. S., "Return to Control," *Quality Progress*, May 1976, p. 20.

2. Consulting experience of one of the authors.

3. Matsumoto, Y. et al., "*SWB System: A Software Factory*," in *Software Engineering Environments*, H. Huenke (editor) North-Holland Publishing Company, New York, 1981, p. 307.

4. Remus, H., "Planning and Measuring Program Implementation," in *Software Engineering Environments*, op. cit.

5. Carpenter Jr., C. L. and Murine, G. E. "Measuring Software Product Quality," *Quality Progress*, May 1984, pp. 16–20.

6. Remus, H., op. cit.

18.13 SELECTED BIBLIOGRAPHY

American National Standards Institute and American Society for Quality Control (ANSI/ASQC), *Generic Guidelines for Quality Systems*, Standard Z1.15, 1980, America Society for Quality Control, Milwaukee.

Amren, S. C., "Installing a Complete QC System," *Quality Progress*, August 1974, p. 14.

Caplan, Frank, *The Quality System: A Sourcebook for Managers and Engineers*, Chilton Book Company, Radnor, Pa., 1980.

Ilminen, Gary R., "PERT Your Program," *Quality Progress*, March 1982, p. 12.

Juran, Joseph M. and Gryna Jr., Frank M., *Quality Planning and Analysis*, 2nd ed., McGraw-Hill Book Company, New York, 1980.

Juran, Joseph M., "Then and Now in Quality Control—Standardization and Quality," *Quality Progress*, February 1975, p. 4.

MacDonald, B. A., "List of Quality Standards, Specifications and Related Documents," *Quality Progress*, September 1976, p. 30.

Mellin, Barbara J., "Sheraton's Quality Improvement Program," *Quality Progress,* December 1977, p. 12.

Mills, Charles A., "Interactions of Industrial Integration and Standardization," *Quality Progress,* February 1979, p. 22.

Sayle, A. J., "Impact of New Technology on QA Standards," *Quality Assurance,* Vol. 9, No. 3, September 1983, pp. 67–9.

Troxell, Joseph R., "Standards for Quality Control in the Service Industry," *Quality Progress,* January 1979, p. 32.

19 CHAPTER

QUALITY AUDIT AND CORRECTIVE ACTIONS

Audits are a fairly frequent event in an enterprise. Depending on the circumstances, some consider an audit with mixed feelings, some as a welcomed checkup of the financial and operational conditions. All audits have in common that an authorized and qualified person examines documentations and performances against established requirements. For instance, an auditor employed by the national revenue department determines whether or not income tax declarations correspond with accounting records and are based on proper accounting principles. In a similar way, the quality auditor sent by a customer assesses compliance with standards for product or process quality and with standards for the quality control program effectiveness.

Designing and implementing a quality assurance program is best carried out as a special project with a project team in charge. By this time the program has been successfully launched and everyone is actively involved towards attaining quality goals individually and jointly with others. However, the program's actual effectiveness can always be improved, and not just maintained. Special projects of the type already established along with the program, can now be designed and carried out for examining the performance and actual quality assurance. Such projects are usually called *audits* with their purpose and effect aimed at inducing improvements.

People tend to ignore procedures and performance directives after a while, and other forces, in the company or elsewhere, continuously develop and detract from quality assurance approaches and goals. Audits can help management and staff counteract such *decay* of the quality assurance program.

Purposes and potential benefits from well-planned and conducted quality audits are such that they:

1. Assess compliance with quality control procedures and quality program standards.

2. Evaluate decision-making processes for validity.

3. Evaluate quality characteristics of products and processes with regard to specifications of customer or designer, by control of regular inspection.

4. Improve effectiveness of quality management programs.

5. Explore causes of defects, customer complaints, and other problems.

6. Attain formal certification of the quality management program.

7. Guide and motivate staff in quality matters, demonstrate management concern for quality, and create quality awareness.

8. Demonstrate management quality concern to suppliers and customers and gain protection against product liability claims.

9. Introduce some necessary formality and consistency in the small business quality program.

In order to attain the potential benefits from quality audits, the projects must be properly planned and executed. Otherwise, such audits can detract from quality assurance, rather than promoting it. For instance, audits conducted by insufficiently trained and unqualified persons often create resistance to constructively searching for weaknesses and improvement. A fear of being blamed for noncompliance and other shortcomings dominates and distracts from the actual purpose of such audits. Punitively oriented audits, frequently carried out by many customers, disturb quality assurance in normal operations and become a costly nuisance for all concerned, including the auditing agency itself.

In the following we shall outline the essentials of sound auditing of quality assurance program implementation and maintenance. Audit concepts and practices will be described with reference to the major standards and recent developments. A quality audit system consisting of interrelated audit programs and projects and auditing for *real effectiveness of quality assurance* will complete the brief survey of this modern management tool.

19.1 AUDIT CONCEPTS AND PRACTICES

A *quality audit* is defined in *standards* as *a systematic and independent examination of the effectiveness of quality assurance programs or its parts, including quality of products and processes*. The purpose of audits in inducing improvements is demonstrated by the diagram in Figure 19.1. The qualified auditor examines conclusive facts and figures as *evidence* for *satisfactory compliance* with required standards. However, mere adherence to product, process, and performance standards does not necessarily mean effective quality assurance. Assessment of the standard in view of changed conditions, needs, and actual adaptations in performances might lead to a review of the standard itself. The examination and judgment is to inform management about possible improvements in quality assurance and thus, to actively contribute to required and satisfactory quality of products and services.

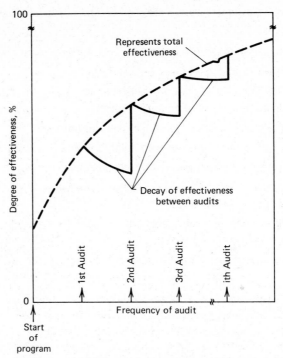

FIGURE 19.1 Illustration of the purpose and effect of audit. Source: Willborn, W. O., "A Generic Guideline for Quality Audit." *ASQC Quality Congress Transactions*, Boston, 1983, pp. 327–30. Copyright American Society for Quality Control Inc. Reprinted by permission.

Standards for an audit, as listed in Figure 19.2, describe basic similarities of financial, operational, and special technical audits, such as quality audits. This uniformity facilitates valuable coordination and cooperation in auditing, and increases the practical usefulness of this important management information and control device. Standards clearly delineate audits from surveys, consultations, surveillances, and inspections. Reference to and application of these standards avoids misunderstandings, and even conflicts, particularly in the case of

1. Canadian Standards Association, Standard on Quality Audits, Q 395.
2. ANSI/ASME, Qualification of Quality Assurance Program Audit Personnel for Nuclear Power Plants, N45.2.23.
3. ANSI/ASME, Requirements for Auditing of Quality Assurance Programs for Nuclear Power Plants, N45.2.12.
4. American Institute of Certified Public Accountants, Codification of Statements on Auditing Standards, 1976.
5. The Institute of Internal Auditors, Inc.; Standards for the Professional Practice of Internal Auditing, 1978.
6. Government of Canada, Standards for Internal Financial Audit, Office of the Comptroller General, 1978.
7. United States General Accounting Office; Standards for Audit of Governmental Organizations, Programs, Activities and Functions; The Comptroller General of the United States, 1972.

FIGURE 19.2 Examples of a few audit standards.

external audits by customers or third parties, such as program registration agencies. The typical content of a *generic quality audit standard* is listed in Figure 19.3.

An audit for attaining valid and reliable results must be properly planned by a qualified and competent auditor. As in any other project, the specific audit assignment and objective determines the details of approach, team compositions, resource requirements, time schedules, and reporting modes. The topical layout of a standard is shown in Figure 19.3. The schematic in Figure 19.4 shows the logical elements and phases in an audit.

The client, as the initiator of an audit, plays a key role, as the client is the main user of audit results. In any audit there are three different cooperating parties: the client (often senior management of the company), the auditor, and the audit subject. Benefits can and should also be derived by the audit subject when the audit is explicitly and effectively oriented to maintain and enhance quality assurance in a constructive manner. The quality of an audit depends largely on the proficiency of the auditor and the support given by senior management.

CONTENTS

FIGURE 19.3 A typical content of a generic quality audit standard. Source: "Quality Audit," CAN3-Q395-81, Reprinted by permission of the Canadian Standard Association, Rexdale, Ontario.

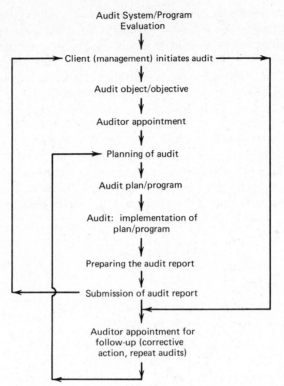

FIGURE 19.4   Elements and phases in an audit program.

Types of Audits

The features of a sound quality audit based on the applicable audit standard should not vary in principle among different kinds of audits. Mainly depending on the object and objective of an audit, the following major types of quality audits can be distinguished.

1. *Internal and external,* depending on who the auditor is; external audits are conducted by an outsider to the company, such as a major customer. Results of the audit are often shared by the company where the external audit was conducted.

2. *System, product, process, location, and organization audits;* product/process audits require respective technological expertise of the auditor.

3. *Baseline and regular audits,* with the first baseline audit usually being more comprehensive and intensive. Regular audits can be augmented by special ad hoc audits for reasons such as major defects, changes, and resource availabilities.

4. *Special and comprehensive audits;* special audits, such as a quality audit are restricted while comprehensive audits involve other objects such as accounting, operations, marketing, etcetera. Compliance with generic audit standards facilitates simultaneous and joint audits.

Principles of Sound Auditing

Some principles of sound auditing that must be observed in order to gain the intended results and benefits of quality assurance are:

1. The auditor should be qualified and independent.

2. The purpose and objective of the audit should be clarified and approved.

3. The audit should be planned and adequately prepared.

4. The persons in charge of activities to be audited should be properly informed before, during, and after the audit.

5. The audit plan and final report should be in writing.

6. The auditor should followup (reaudit) remedial action.

7. Appraisal against standards should be objective, factual, and whenever feasible, quantitative.

8. The audit must not unduly interfere with ongoing operations.

9. The frequency of audits should vary with actual needs; so should the intensity and extent of the audit.

10. Working papers and other documents of the audit should be kept in proper form and order.

11. Sampling for gathering evidence should be unbiased and reliable (sufficiently large samples).

Audit–Auditee Interface

One major problem and issue with auditing is the disturbing and costly multiplicity of audits conducted, mainly by different major customers. Standards, once applied, alleviate the impediment by systematizing and rationalizing audits to some extent. Third party auditing and program registration will bring further improvements with benefits for all concerned. Figure 19.5 describes such an audit system, which is by nature more independent and professionally sound than those audits that are performed by customers. Quality audits usually do not substitute for government inspection of compliance with legally imposed codes. Nevertheless, some cost-effective cooperation will evolve in the future via general integration in comprehensive auditing and program registration.

While audit standards reflect current auditing practice to a great extent, management and quality control specialists have created different quality auditing forms and approaches. The following is an audit plan for the quality program example described earlier (Chapter 15). Project concept and approach, as applied to the program design and implementation (compare Figure 14.7) is here further applied for the auditing and maintenance of the quality program. Such concise and systematic planning helps to ensure expected and required quality of quality audits. A listing of activities is followed by a graph showing interdependencies and time estimates. This data base, possibly with cost and resource data, allows the use of available computerized project information systems. The listing of activities for auditing applicable to the PERT Network of Figure 14.7 was given in Section 14.8.

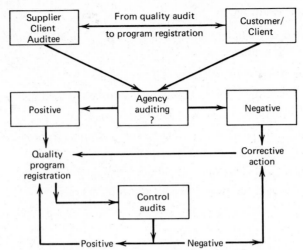

FIGURE 19.5 Audit–auditee interface: from quality audit to program registration.

19.2 AUDITING SYSTEM

Quality audits should be considered by management as an integral component of any quality assurance program, whether simple or complex. The following auditing system assumes a large-scope quality assurance program. However, with some modification it is applicable to simple and small programs. For instance, in small businesses certain auditing tasks can be assigned to operational departments or functions as *self-audits*.

The advanced and effective quality auditing system delineates and integrates various regular and ad hoc audit projects by means of long-term and annual audit programs. As shown in Figure 19.6, each audit project is divided into individual audit elements, each with a particular object for auditing and respective auditor assignment and working papers. In accordance with principles of objective oriented system management, the audit programs and projects have specific interrelated objects, objectives, and auditing bases and approaches. The objects and objectives are only determined once the need and potential benefit of quality audits is firmly established and approved.

Figure 19.7 demonstrates a systematic assessment of needs for a quality audit that is readily adoptable. Once all potential audit objects, such as the quality assurance function, products, processes, branches, departments, and so on are listed as potential audit objects, management can assess the actual need for auditing.

Audit System Flowcharts

The flowchart in Figure 19.8 shows further steps in preparing a long-term and annual audit program. These programs define required audit projects, rank them in priority, assign these to project leaders, and include a time schedule. Once senior management has approved the program and the associated budget, planning of the individual audit projects follows.

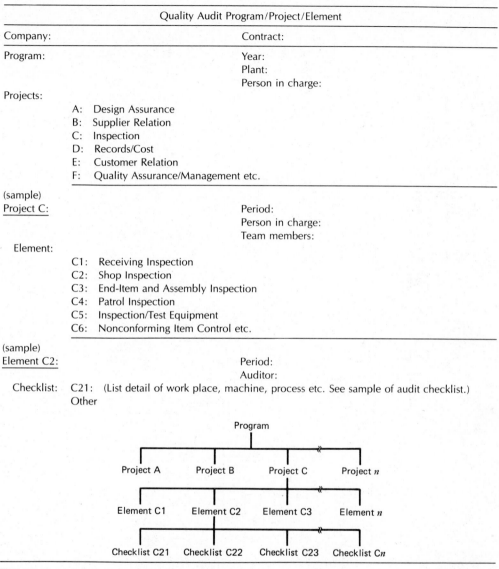

Quality Audit Program/Project/Element

Company: Contract:

Program: Year:
                                           Plant:
                                           Person in charge:

Projects:

A: Design Assurance
B: Supplier Relation
C: Inspection
D: Records/Cost
E: Customer Relation
F: Quality Assurance/Management etc.

(sample)
Project C: Period:
                                           Person in charge:
                                           Team members:

Element:

C1: Receiving Inspection
C2: Shop Inspection
C3: End-Item and Assembly Inspection
C4: Patrol Inspection
C5: Inspection/Test Equipment
C6: Nonconforming Item Control etc.

(sample)
Element C2:   Period:
                                           Auditor:

Checklist: C21: (List detail of work place, machine, process etc. See sample of audit checklist.)
            Other

FIGURE 19.6   Subdivision of audit program into projects and elements.

Audit Project Planning and Execution

An audit project has a particular audit object, objective, and respective demands on the size, technical expertise, and time and cost requirements of the project or audit team. Figure 19.9 is the flowchart for audit project planning and execution. The person in charge of the audit programing will initiate and advise in all the projects. The project team leader, once sufficient qualification is confirmed, supervises the individual auditor, who is a member of the audit team. The qualification of the auditors must also be assured carefully before audit elements

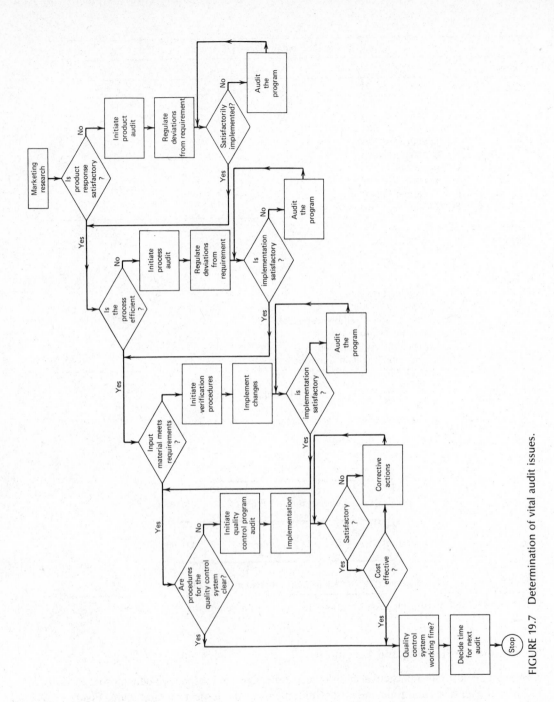

FIGURE 19.7 Determination of vital audit issues.

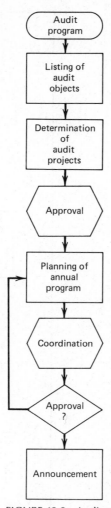

FIGURE 19.8   Audit system flowchart.

are assigned for further detailed planning. The project team will conduct several meetings in preparation for the audit execution.

Data on the audit object have to be compiled, questionnaires and other working papers must be prepared and compared, evidential material must be defined, and priorities must be set. This requires much supervisory support and coordination. The project leader will normally approve the working papers of the auditor as these become official documents later.

Cooperation with the audit subject during the project planning phase helps to ensure audit effectiveness. For instance, the scheduling must allow the subject to devote sufficient time and competent staff to the actual examination of quality assurance operations. Surprise audits should only be carried out when sufficient evidence is otherwise not attainable. During the audit the project leader will conduct various meetings with the audit subject and the responsible management, mainly in order to communicate requirements and inform personnel about major results.

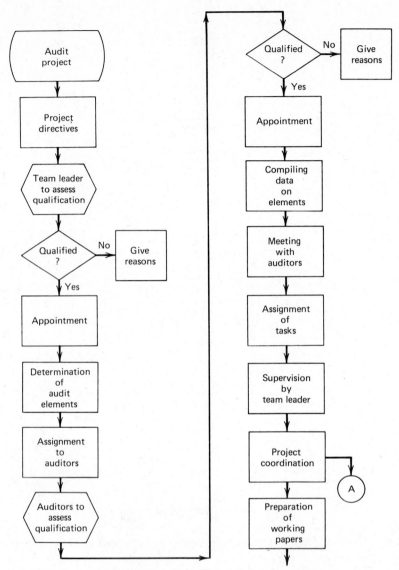

FIGURE 19.9   A comprehensive flow diagram for audit project.

Audit Elements

While the project leader acts as the supervisor and coordinator, individual auditors conduct the actual examination of audit elements for which they have prepared the working papers. Figure 19.10 shows the operational sequence of this core activity in any audit. Whenever the auditor has evidence of a significant deficiency in quality assurance and deviation from procedures and other performance standards, a special report on the facts will be formally made and forwarded to the project leader. Such reports, once properly reviewed and communicated, will be included in the final audit report.

After careful review of the facts with the audit subject, the project leader prepares and

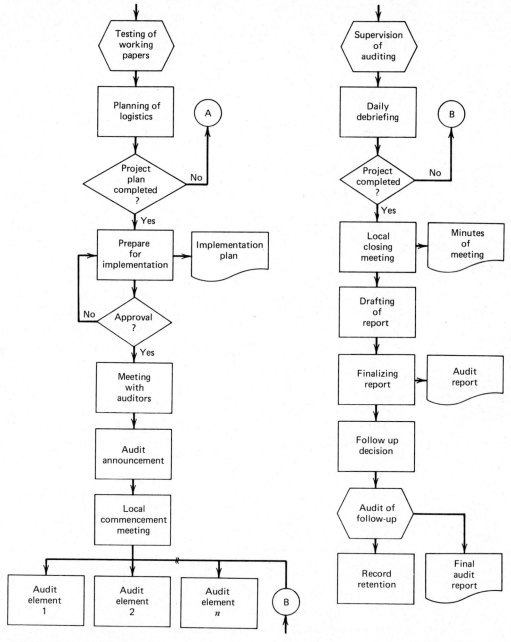

FIGURE 19.9   (*Continued*)

submits the audit report to the management staff that initiated the audit and employed the auditor. An auditor, not being a management consultant, cannot decide on corrective action. Such a decision must be left to the managers in charge. The report should not be unduly delayed, so that remedial actions can be taken and audited accordingly.

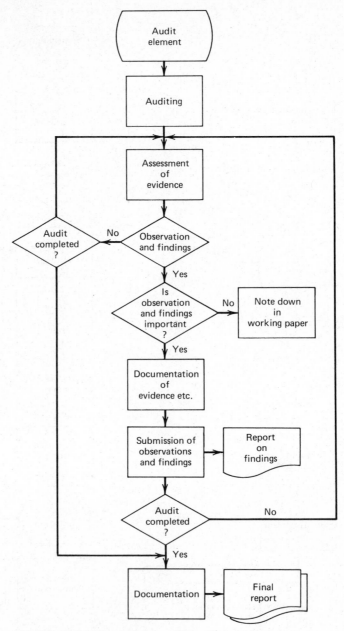

FIGURE 19.10 Flowchart for auditing of elements.

Audit Reports and Documents

The continuous cycle of audit programs, projects, and elements closes with a summarizing annual report on major results of quality audits. All audit documents and reports become part of the documentary evidence with which the company can prove the existence of a sound quality assurance program. Examples of audit documentary forms are shown in Figure 19.11.

The audit system consisting of interrelated programs, projects, and elements, facilitates

Company XYZ	Quality Audit Form No. 1	Year _____

<table>
<tr><td colspan="5">Checklist for Determining Audit Need</td></tr>
<tr><td colspan="5">Note: To be completed by January 15th of each year. Questions and check points can be added. A positive answer to a question indicates a need for an audit.</td></tr>
<tr><td>Question</td><td>Yes</td><td>No</td><td>N/A</td><td>Remarks</td></tr>
<tr><td>(sample questions)

Have major changes occurred in:

   product quality specification
   process quality specification
   vendors
   distributors
   production staff
   management
   other with quality impact (specify)?

Were complaints concerning quality received from:

   production staff
   distributors
   customers?

Were corrective action requirements required by auditors?</td><td></td><td></td><td></td><td></td></tr>
</table>

FIGURE 19.11   Examples of audit documentary forms.

direct involvement of senior management, quality assurance specialists, and supervisors and operators. The programing is usually assigned to the quality assurance manager, while project management is often delegated to competent and independent supervisors. An important requisite is that individual auditors must have adequate independence, technical expertise, and evidence of quality assurance. The proper interplay between senior management as audit client and user, project leaders, and individual auditors, as embodied in the comprehensive audit system, not only maintains quality assurance effectiveness but allows wide participation in and benefiting from quality auditing itself.

19.3   AUDITING FOR REAL PROGRAM EFFECTIVENESS

No quality assurance program nor auditing of it, can guarantee perfect quality of products and services. Such results are neither expected nor can they be delivered by the best qualified and supported auditor.

From mere checking of compliance with standards, a high degree of program and audit effectiveness cannot be concluded. A compliance audit, no matter how systematic and prescribed, cannot be fully satisfactory, when the real effectiveness of a quality assurance program needs to be examined and attained. Assuring quality of products and services does

Company XYZ	Quality Audit Form No. 2	Year_____

AUDIT PROGRAM

Program Supervisor:

Program Objectives:

Program Activities	Dates		Remarks
	Start	Completion	
A. Internal Audits			
Discuss and determine audit projects.			
Select and appoint project supervisors.			
Conduct supervisor orientation.			
Initiate and supervise projects.			
Review audit results.			
Conduct meetings			
Audit projects are:			
B. External Audits			

Results:

a) Corrective Action

Area:	Report No.	Completed/not completed
b) Others		

Comments for next audit program:

Signature:

FIGURE 19.11 (*Continued*)

not rest entirely on adherence to required and imposed performance standards. The standard itself can be deficient and invalid, because a review of the standard lags behind actual developments. New product designs, production processes, staff, technological advancement, impact of competition in the markets, and socioeconomic growth and fluctuations are all factors causing disturbances and instabilities for companies and their quality assurance efforts and commitments.

An adaptation of standards can be brought about through auditing for *real effectiveness*. Described conditions of instability being considerable uncertainty in preventing and detecting major defects, their causes, and the necessary remedial actions. Procedures and control programs can only be gradually perfected. The general ambivalence demands learning,

Company XYZ	Quality Audit Form No. 3	Year_____

AUDIT PROJECT

Project Supervisor:

Team Members:

Objective:

Project Activities	Dates		Remarks
	Start	Completion	
(List of activities except interviews)			

Audit Elements (interview)

Procedure Manual	Reference Standard	Checklist No.	Auditor	Auditee	Date	C/A No.

General Comments:

Corrective Action No.   Signature:

Company XYZ	Quality Audit Forms No. 4	Year_____

Checklist for Interview

Note: Auditor can add questions to the list but may not delete any.

Audit Element Description:

Procedure References: Quality Manual: Standard:

Auditor: Auditee:

Date: Place:

Reference/Aim	Question	Yes	No	N/A	Remarks
	(see content of published checklists in references; answers to questions should not be ranked; it is either go or no-go)				

General Comments:

Signature:

FIGURE 19.11   (*Continued*)

Company XYZ	Quality Audit Form No. 5	Year_____
	CORRECTIVE ACTION REPORT	

Project:

Checklist No.:

Auditor: Auditee:

Date of Audit:

Procedure Reference: Quality Manual: Standard:

Description of noncompliance and/or deficiency:

Date for completion of corrective action:

Auditor's signature: Auditee's signature:

Corrective action description:

Auditee's signature: Date:

Reaudit Result:

Auditor's signature: Date:

FIGURE 19.11   (*Continued*)

testing, and cautious decision making. In order to resolve the negative impact on quality through unpredicted and unpredictable forces and events, there must be much reliance on the staff's ability to react quickly and correctly. Insisting on adherence to directives for incorrect or ineffectual action can create considerable harm and costs. Overruling such directives through adaptation, or complementing them, is necessary and justified.

For auditors, judgement on such adaptations is difficult, as it is an essential component of the audit for real quality assurance program effectiveness. The new concept of a *total quality audit,* which is similar to that of *comprehensive auditing,* prescribes the examination of both adherence and adaptive controls, as they exist in any company and quality assurance program.

Figure 19.12 illustrates the interrelationship of adherence and adaptive controls. The relative scope of these two zones will vary depending on the degrees of operational and quality assurance stability. The overlapping shaded areas indicate partial adherence and adaptations.

When auditors explicitly or implicitly include examination of adaptive controls in their examination, they must observe utmost care in assessing and accepting these. Four main categories in such adaptive decisions and practices should be distinguished.[1]

1. *Practices complementing documented procedures;* that is, immediate remedial response when process breakdown occurs without a known cause, making emergency measures necessary.

2. *Practices without formal procedures;* as when a company does not have formal procedures for ensuring consideration of quality control during product design. However, the quality control manager participates in such design activities informally.

FIGURE 19.12   A total quality audit concept. Source: Sinha, Madhav N. and Willborn, Walter O., "Total Quality Audit—A New Approach," *ASQC Quality Congress Transaction*, Boston 198, pp. 197–201. Copyright American Society for Quality Control, Inc. Reprinted by permission.

3. *Ad hoc problem solving;* that is, when a customer informs the company about a serious failure and immediate rectification is necessary. A formal response might follow.

4. *Negative adaptations;* for example, a design change might create inconvenience to a supplier, who then reacts by accelerating production and deliveries in the traditional way before the changes have been formalized through documentation.

As has been pointed out, total quality audits for real quality assurance effectiveness extend beyond the mere compliance audit to all forms of adaptations and respective quality assurance practices. The auditor will have to observe and judge all four kinds of adaptive controls and report those with significant quality impact. Table 19.1 compares the factors responsible as

TABLE 19.1
Major Elements of Adherence and Adaptive Controls

Elements	Component of Control System	
	Adherence	Adaptation
Production condition	Stable	Unstable
Change of policy	Long term	Short term
Causes of defects	Largely known	Largely unknown
Control procedures	Specific	General
Information base	Historical data	Current data
Staff response	Select and follow procedure; adopt	Select and complement procedure; adopt
Management response	Stress adherence in short-term	Delegate authority and monitor responses
Communication	Formal	Formal and informal
Style of management control	Authoritative	Participative
Quality awareness factor	Constant	Positive and self-perpetuating
Management demands	Fixed, maintaining of status quo	Variable, progressive

Source: Sinha, M. N. and Willborn, W. O., "Auditing for Quality Control," *J. of Int. Auditor*, October 1983, p. 18. Reprinted by permission of the Institute of Internal Auditors, Florida.

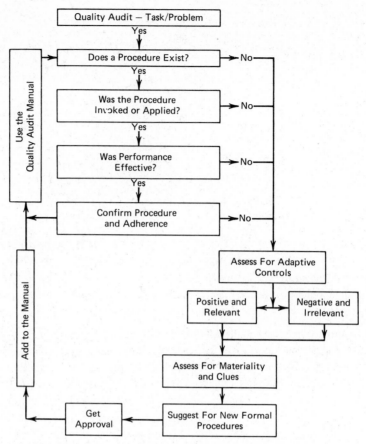

FIGURE 19.13 A simple scheme for auditing a specific control task. Source: Sinha, Madhav N. and Willborn, Walter O., "Auditing for Quality Control," *J. of Internal Auditors,* October 1983, p. 18. Reprinted by permission of the Institute of Internal Auditors, Florida.

major elements in *adherence* and *adaptive* types of controls. The scheme in Figure 19.13 aids in detecting and analyzing adaptive controls. Remedial actions, such as revising the procedures and performance standards, will have to rest with management in charge of the operation.

With senior management taking a more active role in quality assurance, quality auditing has gained in importance for communicating and inducing sound quality assurance programs. Auditing as a management tool is not the prerogative of large organizations.

19.4 SAMPLING FROM EVIDENCE

The audit by nature must sample from a more or less finite domain (population) of evidence and subsequent inference (judgement) on the characteristic or acceptability of the audit object (i.e., correctness of inspection records). Whenever the auditor cannot rely on mere subjective appraisal, on a sample that is not statistically predetermined, or on other *evidence,* the sampling methods provide *objective evidence.* An auditor can never claim to find deficiencies through examination and testing with absolute certainty, nor can he or she be expected to. However, the auditor is held responsible for choosing the proper auditing method. In many situations, the only adequate method is statistical sampling, as will be briefly explained in the following.

Working papers and, in particular, checklists for auditing elements, are not complete without sampling having been considered. In case sampling is applied, proper determination of the sample size, the decision criteria for acceptance, and the selection mode and testing procedure become important. The sample to be drawn from the population in conjunction with auditing is determined through the following decisions.

1. Definition and delineation of the finite populations and the characteristic(s) to be tested.

2. Selection of sampling method: random and judgement sampling.

3. The level of error (nonconformance, deficiency, etc.) acceptable.

4. The confidence limits for the inference or judgement.

It is implied that the statistical distribution and respective parameters are estimated with valid exploratory samples. Sampling in auditing is identical to acceptance sampling in statistical quality control. However, there is also a method called *discovery sampling.* The Quality Audit Standard Q-395 of the Canadian Standards Association suggests the use of the following formula for determining the sample size, n as;

$$n = \frac{\log (1 - C),}{\log P}$$

where P is the desired acceptable performance level, C is the desired probability of finding an error (or a defective item) if the performance level is less than desired (expressed as a binomial term), and *log* is the natural logarithm. For instance, if $C = 0.8$ and $P = 0.9$, the sample size is 15.3.

When an auditor finds two errors in a sample of 20, it can be taken from Table 19.2 with 95% confidence that the actual performance level is not worse than 73%. The same table can be used for determining the sample size for an acceptable number of errors. The lower the number of errors, the higher the performance level; the higher the performance level for any level of error, the higher the sample size. Acceptance performance level $= 1 -$ acceptable error rate.

19.5 QUALITY AUDITS AND COMPUTERS

Computers, singularly and in integrated networks, are used to perform audit tasks. Auditors must also examine computer-based information and production systems.

In computer-assisted auditing, computers are programmed to select, compile, analyze, trace, and transform data, to store data, and link communication lines of reporting, decision making, and operation. The auditor must design, adopt, direct, and manually complement the computerized information system in order to examine and report properly. Using the

TABLE 19.2
Audit Sampling

Confidence Level	Sample size to include one or more errors in the sample				
	Acceptable Performance Levels[a]				
	0.5	0.6	0.7	0.8	0.9
0.6	1	2	3	4	9
0.8	2	3	5	7	15
0.9	3	5	6	10	22
0.999	10	14	19	31	66

Sample Size	Number of Errors	Confidence Level		
		0.90	0.95	0.99
10	0	79	74	63
	1	68	61	50
	2	57	51	39
20	0	89	86	79
	1	83	79	71
	2	77	73	65
100	0	97	97	95
	1	96	95	93
	2	95	94	92

[a]The acceptable performance level is an arbitrary minimum acceptable level selected in order to determine a valid sampling plan. Audit findings show that the actual performance level is no worse than the percentage in the table for the confidence levels shown.

Source: Mills, Chas. A., "The Risk Factor in the Quality Audit," *ASQC Technical Conference Transaction, Atlanta,* 1980, p. 459. Copyright © American Society for Quality Control, Inc. Reprinted by permission.

computer for audit tasks is related to, but not identical with, auditing the quality aspects of *electronic data processing* (EDP) systems.

Quality auditors are particularly concerned about the quality of software and its impact on decisions, performances, and general quality assurance. Electronic data processing audits demand additional knowledge and expertise, although basic principles and methods of auditing a quality assurance program do also apply in these technologically advanced and complex environments.

The Impact of Computers on Auditing

The objective of quality audits is the proficient and independent examination of compliance and effectiveness. The computer is allowed to delegate basically simple audit tasks of data selection, compiling, et cetera, and thus to concentrate on the design of audit plans and activity, and so the interpretation of evidence and performance indicators. Scope, frequency, and intensity of audits increase with computer application. Subsequently the *real effectiveness* of the quality assurance program should also increase.

Greater audit productivity is attained through auditors being trained and able to program the computer, select and assess the software, and to incorporate computer assistance into the audit plan and implementation.

When auditing an EDP system, as a component of a quality assurance system, or auditing the quality assurance program for the EDP system itself, the audit tasks are technologically complex, simply because quality of software and software performance has many different attributes and metrics. Quality auditors must be able to communicate with the computer and computer system in programming languages, but also communicate with the various specialists as well.

The computer has brought a wealth of information to managers and operators, and has also automated processing systems. Information and decision flow have become complex and very sensitive to errors. Quality and reliability of such information expands in importance, the more impersonal communcations and decision making becomes. Consequently, the role of auditors for assuring the quality of such information and the system that produces it has also become more important.

While managers have much control information at their fingertips, they need the auditors to provide quality assurance and to verify the integrity of the software. As more and more personal computers are complementing the mainframe system, clerical and production systems become computerized, along with CAD/CAM, robots, material requirement planning (MRP), and so on. Auditors must examine these processes continuously around and through the computer. However, as the computer has changed work and work relationships, it remains prone to error in its service, because programers cannot predict all possible applications and errors and the computer has limited artificial intelligence.

Auditors using the computer in their regular audits of quality assurance programs, also develop the expertise for examining EDP systems and respective special software quality assurance programs. For auditors and the auditing profession, computers have brought considerable advancement, challenge, and opportunity.

Computer-assisted Auditing of Quality Assurance

Using the computer as auditor enhances proficiency, allows for cooperation with other audit teams and specialists, and helps to improve the quality and usefulness of audits. Auditors of quality assurance systems are normally familiar with software for quality assurance and control, which they can directly apply for their own auditing purposes. They are applicable because much of auditing is sampling, comparing, and analysis of deviations. Table 19.3 lists some examples of computer use in auditor training, audit programing, preparation and application of working papers, reporting, and audit administration.

With the greater application of computers in auditing comes far reaching standardization of auditing routines. This not only brings convenience but permits less-skilled and qualified staff to participate in auditing, possibly in the form of self-auditing. The audit function can thus be more broadly based in an organization, with the quality auditor acting predominantly as audit planner, facilitator, and team leader. Auditing, once assisted through user-oriented software, provides excellent learning experiences to supervisory staff and is a most meaningful involvement and participation in quality assurance. Through computer-assisted self-audits, weaknesses can be spotted and corrected earlier, regular audits can be prepared, and audit follow-up can be implemented and verified.

Auditing of EDP Quality Assurance

For the quality auditor, the EDP system and its respective software is mainly a component and subsystem of the overriding quality assurance system. The audit will normally be limited

TABLE 19.3
Computer Use in Audits

Audit Phase	Auditor Task	Type of Software Function
Auditor qualification	Basic training, familiarization with specific audit project	Learning programs in interactive mode; for example, PLATO programs retrieval of previous audit reports; scanning of stored system graphics and narratives; interactive mode.
Audit programing	Forecasting of needs for audit, Long-term scheduling, budgeting, resource planning, and deployment	Forecasting models, analysis of system records, changes, potential weaknesses and risks. Project management packages, PERT/CPM techniques, simulation, Gantt charts, spread sheets.
Audit planning	Review of QA program	Retrieval of system description and stored records, standardized computer stored checklists used in interactive mode.
	Working paper preparation, questionnaires	Retrieval of working papers and reports from previous audit, adoption of standardized working papers, storing final working papers for retrieval and interactive mode during audit implementation.
	Auditor and subject scheduling and communication	Gantt charts coordination, electronic mail.
Audit implementation	Working paper application	Remote retrieval, completion, storing, comparison with standards, interactive mode.
	Sampling	Generalized audit software packages, quality audit and quality control software.
	Follow-up, clues	Computer search, scanning, comparing, calculating, confirming records, retracting transactions and reports.
	Processing observations	Analyzing deviation and error, compiling facts from record and data files, preparing standardized report (work processing), electronic mail for communication and alert.
Audit report	Summarizing observations and recommendations	Compilation from audit record file, use of standardized report forms (word processing), telecommunication of draft and attaining of report feedbacks.
Audit follow-up	Audit of corrective action	Retrieval of audit record and report, use of audit working papers in interactive mode, updating audit record.
Audit record retention	Completion of audit, storing of audit records	Reviewing and finalizing of the audit file. Sorting audit plan, working papers, and audit report in master file.

to compliance with special quality assurance procedure and its relationship to quality assurance in the company. It is not a comprehensive audit of the EDP system.

As a preliminary step, the auditor reviews the existence of the EDP quality assurance procedure and special program. Such a formal control system is the prerequisite for substantive and detailed testing.

For this introductory and general review, familiar generic quality assurance program standards, together with special EDP and software quality assurance standards provide valuable guidance. Foldes[2] outlines the respective objects in an EDP system for each major attribute of a quality assurance program. The preliminary review should include the organization of the quality assurance group, as well as quality controls of hardware and software.

The EDP quality control function is usually in charge of the planning, implementation, and maintenance of hardware and software testing, safeguards, and error definition, detection, messages, processing, prevention, and communication. It also initiates and controls acquisitions, changes, and strategies, with regard to the important quality aspects. An auditor must expect that quality control procedures are well-described and integrated in a quality manual, so that compliance and effectiveness of operations and services can be examined.

Observations by a quality auditor in a *preaudit review* are, for example:

1. Major quality assurance attributes, practices, or procedures that deviate from applicable standards.

2. Functions of system analysts, programers, computer operators, data conversion operators, librarians, and quality controllers that are insufficiently defined and delineated.

3. Hardware configuration, operation, maintenance, operator training, and safeguards that deviate from standards of safety and integrity of data handling, and assurance of communication linkages if they are not providing proper information.

4. Whether software fails to meet its intended purpose and produces false, inadequate, inconsistent, or redundant messages and information. To see if changes are made in an uncontrolled fashion without proper controls and verifications.

Testing of EDP quality assurance can be performed around the computer or through the computer. For all practical purposes, software must be tested with direct and proficient computer use in the intended environment.

Audit techniques through the computer can use regular EDP system hardware and software, or separate systems especially designed for audits. The given conditions and audit objectives determine the appropriate testing method.

The auditor can use *test decks* on the regular system and then compare the predetermined and expected outcome with the actually obtained one. Controls in the software can also be tested with spiked errors in the test deck. A clear understanding of the data processing system and how it relates to audit objectives is essential for the development of useful test decks. The test plan must identify each test point and the controls and characteristics to be tested. When the auditor uses a client's test deck after having verified its acceptability, test decks should be used with utmost care, particularly when applied with real data and records. Tagging and tracing of real data and data processing can also interfere unduly with ongoing operations. For simple tests and audit trails, a manual approach around the computer might suffice. In order to avoid contamination of real records and communications in high-risk environments, the auditor will prefer specially written audit programs and will apply these separately.

TABLE 19.4
Table of Contents of CSA Preliminary Standard Q396·1–1982
for Software Quality Assurance

Contents

Page

Source: Preliminary CSA Standard Q396·1–1982, *Software Quality Assurance Program—Part I*, p. 3. Reprinted by permission of the Canadian Standard Association, Rexdale, Ontario.

The trade-off for parallel audit programs is the cost of designing and implementing special operational facilities and staff. The parallel simulation and testing can take other forms than simple partial testing programs in order to complete testing facilities and systems. In all cases, it is the output at critical and predetermined test points that must be compared with the output in the regular system.

The auditor might decide to adopt *generalized audit software packages.* These are relatively easy to use, less costly, and usually widely approved and standardized. The auditor, however, will have to carefully verify their applicability, and might have to complement, if not adjust, such packages. This special software for auditing can be used for auditing both quality assurance programs and the EDP system's quality and reliability. It does sampling, statistical analysis, report preparation, retracing, scanning, attaining confirmations, and word processing tasks.

In complex computer related tasks, auditors might require, and be given, special technological assistance. Auditing aids are in particular standards and guidelines. The *American Society for Quality Control* and the *EDP Auditors Association* provide assistance through guidebooks and articles. Many articles on computer auditing can also be found in journals such as *Internal Auditor, Computer World,* and *Business Week.* Textbooks on auditing do also include chapters on EDP auditing. Table 19.4 lists the topics covered in the Canadian Standard for *Software Quality Assurance Program, Part 1,* CSA Preliminary Standard Q396.1–1982. This layout is similar to the ANSI/IEEE 730–1981 standard for *Software Quality Assurance Plans.*

19.6 KEY TERMS AND CONCEPTS

A **quality audit** is a systematic and independent examination of the effectiveness of a quality assurance program or its parts, including quality products and processes. Effectiveness implies adequate compliance with applicable standards.

Types of audits, such as internal or external, system, product, or process audits are often distinguished in practice. The difference rests in the object of the audit and, to a lesser degree, in the auditor's qualifications and approach.

Principles of auditing as described in audit standards and guidelines, stress auditor qualification and independence, clarity in audit objective and scope, adequacy of audit plans and working papers, suitability of evidential material and analysis, validity and reliability of observations and reports, and initiation and follow-up of improvement in compliance with standards and general effectiveness.

Audit standards and guidelines are documents designed to assure the quality and reliability of audits.

They support auditor training and performance and allow objective evaluation of an audit.

Audit organization, programs, and plans are carriers and administrative or managerial frameworks for audits. The organization is a special audit function or unit, assigned with audit tasks. Its status in the organization must support auditing independence and effectiveness. Audit programs are prepared by the auditing function, showing coverage of needs for auditing in the short- and longer-term. Audit plans or specific projects cover specific audits, and are assigned to auditor or auditor teams with a team leader.

Audit working papers describe individual examination elements and, when properly planned, guide the auditor during the audit. Working papers are an aid and should not keep auditors from observing and tracing symptoms of noncompliance and ineffectiveness.

Quality assurance of audits is of particular importance to quality auditors, as they are to give leadership. Complying with standards, observing professional conduct, participating in developing effective audits and quality assurance programs, supporting senior management, and cooperating with other audit groups all indicate a qualified auditor and quality of auditing.

Auditing for real effectiveness requires the auditor not only to examine compliance with standards, but to also consider actual adaptation to need and opportunity for improved quality assurance and quality.

19.7 DISCUSSION AND REVIEW QUESTIONS

(19–1) Define and compare *quality audit* and *inspection.*

(19–2) *The purpose of an audit is not merely to evaluate compliance with a standard.* Explain.

(19–3) *There is no basic difference between a financial audit and a quality audit.* Explain and discuss.

(19–4) *An audit is a project with several project phases.* Explain and relate to project management.

(19–5) *Project management methods can be applied for effective auditing.* Explain.

(19–6) *Standards for quality auditing serve an important role.* Explain and discuss.

(19–7) *There are several different types of quality audits, each having a particular purpose, but the basic auditing approach does not vary.* Explain and discuss.

(19–8) *An auditor should follow certain principles for sound and effective auditing.* Explain.

(19–9) *An auditor must have certain qualifications.* Explain.

(19–10) *An appropriate audit system consists of a structure of audit programs, projects, and elements.* Explain.

(19–11) *The working papers of an auditor are an important auditing aid and documentation.* Explain.

(19–12) *Quality assurance concepts and methods, including statistical quality control techniques, can be applied to auditing.* Explain.

(19–13) *Auditing for real effectiveness of a quality assurance program considers adherence and adaptation.* Explain and discuss.

(19–14) *In principle, an audit should be announced.* Discuss.

(19–15) *An auditor should not express an opinion nor make recommendations.* Explain and discuss.

(19–16) *An auditor should not rely on the results of other audits performed by other auditors.* Discuss.

(19–17) *Financial and quality auditors can and should cooperate, and possibly conduct joint audits.* Discuss.

(19–18) *Self-audits can help to improve quality assurance.* Explain.

19.8 PROBLEMS

(19–1) The hospital in a small community has had several incidents of unacceptable supplies of food procured from local dealers. Fortunately, receiving inspection has so far detected unacceptable deliveries, but considerable uneasiness prevails. Therefore, the hospital board and administration have ordered the auditing of suppliers' quality assurance programs, if there are any. Only those suppliers with positive audit reports will be kept on the list of qualified suppliers.

   a. Prepare an audit plan outlining all steps to be taken.

   b. Prepare a checklist for evaluating the suppliers' inspection procedures and records.

   c. Prepare a working paper for auditing the suppliers' end-item inspection.

   d. Prepare a form for reporting on observations, such as noncompliances.

(19–2) You are the auditor in an audit team to audit nonconformance control procedures in a shoe manufacturing company.

    a. List normal deficiencies that you would expect as a probable occurence.

    b. Prepare the working paper for this audit task. Later, with more information, these working papers can be further detailed.

    c. Outline the audit approach to be taken.

(19–3) With reference to Problem 19–2, assume that you have found that a rejected lot of shoes is not properly tagged nor placed in the predesigned area for nonconformances.

    a. What procedure do you follow?

    b. If this nonconformance is disputed by the supervisor in charge, what do you do? What shouldn't you do?

(19–4) You are appointed as the audit team leader for auditing a precision tool company that is a supplier of the government's military branch. The audit is a joint project for auditors of the quality assurance program, product process, financial control, and personnel.

    a. Prepare a checklist for selecting members of your audit team. This list is to assure proper qualification of the auditors and of the team as a whole.

    b. Outline the essential auditing and operating principles to ensure that the team functions properly and attains the audit objectives effectively.

    c. Prepare an agenda for the first meeting of the audit team.

(19–5) Audit standards have become an important aid to auditors.

    a. List information you, as an auditor of a quality assurance program, would expect from such a standard, and give reasons.

    b. Prepare a plan for using such a standard to draft a general procedure for internal auditing that can later be incorporated in the quality manual.

(19–6) Problem for audit sampling:

    a. Assume that the performance level for an inspection station is set at 0.8, and the desired probability of finding an error if the performance level is less than desired is also 0.8. What should the sample size be?

    b. An auditor samples 10 pages in an inspection record and finds two errors. What is the percentage that shows that the actual performance level is no worse at a confidence level of 95%? What would this percentage express, deducted from 100%?

(19–7) Assume that you want to introduce an internal audit program in a company of your choice. Write a statement for management, describing the purpose, main features, and implementation procedure. Be systematic, brief, concise, and in compliance with rules of effective management reporting.

(19–8) Prepare questions you would like to discuss during the upcoming preaudit sessions.

19.9 NOTES

1. The preceding discussion is based on a paper by Sinha, M. N. and Willborn, W. O., "Auditing for Quality Control," *Journal of Internal Auditors*, October 1983, p. 18.

2. Foldes, Violet, "Software Quality Assurance in the Manufacturing Environment," *Forum 27*, Toronto Section, American Society for Quality Control, 1981.

19.10 SELECTED BIBLIOGRAPHY

Barstow, T. S. and Lambert, J. E., "Auditing A Quality Program With the Help of Statistics," *Quality Progress*, 1982, p. 30.

Cadmus, Bradford, *Operational Auditing Handbook*, Institute of Internal Auditors, Almonte Springs, Fl., 1960.

Canadian Standards Association, Standard Q 395. *Quality Audits*, Rexdale, Ontario, 1981.

Golomski, W. A., "Operational Auditing," *Quality Progress*, September 1978, p. 34.

Johnson, Marvin L., *Quality Assurance Program Evaluation,* Revised edition, Stockton Trade Press, Inc., West Covina, Ca., 1970.

Marguglio, B. W., "Quality Systems Audit," *Industrial Quality Control,* July 1963, p. 12.

Mills, Charles A., "In-Plant Quality Audit," *Quality Progress,* October 1976, p. 22.

Polimeni, R. S., "The Operational Audit of Quality Control," *Journal of Internal Auditor,* January–February 1975, p. 37.

Rusk, J. C., *Quality Assurance System Audit,* General Atomic Company, San Diego, 1977.

Sinha, Madhav N. and Willborn, Walter, "Auditing for Quality Control," *Journal of Internal Auditor,* October 1983, p. 18.

Wachniak, Ray, "Managerial Control—The Quality Audit," *Quality Progress,* November 1975, p. 22.

Willborn, Walter, *Compendium of Audit Standards,* American Society for Quality Control, Milwaukee, 1983.

Willborn, Walter, "Quality Audits in Support of Small Business," *Quality Progress,* April 1979, p. 34.

PART IV

AN EPILOGUE

Successful management of quality assurance will benefit the wider community as well as individual businesses. Quality of products and services contribute to the quality of life. The community at large supports the advancement of quality assurance in many ways. Research and education in the public sphere is more important than ever before, because of the growing importance of quality assurance. New knowledge, skills, and further research is needed.

The final chapter of this book is mainly a look at the new and upcoming challenges and opportunities in this field. The competition for quality and quality assurance will be as demanding as in the past. Commitment to quality for good workmanship and to proficient management and government will help to seize the opportunities in the future.

CHAPTER 20

PUBLIC CONCERN IN QUALITY

Consumer restiveness concerning quality, and product liability claims in cases of unsatisfactory quality exerts pressure on managers. A dissatisfied customer presents a challenge for improvement and an opportunity to take remedial action. Without such resistance to poor quality neither challenge nor opportunity would develop. Consumers would have to be content with minimal quality if they did not have the choices that competition offers them. Ultimately, demand for quality assurance depends on the actual economic condition in a society and the environment of the individual enterprises in the marketplace that make up the standard of living.

With increasing income and a higher standard of living, customer demands for quality and quality assurance have also risen. At the same time, however, the capacity to satisfy expected quality has not kept pace with such demands. Possibly people's demands as consumers are higher than what producers are prepared to produce. Unemployment and inflation are caused in part by poor quality of products and services and the inability to meet demands for more and higher value outputs.

Competition with quality in the marketplace determines the success or failure of enterprises, and thus determines their growth or departure. Governments are called upon by troubled companies and industries to intervene in the free competition of the marketplace. Public regulations and control of independent management increases continuously. Public welfare seems to justify this direct involvement only in industries where excessive cut-throat competition would undermine quality assurance. The Nuclear Regulatory Commission (NRC), Food and Drug Administration, and Air Safety Board are examples of agencies that have a great influence on quality assurance by producers and distributors alike. In essence, these agencies of public control augment individual consumer pressure for quality assurance. For the producers, governmental control and regulation generally also produce a positive impact, in that they induce, guide, and support proper quality assurance in companies.

Therefore, quality assurance, as undertaken within organizations has a close interrelationship with similar activities in the supracompany and thus in the general public domain.

The trend to establish integrated and comprehensive quality assurance programs within companies has gradually continued to embrace service industries and government itself. Government agencies are becoming more and more challenged by themselves and the public at large with regard to their own services and quality assurance.

In this last chapter, the challenge and opportunity to develop a total quality assurance program that extends beyond private enterprises and households into the public domain will be discussed. This chapter concludes the cycle from the quality of the product or service, to the *quality of life*.

20.1 QUALITY COMPETITION

Where there is a challenge, there is the status quo to be questioned and an improvement to be sought. Quality assurance has been the response to the challenge of more traditional, narrowly conceived quality control and inspection. Now it has evolved into more adequately integrated and rationalized quality assurance programs, involving all employees and business partners in an organization. This challenge arose through the dynamics of free enterprise and created opportunities for further advancement. Standards were prepared, specialists attained a broader concept and outlook, and managers become involved with their leadership. The *Japanese syndrome,* along with socioeconomic difficulties, induced in a wider quality awareness. Society at large has come to accept the related challenge and opportunities. The question is, what are the opportunities that can be seized and what actions in the community are needed. For free enterprise that ensures quality of life in society, a blend of competition and cooperation can best master all challenges and opportunities that lie ahead for quality and productivity improvements.

Adherence and Adaptation

In the short run, quality assurance is restricted to those products and services that the producer is able to produce and supply. Customers' quality demands and expectations are largely, but not entirely, specified. Producers, with their respective capacity for quality, eventually compete with others to satisfy this quality demand. The producer either predicts quality requirements through market research, or receives these quality specifications directly from the individual customer. In both cases, however, the producer must verify that the existing capacity to produce matches the given specifications.

Quality assurance means to adhere to the specification. This adherence to customers' quality demands and respective quality specification is assumed to be satisfactory and to attract more customers, to increase the market share, and to out-compete those who do not attain the same degree of adherence.

However, mere adherence and compliance to formal quality specification does not suffice for successful competition in the long run. Customers' demand and producers' capacity in quality terms are not static, and the producer therefore needs to review quality programs and strategy all the time. As shown in Figure 20.1, consumer demand and producer capacity can be thought of as divided into two parts: the actual capacity and the potential capacity. In the short-term perspective, quality assurance seeks to match demand with actual quality output. This is the goal in the immediate perspective of the individual consumer–producer interrelationship. The short-term orientation is similarly true for the consumer–producer relationship in the entire economy and community. In the long run, however, competition among producers is directed towards potential demand and supply, and thus towards seizing opportunities.

Potential Consumer Demands for Quality

Consumer demands and expectations for quality, individually and collectively, fluctuate either randomly, seasonally, or cyclically, according to general living standards and socio-economic conditions. Consumer quality expectations nevertheless escape the attention of producers to a certain degree because quantification and specification have an infinite improvement potential.

Figure 2.1 of Chapter 2 described consumer quality expectations in the light of quality management strategies. Desires for quality characteristics in products and services are practically innumerable among individuals and groups of consumers. All these tangible and intangible desires for quality cannot be satisfied because of many constraints, such as income, traditions, conventions, production capacity, and price to quality relationships. In a consumption oriented society, these desires for quality are closely related with the quality of life that is to be cultivated and created.

Desires are divided into those that are attainable and those that are not. Desires that are attainable in the mind of consumers become requirements for quality. These then, are the more specified expectations, because the consumer feels these requirements are in line with the price paid and the living conditions that can be afforded.

Quality assurance, as normally expected by consumers, extends beyond the boundary of specified and required quality into the unspecified zone. Moreover, in their drive to raise

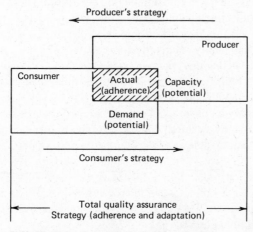

FIGURE 20.1 Organization chart of Temro.

their material quality of life, consumers tend to transgress in their requirements from the attainable and affordable into the unattainable. This is particularly true in cases of services provided by public institutions, such as health, education, social support services, and so on. The individually experienced and demanded quality of life is a psychological phenomenon and depends on the recognition of what the consumer perceives as attainable at a given time and place. There is no doubt that quality assurance will be more aggressively demanded and expected, as more and better products and services are felt to be attainable. The quality of a car that is attainable for the price in the marketplace is the quality benchmark; not what the consumer may specify, in more or less technical terms.

Potential Production Capacity for Quality

The capacity to produce and supply quality that matches a variable and dynamic potential demand is restricted through available human, technological, and natural resources. Still, the potential for quality improvement and for meeting consumer expectations is always larger than the actually exploited and utilized one. People, as producers, have inborn ingenuity, which can be further developed through education, practice, training, and many other challenges and opportunities.

Like consumers, individual producers and workers have numerous aspirations and desires. Work is a means for satisfying these in a creative manner. Not all such expectations of work can be fulfilled in a lifetime because of obvious limitations. However, as long as the chances of work satisfaction are fair and opportunities for proper workplaces and careers are given, then the producer, individually and collectively, is challenged and basically motivated to meet the performance standards.

Within the production section of an economy, and within the reality of the individual workplace, producers must remain in competition by utilizing advancing technology. All production systems must be led to determine the maximum production capacity for quality. Capacity to produce quality and quality assurance is different in human and nonhuman productivity elements, but is always driven and developed by internal systems that managers create as producers.

Quality of life, in individual cases, diminishes with loss of work or vocational opportunities. Technological progress has often been cited as having positive and negative impacts. Through adaptation, producers turn technological potential into enhanced capacity for quality. Figure 2.1(*b*) depicts the sphere of a producer who competes with other producers to satisfy complex consumer requirements for quality. Here, the production capacity for quality is basically determined by the feasible technology, the potential educational and skill level of employees, and natural endowments. Managers will seek to utilize this potential optimally in meeting all goals of the private or public enterprise.

Unfortunately, the actually employed and utilized human resources and technology remain less than the potential in many business enterprises. The higher the proportion of actually employed capacity to the employable and feasible, the more developed and advanced become the production system in companies, and in the economy at large. However, this ratio varies among individual production units. Some are investing more in training for quality than others; some employ modern technology and microelectronics more aggressively and develop this potential further than others. Some apply modern management concepts and methods while others cling to traditional approaches.

Feasible production resources are not always attainable and available, even if one is willing and able to carry the cost. The *technology transfer* is not sufficiently advanced. The educational and training institutions do not emphasize quality of performance and seldom develop knowledge and skills to the greatest extent possible. Both of these factors hinder improving production capacity for quality.

The higher the proportion of utilized technology and human skills within the potential, the greater is the competitive strength, the higher the productivity, and the more effective is quality assurance. The fact that they are *available,* however, does not mean that the resources are effectively *utilized* or *actually employed.* A person might have the knowledge, skill, energy, and talents needed, but might not really employ these for quality performance. Motivation for improved workmanship within the actually available potential is, again, a psychological phenomenon. Motivating a worker just to meet existing performance standards and specified quality benchmarks still might not fully utilize and challenge the individuals and groups in the production process. As with the consumer, where attainable and required quality was not fully specified and therefore hard to be assured, the situation is the same in the production section. Attainable and available human and technological capacity might not be actually utilized for attaining quality. Therefore, challenges and opportunities remain to marshall this already potentially available, but not employed, resource for quality improvement.

Quality of Work Life

Quality of work life, as perceived by workers, depends on the proportion of actually challenged and employed ability to create quality. A worker's perception of the quality of work and the workplace remains instrumental for managements' efforts to actually utilize and enhance the existing potential for quality and quality assurance.

Consumer and Producer Interface

In a society, consumers and producers are more or less the same people. *Quality of life* is measured by the extent to which consumer requirements are actually fulfilled, and also by the degree to which the job offers challenges and opportunities for improving the standard of performance. Quality assurance in this wider perspective of potentials, takes on a dynamic image. "Quality competition" is probably the most fitting descriptive term. Modern quality assurance must dynamically compete in order to meet all potential challenges. It must compete internally for reduction of wastes and externally for consumer satisfaction.

Static, that is, more conventional quality control or inspection, sets standards in view of what is actually explicitly specified, not what is potentially attainable, feasible, and desired. Human resistance to change and to adaptation to modern methods and technology in the workplace are among the major obstacles that all managers face. On the consumer side, insistence on better quality and adaptation to what is feasible can be expected to remain and enlarge indefinitely. Consequently, potential of quality of life can be lost through weak competition and adaptation by producers, if not by managerial incompetence.

Figure 20.1 indicates this dynamic quality assurance that adapts to the challenges and opportunities in the consumer and producer section. Under the producer strategy the thrust is to the left as illustrated by the figure, and vice versa for the consumer. Total quality assurance, however, always involves both consumer and producer. The consumer sets challenges by demanding more quality and moving requirements and quality specifications from the potential into the actual realm; the arrow points toward the production section and the area for actual quality assurance increases. Consumers would, for instance, demand electronic devices in a car that leading models already offer internationally at the same price. This would challenge competing producers to install such devices into their new designs. Such consumer pressure activates competition and forces the producer to rationalize current production.

Consumers might even insist on producers "getting it right the first time." In order to cope with such new consumer demands, the producer will have to utilize available modern technology, or will have to create new methods through respective research and development. Active consumer demand for better quality induces dealers and producers to transform potential and dormant production capacity into utilization. In Figure 20.1, the arrow for producers pointing towards the consumer section, enlarges the actual section for quality assurance. Forces from both consumer and producers thus create dynamic and competitive quality assurance.

Even when the consumer does not challenge producers with greater quality demands, competition with quality assurance *among producers* will lead them to utilize potential consumer demand aggressively. Free international trade, for example, activates the domestic potential for dynamic adaptation and for expanding the area for specified and standardized quality.

Current motivational approaches to quality improvement cling mainly and merely to compliance with existing standards. What is needed, however, is a systematic encouragement to apply individual and collective human creativity. Potential technology might destroy traditional workplaces, but competition requires changes and adaptation. Associated policies for retraining, job security, promotion, pension plans, and so on, can help to cushion individual hardships. However quality competition should not and does not need to be stifled through adherence to traditional and restricted production processes and respective static quality assurance.

20.2 DYNAMIC TOTAL QUALITY ASSURANCE BY COMPANIES AND COMMUNITIES

Many indicators of economic recession can be traced to current weaknesses in quality assurance and competition in the domestic and world-wide marketplace. Societies, like companies, compete with other societies for the material quality of life. Consumers, as members of a society, can help in the quality competition only by increasing their demand for quality, and thus exerting pressure on the producer. Realizing that their demand for quality is practically unsatiable and unspecifiable in totality, they can contribute to their quality of life by demanding only that which is attainable at the time. Excessive quality demands in relation to price does lead to inflation and unemployment, and thus to economic recession. Demands for more public services tend to exceed what can be afforded in terms of public funds.

It might appear utopian to expect of the individual consumer to curtail material expectations. The living standard is perceived as a traditional norm and remains as an unquestioned requirement by consumers. Austerity is hard to accept in best of times. Still, while realistic consumers, temporarily at least, expect less in terms of goods and services, competition must be kept alive, and thus the chances for betterment. Quality of life will never flourish under conditions of weak quality competition of the producers.

Consumer behavior with quality competition and quality assurance in mind, expands and also restricts actual demand for quality, in order to remain within the attainable sphere of quality. The same principle is shared by producers. Producers will have to move carefully in developing their competitive strength for better quality. Quality of life depends in large measure on the quality of the workplace. Government services and supports, as demanded by private enterprise, would have to be assessed for their actual necessity and effectiveness. Government, along with private business, plays an essential role in utilizing and developing new human and technological productive capacity. Cooperation within the community for strengthening quality assurance without protecting weak and incompetent companies can rationalize and expand quality assurance and strengthen opportunities for quality competition. Associations and institutes organized by producers can often more effectively enhance the general capacity for quality in enterprises than government can.

Finally, the individual worker realizes that not only the company surrounding the workplace is in competition with quality, but that this is also true for the community at large. Having been challenged to participate actively through a quality circle type of participative management, workers and their unions should recognize the challenges and opportunities in the quality competition of the marketplace.

Dynamic total quality assurance that aims at improving the *quality of life* in the community depends on the willingness and ability of all the people involved in the work process to adapt to the competition of companies and communities. The solving of quality problems cooperatively at the workplace might not be sufficient, if benefits are not ultimately shared by the community as well. On the other hand, the community can provide much help in overcoming other problems, such as crime, to improve the quality of life. All people have a stake in quality competition. What is needed is to acknowledge this fact with a more enlightened view and spirit. A community-based movement can and should participate in actual quality assurance and quality competition. Through this movement quality competition, and the eventual success in this competition, must be seen as a challenge and opportunity to enhance the quality of life.

20.3 A STRATEGY FOR EFFECTIVE QUALITY COMPETITION

A dynamic quality assurance movement that involves consumers and producers alike, also calls for political action in the public domain. What must be created is a deeper public awareness of the need to forcefully exploit potentials in demand for quality and productivity accordingly. People at large, as consumers and producers, at home and at the workplace, in public and private business, must be able to link improvement of quality of life with stronger energy and success in the marketplace. Assurance of quality of life is a noncontroversial goal. But in view of the desire for quality in products and services, and the limited resources to satisfy these, quality of life remains hard to achieve. This is particularly true during an economic recession. A strategy for dynamic total quality assurance is to be designed to meet the many challenges and opportunities that quality competition offers.

Specialists in quality assurance and managers in public and private business are called upon to develop and implement this strategy. With their background and expertise in setting goals and respective assurance standards, they are well-prepared to formulate the strategy and policy at the community level, as well as at the company level. Their company-based internal quality assurance programs have close relationships with suppliers, customers, and governments. They compete and cooperate at all times as individual organizations and also jointly, as members of the community. A breakthrough with more forceful quality competition must come under their leadership from companies and community. The strategy can act as a bond by creating cooperation without reducing competition and free enterprise in the community.

The following steps lead to formulation and implementation of a strategy of dynamic total quality assurance in the community.

1. Survey of the existing public and private quality assurance capacity; this survey compiles data on the state of quality control in leading industries in the local, regional, or national community.

2. Survey of existing programs that support quality assurance among the consumers and producers of the community; consumer protection agencies, government support programs, education courses, and so on, indicate the potential for developing strong competitive forces for the dynamic total quality assurance program of the community.

3. Formation of a community wide steering committee for quality improvement in general, and specifically, creation of a total quality assurance program.

4. Setting of community related goals for a total quality assurance program; these goals must be clarified and related to the overriding objective of quality of life improvement, and to subgoals for institutions, associations, and individual industries and enterprises.

5. Determination of individual measures leading to goal achievement together with associated educational, information, financial, support programs.

6. Assigning breakthrough projects in key areas in government services, industries, consumer protection branches, and educational institutions.

7. Conducting a campaign for information and orientation of the public, by means of conduct of meetings and demonstration of pilot projects establishing communication channels; the total quality assurance movement follows similar principles to that of the quality circles, only transposed to the community level.

8. Reviewing responses from the community and adjusting the community-wide program, and possibly already-assigned individual projects; communication must be sought and maintained in both directions, from program steering committee to public, and vice versa.

9. Steering committee reviews the results of the first stage of the program and presents a report for a comprehensive and longer-term program to government and other leaders, such as educational institutions and the local ASQC section.

10. A community total quality assurance agency is to be established. This agency itself must set an example in terms of quality assurance of its service in regard to service design, provision, and delivery. Special audits are to be added as a control and improvement device.

Concepts and methods for total quality assurance in which business and community cooperate and compete, separately and jointly, have still to be fully developed. Contributions for this process will have to come primarily from the social sciences and the disciplines of engineering and management.

Several research projects have been conducted for assessing the current practice and potential for quality cooperation at the local, regional, and national level. These are mainly empirical surveys. Successful competitors, such as the Japanese, have been studied in an amazingly large measure. Practical conclusions of the type "they can do it, why not we?" have been little more than hints as to the necessity of quality assurance. Doubtless, instituting quality circles or the like, does not make for a viable and comprehensive quality movement in companies, nor in communities. The challenge is to create more original and nation-wide suitable strategies and actions.

20.4 PRACTICE OF TOTAL QUALITY ASSURANCE IN COMMUNITIES

Community includes local, regional, and national groups of people. The examples given cannot and do not completely and comprehensively describe interactions between business, community, and governments, in matters of quality assurance. They do, however, show cooperative actions of isolated impact on promoting quality of products and services, of quality assurance, and on the ultimate quality of the community.

Community-wide Quality Survey or Audit

In the Canadian Province of Manitoba, a survey on the state of quality control in manufacturing led to the formation of a steering committee and action by governments, industry, associations, and educational institutions to develop quality assurance at the community level. The survey was called a quality audit because the data compilation was to be followed-up with an action program.[1]

A National Survey of Quality Control in Manufacturing Industry

John Roche conducted a national survey of quality control in manufacturing in Ireland in 1979 and 1980. He wrote: "It is hoped that the report and its recommendations will initiate a national awareness of the importance of quality in our daily lives and lead to recognition that quality has a key role in our economy." Almost 1000 responses (60% response rate) were received, covering over 65% of the total labor force in the manufacturing industry. "The responses show a wide variation in the approaches to product quality maintenance and/ or improvement. . . . There is, however, a general awareness of the importance of product quality and a base of considerable expertise. A variety of agencies and services have assisted in the development of this base."[2]

The above survey objectives were to determine the levels and use of quality assurance, to identify quality strength and deficiencies, to determine usage of quality assessment schemes and standards, and to suggest approaches for consolidating strengths and remedying weaknesses. Questionnaires were used as the main research instrument. Recommendations included the following.

1. Manufacturing firms with well-developed quality systems should make their expertise available to suppliers and educational institutions.

2. Trade unions should consider the statement that the survival of their jobs depends on industry being competitive in producing goods of as high a quality as our competitors and at competitive prices.

3. Unions should become more active in consumer protection.

4. Educational institutions should give recognition to quality control.

5. Governments should promote campaigns to increase general quality awareness.

6. Align industrial policies to quality improvement and encourage manufacturing firms to aim for higher standards of quality.

These and many other recommendations were well substantiated by the analysis of the survey data and responses.

A National Strategy for Quality

A 1978 consultative document by the Department of Prices and Consumer Protection of the United Kingdom examines the need for developing and coordinating a national strategy for quality.[3] "The prime responsibility for the quality of goods and services rests with industry and commerce. Government is considered to have a supporting role; such as influencing national attitudes and motivation, seeing that adequate training and educational facilities exist, encouraging cross-fertilization between different sectors of industry. . . ." The report further explains the need for an improvement strategy to be studied in detail, and remedial comprehensive actions are outlined. It is recommended that the strategy should be formulated by main groups among industry, commerce, consumers, and government, with priority given to promote greater quality awareness and international quality competition, to facilitate modern quality management systems, and "to ensure that national arrangements for quality assurance specifications, testing and certification of goods meet world market needs." The following quotation is a statement for a national quality campaign arising out of this strategy document.

> Quality of design, production and marketing wins markets. Only satisfied customers will repeat orders and make British goods and services their first choice.
> Responsibility for achieving competitive quality rests squarely with top management. But everyone involved in industry must recognise that quality is their business too.
> The National Quality Campaign makes quality a national objective. The Government is contributing to this by offering practical help to firms, developing training and encouraging certification. The Government will also promote quality through its own purchasing decisions.
> I believe the drive for quality will appeal to the good sense of the British people. Pride in quality must become the hallmark of British enterprise. I hope the National Quality Campaign will receive the most enthusiastic and wisespread support.[4]

Governments and Media Influence on Total Quality Assurance

In the United States, government departments and regulatory agencies have taken on some leadership in demanding and promoting quality assurance standards. The following quotations and selections from the public press illustrate this intense, and, on the whole, constructive interplay between business, government, the press, and individual citizen in matters of quality assurance.

> Upon becoming Chairman of the Nuclear Regulatory Commission I was struck by the disparity in quality assurance effectiveness among the utilities constructing nuclear power plants. Some were doing a good job. Others were not doing too well.[5]

As NRC Chairman I became acquainted, firsthand, with the stunning fact that—at least among some of the utilities building nuclear power plants—quality assurance programs were accorded a low priority, and at times quality assurance personnel were ignored, some even abused. . .[5]

With a fuller understanding of QA responsibility in the rank and file, and the commitment I have spoken of at the management level, I hope that we will begin to see QA-consciousness working as an integral part of nuclear project planning and execution.[5]

These evaluations include careful scrutiny of the effectiveness of quality assurance programs. Independent design reviews are also encouraged. Seminars for senior management, for impressing the importance of quality assurance, are being sponsored by the NRC and the industries. Other seminars are being directed to training of supervisory and operational personnel.

An important example of a regulatory agency quality assurance support program is the Food and Drug Administration (FDA) and its Good Manufacturing Practices (GMP). The Food, Drug, and Cosmetic Act provides the regulatory body with ''an authority to promulgate regulations for the efficient enforcement of this Act. . .'' (Section 701 [a]). the first GMP, issued in 1963, was extended in subsequent years from drugs to laboratory, clinical, and manufacturing practices in cosmetics and food. These various GMPs are guides that are closely related to the generic quality assurance program standards and further explain these for specific quality-sensitive products and industries. Good Manufacturing Practices seek documented procedures and programs, training, process control, auditing, and establishment of several other measures.

Dynamic total quality assurance thus brings further development of the GMPs by involving senior management and workers more directly within the company, and consumers within the community. The objective of assuring quality and safety remains unchanged, but GMPs have to become a more explicit and strategy-integrated competitive weapon that is visibly instrumental for quality of life improvement. This example stands for many other U.S. regulatory bodies.

If we really want quality in the American product, we've got to make some changes. The angry consumer is the best weapon we've got. When you get burnt by a product, let the maker know in no uncertain terms that you won't tolerate it. Industry has to change as well. The existing tooling, procedures, systems, methodologies, and policies are no longer adequate. We need new concepts, new ideas, and commitment to quality in substance, not in quality departments and inspection plans and sloganeering.[6]

20.5 ECONOMIC GROWTH THROUGH QUALITY IMPROVEMENT

In the maze of the arguments around high or low interest rates, inflation, and unemployment, we lose sight of the most simple facts and relationships. One such fact is that an all-around improvement of quality of performance, individually and collectively, would make us a stronger competitor in the domestic and international marketplace. The simple reason is that more satisfied customers build pride and confidence in the producer, reward good workmanship, and strengthen the enterprising spirit. Our international competitors, through effective enterprise and hard work, have built a quality image that casts a dark shadow over us as competitors.

If they can do it, why not we? We do not need to belabor the need for quality improvement. The conditions, such as meticulous production planning, absolute ''thinking in of quality,'' and cooperation in the workplace, are all part of the managerial system. However, looking for excuses for our failure to compete more strongly and blaming others is counterproductive. The lowering of interest rates by itself will not help, if this measure is not accompanied by a new enterprising spirit for improved quality of performance. On the contrary, low interest rates can reduce the pressure for moving in the direction of decisive quality improvement.

Asking for better performance in management, government, on the production floor, and in the classroom has become very difficult. Why? People normally perceive such a request as meaning more and harder work, and no one likes to be pushed. For instance, the demand for higher productivity for most of us means more output per unit. This understanding is only partly correct. Better quality of performance and high productivity also mean to work smarter, more creatively, under less pressure, and with more enjoyment and satisfaction. Quality of the workplace, of workmanship, and of management, does lead directly for all concerned, the consumer and the producer, to higher quality of life.

People have difficulty in understanding more abstract economic and political interrelationships, but still grasp the importance of quality. Concerted action for general quality improvement, as is now urgently needed, requires leadership and public awareness, as well as mutual trust. The following are measures to be taken by governments, companies, managers and households. Some of these measures can have immediate results in the direction for improved quality, others will only effect a gradual, long-term change. However, all these measures support general economic and social policy and help to strengthen free enterprise as the motor of economic revival.

1. Relate and explain economic policy matters with regard to incentives for enhanced productivity and quality. Any program within governments must be controlled for its own efficiency and effectiveness. New audit programs, such as those developed and adopted by the federal government, are probably the most forceful and useful instrument in this direction.

2. Our governments, as purchasers, should further develop their quality assurance branches, so that they regain the status that they enjoyed during the 1970s. Procurement branches can particularly help small and new enterprises in many ways in establishing quality control programs. The government standards for such programs and audit services have always been extremely useful as long as managerial responsibilities of the supplier remained undiminished.

3. The receivers of government services, by insisting on adequate quality, help to avoid waste and misappropriation of public funds and resources. Naturally, many services become redundant with changing economic and social developments. Here again, management and productivity audits can control, if the responsible branch does not take corrective action by itself.

4. Governments, as a major employer, provides leadership when it demonstrates quality of performance by planning and control. There is basically no reason for relative job security having a negative impact on quality of performance.

The above discussion on economic revival should also be directed to the reasons for current business failures, unemployment, and inflation. Revival can mainly come from new individual and collective endeavors for better performance, workmanship, and enterprise. This sounds very idealistic. Yet, at times when the real world is depressed and depressing, ideals can act as a mainspring for revival. Ideals express hopes, desires, and expectations. They *do* move people. Revival means "moving." The enterprising spirit of the employed and of employers in small and large companies is a powerful force for economic revival.

The thrust for free and vigorous enterprise weakens in periods of high employment and income. Low risks for business failure and unemployment tend to reduce pressures for pioneering enterprise. When difficulties and failure set in, causes are sought elsewhere, and not in our own back yard. The government, for instance, is blamed because of high interest rate policies, the unions are blamed for unrelenting high wage demands, the workers are blamed for low productivity, the management for incompetence, and so on.

The ideal would be that everyone would seek some cause for individual and collective failure in their own sphere. What can I do; what can my company do; my union, my organization, my institution? That seems to be a key question; doubtless, a very idealistic one. Raising this question, however, can revive the individual enterprising spirit and can build trust in the future.

The answer to the key question is as simple as the question itself: create quality improvement in the market and at the workplaces with enterprising vigor. An economic depression leads to searching for ways of revival. When the quality of our competitor's performance exceeds our own, we have practically no other alternative than to learn and gain strength. The economic revival will only come when quality of products and services improve.

20.6 NOTES

1. Willborn, W., "Quality Audit in Support of Small Business," *Quality Progress,* April 1979, p. 34.

2. Roche, John, report on the "National Survey of Quality Control in Manufacturing Industry," 1981, National Board of Science and Technology, Dublin, Ireland.

3. "A National Strategy for Quality—A Consultive Document," issued by the Department of Prices and Consumer Protection, Millbank, London, U.K., December 1978.

4. Butcher, John, "Britain's National Campaign for Quality," *Quality Progress,* November 1983, p. 39.

5. Palladino, Nunzio, "Quality Assurance: Watchword For the Nuclear Industry," *Quality Progress,* January 1983, p. 19.

6. *Business Week,* Readers Report, November 29, 1982, p. 7.

APPENDIXES

A SAMPLE JOB DESCRIPTION FOR A MANAGER OF QUALITY CONTROL

Company: Winnipeg Manufacturing, Winnipeg
Job Title: Quality Control Manager
Department: Quality Assurance
Section: Quality Assurance
Title of Immediate Supervisor: Vice-President, Operations

I. GENERAL INSTRUCTIONS

In filling out section III, it is requested that you describe what is done as well as how and why it is done, to achieve the assigned tasks. The order in which the tasks are to be listed can be either in logical sequence or by rank of importance. The approximate percentage of time spent on each task should be indicated in the right-hand margin.

In itemizing the duties and responsibilities, care should be taken to ensure that these four elements are adequately reported.

A. Kind and Variety of Work

The purpose of this subsection is to describe the essential duties of the job. It should show the different kinds of work involved and the extent to which they are similar or related.

B. Recommendations, Decisions

The points to cover here are the recommendations and decisions arising out of performing the duties. The statements should reveal the nature and extent of the responsibilities, such as in overall company's business, system, technical areas.

C. Contacts

This covers the contacts with other persons, both inside and outside the work unit that the duties and responsibilities entail. Here, the objectives of such contacts, and the frequency and level at which they are made is significantly important for mention.

D. Errors

This subsection has to do with the extent or degree of responsibility for losses to the organization that may result from mistakes in performance.

II. JOB SUMMARY

Briefly indicate purpose and scope of job. This would consist, for instance, of the following points. Under general supervision, audits all incoming and outgoing material for conformance to quality standards and customer's requirements. Releases satisfactory material. Recommends if nonconforming material is fit for use and can be given normal release. Recommends stop shipment when discrepancies are significant. Recommends special shipment release in certain cases. Occasionally audits packaging and shipping methods in other factory areas. (The written job summary package can be broken down in main areas of responsibilities).

III. DUTIES AND RESPONSIBILITIES

A. Kind and Variety of Work

1. Audits all incoming material evaluation systems and small orders just before shipment to ensure conformance to quality workmanship, standards, and sales order equipment lists. Ensures that the material is correctly identified by model, issue, and option. Completes periodic written audits. Releases only satisfactory material. Recommends whether nonconforming items may be released directly, or on special shipment release. Recommends appropriate corrective action, including scrap and rework.

2. Investigates causes of nonconformance and initiates corrective action. Provides follow-up to ensure that corrective action is effective.

3. Monitors and reports on effectiveness of quality assurance programs.

4. Assists in other quality assurance areas as required. Performs other duties as delegated by top management.

B. Recommendations, Decisions

Recommends, for example, rework or scrap of nonconforming material

1. Release of material where degree of nonconformance is significant, but material may be fit for use.

2. Stop shipment as appropriate.

3. Action to correct basic cause of nonconformance. Decides:

4. Release of conforming material, or with defects judged to be of minor significance.

5. Priorities of own work, through the subfunctional areas of quality engineering, process-control engineering, and quality information systems, within dictates of shipping schedule.

C. Contacts

1. Frequent: Factory personnel up to foreman level and top management, concerning schedules, corrective action, quality improvement projects, and related matters.

2. Regular: Production personnel concerning problems in their associated corrective action.

3. Occasional: Customer's representatives concerning acceptability of material, liability risks, safety, and other similar characteristics.

D. Errors

1. In judging quality: Recommending shipment or rework, or other actions that could result in added factory costs, shipping delays, or escaped notice of poor quality, resulting in customer dissatisfaction, increased warranty costs, or other problems.

2. In dealing with production areas: Matters that could result in disputes, and possibly affect morale and efficiency.

3. In dealing with customer representatives: The matters that could result in additional inspection and rework costs and possibly loss of company image.

IV. SUPERVISION EXERCISED

    A. Number supervised directly.

    B. Number supervised indirectly.

    C. Kind of supervision; indicate what planning, allocating and checking of work is involved, whether the incumbent trains and instructs staff, and what kind of recommendations he or she makes regarding hiring, promotions, discipline, and so on.

V. SUPERVISION RECEIVED

Indicate what guidelines are available; for example, instruction, reference material, established procedures memos from top management, and the degree to which independent action is required.

VI. QUALIFICATIONS REQUIRED

Indicate minimum qualifications and ones preferred for the job.

A. Formal Education or Equivalent

For example, graduation with a degree in metallurgical engineering from a recognized university with ASQC quality/reliability engineering certification.

B. Experience

Two years from graduation.

C. Special Skills

Ability in trouble-shooting, conducting meetings and seminars for training. Must be able to learn to use specialized measuring equipment.

D. Other

Knowledge of national and international quality standards and practices is an asset but not a requirement.

APPENDIX B

PROBABILITY TABLES

TABLE A1
Table of Individual Binomial Probability Terms

				p		
n	r	0.10	0.20	0.30	0.40	0.50
2	0	0.8100	0.6400	0.4900	0.3600	0.2500
2	1	0.1800	0.3200	0.4200	0.4800	0.5000
2	2	0.0100	0.0400	0.0900	0.1600	0.2500
3	0	0.7290	0.5120	0.3430	0.2160	0.1250
3	1	0.2430	0.3840	0.4410	0.4320	0.3750
3	2	0.0270	0.0960	0.1890	0.2880	0.3750
3	3	0.0010	0.0080	0.0270	0.0640	0.1250
4	0	0.6561	0.4096	0.2401	0.1296	0.0625
4	1	0.2916	0.4096	0.4116	0.3456	0.2500
4	2	0.0486	0.1536	0.2646	0.3456	0.3750
4	3	0.0036	0.0256	0.0756	0.1536	0.2500
4	4	0.0001	0.0016	0.0081	0.0254	0.0625
5	0	0.5905	0.3277	0.1681	0.0778	0.0312
5	1	0.3280	0.4096	0.3602	0.2592	0.1562
5	2	0.0729	0.2049	0.3087	0.3456	0.3125
5	3	0.0081	0.0512	0.1323	0.2304	0.3125
5	4	0.0004	0.0064	0.0284	0.0768	0.1562
5	5	0.0001	0.0003	0.0023	0.0102	0.0312

TABLE A1
Table of Individual Binomial Probability Terms (*Continued*)

n	r	p 0.10	0.20	0.30	0.40	0.50
6	0	0.5314	0.2621	0.1176	0.0467	0.0156
6	1	0.3543	0.3932	0.3025	0.1866	0.0938
6	2	0.0984	0.2458	0.3241	0.3110	0.2344
6	3	0.0146	0.0819	0.1852	0.2765	0.3125
6	4	0.0012	0.0154	0.0595	0.1382	0.2344
6	5	0.0001	0.0015	0.0102	0.0369	0.0938
6	6		0.0001	0.0009	0.0041	0.0156
7	0	0.4783	0.2097	0.0824	0.0280	0.0078
7	1	0.3720	0.3670	0.2471	0.1306	0.0547
7	2	0.1240	0.2753	0.3177	0.2613	0.1641
7	3	0.0230	0.1147	0.2269	0.2903	0.2734
7	4	0.0026	0.0287	0.0972	0.1935	0.2734
7	5	0.0002	0.0043	0.0250	0.0774	0.1641
7	6		0.0004	0.0036	0.0172	0.0547
7	7			0.0001	0.0017	0.0078
8	0	0.4305	0.1678	0.0576	0.0168	0.0039
8	1	0.3826	0.3355	0.1977	0.0896	0.0312
8	2	0.1488	0.2936	0.2965	0.2090	0.1094
8	3	0.0331	0.1468	0.2541	0.2787	0.2188
8	4	0.0046	0.0459	0.1361	0.2323	0.2734
8	5	0.0005	0.0092	0.0467	0.1239	0.2188
8	6		0.0011	0.0100	0.0413	0.1094
8	7		0.0001	0.0012	0.0079	0.0312
8	8			0.0001	0.0006	0.0039
9	0	0.3874	0.1342	0.0404	0.0101	0.0020
9	1	0.3874	0.3020	0.1556	0.0606	0.0176
9	2	0.1722	0.3020	0.2668	0.1612	0.0703
9	3	0.0446	0.1762	0.2668	0.2508	0.1641
9	4	0.0074	0.0661	0.1715	0.2508	0.2461
9	5	0.0008	0.0165	0.0735	0.1672	0.2461
9	6	0.0001	0.0028	0.0210	0.0743	0.1641
9	7		0.0003	0.0039	0.0212	0.0703
9	8			0.0004	0.0035	0.0176
9	9			0.0001	0.0004	0.0020
10	0	0.3487	0.1074	0.0282	0.0060	0.0010
10	1	0.3874	0.2684	0.1211	0.0403	0.0098
10	2	0.1937	0.3020	0.2335	0.1209	0.0439
10	3	0.0574	0.2013	0.2668	0.2150	0.1172
10	4	0.0112	0.0881	0.2001	0.2508	0.2051
10	5	0.0015	0.0264	0.1029	0.2007	0.2461

Source: adapted from "Tables of Binomial Probability Distribution," U.S. Department of Commerce, Applied Mathematics Series 6, issued January 1950.

TABLE A2
Table of Cumulative Binomial Probabilities

n	r	p				
		0.10	0.20	0.30	0.40	0.50
2	0	1.000	1.000	1.000	1.000	1.000
2	1	0.190	0.360	0.510	0.640	0.750
2	2	0.010	0.040	0.090	0.160	0.250
3	0	1.000	1.000	1.000	1.000	1.000
3	1	0.271	0.488	0.657	0.784	0.875
3	2	0.028	0.104	0.216	0.352	0.500
3	3	0.001	0.008	0.027	0.064	0.125
4	0	1.000	1.000	1.000	1.000	1.000
4	1	0.344	0.590	0.760	0.870	0.938
4	2	0.052	0.181	0.348	0.525	0.688
4	3	0.004	0.027	0.084	0.179	0.312
4	4		0.002	0.008	0.026	0.062
5	0	1.000	1.000	1.000	1.000	1.000
5	1	0.410	0.672	0.832	0.922	0.969
5	2	0.081	0.263	0.472	0.663	0.812
5	3	0.009	0.058	0.163	0.317	0.500
5	4		0.007	0.031	0.087	0.188
5	5			0.002	0.010	0.031
6	0	1.000	1.000	1.000	1.000	1.000
6	1	0.469	0.738	0.882	0.953	0.984
6	2	0.114	0.345	0.580	0.767	0.891
6	3	0.016	0.099	0.257	0.456	0.656
6	4	0.001	0.017	0.070	0.179	0.344
6	5		0.002	0.011	0.041	0.109
6	6			0.001	0.004	0.016
7	0	1.000	1.000	1.000	1.000	1.000
7	1	0.522	0.790	0.918	0.972	0.992
7	2	0.150	0.423	0.672	0.841	0.938
7	3	0.026	0.148	0.353	0.580	0.773
7	4	0.003	0.033	0.126	0.290	0.500
7	5		0.005	0.029	0.096	0.227
7	6			0.004	0.019	0.062
7	7				0.002	0.008
8	0	1.000	1.000	1.000	1.000	1.000
8	1	0.570	0.832	0.942	0.983	0.996
8	2	0.187	0.497	0.745	0.894	0.965
8	3	0.038	0.203	0.448	0.685	0.855
8	4	0.005	0.056	0.194	0.406	0.637
8	5		0.010	0.058	0.174	0.363
8	6		0.001	0.011	0.050	0.145
8	7			0.001	0.009	0.035
8	8				0.001	0.004

TABLE A2
Table of Cumulative Binomial Probabilities (*Continued*)

n	r	p				
		0.10	0.20	0.30	0.40	0.50
9	0	1.000	1.000	1.000	1.000	1.000
9	1	0.613	0.866	0.960	0.990	0.998
9	2	0.225	0.564	0.804	0.928	0.980
9	3	0.053	0.262	0.537	0.768	0.910
9	4	0.008	0.086	0.270	0.517	0.746
9	5	0.001	0.020	0.099	0.267	0.500
9	6		0.003	0.025	0.099	0.254
9	7			0.004	0.025	0.090
9	8				0.004	0.020
9	9					0.002
10	0	1.000	1.000	1.000	1.000	1.000
10	1	0.651	0.893	0.972	0.994	0.999
10	2	0.261	0.624	0.851	0.954	0.989
10	3	0.070	0.322	0.617	0.833	0.945
10	4	0.013	0.121	0.350	0.618	0.828
10	5	0.002	0.033	0.150	0.367	0.623

Source: adapted from Ordnance Corps Pamphlet ORDP 20-1, Tables of Cumulative Binomial Probabilities, Sept. 1952.

TABLE B
Poisson Probability Distribution

r \ np	0.10	0.20	0.30	0.40	0.50	0.60	0.70	0.80	0.90	1.00
0	.9048	.8187	.7408	.6703	.6065	.5488	.4966	.4493	.4066	.3679
1	.0905	.1637	.2222	.2681	.3033	.3293	.3476	.3595	.3659	.3679
2	.0045	.0164	.0333	.0536	.0758	.0988	.1217	.1438	.1647	.1839
3	.0002	.0011	.0033	.0072	.0126	.0198	.0284	.0383	.0494	.0613
4	.0000	.0001	.0003	.0007	.0016	.0030	.0050	.0077	.0111	.0153
5	.0000	.0000	.0000	.0001	.0002	.0004	.0007	.0012	.0020	.0031
6	.0000	.0000	.0000	.0000	.0000	.0000	.0001	.0002	.0003	.0005
7	.0000	.0000	.0000	.0000	.0000	.0000	.0000	.0000	.0000	.0001

r \ np	1.10	1.20	1.30	1.40	1.50	1.60	1.70	1.80	1.90	2.00
0	.3329	.3012	.2725	.2466	.2231	.2019	.1827	.1653	.1496	.1353
1	.3662	.3614	.3543	.3452	.3347	.3230	.3106	.2975	.2842	.2707
2	.2014	.2169	.2303	.2417	.2510	.2584	.2640	.2678	.2700	.2707
3	.0738	.0867	.0998	.1128	.1255	.1378	.1496	.1607	.1710	.1804
4	.0203	.0260	.0324	.0395	.0471	.0551	.0636	.0723	.0812	.0902
5	.0045	.0062	.0084	.0111	.0141	.0176	.0216	.0260	.0309	.0361
6	.0008	.0012	.0018	.0026	.0035	.0047	.0061	.0078	.0098	.0120
7	.0001	.0002	.0003	.0005	.0008	.0011	.0015	.0020	.0027	.0034
8	.0000	.0000	.0001	.0001	.0001	.0002	.0003	.0005	.0006	.0009
9	.0000	.0000	.0000	.0000	.0000	.0000	.0001	.0001	.0001	.0002

r \ np	2.10	2.20	2.30	2.40	2.50	2.60	2.70	2.80	2.90	3.00
0	.1225	.1108	.1003	.0907	.0821	.0743	.0672	.0606	.0550	.0498
1	.2572	.2438	.2306	.2177	.2052	.1931	.1815	.1703	.1596	.1494
2	.2700	.2681	.2652	.2613	.2565	.2510	.2450	.2384	.2314	.2240
3	.1890	.1966	.2033	.2090	.2138	.2176	.2205	.2225	.2237	.2240
4	.0992	.1082	.1169	.1254	.1336	.1414	.1488	.1557	.1622	.1680
5	.0417	.0476	.0538	.0602	.0668	.0735	.0804	.0872	.0940	.1008
6	.0146	.0174	.0206	.0241	.0278	.0319	.0362	.0407	.0455	.0504
7	.0044	.0055	.0068	.0083	.0099	.0118	.0139	.0163	.0188	.0216
8	.0011	.0015	.0019	.0025	.0031	.0038	.0047	.0057	.0068	.0081
9	.0003	.0004	.0005	.0007	.0009	.0011	.0014	.0018	.0022	.0027
10	.0001	.0001	.0001	.0002	.0002	.0003	.0004	.0005	.0006	.0008
11	.0000	.0000	.0000	.0000	.0000	.0001	.0001	.0001	.0002	.0002
12	.0000	.0000	.0000	.0000	.0000	.0000	.0000	.0000	.0000	.0001

r \ np	3.10	3.20	3.30	3.40	3.50	3.60	3.70	3.80	3.90	4.00
0	.0450	.0408	.0369	.0334	.0302	.0273	.0247	.0224	.0202	.0183
1	.1397	.1304	.1217	.1135	.1057	.0984	.0915	.0850	.0789	.0733

TABLE B
Poisson Probability Distribution (*Continued*)

r	np 3.10	3.20	3.30	3.40	3.50	3.60	3.70	3.80	3.90	4.00
2	.2165	.2067	.2008	.1929	.1850	.1771	.1692	.1615	.1539	.1465
3	.2237	.2226	.2209	.2186	.2158	.2125	.2087	.2046	.2001	.1954
4	.1733	.1781	.1823	.1858	.1888	.1912	.1931	.1944	.1951	.1954
5	.1075	.1140	.1203	.1264	.1322	.1377	.1429	.1477	.1522	.1563
6	.0555	.0608	.0662	.0716	.0771	.0826	.0881	.0936	.0989	.1042
7	.0246	.0278	.0312	.0348	.0385	.0425	.0466	.0508	.0551	.0595
8	.0095	.0111	.0129	.0148	.0169	.0191	.0215	.0241	.0269	.0298
9	.0033	.0040	.0047	.0056	.0066	.0076	.0089	.0102	.0116	.0132
10	.0010	.0013	.0016	.0019	.0023	.0028	.0033	.0039	.0045	.0053
11	.0003	.0004	.0005	.0006	.0007	.0009	.0011	.0013	.0016	.0019
12	.0001	.0001	.0001	.0002	.0002	.0003	.0003	.0004	.0005	.0006
13	.0000	.0000	.0000	.0000	.0001	.0001	.0001	.0001	.0002	.0002
14	.0000	.0000	.0000	.0000	.0000	.0000	.0000	.0000	.0000	.0001

r	np 4.10	4.20	4.30	4.40	4.50	4.60	4.70	4.80	4.90	5.00
0	.0166	.0150	.0136	.0123	.0111	.0101	.0091	.0082	.0074	.0067
1	.0679	.0630	.0583	.0540	.0500	.0462	.0427	.0395	.0365	.0337
2	.1393	.1323	.1254	.1188	.1125	.1063	.1005	.0948	.0894	.0842
3	.1904	.1852	.1798	.1743	.1687	.1631	.1574	.1517	.1460	.1404
4	.1951	.1944	.1933	.1917	.1898	.1875	.1849	.1820	.1789	.1755
5	.1600	.1633	.1662	.1687	.1708	.1725	.1738	.1747	.1753	.1755
6	.1093	.1143	.1191	.1237	.1281	.1323	.1362	.1398	.1432	.1462
7	.0640	.0686	.0732	.0778	.0824	.0869	.0914	.0959	.1002	.1044
8	.0328	.0360	.0393	.0428	.0463	.0500	.0537	.0575	.0614	.0653
9	.0150	.0168	.0188	.0209	.0232	.0255	.0281	.0307	.0334	.0363
10	.0061	.0071	.0081	.0092	.0104	.0118	.0132	.0147	.0164	.0181
11	.0023	.0027	.0032	.0037	.0043	.0049	.0056	.0064	.0073	.0082
12	.0008	.0009	.0011	.0013	.0016	.0019	.0022	.0026	.0030	.0034
13	.0002	.0003	.0004	.0005	.0006	.0007	.0008	.0009	.0011	.0013
14	.0001	.0001	.0001	.0001	.0002	.0002	.0003	.0003	.0004	.0005
15	.0000	.0000	.0000	.0000	.0001	.0001	.0001	.0001	.0001	.0002

r	np 5.10	5.20	5.30	5.40	5.50	5.60	5.70	5.80	5.90	6.00
0	.0061	.0055	.0050	.0045	.0041	.0037	.0033	.0030	.0027	.0025
1	.0311	.0287	.0265	.0244	.0225	.0207	.0191	.0176	.0162	.0149
2	.0793	.0746	.0701	.0659	.0618	.0580	.0544	.0509	.0477	.0446
3	.1348	.1293	.1239	.1185	.1133	.1082	.1033	.0985	.0938	.0892
4	.1719	.1681	.1641	.1600	.1558	.1515	.1472	.1428	.1383	.1339
5	.1753	.1748	.1740	.1728	.1714	.1697	.1678	.1656	.1632	.1606
6	.1490	.1515	.1537	.1555	.1571	.1584	.1594	.1601	.1605	.1606
7	.1086	.1125	.1163	.1200	.1234	.1267	.1298	.1326	.1353	.1377

TABLE B
Poisson Probability Distribution (*Continued*)

r \ np	5.10	5.20	5.30	5.40	5.50	5.60	5.70	5.80	5.90	6.00
8	.0692	.0731	.0771	.0810	.0849	.0887	.0925	.0962	.0998	1033
9	.0392	.0423	.0454	.0486	.0519	.0552	.0586	.0620	.0654	.0688
10	.0200	.0220	.0241	.0262	.0285	.0309	.0334	.0359	.0386	.0413
11	.0093	.0104	.0116	.0129	.0143	.0157	.0173	.0190	.0207	.0225
12	.0039	.0045	.0051	.0058	.0065	.0073	.0082	.0092	.0102	.0113
13	.0015	.0018	.0021	.0024	.0028	.0032	.0036	.0041	.0046	.0052
14	.0006	.0007	.0008	.0009	.0011	.0013	.0015	.0017	.0019	.0022
15	.0002	.0002	.0003	.0003	.0004	.0005	.0006	.0007	.0008	.0009
16	.0001	.0001	.0001	.0001	.0001	.0002	.0002	.0002	.0003	.0003
17	.0000	.0000	.0000	.0000	.0000	.0001	.0001	.0001	.0001	.0001

r \ np	6.10	6.20	6.30	6.40	6.50	6.60	6.70	6.80	6.90	7.00
0	.0022	.0020	.0018	.0017	.0015	.0014	.0012	.0011	.0010	.0009
1	.0137	.0126	.0116	.0106	.0098	.0090	.0082	.0076	.0070	.0064
2	.0417	.0390	.0364	.0340	.0318	.0296	.0276	.0258	.0240	.0223
3	.0848	.0806	.0765	.0726	.0688	.0652	.0617	.0584	.0552	.0521
4	.1294	.1249	.1205	.1161	.1118	.1076	.1034	.0992	.0952	.0912
5	.1579	.1549	.1519	.1487	.1454	.1420	.1385	.1349	.1314	.1277
6	.1605	.1601	.1595	.1586	.1575	.1562	.1546	.1529	.1511	.1490
7	.1399	.1418	.1435	.1450	.1462	.1472	.1480	.1486	.1489	.1490
8	.1066	.1099	.1130	.1160	.1188	.1215	.1240	.1263	.1284	.1304
9	.0723	.0757	.0791	.0825	.0858	.0891	.0923	.0954	.0985	.1014
10	.0441	.0469	.0498	.0528	.0558	.0588	.0618	.0649	.0679	.0710
11	.0244	.0265	.0285	.0307	.0330	.0353	.0377	.0401	.0426	.0452
12	.0124	.0137	.0150	.0164	.0179	.0194	.0210	.0227	.0245	.0263
13	.0058	.0065	.0073	.0081	.0089	.0099	.0108	.0119	.0130	.0142
14	.0025	.0029	.0033	.0037	.0041	.0046	.0052	.0058	.0064	.0071
15	.0010	.0012	.0014	.0016	.0018	.0020	.0023	.0026	.0029	.0033
16	.0004	.0005	.0005	.0006	.0007	.0008	.0010	.0011	.0013	.0014
17	.0001	.0002	.0002	.0002	.0003	.0003	.0004	.0004	.0005	.0006
18	.0000	.0001	.0001	.0001	.0001	.0001	.0001	.0002	.0002	.0002
19	.0000	.0000	.0000	.0000	.0000	.0000	.0001	.0001	.0001	.0001

r \ np	7.10	7.20	7.30	7.40	7.50	7.60	7.70	7.80	7.90	8.00
0	.0008	.0007	.0007	.0006	.0006	.0005	.0005	.0004	.0004	.0003
1	.0059	.0054	.0049	.0045	.0041	.0038	.0035	.0032	.0029	.0027
2	.0208	.0194	.0180	.0167	.0156	.0145	.0134	.0125	.0116	.0107
3	.0492	.0464	.0438	.0413	.0389	.0366	.0345	.0324	.0305	.0286
4	.0874	.0836	.0799	.0764	.0729	.0696	.0663	.0632	.0602	.0573
5	.1241	.1204	.1167	.1130	.1094	.1057	.1021	.0986	.0951	.0916
6	.1468	.1445	.1420	.1394	.1367	.1339	.1311	.1282	.1252	.1221

np r	7.10	7.20	7.30	7.40	7.50	7.60	7.70	7.80	7.90	8.00
7	.1489	.1486	.1481	.1474	.1465	.1454	.1442	.1428	.1413	.1396
8	.1321	.1337	.1351	.1363	.1373	.1381	.1388	.1392	.1395	.1396
9	.1042	.1070	.1096	.1121	.1144	.1167	.1187	.1207	.1224	.1241
10	.0740	.0770	.0800	.0829	.0858	.0887	.0914	.0941	.0967	.0993
11	.0478	.0504	.0531	.0558	.0585	.0613	.0640	.0667	.0695	.0722
12	.0283	.0303	.0323	.0344	.0366	.0388	.0411	.0434	.0457	.0481
13	.0154	.0168	.0181	.0196	.0211	.0227	.0243	.0260	.0278	.0296
14	.0078	.0086	.0095	.0104	.0113	.0123	.0134	.0145	.0157	.0169
15	.0037	.0041	.0046	.0051	.0057	.0062	.0069	.0075	.0083	.0090
16	.0016	.0019	.0021	.0024	.0026	.0030	.0033	.0037	.0041	.0045
17	.0007	.0008	.0009	.0010	.0012	.0013	.0015	.0017	.0019	.0021
18	.0003	.0003	.0004	.0004	.0005	.0006	.0006	.0007	.0008	.0009
19	.0001	.0001	.0001	.0002	.0002	.0002	.0003	.0003	.0003	.0004
20	.0000	.0000	.0001	.0001	.0001	.0001	.0001	.0001	.0001	.0002
21	.0000	.0000	.0000	.0000	.0000	.0000	.0000	.0000	.0001	.0001

np r	8.10	8.20	8.30	8.40	8.50	8.60	8.70	8.80	8.90	9.00
0	.0003	.0003	.0002	.0002	.0002	.0002	.0002	.0002	.0001	.0001
1	.0025	.0023	.0021	.0019	.0017	.0016	.0014	.0013	.0012	.0011
2	.0100	.0092	.0086	.0079	.0074	.0068	.0063	.0058	.0054	.0050
3	.0269	.0252	.0237	.0222	.0208	.0195	.0183	.0171	.0160	.0150
4	.0544	.0517	.0491	.0466	.0443	.0420	.0398	.0377	.0357	.0337
5	.0882	.0849	.0816	.0784	.0752	.0722	.0692	.0663	.0635	.0607
6	.1191	.1160	.1128	.1097	.1066	.1034	.1003	.0972	.0941	.0911
7	.1378	.1358	.1338	.1317	.1294	.1271	.1247	.1222	.1197	.1171
8	.1395	.1392	.1388	.1382	.1375	.1366	.1356	.1344	.1332	.1318
9	.1256	.1269	.1280	.1290	.1299	.1306	.1311	.1315	.1317	.1318
10	.1017	.1040	.1063	.1084	.1104	.1123	.1140	.1157	.1172	.1186
11	.0749	.0776	.0802	.0828	.0853	.0878	.0902	.0925	.0948	.0970
12	.0505	.0530	.0555	.0579	.0604	.0629	.0654	.0679	.0703	.0728
13	.0315	.0334	.0354	.0374	.0395	.0416	.0438	.0459	.0481	.0504
14	.0182	.0196	.0210	.0225	.0240	.0256	.0272	.0289	.0306	.0324
15	.0098	.0107	.0116	.0126	.0136	.0147	.0158	.0169	.0182	.0194
16	.0050	.0055	.0060	.0066	.0072	.0079	.0086	.0093	.0101	.0109
17	.0024	.0026	.0029	.0033	.0036	.0040	.0044	.0048	.0053	.0058
18	.0011	.0012	.0014	.0015	.0017	.0019	.0021	.0024	.0026	.0029
19	.0005	.0005	.0006	.0007	.0008	.0009	.0010	.0011	.0012	.0014
20	.0002	.0002	.0002	.0003	.0003	.0004	.0004	.0005	.0005	.0006
21	.0001	.0001	.0001	.0001	.0001	.0002	.0002	.0002	.0002	.0003
22	.0000	.0000	.0000	.0000	.0001	.0001	.0001	.0001	.0001	.0001

TABLE B
Poisson Probability Distribution (*Continued*)

r \ np	9.10	9.20	9.30	9.40	9.50	9.60	9.70	9.80	9.90	10.00
0	.0001	.0001	.0001	.0001	.0001	.0001	.0001	.0001	.0001	.0000
1	.0010	.0009	.0009	.0008	.0007	.0007	.0006	.0005	.0005	.0005
2	.0046	.0043	.0040	.0037	.0034	.0031	.0029	.0027	.0025	.0023
3	.0140	.0131	.0123	.0115	.0107	.0100	.0093	.0087	.0081	.0076
4	.0319	.0302	.0285	.0269	.0254	.0240	.0226	.0213	.0201	.0189
5	.0581	.0555	.0530	.0506	.0483	.0460	.0439	.0418	.0398	.0378
6	.0881	.0851	.0822	.0793	.0764	.0736	.0709	.0682	.0656	.0631
7	.1145	.1118	.1091	.1064	.1037	.1010	.0982	.0955	.0928	.0901
8	.1302	.1286	.1269	.1251	.1232	.1212	.1191	.1170	.1148	.1126
9	.1317	.1315	.1311	.1306	.1300	.1293	.1284	.1274	.1263	.1251
10	.1198	.1210	.1219	.1228	.1235	.1241	.1245	.1249	.1250	.1251
11	.0991	.1012	.1031	.1049	.1067	.1083	.1098	.1112	.1125	.1137
12	.0752	.0776	.0799	.0822	.0844	.0866	.0888	.0908	.0928	.0948
13	.0526	.0549	.0572	.0594	.0617	.0640	.0662	.0685	.0707	.0729
14	.0342	.0361	.0380	.0399	.0419	.0439	.0459	.0479	.0500	.0521
15	.0208	.0221	.0235	.0250	.0265	.0281	.0297	.0313	.0330	.0347
16	.0118	.0127	.0137	.0147	.0157	.0168	.0180	.0192	.0204	.0217
17	.0063	.0069	.0075	.0081	.0088	.0095	.0103	.0111	.0119	.0128
18	.0032	.0035	.0039	.0042	.0046	.0051	.0055	.0060	.0065	.0071
19	.0015	.0017	.0019	.0021	.0023	.0026	.0028	.0031	.0034	.0037
20	.0007	.0008	.0009	.0010	.0011	.0012	.0014	.0015	.0017	.0019
21	.0003	.0003	.0004	.0004	.0005	.0006	.0006	.0007	.0008	.0009
22	.0001	.0001	.0002	.0002	.0002	.0002	.0003	.0003	.0004	.0004
23	.0000	.0001	.0001	.0001	.0001	.0001	.0001	.0001	.0002	.0002
24	.0000	.0000	.0000	.0000	.0000	.0000	.0000	.0001	.0001	.0001

r \ np	11.	12.	13.	14.	15.	16.	17.	18.	19.	20.
0	.0000	.0000	.0000	.0000	.0000	.0000	.0000	.0000	.0000	.0000
1	.0002	.0001	.0000	.0000	.0000	.0000	.0000	.0000	.0000	.0000
2	.0010	.0004	.0002	.0001	.0000	.0000	.0000	.0000	.0000	.0000
3	.0037	.0018	.0008	.0004	.0002	.0001	.0000	.0000	.0000	.0000
4	.0102	.0053	.0027	.0013	.0006	.0003	.0001	.0001	.0000	.0000
5	.0224	.0127	.0070	.0037	.0019	.0010	.0005	.0002	.0001	.0001
6	.0411	.0255	.0152	.0087	.0048	.0026	.0014	.0007	.0004	.0002
7	.0646	.0437	.0281	.0174	.0104	.0060	.0034	.0019	.0010	.0005
8	.0888	.0655	.0457	.0304	.0194	.0120	.0072	.0042	.0024	.0013
9	.1085	.0874	.0661	.0473	.0324	.0213	.0135	.0083	.0050	.0029
10	.1194	.1048	.0859	.0663	.0486	.0341	.0230	.0150	.0095	.0058
11	.1194	.1144	.1015	.0844	.0663	.0496	.0355	.0245	.0164	.0106
12	.1094	.1144	.1099	.0984	.0829	.0661	.0504	.0368	.0259	.0176
13	.0926	.1056	.1099	.1060	.0956	.0814	.0658	.0509	.0378	.0271

TABLE B
Poisson Probability Distribution (*Continued*)

r \ np	11.	12.	13.	14.	15.	16.	17.	18.	19.	20.
14	.0728	.0905	.1021	.1060	.1024	.0930	.0800	.0655	.0514	.0387
15	.0534	.0724	.0885	.0989	.1024	.0992	.0906	.0786	.0650	.0516
16	.0367	.0543	.0719	.0866	.0960	.0992	.0963	.0884	.0772	.0646
17	.0237	.0383	.0550	.0713	.0847	.0934	.0963	.0936	.0863	.0760
18	.0145	.0256	.0397	.0554	.0706	.0830	.0909	.0936	.0911	.0844
19	.0084	.0161	.0272	.0409	.0557	.0699	.0814	.0887	.0911	.0888
20	.0046	.0097	.0177	.0286	.0418	.0559	.0692	.0798	.0866	.0888
21	.0024	.0055	.0109	.0191	.0299	.0426	.0560	.0684	.0783	.0846
22	.0012	.0030	.0065	.0121	.0204	.0310	.0433	.0560	.0676	.0769
23	.0006	.0016	.0037	.0074	.0133	.0216	.0320	.0438	.0559	.0669
24	.0003	.0008	.0020	.0043	.0083	.0144	.0226	.0329	.0442	.0557
25	.0001	.0004	.0010	.0024	.0050	.0092	.0154	.0237	.0336	.0446
26	.0000	.0002	.0005	.0013	.0029	.0057	.0101	.0164	.0246	.0343
27	.0000	.0001	.0002	.0007	.0016	.0034	.0063	.0109	.0173	.0254
28	.0000	.0000	.0001	.0003	.0009	.0019	.0038	.0070	.0117	.0181
29	.0000	.0000	.0001	.0002	.0004	.0011	.0023	.0044	.0077	.0125
30	.0000	.0000	.0000	.0001	.0002	.0006	.0013	.0026	.0049	.0083
31	.0000	.0000	.0000	.0000	.0001	.0003	.0007	.0015	.0030	.0054
32	.0000	.0000	.0000	.0000	.0001	.0001	.0004	.0009	.0018	.0034
33	.0000	.0000	.0000	.0000	.0000	.0001	.0002	.0005	.0010	.0020
34	.0000	.0000	.0000	.0000	.0000	.0000	.0001	.0002	.0006	.0012
35	.0000	.0000	.0000	.0000	.0000	.0000	.0000	.0001	.0003	.0007
36	.0000	.0000	.0000	.0000	.0000	.0000	.0000	.0001	.0002	.0004
37	.0000	.0000	.0000	.0000	.0000	.0000	.0000	.0000	.0001	.0002
38	.0000	.0000	.0000	.0000	.0000	.0000	.0000	.0000	.0000	.0001
39	.0000	.0000	.0000	.0000	.0000	.0000	.0000	.0000	.0000	.0001

Probability of r occurrences of event that has average number of occurrences equal to np. The Poisson distribution is $P_{(r)} = (np)^r e^{-np}/r!$. Example: if $r = 2$, $np = 0.40$, $P = 0.0536$.

Source: Adapted with permission from, Stockton, J. R. and Clark, C. T.; *Introduction to Business and Economic Statistics,* 5th ed; South-Western Publishing Co; Cincinnati, Ohio, copyright © 1975.

TABLE C
Cumulative Poisson Probabilities

$$P_{(x \le r/np)} = \sum_{x=0}^{x=r} \frac{(np)^x}{x!} \times e^{-np}$$

np \ r	0	1	2	3	4	5	6	7	8	9
0.02	.980	1.000								
0.04	.961	.999	1.000							
0.06	.942	.998	1.000							
0.08	.923	.997	1.000							
0.10	.905	.995	1.000							
0.15	.861	.990	.999	1.000						
0.20	.819	.982	.999	1.000						
0.25	.779	.974	.998	1.000						
0.30	.741	.963	.996	1.000						
0.35	.705	.951	.994	1.000						
0.40	.670	.938	.992	.999	1.000					
0.45	.638	.925	.989	.999	1.000					
0.50	.607	.910	.986	.998	1.000					
0.55	.577	.894	.982	.998	1.000					
0.60	.549	.878	.977	.997	1.000					
0.65	.522	.861	.972	.996	.999	1.000				
0.70	.497	.844	.966	.994	.999	1.000				
0.75	.472	.827	.959	.993	.999	1.000				
0.80	.449	.809	.953	.991	.999	1.000				
0.85	.427	.791	.945	.989	.998	1.000				
0.90	.407	.772	.937	.987	.998	1.000				
0.95	.387	.754	.929	.984	.997	1.000				
1.00	.368	.736	.920	.981	.996	.999	1.000			
1.1	.333	.699	.900	.974	.995	.999	1.000			
1.2	.301	.663	.879	.966	.992	.998	1.000			
1.3	.273	.627	.857	.957	.989	.998	1.000			
1.4	.247	.592	.833	.946	.986	.997	.999	1.000		
1.5	.223	.558	.809	.934	.981	.999	1.000			
1.6	.202	.525	.783	.921	.976	.994	.999	1.000		
1.7	.183	.493	.757	.907	.970	.992	.998	1.000		
1.8	.165	.463	.731	.891	.964	.990	.997	.999	1.000	
1.9	.150	.434	.704	.875	.956	.987	.997	.999	1.000	
2.0	.135	.406	.677	.857	.947	.983	.995	.999	1.000	

TABLE C
Cumulative Poisson Probabilities (*Continued*)

r np	0	1	2	3	4	5	6	7	8	9
2.2	.111	.355	.623	.819	.928	.975	.993	.998	1.000	
2.4	.091	.308	.570	.779	.904	.964	.988	.997	.999	1.000
2.6	.074	.267	.518	.736	.877	.951	.983	.995	.999	1.000
2.8	.061	.231	.469	.692	.848	.935	.976	.992	.998	.999
3.0	.050	.199	.423	.647	.815	.916	.966	.988	.996	.999
3.2	.041	.171	.380	.603	.781	.895	.955	.983	.994	.998
3.4	.033	.147	.340	.558	.744	.871	.942	.977	.992	.997
3.6	.027	.126	.303	.515	.706	.844	.927	.969	.988	.996
3.8	.022	.107	.269	.473	.668	.816	.909	.960	.984	.994
4.0	.018	.092	.238	.433	.629	.785	.889	.949	.979	.992
4.2	.015	.078	.210	.395	.590	.753	.867	.936	.972	.989
4.4	.012	.066	.185	.359	.551	.720	.844	.921	.964	.985
4.6	.010	.056	.163	.326	.513	.686	.818	.905	.955	.980
4.8	.008	.048	.143	.294	.476	.651	.791	.887	.944	.975
5.0	.007	.040	.125	.265	.440	.616	.762	.867	.932	.968
5.2	.006	.034	.109	.238	.406	.581	.732	.845	.918	.960
5.4	.005	.029	.095	.213	.373	.546	.702	.822	.903	.951
5.6	.004	.024	.082	.191	.342	.512	.670	.797	.886	.941
5.8	.003	.021	.072	.170	.313	.478	.638	.771	.867	.929
6.0	.021	.017	.062	.151	.285	.446	.606	.744	.847	.916

np	10	11	12	13	14	15	16
2.8	1.000						
3.0	1.000						
3.2	1.000						
3.4	.999	1.000					
3.6	.999	1.000					
3.8	.998	.999	1.000				
4.0	.997	.999	1.000				
4.2	.996	.999	1.000				
4.4	.994	.998	.999	1.000			
4.6	.992	.997	.999	1.000			
4.8	.990	.996	.999	1.000			
5.0	.986	.995	.998	.999	1.000		
5.2	.982	.993	.997	.999	1.000		
5.4	.977	.990	.996	.999	1.000		
5.6	.972	.988	.995	.998	.999	1.000	
5.8	.965	.984	.993	.997	.999	1.000	
6.0	.957	.980	.991	.996	.999	.999	1.000

r np	0	1	2	3	4	5	6	7	8	9
6.2	.002	.015	.054	.134	.259	.414	.574	.716	.826	.902

TABLE C
Cumulative Poisson Probabilities (*Continued*)

np \ r	0	1	2	3	4	5	6	7	8	9
6.4	.002	.012	.046	.119	.235	.384	.542	.687	.803	.886
6.6	.001	.010	.040	.105	.213	.355	.511	.658	.780	.869
6.8	.001	.009	.034	.093	.192	.327	.480	.628	.755	.850
7.0	.001	.007	.030	.082	.173	.301	.450	.599	.729	.830
7.2	.001	.006	.025	.072	.156	.276	.420	.569	.703	.810
7.4	.001	.005	.022	.063	.140	.253	.392	.539	.676	.788
7.6	.001	.004	.019	.055	.125	.231	.365	.510	.648	.765
7.8	.000	.004	.016	.048	.112	.210	.338	.481	.620	.741
8.0	.000	.003	.014	.042	.100	.191	.313	.453	.593	.717
8.5	.000	.002	.009	.030	.074	.150	.256	.386	.523	.653
9.0	.000	.001	.006	.021	.055	.116	.207	.324	.456	.587
9.5	.000	.001	.004	.015	.040	.089	.165	.269	.392	.522
10.0	.000	.000	.003	.010	.029	.067	.130	.220	.333	.458

	10	11	12	13	14	15	16	17	18	19
6.2	.949	.975	.989	.995	.998	.999	1.000			
6.4	.939	.969	.986	.994	.997	.999	1.000			
6.6	.927	.963	.982	.992	.997	.999	.999	1.000		
6.8	.915	.955	.978	.990	.996	.998	.999	1.000		
7.0	.901	.947	.973	.987	.994	.998	.999	1.000		
7.2	.887	.937	.967	.984	.993	.997	.999	.999	1.000	
7.4	.871	.926	.961	.980	.991	.996	.998	.999	1.000	
7.6	.854	.915	.954	.976	.989	.995	.998	.999	1.000	
7.8	.835	.902	.945	.971	.986	.993	.997	.999	1.000	
8.0	.816	.888	.936	.966	.983	.992	.996	.998	.999	1.000
8.5	.763	.849	.909	.949	.973	.986	.993	.997	.999	.999
9.0	.706	.803	.876	.926	.959	.978	.989	.995	.998	.999
9.5	.645	.752	.836	.898	.940	.967	.982	.991	.996	.998
10.0	.583	.697	.792	.864	.917	.951	.973	.986	.993	.997

	20	21	22
8.5	1.000		
9.0	1.000		
9.5	.999	1.000	
10.0	.998	.999	1.000

np \ r	0	1	2	3	4	5	6	7	8	9
10.5	.000	.000	.002	.007	.021	.050	.102	.179	.279	.397
11.0	.000	.000	.001	.005	.015	.038	.079	.143	.232	.341
11.5	.000	.000	.001	.003	.011	.028	.060	.114	.191	.289
12.0	.000	.000	.001	.002	.008	.020	.046	.090	.155	.242
12.5	.000	.000	.000	.002	.005	.015	.035	.070	.125	.201
13.0	.000	.000	.000	.001	.004	.011	.026	.054	.100	.166

TABLE C
Cumulative Poisson Probabilities (*Continued*)

np \ r	0	1	2	3	4	5	6	7	8	9
13.5	.000	.000	.000	.001	.003	.008	.019	.041	.079	.135
14.0	.000	.000	.000	.000	.002	.006	.014	.032	.062	.109
14.5	.000	.000	.000	.000	.001	.004	.010	.024	.048	.088
15.0	.000	.000	.000	.000	.001	.003	.008	.018	.037	.070

np \ r	10	11	12	13	14	15	16	17	18	19
10.5	.521	.639	.742	.825	.888	.932	.960	.978	.988	.994
11.0	.460	.579	.689	.781	.854	.907	.944	.968	.982	.991
11.5	.402	.520	.633	.733	.815	.878	.924	.954	.974	.986
12.0	.347	.462	.576	.682	.772	.844	.899	.937	.963	.979
12.5	.297	.406	.519	.628	.725	.806	.869	.916	.948	.969
13.0	.252	.353	.463	.573	.675	.764	.835	.890	.930	.957
13.5	.211	.304	.409	.518	.623	.718	.798	.861	.908	.942
14.0	.176	.260	.358	.464	.570	.669	.756	.827	.883	.923
14.5	.145	.220	.311	.413	.518	.619	.711	.790	.853	.901
15.0	.118	.185	.268	.363	.466	.568	.664	.749	.819	.875

np \ r	20	21	22	23	24	25	26	27	28	29
10.5	.997	.999	.999	1.000						
11.0	.995	.998	.999	1.000						
11.5	.992	.996	.998	.999	1.000					
12.0	.988	.994	.997	.999	.999	1.000				
12.5	.983	.991	.995	.998	.999	.999	1.000			
13.0	.975	.986	.992	.996	.998	.999	1.000			
13.5	.965	.980	.989	.994	.997	.998	.999	1.000		
14.0	.952	.971	.983	.991	.995	.997	.999	.999	1.000	
14.5	.936	.960	.976	.986	.992	.996	.998	.999	.999	1.000
15.0	.917	.947	.967	.981	.989	.994	.997	.998	.999	1.000

np \ r	4	5	6	7	8	9	10	11	12	13
16	.000	.001	.004	.010	.022	.043	.077	.127	.193	.275
17	.000	.001	.002	.005	.013	.026	.049	.085	.135	.201
18	.000	.000	.001	.003	.007	.015	.030	.055	.092	.143
19	.000	.000	.001	.002	.004	.009	.018	.035	.061	.098
20	.000	.000	.000	.001	.002	.005	.011	.021	.039	.066
21	.000	.000	.000	.000	.001	.003	.006	.013	.025	.043
22	.000	.000	.000	.000	.001	.002	.004	.008	.015	.028
23	.000	.000	.000	.000	.000	.001	.002	.004	.009	.017
24	.000	.000	.000	.000	.000	.000	.001	.003	.005	.011
25	.000	.000	.000	.000	.000	.000	.001	.001	.003	.006

np \ r	14	15	16	17	18	19	20	21	22	23
16	.368	.467	.566	.659	.742	.812	.868	.911	.942	.963
17	.281	.371	.468	.564	.655	.736	.805	.861	.905	.937

TABLE C
Cumulative Poisson Probabilities (*Continued*)

	14	15	16	17	18	19	20	21	22	23
18	.208	.287	.375	.469	.562	.651	.731	.799	.855	.899
19	.150	.215	.292	.378	.469	.561	.647	.725	.793	.849
20	.105	.157	.221	.297	.381	.470	.559	.644	.721	.787
21	.072	.111	.163	.227	.302	.384	.471	.558	.640	.716
22	.048	.077	.117	.169	.232	.306	.387	.472	.556	.637
23	.031	.052	.082	.123	.175	.238	.310	.389	.472	.555
24	.020	.034	.056	.087	.128	.180	.243	.314	.392	.473
25	.012	.022	.038	.060	.092	.134	.185	.247	.318	.394

	24	25	26	27	28	29	30	31	32	33
16	.978	.987	.993	.996	.998	.999	.999	1.000		
17	.959	.975	.985	.991	.995	.997	.999	.999	1.000	
18	.932	.955	.972	.983	.990	.994	.997	.998	.999	1.000
19	.893	.927	.951	.969	.980	.988	.993	.996	.998	.999
20	.843	.888	.922	.948	.966	.978	.987	.992	.995	.997
21	.782	.838	.883	.917	.944	.963	.976	.985	.991	.994
22	.712	.777	.832	.877	.913	.940	.959	.973	.983	.989
23	.635	.708	.772	.827	.873	.908	.936	.956	.971	.981
24	.554	.632	.704	.768	.823	.868	.904	.932	.953	.969
25	.473	.553	.629	.700	.763	.818	.863	.900	.929	.950

	34	35	36	37	38	39	40	41	42	43
19	.999	1.000								
20	.999	.999	1.000							
21	.997	.998	.999	.999	1.000					
22	.994	.996	.998	.999	.999	1.000				
23	.988	.993	.996	.997	.999	.999	1.000			
24	.979	.987	.992	.995	.997	.998	.999	.999	1.000	
25	.966	.978	.985	.991	.994	.997	.998	.999	.999	1.000

Source: Adapted with permission from Grant, E. L. and Leavenworth, R. S., *Statistical Quality Control,* 4th ed., McGraw-Hill Book Company, New York, copyright © 1972.

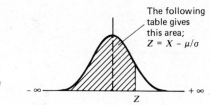

The following table gives this area; $Z = X - \mu/\sigma$

TABLE D
Cumulative Probabilities of the Normal Distribution (Areas under the Standardized Normal Curves from $-\infty$ to Z)[a]

Z	0.09	0.08	0.07	0.06	0.05	0.04	0.03	0.02	0.01	0.00
−3.0	0.00100	0.00104	0.00107	0.00111	0.00114	0.00118	0.00122	0.00126	0.00131	0.00135
−2.9	0.0014	0.0014	0.0015	0.0015	0.0016	0.0016	0.0017	0.0017	0.0018	0.0019
−2.8	0.0019	0.0020	0.0021	0.0021	0.0022	0.0023	0.0023	0.0024	0.0025	0.0026
−2.7	0.0026	0.0027	0.0028	0.0029	0.0030	0.0031	0.0032	0.0033	0.0034	0.0035
−2.6	0.0036	0.0037	0.0038	0.0039	0.0040	0.0041	0.0043	0.0044	0.0045	0.0047
−2.5	0.0048	0.0049	0.0051	0.0052	0.0054	0.0055	0.0057	0.0059	0.0060	0.0062
−2.4	0.0064	0.0066	0.0068	0.0069	0.0071	0.0073	0.0075	0.0078	0.0080	0.0082
−2.3	0.0084	0.0087	0.0089	0.0091	0.0094	0.0096	0.0099	0.0102	0.0104	0.0107
−2.2	0.0110	0.0113	0.0116	0.0119	0.0122	0.0125	0.0129	0.0132	0.0136	0.0139
−2.1	0.0143	0.0146	0.0150	0.0154	0.0158	0.0162	0.0166	0.0170	0.0174	0.0179
−2.0	0.0183	0.0188	0.0192	0.0197	0.0202	0.0207	0.0212	0.0217	0.0222	0.0228
−1.9	0.0233	0.0239	0.0244	0.0250	0.0256	0.0262	0.0268	0.0274	0.0281	0.0287
−1.8	0.0294	0.0301	0.0307	0.0314	0.0322	0.0329	0.0336	0.0344	0.0351	0.0359
−1.7	0.0367	0.0375	0.0384	0.0392	0.0401	0.0409	0.0418	0.0427	0.0436	0.0446
−1.6	0.0455	0.0465	0.0475	0.0485	0.0495	0.0505	0.0516	0.0526	0.0537	0.0548
−1.5	0.0559	0.0571	0.0582	0.0594	0.0606	0.0618	0.0630	0.0643	0.0655	0.0668
−1.4	0.0681	0.0694	0.0708	0.0721	0.0735	0.0749	0.0764	0.0778	0.0793	0.0808
−1.3	0.0823	0.0838	0.0853	0.0869	0.0885	0.0901	0.0918	0.0934	0.0951	0.0968
−1.2	0.0985	0.1003	0.1020	0.1038	0.1057	0.1075	0.1093	0.1112	0.1131	0.1151
−1.1	0.1170	0.1190	0.1210	0.1230	0.1251	0.1271	0.1292	0.1314	0.1335	0.1357
−1.0	0.1379	0.1401	0.1423	0.1446	0.1469	0.1492	0.1515	0.1539	0.1562	0.1587
−0.9	0.1611	0.1635	0.1660	0.1685	0.1711	0.1736	0.1762	0.1788	0.1814	0.1841
−0.8	0.1867	0.1894	0.1922	0.1949	0.1977	0.2005	0.2033	0.2061	0.2090	0.2119
−0.7	0.2148	0.2177	0.2207	0.2236	0.2266	0.2297	0.2327	0.2358	0.2389	0.2420
−0.6	0.2451	0.2483	0.2514	0.2546	0.2578	0.2611	0.2643	0.2676	0.2709	0.2743
−0.5	0.2776	0.2810	0.2843	0.2877	0.2912	0.2946	0.2981	0.3015	0.3050	0.3085
−0.4	0.3121	0.3156	0.3192	0.3228	0.3264	0.3300	0.3336	0.3372	0.3409	0.3446
−0.3	0.3483	0.3520	0.3557	0.3594	0.3632	0.3669	0.3707	0.3745	0.3783	0.3821
−0.2	0.3859	0.3897	0.3936	0.3974	0.4013	0.4052	0.4090	0.4129	0.4168	0.4207
−0.1	0.4247	0.4286	0.4325	0.4364	0.4404	0.4443	0.4483	0.4562	0.4562	0.4602
−0.0	0.4641	0.4681	0.4721	0.4761	0.4801	0.4840	0.4880	0.4920	0.4960	0.5000

TABLE D
Cumulative Probabilities of the Normal Distribution (Areas under the Standardized Normal Curves from $-\infty$ to Z)[a] (*Continued*)

Z	0.09	0.08	0.07	0.06	0.05	0.04	0.03	0.02	0.01	0.00
+0.0	0.5359	0.5319	0.5279	0.5239	0.5199	0.5160	0.5120	0.5080	0.5040	0.5000
+0.1	0.5753	0.5714	0.5675	0.5636	0.5596	0.5557	0.5517	0.5478	0.5438	0.5398
+0.2	0.6141	0.6103	0.6064	0.6026	0.5987	0.5948	0.5910	0.5871	0.5832	0.5793
+0.3	0.6517	0.6480	0.6443	0.6406	0.6368	0.6331	0.6293	0.6255	0.6217	0.6179
+0.4	0.6879	0.6844	0.6808	0.6772	0.6736	0.6700	0.6664	0.6628	0.6591	0.6554
+0.5	0.7224	0.7190	0.7157	0.7123	0.7088	0.7054	0.7019	0.6985	0.6950	0.6915
+0.6	0.7549	0.7517	0.7486	0.7454	0.7422	0.7389	0.7357	0.7324	0.7291	0.7257
+0.7	0.7852	0.7823	0.7794	0.7764	0.7734	0.7704	0.7673	0.7642	0.7611	0.7580
+0.8	0.8133	0.8106	0.8079	0.8051	0.8023	0.7995	0.7967	0.7939	0.7910	0.7881
+0.9	0.8389	0.8365	0.8340	0.8315	0.8289	0.8264	0.8238	0.8212	0.8186	0.8159
+1.0	0.8621	0.8599	0.8577	0.8554	0.8531	0.8508	0.8485	0.8461	0.8438	0.8413
+1.1	0.8830	0.8810	0.8790	0.8770	0.8749	0.8729	0.8708	0.8686	0.8665	0.8643
+1.2	0.9015	0.8997	0.8980	0.8962	0.8944	0.8925	0.8907	0.8888	0.8869	0.8849
+1.3	0.9177	0.9162	0.9147	0.9131	0.9115	0.9099	0.9082	0.9066	0.9049	0.9032
+1.4	0.9319	0.9306	0.9292	0.9279	0.9265	0.9251	0.9236	0.9222	0.9207	0.9192
+1.5	0.9441	0.9429	0.9418	0.9406	0.9394	0.9382	0.9370	0.9357	0.9345	0.9332
+1.6	0.9545	0.9535	0.9525	0.9515	0.9505	0.9495	0.9484	0.9474	0.9463	0.9452
+1.7	0.9633	0.9625	0.9616	0.9608	0.9599	0.9591	0.9582	0.9573	0.9564	0.9554
+1.8	0.9706	0.9699	0.9693	0.9686	0.9678	0.9671	0.9664	0.9656	0.9649	0.9641
+1.9	0.9767	0.9761	0.9756	0.9750	0.9744	0.9738	0.9732	0.9726	0.9719	0.9713
+2.0	0.9817	0.9812	0.9808	0.9803	0.9798	0.9793	0.9788	0.9783	0.9778	0.9773
+2.1	0.9857	0.9854	0.9850	0.9846	0.9842	0.9838	0.9834	0.9830	0.9826	0.9821
+2.2	0.9890	0.9887	0.9884	0.9881	0.9878	0.9875	0.9871	0.9868	0.9864	0.9861
+2.3	0.9916	0.9913	0.9911	0.9909	0.9906	0.9904	0.9901	0.9898	0.9896	0.9893
+2.4	0.9936	0.9934	0.9932	0.9931	0.9929	0.9927	0.9925	0.9922	0.9920	0.9918
+2.5	0.9952	0.9951	0.9949	0.9948	0.9946	0.9945	0.9943	0.9941	0.9940	0.9938
+2.6	0.9964	0.9963	0.9962	0.9961	0.9960	0.9959	0.9957	0.9956	0.9955	0.9953
+2.7	0.9974	0.9973	0.9972	0.9971	0.9970	0.9969	0.9968	0.9967	0.9966	0.9965
+2.8	0.9981	0.9980	0.9979	0.9979	0.9978	0.9977	0.9977	0.9976	0.9975	0.9974
+2.9	0.9986	0.9986	0.9985	0.9985	0.9984	0.9984	0.9983	0.9983	0.9982	0.9981
+3.0	0.99900	0.99896	0.99893	0.99889	0.99886	0.99882	0.99878	0.99874	0.99869	0.99865

[a]Example: when $Z = -1.84$, the probability is 0.0329 of obtaining a value equal to or less than X.

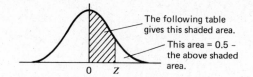

The following table gives this shaded area.

This area = 0.5 − the above shaded area.

Areas Under the Normal Probability Curve From Mean to Z.

Z	.00	.01	.02	.03	.04	.05	.06	.07	.08	.09
0.0	.0000	.0040	.0080	.0120	.0160	.0199	.0239	.0279	.0319	.0359
0.1	.0398	.0438	.0478	.0517	.0557	.0596	.0636	.0675	.0714	.0753
0.2	.0793	.0832	.0871	.0910	.0948	.0987	.1026	.1064	.1103	.1141
0.3	.1179	.1217	.1255	.1293	.1331	.1368	.1406	.1443	.1480	.1517
0.4	.1554	.1591	.1628	.1664	.1700	.1736	.1772	.1808	.1844	.1879
0.5	.1915	.1950	.1985	.2019	.2054	.2088	.2123	.2157	.2190	.2224
0.6	.2257	.2291	.2324	.2357	.2389	.2422	.2454	.2486	.2517	.2549
0.7	.2580	.2611	.2642	.2673	.2704	.2734	.2764	.2794	.2823	.2852
0.8	.2881	.2910	.2939	.2967	.2995	.3023	.3051	.3078	.3106	.3133
0.9	.3159	.3186	.3212	.3238	.3264	.3289	.3315	.3340	.3365	.3389
1.0	.3413	.3438	.3461	.3485	.3508	.3531	.3554	.3577	.3599	.3621
1.1	.3643	.3665	.3686	.3708	.3729	.3749	.3770	.3790	.3810	.3830
1.2	.3849	.3869	.3888	.3907	.3925	.3944	.3962	.3980	.3997	.4015
1.3	.4032	.4049	.4066	.4082	.4099	.4115	.4131	.4147	.4162	.4177
1.4	.4192	.4207	.4222	.4236	.4251	.4265	.4279	.4292	.4306	.4319
1.5	.4332	.4345	.4357	.4370	.4382	.4394	.4406	.4418	.4429	.4441
1.6	.4452	.4463	.4474	.4484	.4495	.4505	.4515	.4525	.4535	.4545
1.7	.4554	.4564	.4573	.4582	.4591	.4599	.4608	.4616	.4625	.4633
1.8	.4641	.4649	.4656	.4664	.4671	.4678	.4686	.4693	.4699	.4706
1.9	.4713	.4719	.4726	.4732	.4738	.4744	.4750	.4756	.4761	.4767
2.0	.4772	.4778	.4783	.4788	.4793	.4798	.4803	.4808	.4812	.4817
2.1	.4821	.4826	.4830	.4834	.4838	.4842	.4846	.4850	.4854	.4857
2.2	.4861	.4864	.4868	.4871	.4875	.4878	.4881	.4884	.4887	.4890
2.3	.4893	.4896	.4898	.4901	.4904	.4906	.4909	.4911	.4913	.4916
2.4	.4918	.4920	.4922	.4925	.4927	.4929	.4931	.4932	.4934	.4936
2.5	.4938	.4940	.4941	.4943	.4945	.4946	.4948	.4949	.4951	.4952
2.6	.4953	.4955	.4956	.4957	.4959	.4960	.4961	.4962	.4963	.4964
2.7	.4965	.4966	.4967	.4968	.4969	.4970	.4971	.4972	.4973	.4974
2.8	.4974	.4975	.4976	.4977	.4977	.4978	.4979	.4979	.4980	.4981
2.9	.4981	.4982	.4982	.4983	.4984	.4984	.4985	.4985	.4986	.4986
3.0	.4987	.4987	.4987	.4988	.4988	.4989	.4989	.4989	.4990	.4990

Degrees of Freedom (dF)	$t_{.005}$	$t_{.025}$	$t_{.05}$	$t_{.10}$	$t_{.20}$
5	4.03	2.57	2.02	1.48	0.92
10	3.17	2.32	1.81	1.37	0.88
15	2.95	2.13	1.75	1.34	0.87
20	2.85	2.09	1.73	1.33	0.86
25	2.79	2.06	1.71	1.32	0.86
30	2.75	2.04	1.70	1.31	0.85
40	2.70	2.02	1.68	1.30	0.85

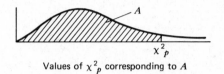

Values of χ^2_p corresponding to A

TABLE F
Distribution of χ^2

DF	$\chi^2_{.005}$	$\chi^2_{.01}$	$\chi^2_{.025}$	$\chi^2_{.05}$	$\chi^2_{.10}$	$\chi^2_{.90}$	$\chi^2_{.95}$	$\chi^2_{.975}$	$\chi^2_{.99}$	$\chi^2_{.995}$
1	0.000039	0.00016	0.00098	0.0039	0.0158	2.71	3.84	5.02	6.63	7.88
2	0.0100	0.0201	0.0506	0.1026	0.2107	4.61	5.99	7.38	9.21	10.60
3	0.0717	0.115	0.216	0.352	0.584	6.25	7.81	9.35	11.34	12.84
4	0.207	0.297	0.484	0.711	1.064	7.78	9.49	11.14	13.28	14.86
5	0.412	0.554	0.831	1.15	1.61	9.24	11.07	12.83	15.09	16.75
6	0.676	0.872	1.24	1.64	2.20	10.64	12.59	14.45	16.81	18.55
7	0.989	1.24	1.69	2.17	2.83	12.02	14.07	16.01	18.48	20.28
8	1.34	1.65	2.18	2.73	3.49	13.36	15.51	17.53	20.09	21.96
9	1.73	2.09	2.70	3.33	4.17	14.68	16.92	19.02	21.67	23.59
10	2.16	2.56	3.25	3.94	4.87	15.99	18.31	20.48	23.21	25.19
11	2.60	3.05	3.82	4.57	5.58	17.28	19.68	21.92	24.73	26.76
12	3.07	3.57	4.40	5.23	6.30	18.55	21.03	23.34	26.22	28.30
13	3.57	4.11	5.01	5.89	7.04	19.81	22.36	24.74	27.69	29.82
14	4.07	4.66	5.63	6.57	7.79	21.06	23.68	26.12	29.14	31.32
15	4.60	5.23	6.26	7.26	8.55	22.31	25.00	27.49	30.58	32.80
16	5.14	5.81	6.91	7.96	9.31	23.54	26.30	28.85	32.00	34.27
18	6.26	7.01	8.23	9.39	10.86	25.99	28.87	31.53	34.81	37.16
20	7.43	8.26	9.59	10.85	12.44	28.41	31.41	34.17	37.57	40.00
24	9.89	10.86	12.40	13.85	15.66	33.20	36.42	39.36	42.98	45.56
30	13.79	14.95	16.79	18.49	20.60	40.26	43.77	46.98	50.89	53.67
40	20.71	22.16	24.43	26.51	29.05	51.81	55.76	59.34	63.69	66.77
60	35.53	37.48	40.48	43.19	46.46	74.40	79.08	83.30	88.38	91.95
120	83.85	86.92	91.58	95.70	100.62	140.23	146.57	152.21	158.95	163.64

Values of χ^2 corresponding to certain selected probabilities (i.e., tail areas under the curve). To illustrate: the probability is 0.95 that a sample with 20 degrees of freedom, taken from a normal distribution, would have $\chi^2 = 31.41$ or smaller.

Source: Adapted with permission from Dixon, W. J. and Massey Jr., F. J., *Introduction to Statistical Analysis,* 3rd ed., McGraw-Hill Book Company, New York, copyright © 1969.

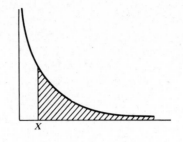

X

TABLE G
Exponential Distribution Values of $e^{-X/\mu}$

$\dfrac{X}{\mu}$	0.00	0.01	0.02	0.03	0.04	0.05	0.06	0.07	0.08	0.09
0.0	1.000	0.9900	0.9802	0.9704	0.9608	0.9512	0.9418	0.9324	0.9231	0.9139
0.1	0.9048	0.8958	0.8860	0.8781	0.8694	0.8607	0.8521	0.8437	0.8353	0.8270
0.2	0.8187	0.8106	0.8025	0.7945	0.7866	0.7788	0.7711	0.7634	0.7558	0.7483
0.3	0.7408	0.7334	0.7261	0.7189	0.7118	0.7047	0.6977	0.6907	0.6839	0.6771
0.4	0.6703	0.6637	0.6570	0.6505	0.6440	0.6376	0.6313	0.6250	0.6188	0.6126
0.5	0.6065	0.6005	0.5945	0.5886	0.5827	0.5769	0.5712	0.5655	0.5599	0.5543
0.6	0.5488	0.5434	0.5379	0.5326	0.5273	0.5220	0.5169	0.5117	0.5066	0.5016
0.7	0.4966	0.4916	0.4868	0.4819	0.4771	0.4724	0.4677	0.4630	0.4584	0.4538
0.8	0.4493	0.4449	0.4404	0.4360	0.4317	0.4274	0.4232	0.4190	0.4148	0.4107
0.9	0.4066	0.4025	0.3985	0.3946	0.3906	0.3867	0.3829	0.3791	0.3753	0.3716

$\dfrac{X}{\mu}$	0.0	0.1	0.2	0.3	0.4	0.5	0.6	0.7	0.8	0.9
1.0	0.3679	0.3329	0.3012	0.2725	0.2466	0.2231	0.2019	0.1827	0.1653	0.1496
2.0	0.1353	0.1225	0.1108	0.1003	0.0907	0.0821	0.0743	0.0672	0.0608	0.0550
3.0	0.0498	0.0450	0.0408	0.0369	0.0334	0.0302	0.0273	0.0247	0.0224	0.0202
4.0	0.0183	0.0166	0.0150	0.0130	0.0123	0.0111	0.0101	0.0091	0.0082	0.0074
5.0	0.0067	0.0061	0.0055	0.0050	0.0045	0.0041	0.0037	0.0033	0.0030	0.0027
6.0	0.0025	0.0022	0.0020	0.0018	0.0017	0.0015	0.0014	0.0012	0.0011	0.0010

Fractional parts of the total area (1.000) under the exponential curve greater than X. To illustrate: if X/μ is 0.45, the probability of occurrence for a value greater than X is 0.6376.

Source: Adapted from S. M. Selby (ed.), *CRC Standard Mathematical Tables,* 17th ed., CRC Press, Cleveland, Ohio, 1969, pp. 201–7.

TABLE H
PV_{sp}, Present Value Factors for Future Single Payments

Years Hence	1%	2%	4%	6%	8%	10%	12%	14%	15%	16%	18%	20%
1	0.990	0.980	0.962	0.943	0.926	0.909	0.893	0.877	0.870	0.862	0.847	0.833
2	0.980	0.961	0.925	0.890	0.857	0.826	0.797	0.769	0.756	0.743	0.718	0.694
3	0.971	0.942	0.889	0.840	0.794	0.751	0.712	0.675	0.658	0.641	0.609	0.579
4	0.961	0.924	0.855	0.792	0.735	0.683	0.636	0.592	0.572	0.552	0.516	0.482
5	0.951	0.906	0.822	0.747	0.681	0.621	0.567	0.519	0.497	0.476	0.437	0.402
6	0.942	0.888	0.790	0.705	0.630	0.564	0.507	0.456	0.432	0.410	0.370	0.335
7	0.933	0.871	0.760	0.665	0.583	0.513	0.452	0.400	0.376	0.354	0.314	0.279
8	0.923	0.853	0.731	0.627	0.540	0.467	0.404	0.351	0.327	0.305	0.266	0.233
9	0.914	0.837	0.703	0.592	0.500	0.424	0.361	0.308	0.284	0.263	0.225	0.194
10	0.905	0.820	0.676	0.558	0.463	0.386	0.322	0.270	0.247	0.227	0.191	0.162
11	0.896	0.804	0.650	0.527	0.429	0.350	0.287	0.237	0.215	0.195	0.162	0.135
12	0.887	0.788	0.625	0.497	0.397	0.319	0.257	0.208	0.187	0.168	0.137	0.112
13	0.879	0.773	0.601	0.469	0.368	0.290	0.229	0.182	0.163	0.145	0.116	0.093
14	0.870	0.758	0.577	0.442	0.340	0.263	0.205	0.160	0.141	0.125	0.099	0.078
15	0.861	0.743	0.555	0.417	0.315	0.239	0.183	0.140	0.123	0.108	0.084	0.065
16	0.853	0.728	0.534	0.394	0.292	0.218	0.163	0.123	0.107	0.093	0.071	0.054
17	0.844	0.714	0.513	0.371	0.270	0.198	0.146	0.108	0.093	0.080	0.060	0.045
18	0.836	0.700	0.494	0.350	0.250	0.180	0.130	0.095	0.081	0.069	0.051	0.038
19	0.828	0.686	0.475	0.331	0.232	0.164	0.116	0.083	0.070	0.060	0.043	0.031
20	0.820	0.673	0.456	0.312	0.215	0.149	0.104	0.073	0.061	0.051	0.037	0.026
21	0.811	0.660	0.439	0.294	0.199	0.135	0.093	0.064	0.053	0.044	0.031	0.022
22	0.803	0.647	0.422	0.278	0.184	0.123	0.083	0.056	0.046	0.038	0.026	0.018
23	0.795	0.634	0.406	0.262	0.170	0.112	0.074	0.049	0.040	0.033	0.022	0.015
24	0.788	0.622	0.390	0.247	0.158	0.102	0.066	0.043	0.035	0.028	0.019	0.013
25	0.780	0.610	0.375	0.233	0.146	0.092	0.059	0.038	0.030	0.024	0.016	0.010

TABLE I
PV_a, Present Value Factor for Annuities

Years (n)	1%	2%	4%	6%	8%	10%	12%	14%	15%	16%	18%	20%
1	0.990	0.980	0.962	0.943	0.926	0.909	0.893	0.877	0.870	0.862	0.847	0.833
2	1.970	1.942	1.886	1.833	1.783	1.736	1.690	1.647	1.626	1.605	1.566	1.528
3	2.941	2.884	2.775	2.673	2.577	2.487	2.402	2.322	2.283	2.246	2.174	2.106
4	3.902	3.808	3.630	3.465	3.312	3.170	3.037	2.914	2.855	2.798	2.690	2.589
5	4.853	4.713	4.452	4.212	3.993	3.791	3.605	3.433	3.352	3.274	3.127	2.991
6	5.795	5.601	5.242	4.917	4.623	4.355	4.111	3.889	3.784	3.685	3.498	3.326
7	6.728	6.472	6.002	5.582	5.206	4.868	4.564	4.288	4.160	4.039	3.812	3.605
8	7.652	7.325	6.733	6.210	5.747	5.335	4.968	4.639	4.487	4.344	4.078	3.837
9	8.566	8.162	7.435	6.802	6.247	5.759	5.328	4.946	4.772	4.607	4.303	4.031
10	9.471	8.983	8.111	7.360	6.710	6.145	5.650	5.216	5.019	4.833	4.494	4.192
11	10.368	9.787	8.760	7.887	7.139	6.495	5.988	5.453	5.234	5.029	4.656	4.327
12	11.255	10.575	9.385	8.384	7.536	6.814	6.194	5.660	5.421	5.197	4.793	4.439
13	12.134	11.343	9.986	8.853	7.904	7.103	6.424	5.842	5.583	5.342	4.910	4.533
14	13.004	12.106	10.563	9.295	8.244	7.367	6.628	6.002	5.724	5.468	5.008	4.611
15	13.865	12.849	11.118	9.712	8.559	7.606	6.811	6.142	5.847	5.575	5.092	4.675
16	14.718	13.578	11.652	10.106	8.851	7.824	6.974	6.265	5.954	5.669	5.162	4.730
17	15.562	14.292	12.166	10.477	9.122	8.022	7.120	6.373	6.047	5.749	5.222	4.775
18	16.398	14.992	12.659	10.828	9.372	8.201	7.250	6.467	6.128	5.818	5.273	4.812
19	17.226	15.678	13.134	11.158	9.604	8.365	7.366	6.550	6.198	5.877	5.316	4.844
20	18.046	16.351	13.590	11.470	9.818	8.514	7.469	6.623	6.259	5.929	5.353	4.870
21	18.857	17.011	14.029	11.764	10.017	8.649	7.562	6.687	6.312	5.973	5.384	4.891
22	19.660	17.658	14.451	12.042	10.201	8.772	7.645	6.743	6.359	6.011	5.410	4.909
23	20.456	18.292	14.857	12.303	10.371	8.883	7.718	6.792	6.399	6.044	5.432	4.925
24	21.243	18.914	15.247	12.550	10.529	8.985	7.784	6.835	6.434	6.073	5.451	4.937
25	22.023	19.523	15.622	12.783	10.675	9.077	7.843	6.873	6.464	6.097	5.467	4.948

APPENDIX C

CONSUMER PRODUCT SAFETY GUIDELINES

Ask yourself the following questions about your business. If you answer "No" to any question, you are missing an important opportunity to find product hazards before they become major problems.

Yes/No

——— Is responsibility assigned within your company for compliance with product safety laws, regulations, and standards?

——— Is safety properly considered along with such other factors as product appearance or production costs?

Do product designs and tests—

——— Comply with existing mandatory safety standards?

——— Make use of voluntary standards when appropriate?

——— Anticipate consumer use patterns and foreseeable misuse?

——— Consider the compatibility of materials and design with the intended operating environment?

——— Are modified designs subjected to the full battery of safety and performance tests?

——— Do testing, inspection, and evaluation continue throughout the production process?

——— Do you keep records that document the steps taken to assure safety and make it possible to trace product hazards to their source?

———— Are there powerful incentives and appropriate training programs for production and purchasing personnel to locate and reject defective goods?

———— Do you provide service technicians and consumers with adequate instructions that explain proper installation, use, maintenance, repair, and disposal of products?

———— Are appropriate precautions, warnings, and antidotes clearly and prominently disclosed and well-explained?

———— Are advertising and point-of-purchase sales representations consistent with safe product use and with manufacturer's warranties?

———— Do you have a system for receiving complaints? For relaying them to your suppliers?

———— Are data from the field (complaints, service reports, and the experience of others in the same business) promptly communicated to the appropriate departments and analyzed for evidence of safety problems?

———— Do you keep track of products that are recalled and the reasons for recalls?

Source: Consumer Product Safety: Responsive Business Approaches to Consumer Needs, U.S. Department of Commerce, Office of Consumer Affairs, 1981.

DEVELOPMENT PLAN FOR SELF-INSPECTION (SI)

NETWORK

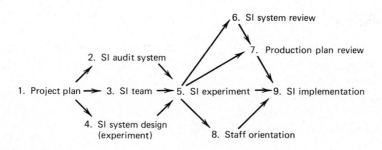

MAJOR ACTIVITIES

1. Project Plan

Feasibility study.
Senior management decision/policy statement.
Quality Assurance prepares SI project plan.
Plan approval and publication.

2. SI Audit System

Appointment of auditor.
Participation in activities under 3 and 4.
Preparation of SI audit plan and worksheets.
Finalizing of plan; integration with regular quality assurance audits.

3. SI Team

Preparation of selection and appointment procedures.
Invitation for application as SI authorized staff.
Orientation meeting.
Formation of a SI team; quality circle approach.
Basic and experimental SI training for SI candidates and for supervisors.
Trial workshop together with quality assurance staff and supervisors.
Selection and formal appointment of SI experimental team.

4. SI System Design

Establish SI design team.
Prepare SI objectives, principles, procedures, manual.
Select SI experimental workplace, station.
Prepare SI procedure and facilities.
Integrate SI procedure with current work procedure and establish new performance standard.
Train SI worker
Test performance against performance standard.
Approve/authorize SI experimental pilot project.

5. SI Experiment

Conduct a meeting with SI auditor (2), SI team (3), and SI experimental subgroup (4).
Prepare SI worker, supervisor, auditor, and quality assurance staff for trial run.
Repeat trial runs with intermediate reviews and adjustments.
Terminate experiment and analyze results.
Prepare a report, including audit report.
Submit to senior management for review and decision on SI.

6. SI System Review

Receive approval and basic guidelines from senior management and quality assurance staff.
Prepare publication on SI experiment and outcome: conduct staff meeting.
Establish SI administrative and operational procedures on a company-wide basis.
Submit to senior management for approval; decision.
Conduct staff meeting(s).
Prepare jointly with auditor final SI audit plan.
Reform SI review team as SI advisory team.

7. Production Plan Review

Determine inspection points, lines, products, work places, staff members etcetera, for possible introduction of SI.
Revise production plan and work methods, assignment for integration with SI.
Train worker and seek SI authorization.
Submit results and seek decision on go-ahead.

8. Staff Orientation

Prepare for several staff meetings and workshops.
Submit plan to SI project team and seek approval from senior management.
Conduct meetings.
Invite and screen application for SI appointment.
Repeat staff meeting for progress report.

9. SI Implementation

Appoint SI worker.
Prepare/review/adjust integrated work procedure and performance standard.
Train worker and conduct trial run.
Analyze results and submit report to SI team for authorization.
Monitor problems, problem solution, and progress.
Conduct an audit and report to senior management.
Conduct a staff meeting for progress report, recognition of SI participants, demonstration of SI benefits.
Prepare and conduct SI regular team meeting(s) with quality circle type of approach. Quality assurance staff and auditors serve as auditor/facilitators.

Note: The above project plan might have to be revised, or even aborted, when results prove the SI system unfeasible. Depending on the availability of staff and budgets or facilities, the plan might have to be adjusted. Careful planning and proceeding with wide participation and continuous communications will avoid possible negative impacts on quality and quality assurance. Moreover, changes at workplaces and in payment systems bear much risk for disharmony and conflicts. The experiment and frequent reporting to staff is designed to assure positive goal achievement.

E APPENDIX

PRODUCT SAFETY CHECKPOINTS

Following are suggested steps for management to follow when establishing policies and procedures for the product safety program.

PRODUCT MANUFACTURE

_____ Establish company policies and procedures

_____ Set up Safety Review Committee

_____ Assign responsibility and lines of authority

_____ Begin product design

      _____ Open product development file
      _____ Establish performance requirements consistent with intended use
      _____ Select raw materials and components
      _____ Prepare written specifications and design drawings
      _____ Document all changes in design
      _____ Sign-off by responsible authority

_____ Conduct prototype testing

      _____ Hire independent lab
      _____ Check for mandatory or voluntary standards

_____ Obtain lab reports, certification, or ''listing''
_____ Document all changes made as a result of testing
_____ Sign-off by responsible authority

_____ Prepare for production

_____ Procure raw materials and components
_____ Develop purchasing specifications and contracts
_____ Inspect materials and components
_____ Document all changes or repairs
_____ Establish procedures for handling and storage
_____ Sign-off by responsible authority

_____ Production quality control

_____ Open quality control file
_____ Train personnel
_____ Put instructions and procedures in writing
_____ Periodically inspect equipment
_____ Inspect and test products from production line
_____ Segregate faulty or non-conforming products
_____ Preserve worker safety

_____ Product distribution

_____ Identify products by lot, batch, serial number, or date of production
_____ Prepare packaging that protects the product and handlers during shipping and storage
_____ Prepare labels and instructions with adequate warnings, list of antidotes, recommended use
_____ Prepare operating instructions covering proper use, anticipated mis-use, installation, maintenance, storage, and disposal
_____ Include instructions for repairs consumers can safely make, and warnings about unsafe repairs
_____ Market test labels and instructions and change if necessary
_____ Sign-off by responsible authority
_____ Prepare advertising and promotions with safety in mind
_____ Demonstrate safe use in advertisements
_____ Emphasize instructions and warnings
_____ Avoid exaggerating safety features and giving user a false sense of security

PRODUCT DISTRIBUTION AND SALE

_____ Contracts between manufacturers, distributors and retailers

_____ Include product description, expected performance, and function

_____ Include guarantees and warranties and certifications

_____ State extent of repairs possible and circumstances in which products must be returned to manufacturer for repair

_____ Include all identification necessary to trace product

_____ Retailers' responsibilities

_____ Train buyers to select products carefully and to be knowledgeable about mandatory and voluntary safety standards

_____ Understand manufacturers' warranties and the circumstances in which retailers can become bound by warranty obligations

_____ Train sales personnel not to remove tags and other safety disclosures from display merchandise, or to discount or contradict manufacturers' safety information. Sales personnel should encourage consumers to read and understand these disclosures

_____ Product servicing

_____ Follow manufacturers' instructions, diagrams, and charts showing correct methods of repair

_____ Return products to manufacturer for major repairs, if manufacturer requests

_____ Train and consider certifying personnel

_____ Promptly report safety problems and potential defects to manufacturer

_____ Customer service and complaints

_____ Establish a system for reporting complaints, injuries, product defects, inadequacies in product literature, and design problems to the manufacturer

_____ Product recall

_____ Agree on procedures in advance with manufacturers, distributors, and retailers

_____ Anticipate the costs of a recall, including repurchase, handling, shipping, and publicity

_____ Keep records by product lot, manufacturer, dates, and purchaser's name, if possible

_____ Help manufacturer keep track of product returns

_____ Make plans for returning products to service centers and for distributing replacement parts

_____ Consider a trial run of recall procedures so response can be prompt and effective

————— Audit procedures

————— Conduct audits to evaluate procedures and evaluate how well they are carried out

————— Maintain the integrity of the recordkeeping system

Source: Excerpted from *Consumer Product Safety: Responsive Business Approaches to Consumer Needs,* U.S. Department of Commerce, Office of Consumer Affairs, 1981.

F APPENDIX

PRODUCT LIABILITY: AN INCREASINGLY IMPORTANT ISSUE

Another critical aspect of producing a new product is the extent to which a businessowner is liable for injury or damage caused by its use. Product liability is one of the potentially most serious of all business liabilities. The trend of the past decade, in most states, has been to abandon the old interpretation which generally did not hold the manufacturer liable for defective products. Today it is increasingly common for the courts to hold the manufacturer, and sometimes the seller, almost totally responsible for injury or property damage due to a defective or unsafe product when used as intended. Overt negligence by the manufacturer during the design, inspection, testing, or instruction process may not have to be proved. Furthermore, misrepresenting the character or quality of a product through advertising is a liable activity if it causes injury or damage.

The current judicial and legislative trend is toward the adoption of a doctrine of strict liability of the manufacturer, and at times even the retailer or wholesaler, for product defects or misuse. It, therefore, is becoming increasingly important for the small businessowner to understand product liability law. The articles list here discuss trends in court decisions, product safety design, product liability and advertising, causes of product failures, and product liability cost estimation.

Goebel, John W. **The Legal Implications of Strict Liability for Marketing—A Mini Course,** in *Increasing Marketing Productivity and Conceptual and Methodological Foundations of Marketing,* Thomas V. Greer, Ed. (Chicago: American Marketing Association, December 1973), pp. 435–346.

The article examines the evolution of the doctrine of strict liability that has been cause for alarm in the marketing field. Potential losses are sizeable and defenses few. While the manufacturer will not necessarily always be responsible for injury, the author notes that "absolute liability" is almost here.

Kuhn, James P. **How to Manage Product Safety,** *Industry Week,* (April 22, 1974), pp. 53–59.

By knowing the similarities and differences between product quality and product safety, the manufacturer can attack the problem of "managing" product safety. Examines the major phases of the product cycle with emphasis on control. A simplified schematic diagram of a feedback system is included along with an eight step approach in setting up an initial system.

Kytle, Rayford P. Jr. **Evaluation for Product Safety Prior to Marketing,** in *Marketing's Contribution to the Firm and Society,* Ronald C. Curham, Ed. (Chicago: American Marketing Association Combined Proceedings, 1974), pp. 361–364.

Today's business climate demands that industry concern itself with product safety. This paper identifies and discusses the various aspects management must concern itself with to product safe products.

Loudenback, Lynn J. and John W. Goebel. **Marketing in the Age of Strict Liability,** *Journal of Marketing,* Vol. 38 (January, 1974), pp. 62–66.

The authors examine the implications of a strict liability doctrine on marketing practices. They also consider potential future changes in the law and its effect on marketing activities.

Loudenback, Lynn J. **Marketing Involvement in Consumer Product Safety Programs,** in *New Marketing for Social and Economic Progress and Marketing's Contributions to the Firm and to the Society,* Ronald C. Curhan, Ed. (Chicago: American Marketing Association, December, 1974), pp. 365–367.

The establishment of the Consumer Product Safety Commission and the adoption of the doctrine of strict liability by our courts has caused business to look more closely at corporate responsiblity for consumer safety. The role of marketing in consumer product safety programs in reviewed and implications for corporate marketing strategy discussed.

Moss, Frank E. **The Manufacturer's Role in Product Safety,** *The Conference Board,* (April, 1974), pp. 30–32.

Senator Moss discusses the emphasis that manufacturers must put on product safety if they are to successfully market products in the future. A "risk based" analysis is presented as one approach to safety and quality control manufacturers might take. The article was written while Senator Moss was chairman of the Consumer Subcommittee of the Senate Committee on Commerce.

Schneider, Lawrence R. **Product Safety Legislation, Old and New,** *The Conference Board,* (April, 1974), pp. 32–37.

The author discusses the impact on both manufacturers and consumers of the consumer Product Safety Act. The recall provisions are given special emphasis as well as briefly discussing the Consumer Product Safety Act and four minor acts now under the jurisdiction of the Consumer Product Safety Division.

Sorensen, Howard C. **Products Liability: The Consumer's Revolt,** *Best's Review,* (property/liability ed.) Vol. 75 (September, 1974), pp. 38–48.

A series of actual court cases involving products liability, strict liability and privity are reviewed. A basis for and proof of negligence is established. Emphasis is put on the growing demands placed on industry by the consumer and the courts.

Varble, Dale L. **Social and Environmental Considerations in New Product Development,** *Journal of Marketing,* Vol. 36 (October, 1972), pp. 11–15.

The process of integrating both social and environmental considerations into new product development is discussed. The advantages of inclusion of social environmental costs into breakeven analysis is overviewed.

Weinstein, Alfred, et al. **Product Liability,** Wiley-Interscience, 605 Third Ave., New York, NY 10016, 1978.

Establishes practical legal and engineering guidelines to help companies avoid costly product liability court cases.

Source: Small Business Bibliography, *Aid* Sb#90, published by U.S. Small Business Administration, Washington, D.C. Bibliography was coauthored by Udell, G. G., University of Oregon, and O'Neill, M. F., of the University of Arizona.

APPENDIX G

LANDMARK CASES IN PRODUCT LIABILITY

MacPherson vs. Buick Motor Co., New York, 1916: A manufacturer is liable for negligently built products that are "reasonably certain to place life and limb in peril," even though consumers do not buy directly from the manufacturer.

Greenman vs. Yuba Power Products Inc., California, 1963: A manufacturer is strictly liable when he sells a product that proves to have a defect that causes injury.

Larson vs. General Motors Corp., U.S. Court of Appeals, 8th Circuit, 1968: When faulty design of a product worsens an injury, a plaintiff may recover damages for the worsened part of the injury, even if the design defect did not cause the injury in the first place.

Cunningham vs. MacNeal Memorial Hospital, Illinois, 1970: It is not a defense to claim that a product (in this case blood infected by hepatitis) could not be made safer by any known technology. This ruling of the Illinois Supreme Court, the only case in which judges squarely refused to consider "state of the art," was reversed by a state statute defining the selling of blood as a service.

Cronin vs. J. B. E. Olson Corp., California, 1972: A product need not be "unreasonably dangerous" to make its manufacturer strictly liable for defective design.

Bexigs vs. Havir Mfg. Co., New Jersey, 1972: If an injury is attributable to the lack of any safety device on a product, the manufacturer cannot base a defense on the contributory negligence of the plaintiff.

Berkabile vs. Brantly Helicopter Corp., Pennsylvania, 1975: Whether the seller could have foreseen a particular injury is irrelevant in a case of strict liability for design defect.

Ault vs. International Harvester Co., California, 1975: Evidence that a manufacturer changed or improved its product line after the manufacture and sale of the particular product that caused an injury may be used to prove design defect.

Micallef vs. Miehle Co., New York, 1976: Evidence that an injured plaintiff obviously knew of a danger inherent in using a product will not defeat his claim if the manufacturer could reasonably have guarded against the danger when designing the product.

Barker vs. Lull Engineering Co., California, 1978: A manufacturer must show that the usefulness of a product involved in an accident outweighs the risks inherent in its design. In this radical ruling, the court shifted the burden of proof in design-defect cases from plaintiff to defendant.

Source: "Landmark Cases in Product Liability," *Business Week,* February 12, 1979, p. 74. Reprinted by special permission, copyright © 1979 by McGraw-Hill, Inc.

APPENDIX H

USEFUL MILITARY STANDARDS AND SPECIFICATIONS

1. Quality Assurance Terms and Definitions — MIL-STD-109
2. Quality Program Requirements — MIL-Q-9858
3. Inspection System Requirements — MIL-I-45208
4. Calibration System Requirements — MIL-C-45662
5. Quality Control System Requirements (Technical Data) — MIL-T-50301
6. Quality Assurance Provisions for Government Agencies — NHB-53004
7. Supplier Quality Assurance Program Requirements — MIL-STD-1535 USAF
8. Sampling Procedures and Methods for Quality Assurance — DSAM 8260.1
9. Handbook of Quality Assurance Forms and Procedures — DSAH 8230.1
10. Magnetic Inspection Units — MIL-M-6867
11. Sampling Procedures and Tables for Inspection by Attributes (International Designations) — MIL-STD-105
12. Sampling Procedures and Tables for Inspection by Variables for Percent Defective — MIL-STD-414
13. Evaluation of Contractors Quality Program — Handbook 50
14. Evaluation of Contractors Inspection System — Handbook 51
15. Evaluation of Contractors Calibration System — Handbook 52
16. Guide for Sampling Inspection — Handbook 53
17. Multi-level Continuous Sampling Procedures and Tables for Inspection by Attributes — Handbook 106
18. Single-level Continuous Sampling Procedures and Tables for Inspection by Attributes — Handbook 107

651

19. Quality Program Provisions for Space System Contractors — NHB 5300.1
20. Quality Program Provisions for Aeronautical and Space System Contractors — NHB 5300.4
21. Inspection System Provisions for Supplier of Space Materials, Parts, Components and Services — NHB 5300.2
22. Control of Manufacturing Supplies — USAF Spec. Bulletin 515
23. Quality Assurance Testing and Inspection of Aircraft Crew Emergency Escape Propellant Equipment — MIL-STD-1550
24. Quality Control of Gaseous and Liquid Aviators Breathing Oxygen at Aircraft Contract Facilities — MIL-STD-1551
25. Quality Standards for Aircraft Pneumatic Tires and Inner Tubes — MIL-STD-698
26. Quality Control of Chemicals — MIL-Q-7640
27. Quality Control General Specifications — MIL-Q-14461
28. Quality of Wood Member Containers — MIL-STD-731
29. Quality Control Requirements General Equipment — MIL-Q-5923
30. Reliability Report — MIL-STD-1304
31. Reliability Prediction — MIL-STD-756
32. Reliability Test Exponential Distribution — MIL-STD-781
33. Reliability Requirements for Weapons Systems (superseded by MIL-R-27542) — MIL-26674
34. Reliability Assurance Program for Electrical Equipment (superseded by MIL-T-27542 USAF) — MIL-R-25717
35. Reliability Assurance Program for Electronic Parts Spec. — MIL-STD-790
36. Reliability Assurance for Production Acceptance of Avionic Equipment General Spec. (superseded by MIL-STD-781A) — MIL-23094
37. Reliability Program for Systems and Equipment Development and Production — MIL-STD-785
38. Reliability Requirements for Design of Electrical Equipment or Systems (superseded by MIL-STD-785) — MIL-R-22256
39. Reliability Index-Determination for Avionic Equipment Control General Spec. — MIL-R-22973
40. Reliability Prediction and Demonstration for Airborne Surveillance Systems (Multipurpose MQM 58A) — MIL-R-55413
41. Reliability and Quality Assurance Requirements for Established Parts General Spec. — MIL-R-38100
42. Reliability Evaluation from Demonstration Data — MIL-STD-757
43. Reliability Requirements for Development of Electrical Subsystems for Equipment Use (MIL-STD-781) — MIL-R-26484
44. Reliability Requirements for Development of Ground Electrical Equipment (superseded by MIL-STD-785) — MIL-R-27070
45. Reliability Requirements for Electrical Ground Checkout Equipment — MIL-R-27173
46. Reliability Requirements for Shipboard Electrical Equipment — MIL-R-22732
47. Reliability Requirements for Shipboard Electronic Equipment — MIL-R-22732
48. Reliability of Production of Electronic Equipment, General Spec. — MIL-R-19610
49. Reliability Prediction of Monolithic Integrated Circuits — MIL-STD-1600
50. Life Cycle Product Quality Program Requirements — OD 46574

APPENDIX I

QUALITY ASSURANCE LITERATURE CLASSIFICATION

830: Design Controls
   831: System Reliability Analysis/Evaluation
   832: Design Reviews
840: Methods of Reliability Analysis
850: Reliability Demonstration/Measurement/Testing
860: Field/Consumer Activity
870: Maintainability

Functional Classification

:00: General

:10: Management   Those functions which direct the activities of an organization

:20: Production   Those functions dealing with a marketable product

:30: Financial   Those functions dealing with monetary aspects of an organization

:40: Procurement   Those functions relating to obtaining material for all organization

:50: Sales and Service   Those functions concerning marketing and servicing the product or service of an organization

:60: Engineering   Those functions establishing the requirements and standards of actual or potential marketable product

:70: Quality   Those functions assuring that product or service conforms to requirements and standards

:80: Industrial Relations   Those functions dealing with the personnel and community relations aspects of an organization.

:90: Management Services   Those functions dealing with the systems and services employed by an organizational complex

Industry and Business Classification

:000—General or Non-Classifiable Establishments
   :000—General
   :099—Non-classifiable
:100—Agriculture, forestry, and fisheries
   :101—Commercial farms
   :102—Non-commercial farms
   :107—Agricultural services and Hunting and grapping
   :108—Forestry
   :109—Fisheries
:200—Mining
   :210—Metal mining
   :211—Anthracite mining
   :212—Bituminous coal and lignite mining
   :213—Crude petroleum and natural gas
   :214—Mining and quarrying of nonmetallic minerals except fuels
:300—Contract construction
:400—Manufacturing
   :419—Ordance and accessories
   :420—Food and kindred products
   :421—Tobacco manufactures
   :422—Textile mill products
   :423—Apparel and other finished products made from fabrics and similar materials
   :424—Lumber and wood products, except furniture
   :425—Furniture and fixtures
   :426—Paper and allied products
   :427—Printing, publishing, and allied industries
   :428—Chemicals and allied products
   :429—Petroleum refining and related industries

:430—Rubber and miscellaneous plastic products
:431—Leather and leather products
:432—Stone, clay and glass products
:433—Primary metal industries
:434—Fabricated metal products, except ordnance, machinery and transportation equipment
:435—Machinery, except electrical
:436—Electrical machinery, equipment, and supplies
:437—Transportation equipment
:438—Professional, scientific, and controlling instruments; photographic and optical goods; watches and clocks
:439—Miscellaneous manufacturing industries
:500—Transportation, communication, electric, gas and sanitary services
:540—Railroad transportation
:541—Local and suburban transit and interurban passenger transportation
:542—Motor freight transportation and warehousing
:544—Water transportation
:545—Transportation by air
:546—Pipe line transportation
:547—Transportation services
:548—Communication
:549—Electric, gas and sanitary services
:600—Wholesale and retail trade
:650—Wholesale trade
:652—Retail trade—building materials, hardware, and farm equipment
:653—Retail trade—general merchandise
:654—Retail trade—food
:655—Automotive dealers and gasoline service stations
:656—Retail trade—apparel and accessories

:657—Retail trade—furniture, home furnishings, and equipment
:658—Retail trade—eating and drinking places
:659—Retail trade—miscellaneous retail stores
:700—Finance, insurance, and real estate
:760—Banking
:761—Credit agencies other than banks
:762—Security and commodity brokers, dealers, exchanges and services
:763—Insurance carriers
:764—Insurance agents, brokers, and service
:765—Real estate
:766—Combinations of real estate, insurance, loans, law offices
:767—Holding and other investment companies
:800—Services
:870—Hotels, rooming houses, camps, and other lodging places
:872—Personal services
:873—Miscellaneous business services
:875—Automobile repair, automobile services, and garages
:876—Miscellaneous repair services
:878—Motion pictures
:879—Amusement and recreation services, except motion pictures
:880—Medical and other health services
:881—Legal services
:882—Educational services
:884—Museums, art galleries, botanical and zoological gardens
:886—Nonprofit membership organizations
:888—Private households

:889—Miscellaneous services
:900—Government
    :991—Federal government
    :992—State government
    :993—Local government
    :994—International government

Source: Adapted from Bureau of Labor Statistics Standard Industrial Classification (SIC).

J APPENDIX

JOB PROFILES IN QUALITY ASSURANCE

INSPECTOR

The inspector's assignment is to determine whether products and materials meet quality standards. He (or she) may have to examine every item that is being produced, or he may sample a few items. Inspectors make measurements, look for visible defects, or may perform chemical tests. Special equipment is often used for these purposes.

The trained inspector learns to do his job well, to be objective in his attitude, and to be consistent in his work. His conscientious performance in interpreting quality standards helps to inspire an atmosphere of cooperation among his fellow workers.

Salaries for inspectors reflect the skill and knowledge required in that industry and company in which they work. In general, inspectors are paid about the same as employees who manufacture and assemble parts and products.

QUALITY OR RELIABILITY TECHNICIAN

The technician is specially trained to use mechanical or laboratory equipment for evaluating quality. In addition, he (or she) has learned how to summarize statistical data that resulted from investigations of quality problems.

The technician is often looked upon as an "engineer in training," although many do not wish to become engineers. Those who work in quality control laboratories usually have specific laboratory training and backgrounds, such as in chemistry or metallurgy.

The technician has at least a high school education and often one or two years of college training. His salary compares favorably with that of other technicians with similar education and experience backgrounds.

The American Society for Quality Control awards a Quality Technician certificate to qualified members who pass a written examination.

QUALITY ENGINEER

The quality engineer is usually a person trained in one of the basic engineering or scientific fields (physics, chemistry, electronics, mechanics, etc.). He (or she) has strengthened this training with additional skills in statistics, quality control techniques, and problem solving. A college degree or its equivalent is often required.

He is usually involved in a wide variety of assignments and works with many people. He is in contact with the design engineer to be sure that quality characteristics are built into the product. He develops sampling plans, inspection and testing procedures, quality control techniques in manufacturing, and he investigates quality problems to identify their causes. He often will work with suppliers and customers, as well as with manufacturing people, in solving those problems.

The quality engineer's knowledge of probability, statistics, and sampling is essential to his work. He must also have the ability to communicate effectively with oral and written reports.

Salaries for quality engineers are generally equal to those of other engineers. There has been a growing demand for good quality engineers. The variety of assignments, the opportunity to introduce new ideas and methods, and the chance to solve problems give the quality engineer a great deal of satisfaction.

The American Society for Quality Control awards a Quality Engineer certificate to qualified members who pass a written examination.

RELIABILITY ENGINEER

The reliability engineer, like the quality engineer, is usually a person trained in one of the basic engineering or scientific fields. Reliability is a special quality characteristic of a product. It is a measure of how well and how long the product will work before it breaks down. The statistical techniques used by the reliability engineer are somewhat different from those used by the quality engineer, although there is a large body of knowledge that they hold in common.

The reliability engineer is usually in close contact with the designers of a product. There is also a need to maintain close touch with suppliers to be certain that good parts will contribute their share to the overall reliability of the finished product. The actual performance of the product in the hands of the customer has to be watched in order to determine if the required degree of reliability has been achieved. For example, the reliability engineer has to know under what condictions a critical automobile part stops working or if a shop tool is used in an unexpected way that could affect its reliability.

The reliability engineer's knowledge of probability, statistics, and sampling is essential to his (or her) work. He must also have the ability to communicate effectively with oral and written reports.

Salaries for reliability engineers are generally equal to those of other engineers. There is a growing demand for good reliability engineers, jsut as there is for good quality engineers. In recent years there has been an active transfer of knowledge and experience from the sophisticated aerospace and defense type industries, where much of the technology of reliability was developed, to commercial applications for goods we see and use every day. Like the quality engineer, the reliability engineer finds his work exciting, challenging, and satisfying.

The American Society for Quality Control awards a Reliability Engineer certificate to qualified members who pass a written examination.

QUALITY OR RELIABILITY MANAGER

Those who are responsible for managing a quality or reliability program in a company have very important jobs. When you think about how everyone's work can affect the quality or reliability of a product, such as designing the product, making the parts, assembling them, packaging, shipping and handling, installing and servicing them, to name only a few, then you can see that the quality or reliability manager is one of the most important members of the management team.

He (or she) has to provide the leadership and technical knowledge that are necessary to plan a total quality program, to see to it that the costs of putting quality into a product are known and controlled, and to coordinate the efforts of the many people who can affect quality. Therefore, this manager must be technically competent, must be objective and data-oriented, must be an administrator, must be able to organize and manage, and must have qualities of leadership.

The manager of a quality or reliability program enjoys the same benefits as any other company department manager does. It is becoming more common to see quality and reliability managers achieve the position of vice president within their companies. He is measured by his ability to satisfy customers through the quality program, to reduce costs, and to contribute to the company's profits.

We are in a new era that is moving such managers from the production and manufacturing floor to the executive suite, because running a total quality program is more like running a total business.

Source: Careers in the Quality Sciences, Copyright American Society for Quality Control Inc., reprinted by permission.

INDEX